Graduate Texts in Mathematics 106

T0184629

For other titles published in this series, go to
http://www.springer.com/series/136

Joseph H. Silverman

The Arithmetic
of Elliptic Curves

Second Edition

With 14 Illustrations

 Springer

Joseph H. Silverman
Department of Mathematics
Brown University
151 Thayer St.
Providence, RI 02912
USA
jhs@math.brown.edu

ISBN   978-1-4419-1858-1   e-ISBN   978-0-387-09494-6
DOI 10.1007/978-0-387-09494-6
Springer Dordrecht Heidelberg London New York

Mathematics Subject Classification (2000): 11-01, 11GXX, 14GXX, 11G05

Springer is part of Springer Science+Business Media (www.springer.com)

Preface to the Second Edition

In the preface to the first edition of this book I remarked on the paucity of introductory texts devoted to the arithmetic of elliptic curves. That unfortunate state of affairs has long since been remedied with the publication of many volumes, among which may be mentioned books by Cassels [43], Cremona [54], Husemöller [118], Knapp [127], McKean et. al [167], Milne [178], and Schmitt et. al [222] that highlight the arithmetic and modular theory, and books by Blake et. al [22], Cohen et. al [51], Hankerson et. al [107], and Washington [304] that concentrate on the use of elliptic curves in cryptography. However, even among this cornucopia of literature, I hope that this updated version of the original text will continue to be useful.

The past two decades have witnessed tremendous progress in the study of elliptic curves. Among the many highlights are the proof by Merel [170] of uniform boundedness for torsion points on elliptic curves over number fields, results of Rubin [215] and Kolyvagin [130] on the finiteness of Shafarevich–Tate groups and on the conjecture of Birch and Swinnerton-Dyer, the work of Wiles [311] on the modularity of elliptic curves, and the proof by Elkies [77] that there exist infinitely many supersingular primes. Although this introductory volume is unable to include proofs of these deep results, it will guide the reader along the beginning of the trail that ultimately leads to these summits.

My primary goals in preparing this second edition, over and above the pedagogical aims of the first edition, are the following:

- Update and expand results and references, especially in Appendix C, which includes a new section on the variation of the trace of Frobenius.

- Add a chapter devoted to algorithmic aspects of elliptic curves, with an emphasis on those features that are used in cryptography.

- Add a section on Szpiro's conjecture and the ABC conjecture.

- Correct, clarify, and simplify the proofs of some results.

- Correct numerous typographical and minor mathematical errors. However, since this volume has been entirely retypeset, I beg the reader's indulgence for any new typos that have been introduced.

- Significantly expand the selection of exercises.

It has been gratifying to see the first edition of this book become a standard text and reference in the subject. In order to maintain backward compatibility of cross-references, I have taken some care to leave the numbering system unchanged. Thus Proposition III.8.1 in the first edition remains Proposition III.8.1 in the second edition, and similarly for Exercise 3.5. New material has been assigned new numbers, and although there are many new exercises, they have been appended to the exercises from the first edition.

Electronic Resources: There are many computer packages that perform computations on elliptic curves. Of particular note are two free packages, Sage [275] and Pari [202], each of which implements an extensive collection of elliptic curve algorithms. For additional links to online elliptic curve resources, and for other material, the reader is invited to visit the *Arithmetic of Elliptic Curves* home page at

$$\texttt{www.math.brown.edu/~jhs/AECHome.html}$$

No book is ever free from error or incapable of being improved. I would be delighted to receive comments, positive or negative, and corrections from you, the reader. You can send mail to me at

$$\texttt{jhs@math.brown.edu}$$

Acknowledgments for the Second Edition

Many people have sent me extensive comments and corrections since the appearance of the first edition in 1986. To all of them, including in particular the following, my deepest thanks: Jeffrey Achter, Andrew Bremner, Frank Calegari, Jesse Elliott, Kirsten Eisenträger, Xander Faber, Joe Fendel, W. Fensch, Alexandru Ghitza, Grigor Grigorov, Robert Gross, Harald Helfgott, Franz Lemmermeyer, Dino Lorenzini, Ronald van Luijk, David Masser, Martin Olsson, Chol Park, Bjorn Poonen, Michael Reid, Michael Rosen, Jordan Risov, Robert Sarvis, Ed Schaefer, René Schoof, Nigel Smart, Jeroen Spandaw, Douglas Squirrel, Katherine Stange, Sinan Unver, John Voight, Jianqiang Zhao, Michael Zieve.

Providence, Rhode Island   JOSEPH H. SILVERMAN
November, 2008

Preface to the First Edition

The preface to a textbook frequently contains the author's justification for offering the public "another book" on a given subject. For our chosen topic, the arithmetic of elliptic curves, there is little need for such an apologia. Considering the vast amount of research currently being done in this area, the paucity of introductory texts is somewhat surprising. Parts of the theory are contained in various books of Lang, especially [135] and [140], and there are books of Koblitz [129] and Robert [210] (the latter now out of print) that concentrate on the analytic and modular theory. In addition, there are survey articles by Cassels [41], which is really a short book, and Tate [289], which is beautifully written, but includes no proofs. Thus the author hopes that this volume fills a real need, both for the serious student who wishes to learn basic facts about the arithmetic of elliptic curves and for the research mathematician who needs a reference source for those same basic facts.

Our approach is more algebraic than that taken in, say, [135] or [140], where many of the basic theorems are derived using complex analytic methods and the Lefschetz principle. For this reason, we have had to rely somewhat more on techniques from algebraic geometry. However, the geometry of (smooth) curves, which is essentially all that we use, does not require a great deal of machinery. And the small price paid in learning a little bit of algebraic geometry is amply repaid in a unity of exposition that, to the author, seems to be lacking when one makes extensive use of either the Lefschetz principle or lengthy, albeit elementary, calculations with explicit polynomial equations.

This last point is worth amplifying. It has been the author's experience that "elementary" proofs requiring page after page of algebra tend to be quite uninstructive. A student may be able to verify such a proof, line by line, and at the end will agree that the proof is complete. But little true understanding results from such a procedure. In this book, our policy is always to state when a result can be proven by such an elementary calculation, indicate briefly how that calculation might be done, and then to give a more enlightening proof that is based on general principles.

The basic (global) theorems in the arithmetic of elliptic curves are the Mordell–Weil theorem, which is proven in Chapter VIII and analyzed more closely in Chapter X, and Siegel's theorem, which is proven in Chapter IX. The reader desiring to reach these results fairly rapidly might take the following path:

I and II (briefly review), III (§§1–8), IV (§§1–6), V (§1)
VII (§§1–5), VIII (§§1–6), IX (§§1–7), X (§§1–6).

This material also makes a good one-semester course, possibly with some time left at the end for special topics. The present volume is built around the notes for such a course, taught by the author at M.I.T. during the spring term of 1983. Of course, there are many other ways to structure a course. For example, one might include all of chapters V and VI, skipping IX and, if pressed for time, X. Other important topics in the arithmetic of elliptic curves, which do not appear in this volume due to time and space limitations, are briefly discussed in Appendix C.

It is certainly true that some of the deepest results in the subject, such as Mazur's theorem bounding torsion over $\mathbb{Q}$ and Faltings' proof of the isogeny conjecture, require many of the resources of modern "SGA-style" algebraic geometry. On the other hand, one needs no machinery at all to write down the equation of an elliptic curve and to do explicit computations with it; so there are many important theorems whose proof requires nothing more than cleverness and hard work. Whether your inclination leans toward heavy machinery or imaginative calculations, you will find much that remains to be discovered in the arithmetic theory of elliptic curves. Happy Hunting!

Acknowledgements

In writing this book, I have consulted a great many sources. Citations have been included for major theorems, but many results that are now considered "standard" have been presented as such. In any case, I can claim no originality for any of the unlabeled theorems in this book, and I apologize in advance to anyone who may feel slighted. The excellent survey articles of Cassels [41] and Tate [289] served as guidelines for organizing the material. (The reader is especially urged to peruse the latter.) In addition to [41] and [289], other sources that were extensively consulted include [135], [139], [186], [210], and [236].

It would not be possible to catalogue all of the mathematicians from whom I learned this beautiful subject, but to all of them, my deepest thanks. I would especially like to thank John Tate, Barry Mazur, Serge Lang, and the "Elliptic Curves Seminar" group at Harvard (1977–1982), whose help and inspiration set me on the road that led to this book. I would also like to thank David Rohrlich and Bill McCallum for their careful reading of the original draft, Gary Cornell and the editorial staff at Springer-Verlag for encouraging me to undertake this project in the first place, and Ann Clee for her meticulous preparation of the manuscript. Finally, I would like to thank my wife, Susan, for her patience and understanding through the turbulent times during which this book was written, and also Deborah and Daniel, for providing much of the turbulence.

Cambridge, Massachusetts   JOSEPH H. SILVERMAN
September, 1985

Acknowledgments for the Second Printing

I would like to thank the following people, who kindly provided corrections that have been incorporated into this second revised printing: Andrew Baker, Arthur Baragar, Wah Keung Chan, Yen-Mei (Julia) Chen, Bob Colemen, Fred Diamond, David Fried, Dick Gross, Ron Jacobwitz, Kevin Keating, Masato Kuwata, Peter Landweber, H.W.

Lenstra Jr., San Ling, Bill McCallum, David Masser, Hwasin Park, Elisabeth Pyle, Ken Ribet, John Rhodes, David Rohrlich, Mike Rosen, Rene Schoof, Udi de Shalit, Alice Silverberg, Glenn Stevens, John Tate, Edlyn Teske, Jaap Top, Paul van Mulbregt, Larry Washington, Don Zagier.

It has unfortunately not been possible to include in this second printing the many important results proven during the past six years, such as the work of Kolyvagin and Rubin on the Birch and Swinnerton-Dyer conjectures (C.16.5) and the finiteness of the Shafarevich–Tate group (X.4.13), Ribet's proof that the conjecture of Shimuara–Taniyama–Weil (C.16.4) implies Fermat's Last Theorem, and recent work of Mestre on elliptic curves of high rank (C §20). The inclusion of such material (and more) will have to await an eventual second edition, so the reader should be aware that some of our general discussion, especially in Appendix C, is out of date. In spite of this obsolescence, it is our hope that this book will continue to provide a useful introduction to the study of the arithmetic of elliptic curves.

Providence, Rhode Island   JOSEPH H. SILVERMAN
August, 1992

Contents

CHAPTER X

CHAPTER XI

APPENDIX A

APPENDIX B

APPENDIX C

Introduction

The study of Diophantine equations, that is, the solution of polynomial equations in integers or rational numbers, has a history stretching back to ancient Greece and beyond. The term *Diophantine geometry* is of more recent origin and refers to the study of Diophantine equations through a combination of techniques from algebraic number theory and algebraic geometry. On the one hand, the problem of finding integer and rational solutions to polynomial equations calls into play the tools of algebraic number theory that describe the rings and fields wherein those solutions lie. On the other hand, such a system of polynomial equations describes an algebraic variety, which is a geometric object. It is the interplay between these two points of view that is the subject of Diophantine geometry.

The simplest sort of equation is linear:

$$aX + bY = c, \qquad a, b, c \in \mathbb{Z}, \qquad a \text{ or } b \neq 0.$$

Such an equation always has rational solutions. It has integer solutions if and only if the greatest common divisor of a and b divides c, and if this occurs, then we can find all solutions using the Euclidean algorithm.

Next in order of difficulty come quadratic equations:

$$aX^2 + bXY + cY^2 + dX + eY + f = 0, \qquad a, \ldots, f \in \mathbb{Z}, \quad a, b \text{ or } c \neq 0.$$

They describe conic sections, and by a suitable change of coordinates *with rational coefficients*, we can transform a given equation into one of the following forms:

$$
\begin{aligned}
AX^2 + BY^2 &= C & &\text{ellipse,} \\
AX^2 - BY^2 &= C & &\text{hyperbola,} \\
AX + BY^2 &= 0 & &\text{parabola.}
\end{aligned}
$$

For quadratic equations we have the following powerful theorem that aids in their solution.

Hasse–Minkowski Theorem 0.1. ([232, IV Theorem 8]) *Let* $f(X, Y) \in \mathbb{Q}[X, Y]$ *be a quadratic polynomial. The equation* $f(X, Y) = 0$ *has a solution* $(x, y) \in \mathbb{Q}^2$ *if and only if it has a solution* $(x, y) \in \mathbb{R}^2$ *and a solution* $(x, y) \in \mathbb{Q}_p^2$ *for every prime p. (Here* $\mathbb{Q}_p$ *is the field of p-adic numbers.)*

In other words, a quadratic polynomial has a solution in $\mathbb{Q}$ if and only if it has a solution in every completion of $\mathbb{Q}$. Hensel's lemma says that checking for solutions in $\mathbb{Q}_p$ is more or less the same as checking for solutions in the finite field $\mathbb{Z}/p\mathbb{Z}$, and this is turn is easily accomplished using quadratic reciprocity. We summarize the steps that go into the Diophantine analysis of quadratic equations.

(1) Analyze the equations over finite fields [quadratic reciprocity].
(2) Use this information to study the equations over complete local fields $\mathbb{Q}_p$ [Hensel's lemma]. (We must also analyze them over $\mathbb{R}$.)
(3) Piece together the local information to obtain results for the global field $\mathbb{Q}$ [Hasse principle].

Where does the geometry appear? Linear and quadratic equations in two variables define curves of genus zero. The above discussion says that we have a fairly good understanding of the arithmetic of such curves. The next simplest case, namely the arithmetic properties of curves of genus one (which are given by cubic equations in two variables), is our object of study in this book. The arithmetic of these so-called *elliptic curves* already presents complexities on which much current research is centered. Further, they provide a standard testing ground for conjectures and techniques that can then be fruitfully applied to the study of curves of higher genus and (abelian) varieties of higher dimension.

Briefly, the organization of this book is as follows. After two introductory chapters giving basic material on algebraic geometry, we start by studying the geometry of elliptic curves over algebraically closed fields (Chapter III). We then follow the program outlined above and investigate the properties of elliptic curves over finite fields (Chapter V), local fields (Chapters VI, VII), and global (number) fields (Chapters VIII, IX, X). Our understanding of elliptic curves over finite and local fields will be fairly satisfactory. However, it turns out that the analogue of the Hasse–Minkowski theorem is false for polynomials of degree greater than 2. This means that the transition from local to global is far more tenuous than in the degree 2 case. We study this problem in some detail in Chapter X. Finally, in Chapter XI we investigate computational aspects of the theory of elliptic curves, especially those that have become important in the field of cryptography.

The theory of elliptic curves is rich, varied, and amazingly vast. The original aim of this book was to provide an essentially self-contained introduction to the basic arithmetic properties of elliptic curves. Even such a limited goal proved to be too ambitious. The material described above is approximately half of what the author had hoped to include. The reader will find a brief discussion and list of references for the omitted topics in Appendix C, about half of which are covered in the companion volume [266] to this book.

Our other goal, that of being self-contained, has been more successful. We have, of course, felt free to state results that every reader should know, even when the proofs are far beyond the scope of this book. However, we have endeavored not to use such results for making further deductions. There are three major exceptions to this general policy. First, we do not prove that every elliptic curve over $\mathbb{C}$ is uniformized by elliptic functions (VI.5.1). This result fits most naturally into a discussion of modular functions, which is one of the omitted topics; it is covered [266, I §4] in the companion volume. Second, we do not prove that over a complete local field, the "nonsingular" points sit with finite index inside the set of all points (VII.6.1). This can be proven by quite explicit polynomial computations (cf. [283]), but they are rather lengthy and have not been included for lack of space. (This result is proven in the companion volume [266, IV §§8, 9].) Finally, in the study of integral points on elliptic curves, we make use of Roth's theorem (IX.1.4) without giving a proof. We include a brief discussion of the proof in (IX §8), and the reader who wishes to see the myriad details can proceed to one of the references listed there.

The prerequisites for reading this book are fairly modest. We assume that the reader has had a first course in algebraic number theory, and thus is acquainted with number fields, rings of integers, prime ideals, ramification, absolute values, completions, etc. The contents of any basic text on algebraic number theory, such as [142, Part I] or [25], should more than suffice. Chapter VI, which deals with elliptic curves over $\mathbb{C}$, assumes a familiarity with the basic principles of complex analysis. In Chapter X, we use a little bit of group cohomology, but just H^0 and H^1. The reader will find in Appendix B the cohomological facts needed to read Chapter X. Finally, since our approach is mainly algebraic, there is the question of background material in algebraic geometry. On the one hand, since much of the theory of elliptic curves can be obtained through the use of explicit equations and calculations, we do not want to require that the reader already know a great deal of algebraic geometry. On the other hand, this being a book on number theory and not algebraic geometry, it would not be reasonable to spend half the book developing from first principles the algebro-geometric facts that we will use. As a compromise, the first two chapters give an introduction to the algebraic geometry of varieties and curves, stating all of the facts that we need, giving complete references, and providing enough proofs so that the reader can gain a flavor for some of the basic techniques used in algebraic geometry.

Numerous exercises have been included at the end of each chapter. The reader desiring to gain a real understanding of the subject is urged to attempt as many as possible. Some of these exercises are (special cases of) results that have appeared in the literature. A list of comments and citations for the exercises may be found on page 461. Exercises with a single asterisk are somewhat more difficult, while two asterisks signal an unsolved problem.

References

Bibliographical references are enclosed in square brackets, e.g., [289, Theorem 6]. Cross-references to theorems, propositions, lemmas, etc., are given in full with the chapter roman numeral or appendix letter, e.g., (IV.3.1) and (B.2.1). Reference to an exercise is given by the chapter number followed by the exercise number, e.g., Exercise 3.6.

Standard Notation

Throughout this book, we use the symbols

$$\mathbb{Z}, \ \mathbb{Q}, \ \mathbb{R}, \ \mathbb{C}, \ \mathbb{F}_q, \ \text{and} \ \mathbb{Z}_\ell$$

to denote the integers, rational numbers, real numbers, complex numbers, a field with q elements, and the ℓ-adic integers, respectively. Further, if R is any ring, then R^* denotes the group of invertible elements of R, and if A is an abelian group, then $A[m]$ denotes the subgroup of A consisting of elements of order dividing m. For a more complete list of notation, see page 467.

Chapter I

Algebraic Varieties

In this chapter we describe the basic objects that arise in the study of algebraic geometry. We set the following notation, which will be used throughout this book.

K a perfect field, i.e., every algebraic extension of K is separable.

$\bar{K}$ a fixed algebraic closure of K.

$G_{\bar{K}/K}$ the Galois group of $\bar{K}/K$.

For this chapter, we also let m and n denote positive integers.

The assumption that K is a perfect field is made solely to simplify our exposition. However, since our eventual goal is to do arithmetic, the field K will eventually be taken to be an algebraic extension of $\mathbb{Q}$, $\mathbb{Q}_p$, or $\mathbb{F}_p$. Thus this restriction on K need not concern us unduly.

For a more extensive exposition of the basic concepts that appear in this chapter, we refer the reader to any introductory book on algebraic geometry, such as [95], [109], [111], or [243].

I.1   Affine Varieties

We begin our study of algebraic geometry with Cartesian (or affine) n-space and its subsets defined by zeros of polynomials.

Definition. *Affine n-space (over K)* is the set of n-tuples

$$\mathbb{A}^n = \mathbb{A}^n(\bar{K}) = \big\{P = (x_1, \ldots, x_n) : x_i \in \bar{K}\big\}.$$

Similarly, the *set of K-rational points of* $\mathbb{A}^n$ is the set

$$\mathbb{A}^n(K) = \big\{P = (x_1, \ldots, x_n) \in \mathbb{A}^n : x_i \in K\big\}.$$

J.H. Silverman, *The Arithmetic of Elliptic Curves, Second Edition*, Graduate Texts in Mathematics 106, DOI 10.1007/978-0-387-09494-6_I,

Notice that the Galois group $G_{\bar{K}/K}$ acts on $\mathbb{A}^n$; for $\sigma \in G_{\bar{K}/K}$ and $P \in \mathbb{A}^n$,

$$P^\sigma = (x_1^\sigma, \ldots, x_n^\sigma).$$

Then $\mathbb{A}^n(K)$ may be characterized by

$$\mathbb{A}^n(K) = \{P \in \mathbb{A}^n : P^\sigma = P \text{ for all } \sigma \in G_{\bar{K}/K}\}.$$

Let $\bar{K}[X] = \bar{K}[X_1, \ldots, X_n]$ be a polynomial ring in n variables, and let $I \subset \bar{K}[X]$ be an ideal. To each such I we associate a subset of $\mathbb{A}^n$,

$$V_I = \{P \in \mathbb{A}^n : f(P) = 0 \text{ for all } f \in I\}.$$

Definition. An (*affine*) *algebraic set* is any set of the form V_I. If V is an algebraic set, the *ideal of* V is given by

$$I(V) = \{f \in \bar{K}[X] : f(P) = 0 \text{ for all } P \in V\}.$$

An algebraic set is *defined over* K if its ideal $I(V)$ can be generated by polynomials in $K[X]$. We denote this by V/K. If V is defined over K, then the *set of K-rational points of* V is the set

$$V(K) = V \cap \mathbb{A}^n(K).$$

Remark 1.1. Note that by the Hilbert basis theorem [8, 7.6], [73, §1.4], all ideals in $\bar{K}[X]$ and $K[X]$ are finitely generated.

Remark 1.2. Let V be an algebraic set, and consider the ideal $I(V/K)$ defined by

$$I(V/K) = \{f \in K[X] : f(P) = 0 \text{ for all } P \in V\} = I(V) \cap K[X].$$

Then we see that V is defined over K if and only if

$$I(V) = I(V/K)\bar{K}[X].$$

Now suppose that V is defined over K and let $f_1, \ldots, f_m \in K[X]$ be generators for $I(V/K)$. Then $V(K)$ is precisely the set of solutions $(x_1, \ldots, x_n)$ to the simultaneous polynomial equations

$$f_1(X) = \cdots = f_m(X) = 0 \quad \text{with } x_1, \ldots, x_n \in K.$$

Thus one of the fundamental problems in the subject of *Diophantine geometry*, namely the solution of polynomial equations in rational numbers, may be said to be the problem of describing sets of the form $V(K)$ when K is a number field.

Notice that if $f(X) \in K[X]$ and $P \in \mathbb{A}^n$, then for any $\sigma \in G_{\bar{K}/K}$,

$$f(P^\sigma) = f(P)^\sigma.$$

Hence if V is defined over K, then the action of $G_{\bar{K}/K}$ on $\mathbb{A}^n$ induces an action on V, and clearly

$$V(K) = \{P \in V : P^\sigma = P \text{ for all } \sigma \in G_{\bar{K}/K}\}.$$

Example 1.3.1. Let V be the algebraic set in $\mathbb{A}^2$ given by the single equation

$$X^2 - Y^2 = 1.$$

Clearly V is defined over K for any field K. Let us assume that $\mathrm{char}(K) \neq 2$. Then the set $V(K)$ is in one-to-one correspondence with $\mathbb{A}^1(K) \smallsetminus \{0\}$, one possible map being

$$\mathbb{A}^1(K) \smallsetminus \{0\} \longrightarrow V(K),$$

$$t \longmapsto \left(\frac{t^2+1}{2t}, \frac{t^2-1}{2t} \right).$$

Example 1.3.2. The algebraic set

$$V : X^n + Y^n = 1$$

is defined over $\mathbb{Q}$. Fermat's last theorem, proven by Andrew Wiles in 1995 [291, 311], states that for all $n \geq 3$,

$$V(\mathbb{Q}) = \begin{cases} \{(1,0),(0,1)\} & \text{if } n \text{ is odd,} \\ \{(\pm 1,0),(0,\pm 1)\} & \text{if } n \text{ is even.} \end{cases}$$

Example 1.3.3. The algebraic set

$$V : Y^2 = X^3 + 17$$

has many $\mathbb{Q}$-rational points, for example

$$(-2,3) \qquad (5234, 378661) \qquad \left(\frac{137}{64}, \frac{2651}{512} \right).$$

In fact, the set $V(\mathbb{Q})$ is infinite. See (I.2.8) and (III.2.4) for further discussion of this example.

Definition. An affine algebraic set V is called an *(affine) variety* if $I(V)$ is a prime ideal in $\bar{K}[X]$. Note that if V is defined over K, it is not enough to check that $I(V/K)$ is prime in $K[X]$. For example, consider the ideal $(X_1^2 - 2X_2^2)$ in $\mathbb{Q}[X_1, X_2]$.

Let V/K be a variety, i.e., V is a variety defined over K. Then the *affine coordinate ring of* V/K is defined by

$$K[V] = \frac{K[X]}{I(V/K)}.$$

The ring $K[V]$ is an integral domain. Its quotient field (field of fractions) is denoted by $K(V)$ and is called the *function field of* V/K. Similarly $\bar{K}[V]$ and $\bar{K}(V)$ are defined by replacing K with $\bar{K}$.

Note that since an element $f \in \bar{K}[V]$ is well-defined up to adding a polynomial vanishing on V, it induces a well-defined function $f : V \to \bar{K}$. If $f(X) \in \bar{K}[X]$ is any polynomial, then $G_{\bar{K}/K}$ acts on f by acting on its coefficients. Hence if V is defined over K, so $G_{\bar{K}/K}$ takes $I(V)$ into itself, then we obtain an action of $G_{\bar{K}/K}$ on $\bar{K}[V]$ and $\bar{K}(V)$. One can check (Exercise 1.12) that $K[V]$ and $K(V)$ are, respectively, the subsets of $\bar{K}[V]$ and $\bar{K}(V)$ fixed by $G_{\bar{K}/K}$. We denote the action of $\sigma \in G_{\bar{K}/K}$ on f by $f \mapsto f^{\sigma}$. Then for all points $P \in V$,

$$\big(f(P)\big)^{\sigma} = f^{\sigma}(P^{\sigma}).$$

Definition. Let V be a variety. The *dimension of* V, denoted by $\dim(V)$, is the transcendence degree of $\bar{K}(V)$ over $\bar{K}$.

Example 1.4. The dimension of $\mathbb{A}^n$ is n, since $\bar{K}(\mathbb{A}^n) = \bar{K}(X_1, \ldots, X_n)$. Similarly, if $V \subset \mathbb{A}^n$ is given by a single nonconstant polynomial equation

$$f(X_1, \ldots, X_n) = 0,$$

then $\dim(V) = n - 1$. (The converse is also true; see [111, I.1.2].) In particular, the examples described in (I.1.3.1), (I.1.3.2), and (I.1.3.3) all have dimension one.

In studying a geometric object, we are naturally interested in whether it looks reasonably "smooth." The next definition formalizes this notion in terms of the usual Jacobian criterion for the existence of a tangent plane.

Definition. Let V be a variety, $P \in V$, and $f_1, \ldots, f_m \in \bar{K}[X]$ a set of generators for $I(V)$. Then V is *nonsingular* (or *smooth*) *at* P if the $m \times n$ matrix

$$\left(\frac{\partial f_i}{\partial X_j}(P) \right)_{\substack{1 \le i \le m \\ 1 \le j \le n}}$$

has rank $n - \dim(V)$. If V is nonsingular at every point, then we say that V is *nonsingular* (or *smooth*).

Example 1.5. Let V be given by a single nonconstant polynomial equation

$$f(X_1, \ldots, X_n) = 0.$$

Then (I.1.4) tells us that $\dim(V) = n - 1$, so $P \in V$ is a singular point if and only if

$$\frac{\partial f}{\partial X_1}(P) = \cdots = \frac{\partial f}{\partial X_n}(P) = 0.$$

Since P also satisfies $f(P) = 0$, this gives $n + 1$ equations for the n coordinates of any singular point. Thus for a "randomly chosen" polynomial f, one would expect V to be nonsingular. We will not pursue this idea further, but see Exercise 1.1.

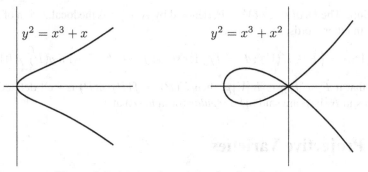

Figure 1.1: A smooth curve and a singular curve.

Example 1.6. Consider the two varieties

$$V_1 : Y^2 = X^3 + X \qquad \text{and} \qquad V_2 : Y^2 = X^3 + X^2.$$

Using (I.1.5), we see that any singular points on V_1 and V_2 satisfy, respectively,

$$V_1^{\text{sing}} : 3X^2 + 1 = 2Y = 0 \qquad \text{and} \qquad V_2^{\text{sing}} : 3X^2 + 2X = 2Y = 0.$$

Thus V_1 is nonsingular, while V_2 has one singular point, namely $(0,0)$. The graphs of $V_1(\mathbb{R})$ and $V_2(\mathbb{R})$ illustrate the difference; see Figure 1.1.

There is another characterization of smoothness, in terms of the functions on the variety V, that is often quite useful. For each point $P \in V$, we define an ideal M_P of $\bar{K}[V]$ by

$$M_P = \{f \in \bar{K}[V] : f(P) = 0\}.$$

Notice that M_P is a maximal ideal, since there is an isomorphism

$$\bar{K}[V]/M_P \longrightarrow \bar{K} \quad \text{given by} \quad f \longmapsto f(P).$$

The quotient M_P/M_P^2 is a finite-dimensional $\bar{K}$-vector space.

Proposition 1.7. *Let V be a variety. A point $P \in V$ is nonsingular if and only if*

$$\dim_{\bar{K}} M_P/M_P^2 = \dim V.$$

PROOF. [111, I.5.1]. (See Exercise 1.3 for a special case.) □

Example 1.8. Consider the point $P = (0,0)$ on the varieties V_1 and V_2 of (I.1.6). In both cases, M_P is the ideal of $\bar{K}[V]$ generated by X and Y, and M_P^2 is the ideal generated by X^2, XY, and Y^2. For V_1 we have

$$X = Y^2 - X^3 \equiv 0 \pmod{M_P^2},$$

so M_P/M_P^2 is generated by Y alone. On the other hand, for V_2 there is no nontrivial relationship between X and Y modulo M_P^2, so M_P/M_P^2 requires both X and Y as generators. Since each V_i has dimension one, (I.1.7) implies that V_1 is smooth at P and V_2 is not.

Definition. The *local ring of V at P*, denoted by $\bar{K}[V]_P$, is the localization of $\bar{K}[V]$ at M_P. In other words,

$$\bar{K}[V]_P = \{F \in \bar{K}(V) : F = f/g \text{ for some } f, g \in \bar{K}[V] \text{ with } g(P) \neq 0\}.$$

Notice that if $F = f/g \in \bar{K}[V]_P$, then $F(P) = f(P)/g(P)$ is well-defined. The functions in $\bar{K}[V]_P$ are said to be *regular* (or *defined*) at P.

I.2   Projective Varieties

Historically, projective space arose through the process of adding "points at infinity" to affine space. We define projective space to be the collection of lines through the origin in affine space of one dimension higher.

Definition. *Projective n-space* (*over K*), denoted by $\mathbb{P}^n$ or $\mathbb{P}^n(\bar{K})$, is the set of all $(n+1)$-tuples

$$(x_0, \ldots, x_n) \in \mathbb{A}^{n+1}$$

such that at least one x_i is nonzero, modulo the equivalence relation

$$(x_0, \ldots, x_n) \sim (y_0, \ldots, y_n)$$

if there exists a $\lambda \in \bar{K}^*$ such that $x_i = \lambda y_i$ for all i. An equivalence class

$$\{(\lambda x_0, \ldots, \lambda x_n) : \lambda \in \bar{K}^*\}$$

is denoted by $[x_0, \ldots, x_n]$, and the individual $x_0, \ldots, x_n$ are called *homogeneous coordinates* for the corresponding point in $\mathbb{P}^n$. The *set of K-rational points in* $\mathbb{P}^n$ is the set

$$\mathbb{P}^n(K) = \{[x_0, \ldots, x_n] \in \mathbb{P}^n : \text{all } x_i \in K\}.$$

Remark 2.1. Note that if $P = [x_0, \ldots, x_n] \in \mathbb{P}^n(K)$, it does not follow that each $x_i \in K$. However, choosing some i with $x_i \neq 0$, it does follow that $x_j/x_i \in K$ for every j.

Definition. Let $P = [x_0, \ldots, x_n] \in \mathbb{P}^n(\bar{K})$. The *minimal field of definition for P* (*over K*) is the field

$$K(P) = K(x_0/x_i, \ldots, x_n/x_i) \quad \text{for any } i \text{ with } x_i \neq 0.$$

The Galois group $G_{\bar{K}/K}$ acts on $\mathbb{P}^n$ by acting on homogeneous coordinates,

$$[x_0, \ldots, x_n]^\sigma = [x_0^\sigma, \ldots, x_n^\sigma].$$

This action is well-defined, independent of choice of homogeneous coordinates, since

$$[\lambda x_0, \ldots, \lambda x_n]^\sigma = [\lambda^\sigma x_0^\sigma, \ldots, \lambda^\sigma x_n^\sigma] = [x_0^\sigma, \ldots, x_n^\sigma].$$

It is not difficult to check that

$$\mathbb{P}^n(K) = \{P \in \mathbb{P}^n : P^\sigma = P \text{ for all } \sigma \in G_{\bar{K}/K}\},$$

and that

$$K(P) = \text{fixed field of } \{\sigma \in G_{\bar{K}/K} : P^\sigma = P\};$$

see Exercise 1.12.

Definition. A polynomial $f \in \bar{K}[X] = \bar{K}[X_0, \ldots, X_n]$ is *homogeneous of degree d* if

$$f(\lambda X_0, \ldots, \lambda X_n) = \lambda^d f(X_0, \ldots, X_n) \quad \text{for all } \lambda \in \bar{K}.$$

An ideal $I \subset \bar{K}[X]$ is *homogeneous* if it is generated by homogeneous polynomials.

Let f be a homogeneous polynomial and let $P \in \mathbb{P}^n$. It makes sense to ask whether $f(P) = 0$, since the answer is independent of the choice of homogeneous coordinates for P. To each homogeneous ideal I we associate a subset of $\mathbb{P}^n$ by the rule

$$V_I = \{P \in \mathbb{P}^n : f(P) = 0 \text{ for all homogeneous } f \in I\}.$$

Definition. A *(projective) algebraic set* is any set of the form V_I for a homogeneous ideal I. If V is a projective algebraic set, the *(homogeneous) ideal of V*, denoted by $I(V)$, is the ideal of $\bar{K}[X]$ generated by

$$\{f \in \bar{K}[X] : f \text{ is homogeneous and } f(P) = 0 \text{ for all } P \in V\}.$$

Such a V is *defined over K*, denoted by V/K, if its ideal $I(V)$ can be generated by homogeneous polynomials in $K[X]$. If V is defined over K, then the *set of K-rational points of V* is the set

$$V(K) = V \cap \mathbb{P}^n(K).$$

As usual, $V(K)$ may also be described as

$$V(K) = \{P \in V : P^\sigma = P \text{ for all } \sigma \in G_{\bar{K}/K}\}.$$

Example 2.2. A *line* in $\mathbb{P}^2$ is an algebraic set given by a linear equation

$$aX + bY + cZ = 0$$

with $a, b, c \in \bar{K}$ not all zero. If, say, $c \neq 0$, then such a line is defined over any field containing a/c and b/c. More generally, a *hyperplane* in $\mathbb{P}^n$ is given by an equation

$$a_0 X_0 + a_1 X_1 + \cdots + a_n X_n = 0$$

with $a_i \in \bar{K}$ not all zero.

Example 2.3. Let V be the algebraic set in $\mathbb{P}^2$ given by the single equation

$$X^2 + Y^2 = Z^2.$$

Then for any field K with $\text{char}(K) \neq 2$, the set $V(K)$ is isomorphic to $\mathbb{P}^1(K)$, for example by the map

$$\mathbb{P}^1(K) \longrightarrow V(K), \qquad [s, t] \longmapsto [s^2 - t^2, 2st, s^2 + t^2].$$

(For the precise definition of "isomorphic," see (I.3.5).)

Remark 2.4. A point of $\mathbb{P}^n(\mathbb{Q})$ has the form $[x_0, \ldots, x_n]$ with $x_i \in \mathbb{Q}$. Multiplying by an appropriate $\lambda \in \mathbb{Q}$, we can clear denominators and common factors from the x_i's. In other words, every $P \in \mathbb{P}^n(\mathbb{Q})$ may be written with homogeneous coordinates $[x_0, \ldots, x_n]$ satisfying

$$x_0, \ldots, x_n \in \mathbb{Z} \quad \text{and} \quad \gcd(x_0, \ldots, x_n) = 1.$$

Note that the x_i's are determined by P up to multiplication by -1.

Thus if an ideal of an algebraic set $V/\mathbb{Q}$ is generated by homogeneous polynomials $f_1, \ldots, f_m \in \mathbb{Q}[X]$, then describing $V(\mathbb{Q})$ is equivalent to finding the solutions to the homogeneous equations

$$f_1(X_0, \ldots, X_n) = \cdots = f_m(X_0, \ldots, X_n) = 0$$

in relatively prime integers $x_0, \ldots, x_n$.

Example 2.5. The algebraic set

$$V : X^2 + Y^2 = 3Z^2$$

is defined over $\mathbb{Q}$. However, $V(\mathbb{Q}) = \emptyset$. To see this, suppose that $[x, y, z] \in V(\mathbb{Q})$ with $x, y, z \in \mathbb{Z}$ and $\gcd(x, y, z) = 1$. Then

$$x^2 + y^2 \equiv 0 \pmod{3},$$

so the fact that -1 is not a square modulo 3 implies that

$$x \equiv y \equiv 0 \pmod{3}.$$

Hence x^2 and y^2 are divisible by 3^2. It follows from the equation for V that 3 also divides z, which contradicts the assumption that $\gcd(x, y, z) = 1$.

This example illustrates a fundamental tool used in the study of Diophantine equations.

> *In order to show that an algebraic set $V/\mathbb{Q}$ has no $\mathbb{Q}$-rational points, it suffices to show that the corresponding homogeneous polynomial equations have no nonzero solutions modulo p for any one prime p (or even for one prime power p^r).*

A more succinct way to phrase this is to say that if $V(\mathbb{Q})$ is nonempty, then $V(\mathbb{Q}_p)$ is nonempty for every p-adic field $\mathbb{Q}_p$. Similarly, $V(\mathbb{R})$ would also be nonempty. One of the reasons that the study of Diophantine equations is so difficult is that the converse to this statement, which is called the *Hasse principle*, does not hold in general. An example, due to Selmer [225, 227], is the equation

$$V : 3X^3 + 4Y^2 + 5Z^3 = 0.$$

One can check that $V(\mathbb{Q}_p)$ is nonempty for every prime p, yet $V(\mathbb{Q})$ is empty. See, e.g., [41, §4] for a proof. Other examples are given in (X.6.5).

Definition. A projective algebraic set is called a (*projective*) *variety* if its homogeneous ideal $I(V)$ is a prime ideal in $\bar{K}[X]$.

It is clear that $\mathbb{P}^n$ contains many copies of $\mathbb{A}^n$. For example, for each $0 \leq i \leq n$, there is an inclusion

$$\phi_i : \mathbb{A}^n \longrightarrow \mathbb{P}^n,$$

$$(y_1, \ldots, y_n) \longmapsto [y_1, y_2, \ldots, y_{i-1}, 1, y_i, \ldots, y_n].$$

We let H_i denote the hyperplane in $\mathbb{P}^n$ given by $X_i = 0$,

$$H_i = \{P = [x_0, \ldots, x_n] \in \mathbb{P}^n : x_i = 0\},$$

and we let U_i be the complement of H_i,

$$U_i = \{P = [x_0, \ldots, x_n] \in \mathbb{P}^n : x_i \neq 0\} = \mathbb{P}^n \smallsetminus H_i.$$

There is a natural bijection

$$\phi_i^{-1} : U_i \longrightarrow \mathbb{A}^n,$$

$$[x_0, \ldots, x_n] \longmapsto \left(\frac{x_0}{x_i}, \frac{x_1}{x_i}, \ldots, \frac{x_{i-1}}{x_i}, \frac{x_{i+1}}{x_i}, \ldots, \frac{x_n}{x_i}, \right).$$

(Note that for any point of $\mathbb{P}^n$ with $x_i \neq 0$, the quantities x_j/x_i are well-defined.) For a fixed i, we will normally identify $\mathbb{A}^n$ with the set U_i in $\mathbb{P}^n$ via the map ϕ_i.

Now let V be a projective algebraic set with homogeneous ideal $I(V) \subset \bar{K}[X]$. Then $V \cap \mathbb{A}^n$, by which we mean $\phi_i^{-1}(V \cap U_i)$ for some fixed i, is an affine algebraic set with ideal $I(V \cap \mathbb{A}^n) \subset \bar{K}[Y]$ given by

$$I(V \cap \mathbb{A}^n) = \{f(Y_1, \ldots, Y_{i-1}, 1, Y_{i+1}, \ldots, Y_n) : f(X_0, \ldots, X_n) \in I(V)\}.$$

Notice that the sets $U_0, \ldots, U_n$ cover all of $\mathbb{P}^n$, so any projective variety V is covered by subsets $V \cap U_0, \ldots, V \cap U_n$, each of which is an affine variety via an appropriate ϕ_i^{-1}. The process of replacing the polynomial $f(X_0, \ldots, X_n)$ with the polynomial $f(Y_1, \ldots, Y_{i-1}, 1, Y_{i+1}, \ldots, Y_n)$ is called *dehomogenization with respect to* X_i.

This process can be reversed. For any $f(Y) \in \bar{K}[Y]$, we define

$$f^*(X_0, \ldots, X_n) = X_i^d f \left(\frac{X_0}{X_i}, \frac{X_1}{X_i}, \ldots, \frac{X_{i-1}}{X_i}, \frac{X_{i+1}}{X_i}, \ldots, \frac{X_n}{X_i} \right),$$

where $d = \deg(f)$ is the smallest integer for which f^* is a polynomial. We say that f^* is the *homogenization of f with respect to* X_i.

Definition. Let $V \subset \mathbb{A}^n$ be an affine algebraic set with ideal $I(V)$, and consider V as a subset of $\mathbb{P}^n$ via

$$V \subset \mathbb{A}^n \xrightarrow{\phi_i} \mathbb{P}^n.$$

The *projective closure of* V, denoted by $\bar{V}$, is the projective algebraic set whose homogeneous ideal $I(\bar{V})$ is generated by

$$\{f^*(X) : f \in I(V)\}.$$

Proposition 2.6. (a) *Let V be an affine variety. Then $\bar{V}$ is a projective variety, and*

$$V = \bar{V} \cap \mathbb{A}^n.$$

(b) *Let V be a projective variety. Then $V \cap \mathbb{A}^n$ is an affine variety, and either*

$$V \cap \mathbb{A}^n = \emptyset \qquad or \qquad V = \overline{V \cap \mathbb{A}^n}.$$

(c) *If an affine (respectively projective) variety V is defined over K, then $\bar{V}$ (respectively $V \cap \mathbb{A}^n$) is also defined over K.*

PROOF. See [111, I.2.3] for (a) and (b). Part (c) is clear from the definitions. □

Remark 2.7. In view of (I.2.6), each affine variety may be identified with a unique projective variety. Notationally, since it is easier to deal with affine coordinates, we will often say "let V be a projective variety" and write down some inhomogeneous equations, with the understanding that V is the projective closure of the indicated affine variety W. The points of $V \smallsetminus W$ are called the *points at infinity* on V.

Example 2.8. Let V be the projective variety given by the equation

$$V : Y^2 = X^3 + 17.$$

This really means that V is the variety in $\mathbb{P}^2$ given by the homogeneous equation

$$\bar{Y}^2 \bar{Z} = \bar{X}^3 + 17\bar{Z}^3,$$

the identification being

$$X = \bar{X}/\bar{Z}, \qquad Y = \bar{Y}/\bar{Z}.$$

This variety has one point at infinity, namely $[0, 1, 0]$, obtained by setting $\bar{Z} = 0$. Thus, for example,

$$V(\mathbb{Q}) = \left\{ (x, y) \in \mathbb{A}^2(\mathbb{Q}) : y^2 = x^3 + 17 \right\} \cup \left\{ [0, 1, 0] \right\}.$$

In (I.1.3.3) we listed several points in $V(\mathbb{Q})$. The reader may verify (Exercise 1.5) that the line connecting any two points of $V(\mathbb{Q})$ intersects V in a third point of $V(\mathbb{Q})$ (provided that the line is not tangent to V). Using this secant line procedure repeatedly leads to infinitely many points in $V(\mathbb{Q})$, although this is by no means obvious. The variety V is an *elliptic curve*, and as such, it provides the first example of the varieties that will be our principal object of study in this book. See (III.2.4) for further discussion of this example.

   Many important properties of a projective variety V may now be defined in terms of the affine subvariety $V \cap \mathbb{A}^n$.

Definition. Let V/K be a projective variety and choose $\mathbb{A}^n \subset \mathbb{P}^n$ such that $V \cap \mathbb{A}^n \neq \emptyset$. The *dimension of V* is the dimension of $V \cap \mathbb{A}^n$.

   The *function field of V*, denoted by $K(V)$, is the function field of $V \cap \mathbb{A}^n$, and similarly for $\bar{K}(V)$. We note that for different choices of $\mathbb{A}^n$, the different $K(V)$ are canonically isomorphic, so we may identify them. (See (I.2.9) for another description of $K(V)$.)

Definition. Let V be a projective variety, let $P \in V$, and choose $\mathbb{A}^n \subset \mathbb{P}^n$ with $P \in \mathbb{A}^n$. Then V is *nonsingular* (or *smooth*) *at* P if $V \cap \mathbb{A}^n$ is nonsingular at P. The *local ring of V at P*, denoted by $\bar{K}[V]_P$, is the local ring of $V \cap \mathbb{A}^n$ at P. A function $F \in \bar{K}(V)$ is *regular* (or *defined*) *at* P if it is in $\bar{K}[V]_P$, in which case it makes sense to evaluate F at P.

Remark 2.9. The function field of $\mathbb{P}^n$ may also be described as the subfield of $\bar{K}(X_0, \ldots, X_n)$ consisting of rational functions $F(X) = f(X)/g(X)$ for which f and g are *homogeneous* polynomials of the *same* degree. Such an expression gives a well-defined function on $\mathbb{P}^n$ at all point P where $g(P) \neq 0$. Similarly, the function field of a projective variety V is the field of rational functions $F(X) = f(X)/g(X)$ such that:

(i) f and g are homogeneous of the same degree;

(ii) $g \notin I(V)$;

(iii) two functions f_1/g_1 and f_2/g_2 are identified if $f_1 g_2 - f_2 g_1 \in I(V)$.

I.3 Maps Between Varieties

In this section we look at algebraic maps between projective varieties. These are maps that are defined by rational functions.

Definition. Let V_1 and $V_2 \subset \mathbb{P}^n$ be projective varieties. A *rational map from V_1 to V_2* is a map of the form

$$f : V_1 \longrightarrow V_2, \qquad \phi = [f_0, \ldots, f_n],$$

where the functions $f_0, \ldots, f_n \in \bar{K}(V_1)$ have the property that for every point $P \in V_1$ at which $f_0, \ldots, f_n$ are all defined,

$$\phi(P) = [f_0(P), \ldots, f_n(P)] \in V_2.$$

If V_1 and V_2 are defined over K, then $G_{\bar{K}/K}$ acts on ϕ in the obvious way,

$$\phi^\sigma(P) = [f_0^\sigma(P), \ldots, f_n^\sigma(P)].$$

Notice that we have the formula

$$\phi(P)^\sigma = \phi^\sigma(P^\sigma) \qquad \text{for all } \sigma \in G_{\bar{K}/K} \text{ and } P \in V_1.$$

If, in addition, there is some $\lambda \in \bar{K}^*$ such that $\lambda f_0, \ldots, \lambda f_n \in K(V_1)$, then ϕ is said to be *defined over K*. Note that $[f_0, \ldots, f_n]$ and $[\lambda f_0, \ldots, \lambda f_n]$ give the same map on points. As usual, it is true that ϕ is defined over K if and only if $\phi = \phi^\sigma$ for all $\sigma \in G_{\bar{K}/K}$; see Exercise 1.12c.

Remark 3.1. A rational map $\phi : V_1 \to V_2$ is not necessarily a well-defined function at every point of V_1. However, it may be possible to evaluate $\phi(P)$ at points P of V_1 where some f_i is not regular by replacing each f_i by gf_i for an appropriate $g \in \bar{K}(V_1)$.

Definition. A rational map

$$\phi = [f_0, \ldots, f_n] : V_1 \longrightarrow V_2$$

is *regular* (or *defined*) at $P \in V_1$ if there is a function $g \in \bar{K}(V_1)$ such that

(i) each gf_i is regular at P;

(ii) there is some i for which $(gf_i)(P) \neq 0$.

If such a g exists, then we set

$$\phi(P) = \big[(gf_0)(P), \ldots, (gf_n)(P)\big].$$

N.B. It may be necessary to take different g's for different points. A rational map that is regular at every point is called a *morphism*.

Remark 3.2. Let $V_1 \subset \mathbb{P}^m$ and $V_2 \subset \mathbb{P}^n$ be projective varieties. Recall (I.2.9) that the functions in $\bar{K}(V_1)$ may be described as quotients of homogeneous polynomials in $\bar{K}[X_0, \ldots, X_m]$ having the same degree. Thus by multiplying a rational map $\phi = [f_0, \ldots, f_n]$ by a homogeneous polynomial that "clears the denominators" of the f_i's, we obtain the following alternative definition:

A *rational map* $\phi : V_1 \to V_2$ is a map of the form

$$\phi = \big[\phi_0(X), \ldots, \phi_n(X)\big],$$

where

(i) the $\phi_i(X) \in \bar{K}[X] = \bar{K}[X_0, \ldots, X_n]$ are homogeneous polynomials, not all in $I(V_1)$, having the same degree;

(ii) for very $f \in I(V_2)$,

$$f\big(\phi_0(X), \ldots, \phi_n(X)\big) \in I(V_1).$$

Clearly, $\phi(P)$ is well-defined provided that some $\phi_i(P) \neq 0$. However, even if all $\phi_i(P) = 0$, it may be possible to alter ϕ so as to make sense of $\phi(P)$. We make this precise as follows:

A rational map $\phi = [\phi_0, \ldots, \phi_n] : V_1 \to V_2$ as above is *regular* (or *defined*) at $P \in V_1$ if there exist homogeneous polynomials $\psi_0, \ldots, \psi_n \in \bar{K}[X]$ such that

(i) $\psi_0, \ldots, \psi_n$ have the same degree;

(ii) $\phi_i \psi_j \equiv \phi_j \psi_i \pmod{I(V_1)}$ for all $0 \le i, j \le n$;

(iii) $\psi(P) \neq 0$ for some i.

If this occurs, then we set

$$\phi(P) = [\psi_0(P), \ldots, \psi_n(P)].$$

As above, a rational map that is everywhere regular is called a *morphism*.

Remark 3.3. Let $\phi = [\phi_0, \ldots, \phi_n] : \mathbb{P}^m \to \mathbb{P}^n$ be a rational map as in (I.3.2), where the $\phi_i \in \bar{K}[X]$ are homogeneous polynomials of the same degree. Since $\bar{K}[X]$ is a unique factorization domain (UFD), we may assume that the ϕ_i's have no common factor. Then ϕ is regular at a point $P \in \mathbb{P}^m$ if and only if some $\phi_i(P) \neq 0$. (Note that $I(\mathbb{P}^m) = (0)$, so there is no way to alter the ϕ_i's.) Hence ϕ is a morphism if and only if the ϕ_i's have no common zero in $\mathbb{P}^m$.

Definition. Let V_1 and V_2 be varieties. We say that V_1 and V_2 are *isomorphic*, and write $V_1 \cong V_2$, if there are morphisms $\phi : V_1 \to V_2$ and $\psi : V_2 \to V_1$ such that $\psi \circ \phi$ and $\phi \circ \psi$ are the identity maps on V_1 and V_2, respectively. We say that V_1/K and V_2/K are *isomorphic over K* if ϕ and ψ can be defined over K. Note that both ϕ and ψ must be morphisms, not merely rational maps.

Remark 3.4. If $\phi : V_1 \to V_2$ is an isomorphism defined over K, then ϕ identifies $V_1(K)$ with $V_2(K)$. Hence for Diophantine problems, it suffices to study any one variety in a given K-isomorphism class of varieties.

Example 3.5. Assume that $\text{char}(K) \neq 2$ and let V be the variety from (I.2.3),

$$V : X^2 + Y^2 = Z^2.$$

Consider the rational map

$$\phi : V \longrightarrow \mathbb{P}^1, \qquad \phi = [X + Z, Y].$$

Clearly ϕ is regular at every point of V except possibly at $[1, 0, -1]$, i.e., at the point where $X + Z = Y = 0$. However, using

$$(X + Z)(X - Z) \equiv -Y^2 \pmod{I(V)},$$

we have

$$\phi = [X + Z, Y] = [X^2 - Z^2, Y(X - Z)] = [-Y^2, Y(X - Z)] = [-Y, X - Z].$$

Thus

$$\phi([1, 0, -1]) = [0, 2] = [0, 1],$$

so ϕ is regular at every point of V, i.e., ϕ is a morphism. One easily checks that the map

$$\psi : \mathbb{P}^1 \longrightarrow V, \qquad \psi = [S^2 - T^2, 2ST, S^2 + T^2],$$

is a morphism and provides an inverse for ϕ, so V and $\mathbb{P}^1$ are isomorphic.

Example 3.6. Using (I.3.3), we see that the rational map

$$\phi : \mathbb{P}^2 \longrightarrow \mathbb{P}^2, \qquad \phi = [X^2, XY, Z^2],$$

is regular everywhere except at the point $[0, 1, 0]$.

Example 3.7. Let V be the variety

$$V : Y^2 Z = X^3 + X^2 Z$$

and consider the rational maps

$$\psi : \mathbb{P}^1 \to V, \qquad\qquad \psi = [(S^2 - T^2)T, (S^2 - T^2)S, T^3],$$
$$\phi : V \to \mathbb{P}^1, \qquad\qquad \phi = [Y, X].$$

Here ψ is a morphism, while ϕ is not regular at $[0, 0, 1]$. Not coincidentally, the point $[0, 0, 1]$ is a singular point of V; see (II.2.1). We emphasize that although the compositions $\phi \circ \psi$ and $\psi \circ \phi$ are the identity map wherever they are defined, the maps ϕ and ψ are not isomorphisms, because ϕ is not a morphism.

Example 3.8. Consider the varieties

$$V_1 : X^2 + Y^2 = Z^2 \qquad \text{and} \qquad V_2 : X^2 + Y^2 = 3Z^2.$$

They are not isomorphic over $\mathbb{Q}$, since $V_2(\mathbb{Q}) = \emptyset$ from (I.2.5), while $V_1(\mathbb{Q})$ contains lots of points. (More precisely, $V_1(\mathbb{Q}) = \mathbb{P}^1(\mathbb{Q})$ from (I.3.5).) However, the varieties V_1 and V_2 are isomorphic over $\mathbb{Q}(\sqrt{3})$, an isomorphism being given by

$$\phi : V_2 \longrightarrow V_1, \qquad \phi = [X, Y, \sqrt{3}\, Z].$$

Exercises

1.1. Let $A, B \in \bar{K}$. Characterize the values of A and B for which each of the following varieties is singular. In particular, as (A, B) ranges over $\mathbb{A}^2$, show that the "singular values" lie on a one-dimensional subset of $\mathbb{A}^2$, so "most" values of (A, B) give a nonsingular variety.
  (a) $V : Y^2 Z + AXYZ + BYZ^2 = X^3$.
  (b) $V : Y^2 Z = X^3 + AXZ^2 + BZ^3$. (You may assume that $\text{char}(K) \neq 2$.)

1.2. Find the singular point(s) on each of the following varieties. Sketch $V(\mathbb{R})$.
  (a) $V : Y^2 = X^3$ in $\mathbb{A}^2$.
  (b) $V : 4X^2 Y^2 = (X^2 + Y^2)^3$ in $\mathbb{A}^2$.
  (c) $V : Y^2 = X^4 + Y^4$ in $\mathbb{A}^2$.
  (d) $V : X^2 + Y^2 = (Z - 1)^2$ in $\mathbb{A}^3$.

1.3. Let $V \subset \mathbb{A}^n$ be a variety given by a single equation as in (I.1.4). Prove that a point $P \in V$ is nonsingular if and only if

$$\dim_{\bar{K}} M_P/M_P^2 = \dim V.$$

(*Hint.* Let $f = 0$ be the equation of V and define the *tangent plane of V at P* by

$$T = \left\{ (y_1, \ldots, y_n) \in \mathbb{A}^n : \sum_{i=1}^{n} \left(\frac{\partial f}{\partial X_i}(P) \right) y_i = 0 \right\}.$$

Show that the map

$$M_P/M_P^2 \times T \longrightarrow \bar{K}, \qquad (g, y) \longmapsto \sum_{i=1}^{n} \left(\frac{\partial g}{\partial X_i}(P) \right) y_i,$$

is a well-defined perfect pairing of $\bar{K}$-vector spaces. Now use (I.1.5).)

1.4. Let $V/\mathbb{Q}$ be the variety

$$V : 5X^2 + 6XY + 2Y^2 = 2YZ + Z^2.$$

Prove that $V(\mathbb{Q}) = \emptyset$.

1.5. Let $V/\mathbb{Q}$ be the projective variety

$$V : Y^2 = X^3 + 17,$$

and let $P_1 = (x_1, y_1)$ and $P_2 = (x_2, y_2)$ be distinct points of V. Let L be the line through P_1 and P_2.
  (a) Show that $V \cap L = \{P_1, P_2, P_3\}$ and express $P_3 = (x_3, y_3)$ in terms of P_1 and P_2. (If L is tangent to V, then P_3 may equal P_1 or P_2.)
  (b) Calculate P_3 for $P_1 = (-1, 4)$ and $P_2 = (2, 5)$.
  (c) Show that if $P_1, P_2 \in V(\mathbb{Q})$, then $P_3 \in V(\mathbb{Q})$.

1.6. Let V be the variety

$$V : Y^2 Z = X^3 + Z^3.$$

Show that the map

$$\phi : V \longrightarrow \mathbb{P}^2, \qquad \phi = [X^2, XY, Z^2],$$

is a morphism. (Notice that ϕ does not give a morphism $\mathbb{P}^2 \to \mathbb{P}^2$.)

1.7. Let V be the variety

$$V : Y^2 Z = X^3,$$

and let ϕ be the map

$$\phi : \mathbb{P}^1 \longrightarrow V, \qquad \phi = [S^2 T, S^3, T^3].$$

  (a) Show that ϕ is a morphism.
  (b) Find a rational map $\psi : V \to \mathbb{P}^1$ such that $\phi \circ \psi$ and $\psi \circ \phi$ are the identity map wherever they are defined.
  (c) Is ϕ an isomorphism?

1.8. Let $\mathbb{F}_q$ be a finite field with q elements and let $V \subset \mathbb{P}^n$ be a variety defined over $\mathbb{F}_q$.
  (a) Prove that the qth-power map

$$\phi = [X_0^q, \ldots, X_n^q]$$

is a morphism $\phi : V \to V$. It is called the *Frobenius morphism*.
  (b) Prove that ϕ is one-to-one and onto.
  (c) Prove that ϕ is not an isomorphism.

(d) Prove that $V(\mathbb{F}_q) = \{P \in V : \phi(P) = P\}$.

1.9. If $m > n$, prove that there are no nonconstant morphisms $\mathbb{P}^m \to \mathbb{P}^n$. (*Hint.* Use the dimension theorem [111, I.7.2].)

1.10. For each prime $p \geq 3$, let $V_p \subset \mathbb{P}^2$ be the variety given by the equation

$$V_p : X^2 + Y^2 = pZ^2.$$

(a) Prove that V_p is isomorphic to $\mathbb{P}^1$ over $\mathbb{Q}$ if and only if $p \equiv 1 \pmod 4$.
(b) Prove that for $p \equiv 3 \pmod 4$, no two of the V_p's are isomorphic over $\mathbb{Q}$.

1.11. (a) Let $f \in K[X_0, \ldots, X_n]$ be a homogeneous polynomial, and let

$$V = \{P \in \mathbb{P}^n : f(P) = 0\}$$

be the hypersurface defined by f. Prove that if a point $P \in V$ is singular, then

$$\frac{\partial f}{\partial X_0}(P) = \cdots = \frac{\partial f}{\partial X_n}(P) = 0.$$

Thus for hypersurfaces in projective space, we can check for smoothness using homogeneous coordinates.
(b) Let $n \geq 1$, and let $W \subset \mathbb{P}^n$ be a smooth algebraic set, each of whose component varieties has dimension $n - 1$. Prove that W is a variety. (*Hint.* First use Krull's Hauptidealsatz [8, page 122], [73, Theorem 10.1], to show that W is the zero of a single homogeneous polynomial.)

1.12. (a) Let V/K be an affine variety. Prove that

$$K[V] = \{f \in \bar{K}[V] : f^\sigma = f \text{ for all } \sigma \in G_{\bar{K}/K}\}.$$

(*Hint.* One inclusion is clear. For the other, choose some polynomial $F \in \bar{K}[X]$ with $F \equiv f \pmod{I(V)}$. Show that the map $G_{\bar{K}/K} \to I(V)$ defined by $\sigma \mapsto F^\sigma - F$ is a 1-cocycle; see (B §2). Now use (B.2.5a) to conclude that there exists a $G \in I(V)$ such that $F + G \in K[X]$.)
(b) Prove that

$$\mathbb{P}^n(K) = \{P \in \mathbb{P}^n(\bar{K}) : P^\sigma = P \text{ for all } \sigma \in G_{\bar{K}/K}\}.$$

(*Hint.* Write $P = [x_0, \ldots, x_n]$. If $P = P^\sigma$, then there is a $\lambda_\sigma \in \bar{K}^*$ such that $x_i^\sigma = \lambda_\sigma x_i$ for all $0 \leq i \leq n$. Show that the map $\sigma \mapsto \lambda_\sigma$ gives a 1-cocycle from $G_{\bar{K}/K}$ to $\bar{K}^*$. Now use Hilbert's Theorem 90 (B.2.5b) to find an $\alpha \in \bar{K}^*$ such that $[\alpha x_0, \ldots, \alpha x_n] \in \mathbb{P}^n(K)$.)
(c) Let $\phi : V_1 \to V_2$ be a rational map of projective varieties. Prove that ϕ is defined over K if and only if $\phi^\sigma = \phi$ for every $\sigma \in G_{\bar{K}/K}$. (*Hint.* Use (a) and (b).)

Chapter II

Algebraic Curves

In this chapter we present basic facts about algebraic curves, i.e., projective varieties of dimension one, that will be needed for our study of elliptic curves. Actually, since elliptic curves are curves of genus one, one of our tasks will be to define the genus of a curve. As in Chapter I, we give references for those proofs that are not included. There are many books in which the reader will find more material on the subject of algebraic curves, for example [111, Chapter IV], [133], [180], [243], [99, Chapter 2], and [302].

We recall the following notation from Chapter I that will be used in this chapter. Here C denotes a curve and $P \in C$ is a point of C.

C/K	C is defined over K.
$\bar{K}(C)$	the function field of C over $\bar{K}$.
$K(C)$	the function field of C over K.
$\bar{K}[C]_P$	the local ring of C at P.
M_P	the maximal ideal of $\bar{K}[C]_P$.

II.1 Curves

By a *curve* we will always mean a projective variety of dimension one. We generally deal with curves that are smooth. Examples of smooth curves include $\mathbb{P}^1$, (I.2.3), and (I.2.8). We start by describing the local rings at points on a smooth curve.

Proposition 1.1. *Let C be a curve and $P \in C$ a smooth point. Then $\bar{K}[C]_P$ is a discrete valuation ring.*

PROOF. From (I.1.7), the vector space M_P/M_P^2 is a one-dimensional vector space over the field $\bar{K} = \bar{K}[C]_P/M_P$. Now use [8, Proposition 9.2] or Exercise 2.1.   □

Definition. Let C be a curve and $P \in C$ a smooth point. The (*normalized*) *valuation on $\bar{K}[C]_P$* is given by

J.H. Silverman, *The Arithmetic of Elliptic Curves, Second Edition*, Graduate Texts in Mathematics 106, DOI 10.1007/978-0-387-09494-6_II,
© Springer Science+Business Media, LLC 2009

$$\mathrm{ord}_P : \bar{K}[C]_P \longrightarrow \{0, 1, 2, \ldots\} \cup \{\infty\},$$
$$\mathrm{ord}_P(f) = \sup\{d \in \mathbb{Z} : f \in M_P^d\}.$$

Using $\mathrm{ord}_P(f/g) = \mathrm{ord}_P(f) - \mathrm{ord}_P(g)$, we extend ord_P to $\bar{K}(C)$,

$$\mathrm{ord}_P : \bar{K}(C) \longrightarrow \mathbb{Z} \cup \infty.$$

A *uniformizer for C at P* is any function $t \in \bar{K}(C)$ with $\mathrm{ord}_P(t) = 1$, i.e., a generator for the ideal M_P.

Remark 1.1.1. If $P \in C(K)$, then it is not hard to show that $K(C)$ contains uniformizers for P; see Exercise 2.16.

Definition. Let C and P be as above, and let $f \in \bar{K}(C)$. The *order of f at P* is $\mathrm{ord}_P(f)$. If $\mathrm{ord}_P(f) > 0$, then f has a *zero* at P, and if $\mathrm{ord}_P(f) < 0$, then f has a *pole* at P. If $\mathrm{ord}_P(f) \geq 0$, then f is *regular* (or *defined*) at P and we can evaluate $f(P)$. Otherwise f has a pole at P and we write $f(P) = \infty$.

Proposition 1.2. *Let C be a smooth curve and $f \in \bar{K}(C)$ with $f \neq 0$. Then there are only finitely many points of C at which f has a pole or zero. Further, if f has no poles, then $f \in \bar{K}$.*

PROOF. See [111, I.6.5], [111, II.6.1], or [243, III §1] for the finiteness of the number of poles. To deal with the zeros, look instead at $1/f$. The last statement is [111, I.3.4a] or [243, I §5, Corollary 1]. $\qquad\square$

Example 1.3. Consider the two curves

$$C_1 : Y^2 = X^3 + X \qquad \text{and} \qquad C_2 : Y^2 = X^3 + X^2.$$

(Remember our convention (I.2.7) concerning affine equations for projective varieties. Each of C_1 and C_2 has a single point at infinity.) Let $P = (0, 0)$. Then C_1 is smooth at P and C_2 is not (I.1.6). The maximal ideal M_P of $\bar{K}[C_1]_P$ has the property that M_P/M_P^2 is generated by Y (I.1.8), so for example,

$$\mathrm{ord}_P(Y) = 1, \qquad \mathrm{ord}_P(X) = 2, \qquad \mathrm{ord}_P(2Y^2 - X) = 2.$$

(For the last, note that $2Y^2 - X = 2X^3 + X$.) On the other hand, $\bar{K}[C_2]_P$ is not a discrete valuation ring.

The next proposition is useful in dealing with curves over fields of characteristic $p > 0$. (See also Exercise 2.15.)

Proposition 1.4. *Let C/K be a curve, and let $t \in K(C)$ be a uniformizer at some nonsingular point $P \in C(K)$. Then $K(C)$ is a finite separable extension of $K(t)$.*

PROOF. The field $K(C)$ is clearly a finite (algebraic) extension of $K(t)$, since it is finitely generated over K, has transcendence degree one over K (since C is a curve), and $t \notin K$. Let $x \in K(C)$. We claim that x is separable over $K(t)$.

In any case, x is algebraic over $K(t)$, so it satisfies some polynomial relation

$$\sum a_{ij}t^i x^j = 0, \qquad \text{where} \quad \Phi(T,X) = \sum a_{ij}T^i X^j \in K[X,T].$$

We may further assume that Φ is chosen so as to have minimal degree in X, i.e., $\Phi(t,X)$ is a minimal polynomial for x over $K(t)$. Let $p = \operatorname{char}(K)$. If Φ contains a nonzero term $a_{ij}T^i X^j$ with $j \not\equiv 0 \pmod{p}$, then $\partial \Phi(t,X)/\partial X$ is not identically 0, so x is separable over $K(t)$.

Suppose instead that $\Phi(T,X) = \Psi(T,X^p)$. We proceed to derive a contradiction. The main point to note is that if $F(T,X) \in K[T,X]$ is any polynomial, then $F(T^p,X^p)$ is a pth power. This is true because we have assumed that K is perfect, which implies that every element of K is a pth power. Thus if $F(T,X) = \sum \alpha_{ij}T^i X^j$, then writing $\alpha_{ij} = \beta_{ij}^p$ gives $F(T^p,X^p) = \left(\sum \beta_{ij}T^i X^j\right)^p$.

We regroup the terms in $\Phi(T,X) = \Psi(T,X^p)$ according to powers of T modulo p. Thus

$$\Phi(T,X) = \Psi(T,X^p) = \sum_{k=0}^{p-1} \left(\sum_{i,j} b_{ijk}T^{ip}X^{jp} \right) T^k = \sum_{k=0}^{p-1} \phi_k(T,X)^p T^k.$$

By assumption we have $\Phi(t,x) = 0$. On the other hand, since t is a uniformizer at P, we have

$$\operatorname{ord}_P\!\big(\phi_k(t,x)^p t^k\big) = p \operatorname{ord}_P\!\big(\phi_k(t,x)\big) + k \operatorname{ord}_P(t) \equiv k \pmod{p}.$$

Hence each of the terms in the sum $\sum \phi_k(t,x)^p t^k$ has a distinct order at P, so every term must vanish,

$$\phi_0(t,x) = \phi_1(t,x) = \cdots = \phi_{p-1}(t,x) = 0.$$

But at least one of the $\phi_k(T,X)$'s must involve X, and for that k, the relation $\phi_k(t,x) = 0$ contradicts our choice of $\Phi(t,X)$ as a minimal polynomial for x over $K(t)$. (Note that $\deg_X \phi_k(T,X) \le \frac{1}{p} \deg_X \Phi(T,X)$.) This contradiction completes the proof that x is separable over $K(t)$. $\qquad\square$

II.2 Maps Between Curves

We start with the fundamental result that for smooth curves, a rational map is defined at every point.

Proposition 2.1. *Let C be a curve, let $V \subset \mathbb{P}^N$ be a variety, let $P \in C$ be a smooth point, and let $\phi : C \to V$ be a rational map. Then ϕ is regular at P. In particular, if C is smooth, then ϕ is a morphism.*

PROOF. Write $\phi = [f_0, \ldots, f_N]$ with functions $f_i \in \bar{K}(C)$, and choose a uniformizer $t \in \bar{K}(C)$ for C at P. Let

$$n = \min_{0 \le i \le N} \mathrm{ord}_P(f_i).$$

Then

$$\mathrm{ord}_P(t^{-n} f_i) \ge 0 \quad \text{for all } i \qquad \text{and} \qquad \mathrm{ord}_P(t^{-n} f_j) = 0 \quad \text{for some } j.$$

Hence every $t^{-n} f_i$ is regular at P, and $(t^{-n} f_j)(P) \ne 0$. Therefore ϕ is regular at P.   □

See (I.3.6) and (I.3.7) for examples where (II.2.1) is false if P is not smooth or if C has dimension greater than 1.

Example 2.2. Let C/K be a smooth curve and let $f \in K(C)$ be a function. Then f defines a rational map, which we also denote by f,

$$f : C \longrightarrow \mathbb{P}^1, \qquad P \longmapsto [f(P), 1].$$

From (II.2.1), this map is actually a morphism. It is given explicitly by

$$f(P) = \begin{cases} [f(P), 1] & \text{if } f \text{ is regular at } P, \\ [1, 0] & \text{if } f \text{ has a pole at } P. \end{cases}$$

Conversely, let

$$\phi : C \longrightarrow \mathbb{P}^1, \qquad \phi = [f, g],$$

be a rational map defined over K. Then either $g = 0$, in which case ϕ is the constant map $\phi = [1, 0]$, or else ϕ is the map corresponding to the function $f/g \in K(C)$. Denoting the former map by ∞, we thus have a one-to-one correspondence

$$K(C) \cup \{\infty\} \longleftrightarrow \{\text{maps } C \to \mathbb{P}^1 \text{ defined over } K\}.$$

We will often implicitly identify these two sets.

Theorem 2.3. *Let $\phi : C_1 \to C_2$ be a morphism of curves. Then ϕ is either constant or surjective.*

PROOF. See [111, II.6.8] or [243, I §5, Theorem 4].   □

Let C_1/K and C_2/K be curves and let $\phi : C_1 \to C_2$ be a nonconstant rational map defined over K. Then composition with ϕ induces an injection of function fields fixing K,

$$\phi^* : K(C_2) \longrightarrow K(C_1), \qquad \phi^* f = f \circ \phi.$$

Theorem 2.4. *Let C_1/K and C_2/K be curves.*
 (a) *Let $\phi : C_1 \to C_2$ be a nonconstant map defined over K. Then $K(C_1)$ is a finite extension of $\phi^*(K(C_2))$.*
 (b) *Let $\iota : K(C_2) \to K(C_1)$ be an injection of function fields fixing K. Then there exists a unique nonconstant map $\phi : C_1 \to C_2$ (defined over K) such that $\phi^* = \iota$.*

(c) Let $\mathbb{K} \subset K(C_1)$ be a subfield of finite index containing K. Then there exist a smooth curve C'/K, unique up to K-isomorphism, and a nonconstant map $\phi : C_1 \to C'$ defined over K such that $\phi^* K(C') = \mathbb{K}$.

PROOF. (a) [111, II.6.8].

(b) Let $C_1 \subset \mathbb{P}^N$, and for each i, let $g_i \in K(C_2)$ be the function on C_2 corresponding to X_i/X_0. (Relabeling if necessary, we may assume that C_2 is not contained in the hyperplane $X_0 = 0$.) Then

$$\phi = \big[1, \iota(g_1), \ldots, \iota(g_N)\big]$$

gives a map $\phi : C_1 \to C_2$ with $\phi^* = \iota$. (Note that ϕ is not constant, since the g_i's cannot all be constant and ι is injective.) Finally, if $\psi = [f_0, \ldots, f_N]$ is another map with $\psi^* = \iota$, then for each i,

$$f_i/f_0 = \psi^* g_i = \phi^* g_i = \iota(g_i),$$

which shows that $\psi = \phi$.

(c) See [111, I.6.12] for the case that K is algebraically closed. The general case can be proven similarly, or it may be deduced from the algebraically closed case by examining $G_{\bar{K}/K}$-invariants. $\qquad\square$

Definition. Let $\phi : C_1 \to C_2$ be a map of curves defined over K. If ϕ is constant, we define the *degree of ϕ* to be 0. Otherwise we say that ϕ is a *finite map* and we define its *degree* to be

$$\deg \phi = \big[K(C_1) : \phi^* K(C_2)\big].$$

We say that ϕ is *separable*, *inseparable*, or *purely inseparable* if the field extension $K(C_1)/\phi^* K(C_2)$ has the corresponding property, and we denote the separable and inseparable degrees of the extension by $\deg_s \phi$ and $\deg_i \phi$, respectively.

Definition. Let $\phi : C_1 \to C_2$ be a nonconstant map of curves defined over K. From (II.2.4a) we know that $K(C_1)$ is a finite extension of $\phi^* K(C_2)$. We use the norm map relative to ϕ^* to define a map in the other direction,

$$\phi_* : K(C_1) \longmapsto K(C_2), \qquad \phi_* = (\phi^*)^{-1} \circ N_{K(C_1)/\phi^* K(C_2)} .$$

Corollary 2.4.1. *Let C_1 and C_2 be smooth curves, and let $\phi : C_1 \to C_2$ be a map of degree one. Then ϕ is an isomorphism.*

PROOF. By definition, $\deg \phi = 1$ means that $\phi^* \bar{K}(C_2) = \bar{K}(C_1)$, so ϕ^* is an isomorphism of function fields. Hence from (II.2.5b), corresponding to the inverse map $(\phi^*)^{-1} : \bar{K}(C_1) \xrightarrow{\sim} \bar{K}(C_2)$, there is a rational map $\psi : C_2 \to C_1$ such that $\psi^* = (\phi^*)^{-1}$. Further, since C_2 is smooth, (II.2.1) tells us that ψ is actually a morphism. Finally, since $(\phi \circ \psi)^* = \psi^* \circ \phi^*$ is the identity map on $\bar{K}(C_2)$, and similarly $(\psi \circ \phi)^* = \phi^* \circ \psi^*$ is the identity map on $\bar{K}(C_1)$, the uniqueness assertion of (II.2.4b) implies that $\phi \circ \psi$ and $\psi \circ \phi$ are, respectively, the identity maps on C_2 and C_1. Hence ϕ and ψ are isomorphisms. $\qquad\square$

Remark 2.5. The above result (II.2.4) shows the close connection between (smooth) curves and their function fields. This can be made precise by stating that the following map is an equivalence of categories. (See [111, I §6] for details.)

$$
\begin{bmatrix}
\textit{Objects}: \text{smooth curves} \\
\text{defined over } K \\
\textit{Maps}: \text{nonconstant rational} \\
\text{maps (equivalently} \\
\text{surjective morphisms)} \\
\text{defined over } K
\end{bmatrix}
\rightsquigarrow
\begin{bmatrix}
\textit{Objects}: \text{finitely generated} \\
\text{extensions } \mathbb{K}/K \text{ of} \\
\text{transcendence degree one with} \\
\mathbb{K} \cap \bar{K} = K \\
\textit{Maps}: \text{field injections fixing } K
\end{bmatrix}
$$

$$C/K \rightsquigarrow K(C)$$

$$\phi : C_1 \to C_2 \rightsquigarrow \phi^* : K(C_2) \to K(C_1)$$

Example 2.5.1. *Hyperelliptic Curves.* We assume that $\operatorname{char}(K) \neq 2$. We choose a polynomial $f(x) \in K[x]$ of degree d and consider the *affine* curve C_0/K given by the equation

$$C_0 : y^2 = f(x) = a_0 x^d + a_1 x^{d-1} + \cdots + a_d.$$

Suppose that the point $P = (x_0, y_0) \in C_0$ is singular. Then

$$2y_0 = f'(x_0) = 0,$$

which means that $y_0 = 0$ and x_0 is a double root of $f(x)$. Hence, if we assume that $\operatorname{disc}(f) \neq 0$, then the affine curve $y^2 = f(x)$ will be nonsingular.

If we treat C_0 as a curve in $\mathbb{P}^2$ by homogenizing its affine equation, then one easily checks that the point(s) at infinity are singular whenever $d \geq 4$. On the other hand, (II.2.4c) assures us that there exists some smooth projective curve C/K whose function field equals $K(C_0) = K(x, y)$. The problem is that this smooth curve is not a subset of $\mathbb{P}^2$.

For example, consider the case $d = 4$. (See also Exercise 2.14.) Then C_0 has an affine equation

$$C_0 : y^2 = a_0 x^4 + a_1 x^3 + a_2 x^2 + a_3 x + a_4.$$

We define a map

$$[1, x, y, x^2] : C_0 \longrightarrow \mathbb{P}^3.$$

Letting $[X_0, X_1, X_2, X_3] = [1, x, y, x^2]$, the ideal of the image clearly contains the two homogeneous polynomials

$$F = X_3 X_0 - X_1^2,$$
$$G = X_2^2 X_0^2 - a_0 X_1^4 - a_1 X_1^3 X_0 - a_2 X_1^2 X_0^2 - a_3 X_1 X_0^3 - a_4 X_0^4.$$

However, the zero set of these two polynomials cannot be the desired curve C, since it includes the line $X_0 = X_1 = 0$. So we substitute $X_1^2 = X_0 X_3$ into G and cancel an X_0^2 to obtain the quadratic polynomial

$$H = X_2^2 - a_0 X_3^2 - a_1 X_1 X_3 - a_2 X_0 X_3 - a_3 X_0 X_1 - a_4 X_0^2.$$

We claim that the ideal generated by F and H gives a smooth curve C.

To see this, note first that if $X_0 \neq 0$, then dehomogenization with respect to X_0 gives the affine curve (setting $x = X_1/X_0$, $y = X_2/X_0$, and $z = X_3/X_0$)

$$z = x^2 \qquad \text{and} \qquad y^2 = a_0 z^2 + a_1 xz + a_2 z + a_3 x + a_4.$$

Substituting the first equation into the second gives us back the original curve C_0. Thus $C_0 \cong C \cap \{X_0 \neq 0\}$.

Next, if $X_0 = 0$, then necessarily $X_1 = 0$, and then $X_2 = \pm\sqrt{a_0}\, X_3$. Thus C has two points $[0, 0, \pm\sqrt{a_0}, 1]$ on the hyperplane $X_0 = 0$. (Note that $a_0 \neq 0$, since we have assumed that $f(x)$ has degree exactly four.) To check that C is nonsingular at these two points, we dehomogenize with respect to X_3, setting $u = X_0/X_3$, $v = X_1/X_3$, and $w = X_2/X_3$. This gives the equations

$$u = v^2 \qquad \text{and} \qquad w^2 = a_0 + a_1 v + a_2 u + a_3 uv + a_4 u^2,$$

from which we obtain the single affine equation

$$w^2 = a_0 + a_1 v + a_2 v^2 + a_3 v^3 + a_4 v^4.$$

Again using the assumption that the polynomial $f(x)$ has no double roots, we see that the points $(v, w) = \left(0, \pm\sqrt{a_0}\right)$ are nonsingular.

We summarize the preceding discussion in the following proposition, which will be used in Chapter X.

Proposition 2.5.2. *Let $f(X) \in K[x]$ be a polynomial of degree 4 with $\mathrm{disc}(f) \neq 0$. There exists a smooth projective curve $C \subset \mathbb{P}^3$ with the following properties:*
  (i) *The intersection of C with $\mathbb{A}^3 = \{X_0 \neq 0\}$ is isomorphic to the affine curve $y^2 = f(x)$.*
  (ii) *Let $f(x) = a_0 x^4 + \cdots + a_4$. Then the intersection of C with the hyperplane $X_0 = 0$ consists of the two points $\left[0, 0, \pm\sqrt{a_0}, 1\right]$.*

We next look at the behavior of a map in the neighborhood of a point.

Definition. Let $\phi : C_1 \to C_2$ be a nonconstant map of smooth curves, and let $P \in C_1$. The *ramification index of ϕ at P*, denoted by $e_\phi(P)$, is the quantity

$$e_\phi(P) = \mathrm{ord}_P\left(\phi^* t_{\phi(P)}\right),$$

where $t_{\phi(P)} \in K(C_2)$ is a uniformizer at $\phi(P)$. Note that $e_\phi(P) \geq 1$. We say that ϕ is *unramified at P* if $e_\phi(P) = 1$, and that ϕ is *unramified* if it is unramified at every point of C_1.

Proposition 2.6. *Let $\phi : C_1 \to C_2$ be a nonconstant map of smooth curves.*
  (a) *For every $Q \in C_2$,*
$$\sum_{P \in \phi^{-1}(Q)} e_\phi(P) = \deg(\phi).$$

(b) *For all but finitely many* $Q \in C_2$,

$$\#\phi^{-1}(Q) = \deg_s(\phi).$$

(c) *Let* $\psi : C_2 \to C_3$ *be another nonconstant map of smooth curves. Then for all* $P \in C_1$,

$$e_{\psi \circ \phi}(P) = e_\phi(P)e_\psi(\phi P).$$

PROOF. (a) Use [111, II.6.9] with $Y = \mathbb{P}^1$ and $D = (0)$, or see [142, Proposition 2], [233, I Proposition 10], or [243, III §2, Theorem 1].

(b) See [111, II.6.8].

(c) Let $t_{\phi P}$ and $t_{\psi \phi P}$ be uniformizers at the indicated points. By definition, the functions

$$t_{\phi P}^{e_\psi(\phi P)} \quad \text{and} \quad \psi^* t_{\psi \phi P}$$

have the same order at $\phi(P)$. Applying ϕ^* and taking orders at P yields

$$\mathrm{ord}_P \left(\phi^* t_{\phi P}^{e_\psi(\phi P)} \right) = \mathrm{ord}_P ((\psi \phi)^* t_{\psi \phi P}),$$

which is the desired result. $\qquad\qquad\qquad\qquad\qquad\qquad\qquad\qquad\qquad\qquad$ □

Corollary 2.7. *A map a* $\phi : C_1 \to C_2$ *is unramified if and only if*

$$\#\phi^{-1}(Q) = \deg(\phi) \qquad \text{for all } Q \in C_2.$$

PROOF. From (II.2.6a), we see that $\#\phi^{-1}(Q) = \deg(\phi)$ if and only if

$$\sum_{P \in \phi^{-1}(Q)} e_\phi(P) = \#\phi^{-1}(Q).$$

Since $e_\phi(P) \geq 1$, this occurs if and only if each $e_\phi(P) = 1$. $\qquad\qquad\qquad$ □

Remark 2.8. The content of (II.2.6) is exactly analogous to the theorems describing the ramification of primes in number fields. Thus let L/K be number fields. Then (II.2.6a) is the analogue of the $\sum e_i f_i = [K : \mathbb{Q}]$ theorem ([142, I, Proposition 21], [233, I, Proposition 10]), while (II.2.6b) is analogous to the fact that only finitely many primes of K ramify in L, and (II.2.6c) gives the multiplicativity of ramification degrees in towers of fields. Of course, (II.2.6) and the analogous results for number fields are both merely special cases of the basic theorems describing finite extensions of Dedekind domains.

Example 2.9. Consider the map

$$\phi : \mathbb{P}^1 \longrightarrow \mathbb{P}^1, \qquad \phi([X, Y]) = [X^3(X - Y)^2, Y^5].$$

Then ϕ is ramified at the points $[0, 1]$ and $[1, 1]$. Further,

$$e_\phi([0, 1]) = 3 \quad \text{and} \quad e_\phi([1, 1]) = 2,$$

so

$$\sum_{P \in \phi^{-1}([0,1])} e_\phi(P) = e_\phi([0, 1]) + e_\phi([1, 1]) = 5 = \deg \phi,$$

which is in accordance with (II.2.6a).

The Frobenius Map

Assume that $\text{char}(K) = p > 0$ and let $q = p^r$. For any polynomial $f \in K[X]$, let $f^{(q)}$ be the polynomial obtained from f by raising each coefficient of f to the qth power. Then for any curve C/K, we can define a new curve $C^{(q)}/K$ as the curve whose homogeneous ideal is given by

$$I(C^{(q)}) = \text{ideal generated by } \{f^{(q)} : f \in I(C)\}.$$

Further, there is a natural map from C to $C^{(q)}$, called the qth-*power Frobenius morphism*, given by

$$\phi : C \longrightarrow C^{(q)}, \qquad \phi([x_0, \ldots, x_n]) = [x_0^q, \ldots, x_n^q].$$

To see that ϕ maps C to $C^{(q)}$, it suffices to show that for every point

$$P = [x_0, \ldots, x_n] \in C,$$

the image $\phi(P)$ is a zero of each generator $f^{(q)}$ of $I(C^{(q)})$. We compute

$$
\begin{aligned}
f^{(q)}(\phi(P)) &= f^{(q)}(x_0^q, \ldots, x_n^q) \\
&= \big(f(x_0, \ldots, x_n)\big)^q && \text{since } \text{char}(K) = p, \\
&= 0 && \text{since } f(P) = 0.
\end{aligned}
$$

Example 2.10. Let C be the curve in $\mathbb{P}^2$ given by the single equation

$$C : Y^2 Z = X^2 + aXZ^2 + bZ^3.$$

Then $C^{(q)}$ is the curve given by the equation

$$C^{(q)} : Y^2 Z = X^2 + a^q XZ^2 + b^q Z^3.$$

The next proposition describes the basic properties of the Frobenius map.

Proposition 2.11. *Let K be a field of characteristic $p > 0$, let $q = p^r$, let C/K be a curve, and let $\phi : C \to C^{(q)}$ be the qth-power Frobenius morphism.*
(a) $\phi^* K(C^{(q)}) = K(C)^q = \{f^q : f \in K(C)\}$.
(b) *ϕ is purely inseparable.*
(c) $\deg \phi = q$.
(*N.B. We are assuming that K is perfect. If K is not perfect, then (b) and (c) remain true, but (a) must be modified.*)

PROOF. (a) Using the description (I.2.9) of $K(C)$ as consisting of quotients f/g of homogeneous polynomials of the same degree, we see that $\phi^* K(C^{(q)})$ is the subfield of $K(C)$ given by quotients

$$\phi^* \left(\frac{f}{g} \right) = \frac{f(X_0^q, \ldots, X_n^q)}{g(X_0^q, \ldots, X_n^q)}.$$

Similarly, $K(C)^q$ is the subfield of $K(C)$ given by quotients

$$\frac{f(X_0,\ldots,X_n)^q}{g(X_0,\ldots,X_n)^q}.$$

However, since K is perfect, we know that every element of K is a qth power, so

$$\left(K[X_0,\ldots,X_n]\right)^q = K[X_0^q,\ldots,X_n^q].$$

Thus the set of quotients $f(X_i^q)/g(X_i^q)$ and the set of quotients $f(X_i)^q/g(X_i)^q$ give the exact same subfield of $K(C)$.

(b) Immediate from (a).

(c) Taking a finite extension of K if necessary, we may assume that there is a smooth point $P \in K(C)$. Let $t \in K(C)$ be a uniformizer at P (II.1.1.1). Then (II.1.4) says that $K(C)$ is separable over $K(t)$. Consider the tower of fields

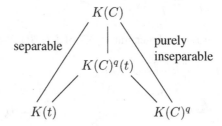

It follows that $K(C) = K(C)^q(t)$, so from (a),

$$\deg \phi = \left[K(C)^q(t) : K(C)^q\right].$$

Now $t^q \in K(C)^q$, so in order to prove that $\deg \phi = q$, we need merely show that $t^{q/p} \notin K(C)^q$. But if $t^{q/p} = f^q$ for some $f \in K(C)$, then

$$\frac{q}{p} = \operatorname{ord}_P(t^{q/p}) = q \operatorname{ord}_P(f),$$

which is impossible, since $\operatorname{ord}_P(f)$ must be an integer. $\square$

Corollary 2.12. *Every map $\psi : C_1 \to C_2$ of (smooth) curves over a field of characteristic $p > 0$ factors as*

$$C_1 \xrightarrow{\ \phi\ } C_1^{(q)} \xrightarrow{\ \lambda\ } C_2,$$

where $q = \deg_i(\psi)$, the map ϕ is the qth-power Frobenius map, and the map λ is separable.

PROOF. Let $\mathbb{K}$ be the separable closure of $\psi^* K(C_2)$ in $K(C_1)$. Then $K(C_1)/\mathbb{K}$ is purely inseparable of degree q, so $K(C_1)^q \subset \mathbb{K}$. From (II.2.11a,c) we have,

$$K(C_1)^q = \phi^*\left(K(C_1^{(q)})\right) \qquad \text{and} \qquad \left[K(C_1) : \phi^*(K(C_1^{(q)}))\right] = q.$$

Comparing degrees, we conclude that $\mathbb{K} = \phi^*(C_1^{(q)})$. We now have a tower of function fields

$$K(C_1) \,/\, \phi^*K(C_1^{(q)}) \,/\, \psi^*K(C_2),$$

and from (II.2.4b), this corresponds to maps

$$C_1 \xrightarrow{\ \phi\ } C_1^{(q)} \xrightarrow{\ \lambda\ } C_2$$
$$\underset{\psi}{\underbrace{\phantom{C_1 \xrightarrow{\ \phi\ } C_1^{(q)}}}}$$

$\square$

II.3   Divisors

The *divisor group of a curve* C, denoted by $\mathrm{Div}(C)$, is the free abelian group generated by the points of C. Thus a divisor $D \in \mathrm{Div}(C)$ is a formal sum

$$D = \sum_{P \in C} n_P(P),$$

where $n_P \in \mathbb{Z}$ and $n_P = 0$ for all but finitely many $P \in C$. The *degree of* D is defined by

$$\deg D = \sum_{P \in C} n_P.$$

The *divisors of degree* 0 form a subgroup of $\mathrm{Div}(C)$, which we denote by

$$\mathrm{Div}^0(C) = \{D \in \mathrm{Div}(C) : \deg D = 0\}.$$

If C is defined over K, we let $G_{\bar{K}/K}$ act on $\mathrm{Div}(C)$ and $\mathrm{Div}^0(C)$ in the obvious way,

$$D^\sigma = \sum_{P \in C} n_P(P^\sigma).$$

Then D is *defined over* K if $D^\sigma = D$ for all $\sigma \in G_{\bar{K}/K}$. We note that if $D = n_1(P_1) + \cdots + n_r(P_r)$ with $n_1, \ldots, n_r \neq 0$, then to say that D is defined over K does *not* mean that $P_1, \ldots, P_r \in C(K)$. It suffices for the group $G_{\bar{K}/K}$ to permute the P_i's in an appropriate fashion. We denote the *group of divisors defined over* K by $\mathrm{Div}_K(C)$, and similarly for $\mathrm{Div}_K^0(C)$.

Assume now that the curve C is smooth, and let $f \in \bar{K}(C)^*$. Then we can associate to f the divisor $\mathrm{div}(f)$ given by

$$\mathrm{div}(f) = \sum_{P \in C} \mathrm{ord}_P(f)(P).$$

This is a divisor by (II.1.2). If $\sigma \in G_{\bar{K}/K}$, then it is easy to see that

$$\mathrm{div}(f^\sigma) = \big(\mathrm{div}(f)\big)^\sigma.$$

In particular, if $f \in K(C)$, then $\mathrm{div}(f) \in \mathrm{Div}_K(C)$.

Since each ord_P is a valuation, the map

$$\mathrm{div} : \bar{K}(C)^* \longrightarrow \mathrm{Div}(C)$$

is a homomorphism of abelian groups. It is analogous to the map that sends an element of a number field to the corresponding fractional ideal. This prompts the following definitions.

Definition. A divisor $D \in \mathrm{Div}(C)$ is *principal* if it has the form $D = \mathrm{div}(f)$ for some $f \in \bar{K}(C)^*$. Two divisors are *linearly equivalent*, written $D_1 \sim D_2$, if $D_1 - D_2$ is principal. The *divisor class group* (or *Picard group*) of C, denoted by $\mathrm{Pic}(C)$, is the quotient of $\mathrm{Div}(C)$ by its subgroup of principal divisors. We let $\mathrm{Pic}_K(C)$ be the subgroup of $\mathrm{Pic}(C)$ fixed by $G_{\bar{K}/K}$. N.B. In general, $\mathrm{Pic}_K(C)$ is not the quotient of $\mathrm{Div}_K(C)$ by its subgroup of principal divisors. But see exericse 2.13 for a case in which this is true.

Proposition 3.1. *Let C be a smooth curve and let $f \in \bar{K}(C)^*$.*
(a) $\mathrm{div}(f) = 0$ *if and only if $f \in \bar{K}^*$.*
(b) $\deg\big(\mathrm{div}(f)\big) = 0$.

PROOF. (a) If $\mathrm{div}(f) = 0$, then f has no poles, so the associated map $f : C \to \mathbb{P}^1$ as defined in (II.2.2) is not surjective. Then (II.2.3) tells us that the map is constant, so $f \in \bar{K}^*$. The converse is clear.
(b) See [111, II.6.10], [243, III 2, corollary to Theorem 1], or (II.3.7). $\qquad\square$

Example 3.2. On $\mathbb{P}^1$, every divisor of degree 0 is principal To see this, suppose that $D = \sum n_P(P)$ has degree 0. Writing $P = [\alpha_P, \beta_P] \in \mathbb{P}^1$, we see that D is the divisor of the function

$$\prod_{P \in \mathbb{P}^1} (\beta_P X - \alpha_P Y)^{n_P}.$$

Note that $\sum n_P = 0$ ensures that this function is in $K(\mathbb{P}^1)$. It follows that the degree map $\deg : \mathrm{Pic}(\mathbb{P}^1) \to \mathbb{Z}$ is an isomorphism. The converse is also true, i.e., if C is a smooth curve and $\mathrm{Pic}(C) \cong \mathbb{Z}$, then C is isomorphic to $\mathbb{P}^1$.

Example 3.3. Assume that $\mathrm{char}(K) \neq 2$. Let $e_1, e_2, e_3 \in \bar{K}$ be distinct, and consider the curve

$$C : y^2 = (x - e_1)(x - e_2)(x - e_3).$$

One can check that C is smooth and that it has a single point at infinity, which we denote by P_∞. For $i = 1, 2, 3$, let $P_i = (e_i, 0) \in C$. Then

$$\mathrm{div}(x - e_i) = 2(P_i) - 2(P_\infty),$$
$$\mathrm{div}(y) = (P_1) + (P_2) + (P_3) - 3(P_\infty).$$

Definition. It follows from (II.3.1b) that the principal divisors form a subgroup of $\mathrm{Div}^0(C)$. We define the *degree-0 part of the divisor class group of C* to be the quotient of $\mathrm{Div}^0(C)$ by the subgroup of principal divisors. We denote this group by $\mathrm{Pic}^0(C)$. Similarly, we write $\mathrm{Pic}_K^0(C)$ for the subgroup of $\mathrm{Pic}^0(C)$ fixed by $G_{\bar{K}/K}$.

Remark 3.4. The above definitions and (II.3.1) may be summarized by saying that there is an exact sequence

$$1 \longrightarrow \bar{K}^* \longrightarrow \bar{K}(C)^* \xrightarrow{\text{div}} \text{Div}^0(C) \longrightarrow \text{Pic}^0(C) \longrightarrow 0.$$

This sequence is the function field analogue of the fundamental exact sequence in algebraic number theory, which for a number field K reads

$$1 \longrightarrow \begin{pmatrix} \text{units} \\ \text{of } K \end{pmatrix} \longrightarrow K^* \longrightarrow \begin{pmatrix} \text{fractional} \\ \text{ideals of } K \end{pmatrix} \longrightarrow \begin{pmatrix} \text{ideal class} \\ \text{group of } K \end{pmatrix} \longrightarrow 1.$$

Let $\phi : C_1 \to C_2$ be a nonconstant map of smooth curves. As we have seen, ϕ induces maps on the function fields of C_1 and C_2,

$$\phi^* : \bar{K}(C_2) \longrightarrow \bar{K}(C_1) \quad \text{and} \quad \phi_* : \bar{K}(C_1) \longrightarrow \bar{K}(C_2).$$

We similarly define maps of divisor groups as follows:

$$\phi^* : \text{Div}(C_2) \longrightarrow \text{Div}(C_1), \qquad\qquad \phi_* : \text{Div}(C_1) \longrightarrow \text{Div}(C_2),$$

$$(Q) \longmapsto \sum_{P \in \phi^{-1}(Q)} e_\phi(P)(P), \qquad\qquad (P) \longmapsto (\phi P),$$

and extend $\mathbb{Z}$-linearly to arbitrary divisors.

Example 3.5. Let C be a smooth curve, let $f \in \bar{K}(C)$ be a nonconstant function, and let $f : C \to \mathbb{P}^1$ be the corresponding map (II.2.2). Then directly from the definitions,

$$\text{div}(f) = f^*\big((0) - (\infty)\big).$$

Proposition 3.6. *Let $\phi : C_1 \to C_2$ be a nonconstant map of smooth curves.*

(a) $\deg(\phi^* D) = (\deg \phi)(\deg D)$ *for all $D \in \text{Div}(C_2)$.*

(b) $\phi^*(\text{div } f) = \text{div}(\phi^* f)$ *for all $f \in \bar{K}(C_2)^*$.*

(c) $\deg(\phi_* D) = \deg D$ *for all $D \in \text{Div}(C_1)$.*

(d) $\phi_*(\text{div } f) = \text{div}(\phi_* f)$ *for all $f \in \bar{K}(C_1)^*$.*

(e) $\phi_* \circ \phi^*$ *acts as multiplication by $\deg \phi$ on $\text{Div}(C_2)$.*

(f) *If $\psi : C_2 \to C_3$ is another such map, then*

$$(\psi \circ \phi)^* = \phi^* \circ \psi^* \quad \text{and} \quad (\psi \circ \phi)_* = \psi_* \circ \phi_*.$$

PROOF. (a) Follows directly from (II.2.6a).

(b) Follows from the definitions and the easy fact (Exercise 2.2) that for all $P \in C_1$,

$$\text{ord}_P(\phi^* f) = e_\phi(P)\,\text{ord}_{\phi P}(f).$$

(c) Clear from the definitions.

(d) See [142, Chapter 1, Proposition 22] or [233, I, Proposition 14].

(e) Follows directly from (II.2.6a).

(f) The first equality follows from (II.2.6c). The second is obvious.   □

Remark 3.7. From (II.3.6) we see that ϕ^* and ϕ_* take divisors of degree 0 to divisors of degree 0, and principal divisors to principal divisors. They thus induce maps

$$\phi^* : \mathrm{Pic}^0(C_2) \longrightarrow \mathrm{Pic}^0(C_1) \qquad \text{and} \qquad \phi_* : \mathrm{Pic}^0(C_1) \longrightarrow \mathrm{Pic}^0(C_2).$$

In particular, if $f \in \bar{K}(C)$ gives the map $f : C \to \mathbb{P}^1$, then

$$\deg \mathrm{div}(f) = \deg f^*\big((0) - (\infty)\big) = \deg f - \deg f = 0.$$

This provides a proof of (II.3.1b)

II.4   Differentials

In this section we discuss the vector space of differential forms on a curve. This vector space serves two distinct purposes. First, it performs the traditional calculus role of linearization. (See (III §5), especially (III.5.2).) Second, it gives a useful criterion for determining when an algebraic map is separable. (See (II.4.2) and its utilization in the proof of (III.5.5).) Of course, the latter is also a familiar use of calculus, since a field extension is separable if and only if the minimal polynomial of each element has a nonzero derivative

Definition. Let C be a curve. The *space of (meromorphic) differential forms* on C, denoted by Ω_C, is the $\bar{K}$-vector space generated by symbols of the form dx for $x \in \bar{K}(C)$, subject to the usual relations:

(i) $d(x + y) = dx + dy$ for all $x, y \in \bar{K}(C)$.

(ii) $d(xy) = x\,dy + y\,dx$ for all $x, y \in \bar{K}(C)$.

(iii) $da = 0$ for all $a \in \bar{K}$.

Remark 4.1. There is, of course, a functorial definition of Ω_C. See, for example, [164, Chapter 10], [111, II.8], or [210, II §3].

Let $\phi : C_1 \to C_2$ be a nonconstant map of curves. The associated function field map $\phi^* : \bar{K}(C_2) \to \bar{K}(C_1)$ induces a map on differentials,

$$\phi^* : \Omega_{C_2} \longrightarrow \Omega_{C_1}, \qquad \phi^*\left(\sum f_i\,dx_i\right) = \sum (\phi^* f_i)d(\phi^* x_i).$$

This map provides a useful criterion for determining when ϕ is separable.

Proposition 4.2. *Let C be a curve.*

(a) *Ω_C is a 1-dimensional $\bar{K}(C)$-vector space.*

(b) *Let $x \in \bar{K}(C)$. Then dx is a $\bar{K}(C)$-basis for Ω_C if and only if $\bar{K}(C)/\bar{K}(x)$ is a finite separable extension.*

(c) *Let $\phi : C_1 \to C_2$ be a nonconstant map of curves. Then ϕ is separable if and only if the map*

$$\phi^* : \Omega_{C_2} \longrightarrow \Omega_{C_1}$$

is injective (equivalently, nonzero).

PROOF. (a) See [164, 27.A,B], [210, II.3.4], or [243, III §4, Theorem 3].

(b) See [164, 27A,B] or [243, III §4, Theorem 4].

(c) Using (a) and (b), choose $y \in \bar{K}(C_2)$ such that $\Omega_{C_2} = \bar{K}(C_2)\,dy$ and such that $\bar{K}(C_2)/\bar{K}(y)$ is a separable extension. Note that $\phi^* \bar{K}(C_2)$ is then separable over $\phi^* \bar{K}(y) = \bar{K}(\phi^* y)$. Now

$$\phi^* \text{ is injective} \iff d(\phi^* y) \neq 0$$
$$\iff d(\phi^* y) \text{ is a basis for } \Omega_{C_1} \text{ (from (a))},$$
$$\iff \bar{K}(C_1)/\bar{K}(\phi^* y) \text{ is separable (from (b))},$$
$$\iff \bar{K}(C_1)/\phi^* \bar{K}(C_2) \text{ is separable},$$

where the last equivalence follows because we already know that $\phi^* \bar{K}(C_2)/\bar{K}(\phi^* y)$ is separable. $\qquad \square$

Proposition 4.3. *Let C be a curve, let $P \in C$, and let $t \in \bar{K}(C)$ be a uniformizer at P.*

(a) *For every $\omega \in \Omega_C$ there exists a unique function $g \in \bar{K}(C)$, depending on ω and t, satisfying*

$$\omega = g\,dt.$$

We denote g by ω/dt.

(b) *Let $f \in \bar{K}(C)$ be regular at P. Then df/dt is also regular at P.*

(c) *Let $\omega \in \Omega_C$ with $\omega \neq 0$. The quantity*

$$\mathrm{ord}_P(\omega/dt)$$

depends only on ω and P, independent of the choice of uniformizer t. We call this value the order *of ω at P and denote it by $\mathrm{ord}_P(\omega)$.*

(d) *Let $x, f \in \bar{K}(C)$ with $x(P) = 0$, and let $p = \mathrm{char}\, K$. Then*

$$\mathrm{ord}_P(f\,dx) = \mathrm{ord}_P(f) + \mathrm{ord}_P(x) - 1, \quad \text{if } p = 0 \text{ or } p \nmid \mathrm{ord}_P(x),$$
$$\mathrm{ord}_P(f\,dx) \geq \mathrm{ord}_P(f) + \mathrm{ord}_P(x), \quad \text{if } p > 0 \text{ and } p \mid \mathrm{ord}_P(x).$$

(e) *Let $\omega \in \Omega_C$ with $\omega \neq 0$. Then*

$$\mathrm{ord}_P(\omega) = 0 \quad \text{for all but finitely many } P \in C.$$

PROOF. (a) This follows from (II.1.4) and (4.2ab).

(b) See [111, comment following IV.2.1] or [210, II.3.10].

(c) Let t' be another uniformizer at P. Then from (b) we see that dt/dt' and dt'/dt are both regular at P, so $\mathrm{ord}_P(dt'/dt) = 0$. The desired result then follows from

$$\omega = g\,dt' = g(dt'/dt)\,dt.$$

(d) Write $x = ut^n$ with $n = \mathrm{ord}_P(x) \geq 1$, so $\mathrm{ord}_P(u) = 0$. Then

$$dx = \left[nut^{n-1} + (du/dt)t^n \right] dt.$$

From (b) we know that du/dt is regular at P. Hence if $n \neq 0$, then the first term dominates, which gives the desired equality

$$\mathrm{ord}_P(f\,dx) = \mathrm{ord}_P(fnut^{n-1}\,dt) = \mathrm{ord}_P(f) + n - 1.$$

On the other hand, if $p > 0$ and $p \mid n$, then the first term vanishes and we find that

$$\mathrm{ord}_P(f\,dx) = \mathrm{ord}_P(f(du/dt)t^n\,dt) \geq \mathrm{ord}_P(f) + n.$$

(e) Choose some $x \in \bar{K}(C)$ such that $\bar{K}(C)/\bar{K}(x)$ is separable and write $\omega = f\,dx$. From [111, IV.2.2a], the map $x : C \to \mathbb{P}^1$ ramifies at only finitely many points of C. Hence discarding finitely many points, we may restrict attention to points $P \in C$ such that

$$f(P) \neq 0, \quad f(P) \neq \infty, \quad x(P) \neq \infty,$$

and the map $x : C \to \mathbb{P}^1$ is unramified at P. The two conditions on x imply that $x - x(P)$ is a uniformizer at P, so

$$\mathrm{ord}_P(\omega) = \mathrm{ord}_P\big(f\,d(x - x(P))\big) = 0.$$

Hence $\mathrm{ord}_P(\omega) = 0$ for all but finitely many P.   □

Definition. Let $\omega \in \Omega_C$. The *divisor associated to ω* is

$$\mathrm{div}(\omega) = \sum_{P \in C} \mathrm{ord}_P(\omega)(P) \in \mathrm{Div}(C).$$

The differential $\omega \in \Omega_C$ is *regular* (or *holomorphic*) if

$$\mathrm{ord}_P(\omega) \geq 0 \qquad \text{for all } P \in C.$$

It is *nonvanishing* if

$$\mathrm{ord}_P(\omega) \leq 0 \qquad \text{for all } P \in C.$$

Remark 4.4. If $\omega_1, \omega_2 \in \Omega_C$ are nonzero differentials, then (II.4.2a) implies that there is a function $f \in \bar{K}(C)^*$ such that $\omega_1 = f\omega_2$. Thus

$$\mathrm{div}(\omega_1) = \mathrm{div}(f) + \mathrm{div}(\omega_2),$$

which shows that the following definition makes sense.

Definition. The *canonical divisor class on C* is the image in $\mathrm{Pic}(C)$ of $\mathrm{div}(\omega)$ for any nonzero differential $\omega \in \Omega_C$. Any divisor in this divisor class is called a *canonical divisor*.

Example 4.5. We are going to show that there are no holomorphic differentials on $\mathbb{P}^1$. First, if t is a coordinate function on $\mathbb{P}^1$, then

$$\mathrm{div}(dt) = -2(\infty).$$

To see this, note that for all $\alpha \in \bar{K}$, the function $t - \alpha$ is a uniformizer at α, so

$$\mathrm{ord}_\alpha(dt) = \mathrm{ord}_\alpha\big(d(t - \alpha)\big) = 0.$$

However, at $\infty \in \mathbb{P}^1$ we need to use a function such as $1/t$ as our uniformizer, so

$$\mathrm{ord}_\infty(dt) = \mathrm{ord}_\infty\left(-t^2\, d\left(\frac{1}{t}\right)\right) = -2.$$

Thus dt is not holomorphic. But now for any nonzero $\omega \in \Omega_{\mathbb{P}^1}$, we can use (II.4.3a) to compute

$$\deg\,\mathrm{div}(\omega) = \deg\,\mathrm{div}(dt) = -2,$$

so ω cannot be holomorphic either.

Example 4.6. Let C be the curve

$$C : y^2 = (x - e_1)(x - e_2)(x - e_3),$$

where we continue with the notation from (II.3.3). Then

$$\mathrm{div}(dx) = (P_1) + (P_2) + (P_3) - 3(P_\infty).$$

(Note that $dx = d(x - e_i) = -x^2\, d(1/x)$.) We thus see that

$$\mathrm{div}(dx/y) = 0.$$

Hence the differential dx/y is both holomorphic and nonvanishing.

II.5 The Riemann–Roch Theorem

Let C be a curve. We put a partial order on $\mathrm{Div}(C)$ in the following way.

Definition. A divisor $D = \sum n_P(P)$ is *positive* (or *effective*), denoted by

$$D \geq 0,$$

if $n_P \geq 0$ for every $P \in C$. Similarly, for any two divisors $D_1, D_2 \in \mathrm{Div}(C)$, we write

$$D_1 \geq D_2$$

to indicate that $D_1 - D_2$ is positive.

Example 5.1. Let $f \in \bar{K}(C)^*$ be a function that is regular everywhere except at one point $P \in C$, and suppose that it has a pole of order at most n at P. These requirements on f may be succinctly summarized by the inequality

$$\mathrm{div}(f) \geq -n(P).$$

Similarly,

$$\mathrm{div}(f) \geq (Q) - n(P)$$

says that in addition, f has a zero at Q. Thus divisorial inequalities are a useful tool for describing poles and/or zeros of functions.

Definition. Let $D \in \mathrm{Div}(C)$. We associate to D the set of functions

$$\mathcal{L}(D) = \{f \in \bar{K}(C)^* : \mathrm{div}(f) \geq -D\} \cup \{0\}.$$

The set $\mathcal{L}(D)$ is a finite-dimensional $\bar{K}$-vector space (see (II.5.2b) below), and we denote its dimension by

$$\ell(D) = \dim_{\bar{K}} \mathcal{L}(D).$$

Proposition 5.2. *Let* $D \in \mathrm{Div}(C)$.
(a) *If* $\deg D < 0$, *then*

$$\mathcal{L}(D) = \{0\} \qquad and \qquad \ell(D) = 0.$$

(b) $\mathcal{L}(D)$ *is a finite-dimensional* $\bar{K}$-*vector space.*
(c) *If* $D' \in \mathrm{Div}(C)$ *is linearly equivalent to* D, *then*

$$\mathcal{L}(D) \cong \mathcal{L}(D'), \qquad and\ so \qquad \ell(D) = \ell(D').$$

PROOF. (a) Let $f \in \mathcal{L}(D)$ with $f \neq 0$. Then (II.3.1b) tells us that

$$0 = \deg \mathrm{div}(f) \geq \deg(-D) = -\deg D,$$

so $\deg D \geq 0$.
(b) See [111, II.5.19] or Exercise 2.4.
(c) If $D = D' + \mathrm{div}(g)$, then the map

$$\mathcal{L}(D) \longrightarrow \mathcal{L}(D'), \qquad f \longmapsto fg$$

is an isomorphism.   □

Example 5.3. Let $K_C \in \mathrm{Div}(C)$ be a canonical divisor on C, say

$$K_C = \mathrm{div}(\omega).$$

Then each function $f \in \mathcal{L}(K_C)$ has the property that

$$\mathrm{div}(f) \geq -\mathrm{div}(\omega), \qquad so \quad \mathrm{div}(f\omega) \geq 0.$$

In other words, $f\omega$ is holomorphic. Conversely, if the differential $f\omega$ is holomorphic, then $f \in \mathcal{L}(K_C)$. Since every differential on C has the form $f\omega$ for some f, we have established an isomorphism of $\bar{K}$-vector spaces,

$$\mathcal{L}(K_C) \cong \{\omega \in \Omega_C : \omega \text{ is holomorphic}\}.$$

The dimension $\ell(K_C)$ of these spaces is an important invariant of the curve C.

We are now ready to state a fundamental result in the algebraic geometry of curves. Its importance, as we will see amply demonstrated in (III §3), lies in its ability to tell us that there are functions on C having prescribed zeros and poles.

Theorem 5.4. (Riemann–Roch) *Let C be a smooth curve and let K_C be a canonical divisor on C. There is an integer $g \geq 0$, called the* genus *of C, such that for every divisor $D \in \mathrm{Div}(C)$,*

$$\ell(D) - \ell(K_C - D) = \deg D - g + 1.$$

PROOF. For a fancy proof using Serre duality, see [111, IV §1]. A more elementary proof, due to Weil, is given in [136, Chapter 1]. □

Corollary 5.5. (a) $\ell(K_C) = g$.
(b) $\deg K_C = 2g - 2$.
(c) *If* $\deg D > 2g - 2$, *then*

$$\ell(D) = \deg D - g + 1.$$

PROOF. (a) Use (II.5.4) with $D = 0$. Note that $\mathcal{L}(0) = \bar{K}$ from (II.1.2), so $\ell(0) = 1$.
(b) Use (a) and (II.5.4) with $D = K_C$.
(c) From (b) we have $\deg(K_C - D) < 0$. Now use (II.5.4) and (II.5.2a). □

Example 5.6. Let $C = \mathbb{P}^1$. Then (II.4.5) says that there are no holomorphic differentials on C, so using the identification from (II.5.3), we see that $\ell(K_C) = 0$. Then (II.5.5a) says that $\mathbb{P}^1$ has genus 0, and the Riemann–Roch theorem reads

$$\ell(D) - \ell(-2(\infty) - D) = \deg D + 1.$$

In particular, if $\deg D \geq -1$, then

$$\ell(D) = \deg D + 1.$$

(See Exercise 2.3b.)

Example 5.7. Let C be the curve

$$C : y^2 = (x - e_1)(x - e_2)(x - e_3),$$

where we continue with the notation of (II.3.3) and (II.4.6). We have seen in (II.4.6) that

$$\mathrm{div}(dx/y) = 0,$$

so the canonical class on C is trivial, i.e., we may take $K_C = 0$. Hence using (II.5.5a) we find that

$$g = \ell(K_C) = \ell(0) = 1,$$

so C has genus one. The Riemann–Roch theorem (II.5.5c) then tells us that

$$\ell(D) = \deg D \qquad \text{provided } \deg D \geq 1.$$

We consider several special cases.

(i) Let $P \in C$. Then $\ell\big((P)\big) = 1$. But $\mathcal{L}\big((P)\big)$ certainly contains the constant functions, which have no poles, so this shows that there are no functions on C having a single simple pole.

(ii) Recall that P_∞ is the point at infinity on C. Then $\ell\big(2(P_\infty)\big) = 2$, and $\{1, x\}$ provides a basis for $\mathcal{L}\big(2(P_\infty)\big)$.

(iii) Similarly, the set $\{1, x, y\}$ is a basis for $\mathcal{L}\big(3(P_\infty)\big)$, and $\{1, x, y, x^2\}$ is a basis for $\mathcal{L}\big(4(P_\infty)\big)$.

(iv) Now we observe that the seven functions $1, x, y, x^2, xy, x^3, y^2$ are all in $\mathcal{L}\big(6(P_\infty)\big)$, but $\ell\big(6(P_\infty)\big) = 6$, so these seven functions must be $\bar{K}$-linearly dependent. Of course, the equation $y^2 = (x - e_1)(x - e_2)(x - e_3)$ used to define C gives an equation of linear dependence among them.

The next result says that if C and D are defined over K, then so is $\mathcal{L}(D)$.

Proposition 5.8. *Let C/K be a smooth curve and let $D \in \mathrm{Div}_K(C)$. Then $\mathcal{L}(D)$ has a basis consisting of functions in $K(C)$.*

PROOF. Since D is defined over K, we have

$$f^\sigma \in \mathcal{L}(D^\sigma) = \mathcal{L}(D) \qquad \text{for all } f \in \mathcal{L}(D) \text{ and all } \sigma \in G_{\bar{K}/K}.$$

Thus $G_{\bar{K}/K}$ acts on $\mathcal{L}(D)$, and the desired conclusion follows from the following general lemma. $\qquad\square$

Lemma 5.8.1. *Let V be a $\bar{K}$-vector space, and assume that $G_{\bar{K}/K}$ acts continuously on V in a manner compatible with its action on $\bar{K}$. Let*

$$V_K = V^{G_{\bar{K}/K}} = \{\mathbf{v} \in V : \mathbf{v}^\sigma = \mathbf{v} \text{ for all } \sigma \in G_{\bar{K}/K}\}.$$

Then

$$V \cong \bar{K} \otimes_K V_K,$$

i.e., the vector space V has a basis of $G_{\bar{K}/K}$-invariant vectors.

PROOF. It is clear that V_K is a K-vector space, so it suffices to show that every $\mathbf{v} \in V$ is a $\bar{K}$-linear combination of vectors in V_K. Let $\mathbf{v} \in V$ and let L/K be a finite Galois extension such that $\mathbf{v}$ is fixed by $G_{\bar{K}/L}$. (The assumption that $G_{\bar{K}/K}$ acts continuously on V means precisely that the subgroup $\{\sigma \in G_{\bar{K}/K} : \mathbf{v}^\sigma = \mathbf{v}\}$ has finite index in K, so we can take L to be the Galois closure of its fixed field.) Let $\{\alpha_1, \ldots, \alpha_n\}$ be a basis for L/K, and let $\{\sigma_1, \ldots, \sigma_n\} = G_{L/K}$. For each $1 \le i \le n$, consider the vector

$$\mathbf{w}_i = \sum_{j=1}^{n} (\alpha_i \mathbf{v})^{\sigma_j} = \mathrm{Trace}_{L/K}(\alpha_i \mathbf{v}).$$

It is clear that $\mathbf{w}_i$ is $G_{\bar{K}/K}$ invariant, so $\mathbf{w}_i \in V_K$. A basic result from field theory [142, III, Proposition 9] says that the matrix $\big(\alpha_i^{\sigma_j}\big)_{1 \le i, j \le n}$ is nonsingular, so each $\mathbf{v}^{\sigma_j}$, and in particular $\mathbf{v}$, is an L-linear combination of the $\mathbf{w}_i$'s. (For a fancier proof, see Exercise 2.12.) $\qquad\square$

We conclude this section with a classic relationship connecting the genera of curves linked by a nonconstant map.

Theorem 5.9. (Hurwitz) *Let* $\phi : C_1 \to C_2$ *be a nonconstant separable map of smooth curves of genera* g_1 *and* g_2, *respectively. Then*

$$2g_1 - 2 \geq (\deg \phi)(2g_2 - 2) + \sum_{P \in C_1} \left(e_\phi(P) - 1 \right).$$

Further, equality holds if and only if one of the following two conditions is true:
  (i) $\mathrm{char}(K) = 0$.
  (ii) $\mathrm{char}(K) = p > 0$ *and* p *does not divide* $e_\phi(P)$ *for all* $P \in C_1$.

PROOF. Let $\omega \in \Omega_C$ be a nonzero differential, let $P \in C_1$, and let $Q = \phi(P)$. Since ϕ is separable, (II.4.2c) tells us that $\phi^* \omega \neq 0$. We need to relate the values of $\mathrm{ord}_P(\phi^* \omega)$ and $\mathrm{ord}_Q(\omega)$. Write $\omega = f\, dt$ with $t \in \bar{K}(C_2)$ a uniformizer at Q. Letting $e = e_\phi(P)$, we have $\phi^* t = us^e$, where s is a uniformizer at P and $u(P) \neq 0, \infty$. Hence

$$\phi^* \omega = (\phi^* f)d(\phi^* t) = (\phi^* f)d(us^e) = (\phi^* f)\left[eus^{e-1} + (du/ds)s^e \right] ds.$$

Now $\mathrm{ord}_P(du/ds) \geq 0$ from (II.4.3b), so we see that

$$\mathrm{ord}_P(\phi^* \omega) \geq \mathrm{ord}_P(\phi^* f) + e - 1,$$

with equality if and only if $e \neq 0$ in K. Further,

$$\mathrm{ord}_P(\phi^* f) = e_\phi(P)\,\mathrm{ord}_Q(f) = e_\phi(P)\,\mathrm{ord}_Q(\omega).$$

Hence adding over all $P \in C_1$ yields

$$\begin{aligned}
\deg \mathrm{div}(\phi^* \omega) &\geq \sum_{P \in C_1} \left[e_\phi(P)\,\mathrm{ord}_{\phi(P)}(\omega) + e_\phi(P) - 1 \right] \\
&= \sum_{Q \in C_2} \sum_{P \in \phi^{-1}(Q)} e_\phi(P)\,\mathrm{ord}_Q(\omega) + \sum_{P \in C_1} \left(e_\phi(P) - 1 \right) \\
&= (\deg \phi)(\deg \mathrm{div}(\omega)) + \sum_{P \in C_1} \left(e_\phi(P) - 1 \right),
\end{aligned}$$

where the last equality follows from (II.2.6a). Now Hurwitz's formula is a consequence of (II.5.5b), which says that on a curve of genus g, the divisor of any nonzero differential has degree $2g - 2$. $\qquad\square$

Exercises

2.1. Let R be a Noetherian local domain that is not a field, let $\mathfrak{M}$ be its maximal ideal, and let $k = R/\mathfrak{M}$ be its residue field. Prove that the following are equivalent:
  (i) R is a discrete valuation ring.

(ii) $\mathfrak{M}$ is principal.

(iii) $\dim_k \mathfrak{M}/\mathfrak{M}^2 = 1$.

(Note that this lemma was used in (II.1.1) to show that on a smooth curve, the local rings $\bar{K}[C]_P$ are discrete valuation rings.)

2.2. Let $\phi : C_1 \to C_2$ be a nonconstant map of smooth curves, let $f \in \bar{K}(C_2)^*$, and let $P \in C_1$. Prove that

$$\text{ord}_P(\phi^* f) = e_\phi(P)\,\text{ord}_{\phi(P)}(f).$$

2.3. Verify directly that each of the following results from the text is true for the particular case of the curve $C = \mathbb{P}^1$.

(a) Prove the two parts of (II.2.6):

(i) $\displaystyle\sum_{P \in \phi^{-1}(Q)} e_\phi(P) = \deg \phi \qquad$ for all $Q \in \mathbb{P}^1$.

(ii) $\#\phi^{-1}(Q) = \deg_s(\phi) \qquad$ for all but finitely many $Q \in \mathbb{P}^1$.

(b) Prove the Riemann–Roch theorem (II.5.4) for $\mathbb{P}^1$.

(c) Prove Hurwitz's theorem (II.5.9) for a nonconstant separable map $\phi : \mathbb{P}^1 \to \mathbb{P}^1$.

2.4. Let C be a smooth curve and let $D \in \text{Div}(C)$. Without using the Riemann–Roch theorem, prove the following statements.

(a) $\mathcal{L}(D)$ is a $\bar{K}$-vector space.

(b) If $\deg D \geq 0$, then

$$\ell(D) \leq \deg D + 1.$$

2.5. Let C be a smooth curve. Prove that the following are equivalent (over $\bar{K}$):

(i) C is isomorphic to $\mathbb{P}^1$.

(ii) C has genus 0.

(iii) There exist distinct points $P, Q \in C$ satisfying $(P) \sim (Q)$.

2.6. Let C be a smooth curve of genus one, and fix a base point $P_0 \in C$.

(a) Prove that for all $P, Q \in C$ there exists a unique $R \in C$ such that

$$(P) + (Q) \sim (R) + (P_0).$$

Denote this point R by $\sigma(P, Q)$.

(b) Prove that the map $\sigma : C \times C \to C$ from (a) makes C into an abelian group with identity element P_0.

(c) Define a map

$$\kappa : C \longrightarrow \text{Pic}^0(C), \qquad P \longmapsto \text{divisor class of } (P) - (P_0).$$

Prove that κ is a bijection of sets, and hence that κ can be used to make C into a group via the rule

$$P + Q = \kappa^{-1}\big(\kappa(P) + \kappa(Q)\big).$$

(d) Prove that the group operations on C defined in (b) and (c) are the same.

2.7. Let $F(X, Y, Z) \in K[X, Y, Z]$ be a homogeneous polynomial of degree $d \geq 1$, and assume that the curve C in $\mathbb{P}^2$ given by the equation $F = 0$ is nonsingular. Prove that

$$\text{genus}(C) = \frac{(d-1)(d-2)}{2}.$$

(*Hint.* Define a map $C \to \mathbb{P}^1$ and use (II.5.9).)

2.8. Let $\phi : C_1 \to C_2$ be a nonconstant separable map of smooth curves.
(a) Prove that $\text{genus}(C_1) \geq \text{genus}(C_2)$.
(b) Prove that if C_1 and C_2 have the same genus g, then one of the following is true:
    (i) $g = 0$.
    (ii) $g = 1$ and ϕ is unramified.
    (iii) $g \geq 2$ and ϕ is an isomorphism.

2.9. Let a, b, c, d be squarefree integers with $a > b > c > 0$, and let C be the curve in $\mathbb{P}^2$ given by the equation
$$C : aX^3 + bY^3 + cZ^3 + dXYZ = 0.$$
Let $P = [x, y, z] \in C$ and let L be the tangent line to C at P.
(a) Show that $C \cap L = \{P, P'\}$ and calculate $P' = [x', y', z']$ in terms of a, b, c, d, x, y, z.
(b) Show that if $P \in C(\mathbb{Q})$, then $P' \in C(\mathbb{Q})$.
(c) Let $P \in C(\mathbb{Q})$. Choose homogeneous coordinates for P and P' that are integers satisfying $\gcd(x, y, z) = 1$ and $\gcd(x', y', z') = 1$. Prove that

$$|x'y'z'| > |xyz|.$$

(Note the strict inequality.)
(d) Conclude that either $C(\mathbb{Q}) = \emptyset$ or else $C(\mathbb{Q})$ is an infinite set.
(e) ** Characterize, in terms of a, b, c, d, whether $C(\mathbb{Q})$ contains any points.

2.10. Let C be a smooth curve. The *support* of a divisor $D = \sum n_P(P) \in \text{Div}(C)$ is the set of points $P \in C$ for which $n_P \neq 0$. Let $f \in \bar{K}(C)^*$ be a function such that $\text{div}(f)$ and D have disjoint supports. Then it makes sense to define

$$f(D) = \prod_{P \in C} f(P)^{n_P}.$$

Let $\phi : C_1 \to C_2$ be a nonconstant map of smooth curves. Prove that the following two equalities are valid in the sense that if both sides are well-defined, then they are equal.
(a) $f(\phi^* D) = (\phi_* f)(D)$ for all $f \in \bar{K}(C_1)^*$ and all $D \in \text{Div}(C_2)$.
(b) $f(\phi_* D) = (\phi^* f)(D)$ for all $f \in \bar{K}(C_2)^*$ and all $D \in \text{Div}(C_1)$.

2.11. Let C be a smooth curve and let $f, g \in \bar{K}(C)^*$ be functions such that $\text{div}(f)$ and $\text{div}(g)$ have disjoint support. (See Exercise 2.10.) Prove *Weil's reciprocity law*

$$f\big(\text{div}(g)\big) = g\big(\text{div}(f)\big)$$

using the following two steps:
(a) Verify Weil's reciprocity law directly for $C = \mathbb{P}^1$.
(b) Now prove it for arbitrary C by using the map $g : C \to \mathbb{P}^1$ to reduce to (a).

2.12. Use the extension of Hilbert's Theorem 90 (B.3.2), which says that

$$H^1\big(G_{\bar{K}/K},\mathrm{GL}_n(\bar{K})\big) = 0,$$

to give another proof of (II.5.8.1).

2.13. Let C/K be a curve.
(a) Prove that the following sequence is exact:

$$1 \longrightarrow K^* \longrightarrow K(C)^* \longrightarrow \mathrm{Div}_K^0(C) \longrightarrow \mathrm{Pic}_K^0(C).$$

(b) Suppose that C has genus one and that $C(K) \neq \emptyset$. Prove that the map

$$\mathrm{Div}_K^0(C) \longrightarrow \mathrm{Pic}_K^0(C)$$

is surjective.

2.14. For this exercise we assume that char $K \neq 2$. Let $f(x) \in K[x]$ be a polynomial of degree $d \geq 1$ with nonzero discriminant, let C_0/K be the affine curve given by the equation

$$C_0 : y^2 = f(x) = a_0 x^d + a_1 x^{d-1} + \cdots + a_{d-1}x + a_d,$$

and let g be the unique integer satisfying $d - 3 < 2g \leq d - 1$.
(a) Let C be the closure of the image of C_0 via the map

$$[1, x, x^2, \ldots, x^{g-1}, y] : C_0 \longrightarrow \mathbb{P}^{g+2}.$$

Prove that C is smooth and that $C \cap \{X_0 \neq 0\}$ is isomorphic to C_0. The curve C is called a *hyperelliptic curve*.

(b) Let

$$f^*(v) = v^{2g+2}f(1/v) = \begin{cases} a_0 + a_1 v + \cdots + a_{d-1}v^{d-1} + a_d v^d & \text{if } d \text{ is even,} \\ a_0 v + a_1 v^2 + \cdots + a_{d-1}v^d + a_d v^{d+1} & \text{if } d \text{ is odd.} \end{cases}$$

Show that C consists of two affine pieces

$$C_0 : y^2 = f(x) \qquad \text{and} \qquad C_1 : w^2 = f^*(v),$$

"glued together" via the maps

$$\begin{array}{ll} C_0 \longrightarrow C_1, & \qquad C_1 \longrightarrow C_0, \\ (x, y) \longmapsto (1/x, y/x^{g+1}), & \qquad (v, w) \longmapsto (1/v, w/v^{g+1}). \end{array}$$

(c) Calculate the divisor of the differential dx/y on C and use the result to show that C has genus g. Check your answer by applying Hurwitz's formula (II.5.9) to the map $[1, x] : C \to \mathbb{P}^1$. (Note that Exercise 2.7 does not apply, since $C \not\subset \mathbb{P}^2$.)
(d) Find a basis for the holomorphic differentials on C. (*Hint.* Consider the set of differential forms $\{x^i\, dx/y : i = 0, 1, 2, \ldots\}$. How many elements in this set are holomorphic?)

2.15. Let C/K be a smooth curve defined over a field of characteristic $p > 0$, and let $t \in K(C)$. Prove that the following are equivalent:
(i) $K(C)$ is a finite separable extension of $K(t)$.
(ii) For all but finitely many points $P \in C$, the function $t - t(P)$ is a uniformizer at P.
(iii) $t \notin K(C)^p$.

2.16. Let C/K be a curve that is defined over K and let $P \in C(K)$. Prove that $K(C)$ contains uniformizers for C at P, i.e., prove that there are uniformizers that are defined over K.

Chapter III

The Geometry of Elliptic Curves

Elliptic curves, our principal object of study in this book, are curves of genus one having a specified base point. Our ultimate goal, as the title of the book indicates, is to study the arithmetic properties of these curves. In other words, we will be interested in analyzing their points defined over arithmetically interesting fields, such as finite fields, local (p-adic) fields, and global (number) fields. However, before doing so we are well advised to study the properties of these curves in the simpler situation of an algebraically closed field, i.e., to study their geometry. This reflects the general principle in Diophantine geometry that in attempting to study any significant problem, it is essential to have a thorough understanding of the geometry before one can hope to make progress on the number theory. It is the purpose of this chapter to make an intensive study of the geometry of elliptic curves over arbitrary algebraically closed fields. (The particular case of elliptic curves over the complex numbers is studied in more detail in Chapter VI.)

We start in the first two sections by looking at elliptic curves given by explicit polynomial equations called Weierstrass equations. Using these explicit equations, we show, among other things, that the set of points of an elliptic curve forms an abelian group, and that the group law is given by rational functions. Then, in Section 3, we use the Riemann–Roch theorem to study arbitrary elliptic curves and to show that every elliptic curve has a Weierstrass equation, so the results from the first two sections in fact apply generally. The remainder of the chapter studies, in various guises, the algebraic maps between elliptic curves. In particular, since the points of an elliptic curve form a group, for each integer m there is a multiplication-by-m map from the curve to itself. It would be difficult to overestimate the importance of these multiplication maps in any attempt to study the arithmetic of elliptic curves, which will explain why we devote so much space to them in this chapter.

J.H. Silverman, *The Arithmetic of Elliptic Curves, Second Edition*, Graduate Texts in Mathematics 106, DOI 10.1007/978-0-387-09494-6_III,
© Springer Science+Business Media, LLC 2009

III.1   Weierstrass Equations

Our primary objects of study are *elliptic curves*, which are curves of genus one having a specified base point. As we will see in (III §3), every such curve can be written as the locus in $\mathbb{P}^2$ of a cubic equation with only one point, the base point, on the line at ∞. Then, after X and Y are scaled appropriately, an elliptic curve has an equation of the form

$$Y^2Z + a_1XYZ + a_3YZ^2 = X^3 + a_2X^2Z + a_4XZ^2 + a_6Z^3.$$

Here $O = [0, 1, 0]$ is the base point and $a_1, \ldots, a_6 \in \bar{K}$. (It will become clear later why the coefficients are labeled in this way.) In this section and the next, we study the curves given by such *Weierstrass equations*, using explicit formulas as much as possible to replace the need for general theory.

To ease notation, we generally write the Weierstrass equation for our elliptic curve using non-homogeneous coordinates $x = X/Z$ and $y = Y/Z$,

$$E : y^2 + a_1xy + a_3y = x^3 + a_2x^2 + a_4x + a_6,$$

always remembering that there is an extra point $O = [0, 1, 0]$ out at infinity. As usual, if $a_1, \ldots, a_6 \in K$, then E is said to be *defined over* K.

If $\operatorname{char}(\bar{K}) \neq 2$, then we can simplify the equation by completing the square. Thus the substitution

$$y \longmapsto \frac{1}{2}(y - a_1x - a_3)$$

gives an equation of the form

$$E : y^2 = 4x^3 + b_2x^2 + 2b_4x + b_6,$$

where

$$b_2 = a_1^2 + 4a_4, \qquad b_4 = 2a_4 + a_1a_3, \qquad b_6 = a_3^2 + 4a_6.$$

We also define quantities

$$b_8 = a_1^2a_6 + 4a_2a_6 - a_1a_3a_4 + a_2a_3^2 - a_4^2,$$
$$c_4 = b_2^2 - 24b_4,$$
$$c_6 = -b_2^3 + 36b_2b_4 - 216b_6,$$
$$\Delta = -b_2^2b_8 - 8b_4^3 - 27b_6^2 + 9b_2b_4b_6,$$
$$j = c_4^3/\Delta,$$
$$\omega = \frac{dx}{2y + a_1x + a_3} = \frac{dy}{3x^2 + 2a_2x + a_4 - a_1y}.$$

One easily verifies that they satisfy the relations

$$4b_8 = b_2b_6 - b_4^2 \qquad \text{and} \qquad 1728\Delta = c_4^3 - c_6^2.$$

If further $\operatorname{char}(\bar{K}) \neq 2, 3$, then the substitution

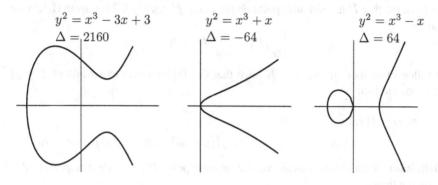

Figure 3.1: Three elliptic curves

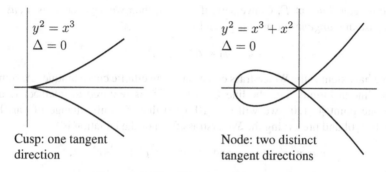

Cusp: one tangent
direction

Node: two distinct
tangent directions

Figure 3.2: Two singular cubic curves.

$$(x, y) \longmapsto \left(\frac{x - 3b_2}{36}, \frac{y}{108} \right)$$

eliminates the x^2 term, yielding the simpler equation

$$E : y^2 = x^3 - 27c_4 x - 54c_6.$$

Definition. The quantity Δ is the *discriminant* of the Weierstrass equation, the quantity j is the *j-invariant* of the elliptic curve, and ω is the *invariant differential* associated to the Weierstrass equation.

Example 1.1. It is easy to graph the real locus of a Weierstrass equation. Some representative examples are shown in Figure 3.1. If $\Delta = 0$, then we will see later (III.1.4) that the curve is singular. Two sorts of behavior can occur, as illustrated in Figure 3.2.

With these singular examples in mind, we consider the general situation. Let $P = (x_0, y_0)$ be a point satisfying a Weierstrass equation

$$f(x, y) = y^2 + a_1 xy + a_3 y - x^3 - a_2 x^2 - a_4 x - a_6 = 0,$$

and assume that P is a singular point on the curve $f(x, y) = 0$. Then from (I.1.5) we have

$$\frac{\partial f}{\partial x}(P) = \frac{\partial f}{\partial y}(P) = 0.$$

It follows that there are $\alpha, \beta \in \bar{K}$ such that the Taylor series expansion of $f(x, y)$ at P has the form

$$f(x, y) - f(x_0, y_0)$$
$$= \big((y - y_0) - \alpha(x - x_0)\big)\big((y - y_0) - \beta(x - x_0)\big) - (x - x_0)^3.$$

Definition. With notation as above, the singular point P is a *node* if $\alpha \neq \beta$. In this case, the lines

$$y - y_0 = \alpha(x - x_0) \qquad \text{and} \qquad y - y_0 = \beta(x - x_0)$$

are the *tangent lines* at P. Conversely, if $\alpha = \beta$, then we say that P is a *cusp*, in which case the *tangent line* at P is given by

$$y - y_0 = \alpha(x - x_0).$$

To what extent is the Weierstrass equation for an elliptic curve unique? Assuming that the line at infinity, i.e., the line $Z = 0$ in $\mathbb{P}^2$, is required to intersect E only at the one point $[0, 1, 0]$, we will see (III.3.1b) that the only change of variables fixing $[0, 1, 0]$ and preserving the Weierstrass form of the equation is

$$x = u^2 x' + r \qquad \text{and} \qquad y = u^3 y' + u^2 s x' + t,$$

where $u, r, s, t \in \bar{K}$ and $u \neq 0$. It is now a simple (but tedious) matter to make this substitution and compute the a_i' coefficients and associated quantities for the new equation. The results are compiled in Table 3.1.

It is now clear why the j-invariant has been so named; it is an invariant of the isomorphism class of the curve, and does not depend on the particular equation chosen. For algebraically closed fields, the converse is true, a fact that we establish later in this section (III.1.4b).

Remark 1.3. As we have seen, if the characteristic of K is different from 2 and 3, then any elliptic curve over K has a Weierstrass equation of a particularly simple kind. Thus any proof that involves extensive algebraic manipulation with Weierstrass equation, for example that of (III.1.4) later in this section, tends to be much shorter if K is so restricted. On the other hand, even if one is primarily interested in characteristic 0, e.g., $K = \mathbb{Q}$, an important tool is the process of reducing the coefficients of an equation modulo p for various primes p, including $p = 2$ and $p = 3$. So even for $K = \mathbb{Q}$, it is important to understand elliptic curves in all characteristics. Consequently, we adopt the following policy. All theorems will be stated for a general Weierstrass equation, but if it makes the proof substantially shorter, we will make the assumption that the characteristic of K is not 2 or 3 and give the proof in that case. Then, in the interest of completeness, we return to these theorems in Appendix A and give the proofs for general Weierstrass equations and arbitrary characteristic.

$$ua'_1 = a_1 + 2s$$
$$u^2a'_2 = a_2 - sa_1 + 3r - s^2$$
$$u^3a'_3 = a_3 + ra_1 + 2t$$
$$u^4a'_4 = a_4 - sa_3 + 2ra_2 - (t+rs)a_1 + 3r^2 - 2st$$
$$u^6a'_6 = a_6 + ra_4 + r^2a_2 + r^3 - ta_3 - t^2 - rta_1$$

$$u^2b'_2 = b_2 + 12r$$
$$u^4b'_4 = b_4 + rb_2 + 6r^2$$
$$u^6b'_6 = b_6 + 2rb_4 + r^2b_2 + 4r^3$$
$$u^8b'_8 = b_8 + 3rb_6 + 3r^2b_4 + r^3b_2 + 3r^4$$

$$u^4c'_4 = c_4$$
$$u^6c'_6 = c_6$$
$$u^{12}\Delta' = \Delta$$

$$j' = j$$
$$u^{-1}\omega' = \omega$$

Table 3.1: Change-of-variable formulas for Weierstrass equations.

Assuming now that the characteristic of K is not 2 or 3, our elliptic curve(s) have Weierstrass equation(s) of the form

$$E : y^2 = x^3 + Ax + B.$$

Associated to this equation are the quantities

$$\Delta = -16(4A^3 + 27B^2) \quad \text{and} \quad j = -1728\frac{(4A)^3}{\Delta}.$$

The only change of variables preserving this form of the equation is

$$x = u^2x' \quad \text{and} \quad y = u^3y' \quad \text{for some } u \in \bar{K}^*;$$

and then

$$u^4A' = A, \quad u^6B' = B, \quad u^{12}\Delta' = \Delta.$$

Proposition 1.4. (a) *The curve given by a Weierstrass equation satisfies:*

    (i) *It is nonsingular if and only if $\Delta = 0$.*
    (ii) *It has a node if and only if $\Delta = 0$ and $c_4 \neq 0$.*
    (iii) *It has a cusp if and only if $\Delta = c_4 = 0$.*

    In cases (ii) *and* (iii)*, there is only the one singular point.*
(b) *Two elliptic curves are isomorphic over $\bar{K}$ if and only if they both have the same j-invariant.*
(c) *Let $j_0 \in \bar{K}$. There exists an elliptic curve defined over $K(j_0)$ whose j-invariant is equal to j_0.*

PROOF. Let E be given by the Weierstrass equation

$$E : f(x,y) = y^2 + a_1xy + a_3y - x^3 - a_2x^2 - a_4x - a_6 = 0.$$

We start by showing that the point at infinity is never singular. Thus we look at the curve in $\mathbb{P}^2$ with homogeneous equation

$$F(X,Y,Z) = Y^2Z + a_1XYZ + a_3YZ^2 - X^3 - a_2X^2Z - a_4XZ^2 - a_6Z^3$$
$$= 0$$

and at the point $O = [0,1,0]$. Since

$$\frac{\partial F}{\partial Z}(O) = 1 \neq 0,$$

we see that O is a nonsingular point of E.

Next suppose that E is singular, say at $P_0 = (x_0, y_0)$. The substitution

$$x = x' + x_0 \qquad y = y' + y_0$$

leaves Δ and c_4 invariant (III.1.2), so without loss of generality we may assume that E is singular at $(0,0)$. Then

$$a_6 = f(0,0) = 0, \qquad a_4 = \frac{\partial f}{\partial x}(0,0) = 0, \qquad a_3 = \frac{\partial f}{\partial y}(0,0) = 0,$$

so the equation for E takes the form

$$E : f(x,y) = y^2 + a_1xy - a_2x^2 - x^3 = 0.$$

This equation has associated quantities

$$c_4 = (a_1^2 + 4a_2)^2 \qquad \text{and} \qquad \Delta = 0.$$

By definition, E has a node (respectively cusp) at $(0,0)$ if the quadratic form $y^2 + a_1xy - a_2x^2$ has distinct (respectively equal) factors, which occurs if and only if the discriminant of this quadratic form satisfies

$$a_1^2 + 4a_2 \neq 0 \qquad \text{(respectively } a_1^2 + 4a_2 = 0\text{).}$$

This proves the "only if" part of (ii) and (iii).

To complete the proof of (i)–(iii), it remains to show that if E is nonsingular, then $\Delta \neq 0$. To simplify the computation, we assume that $\operatorname{char}(K) \neq 2$ and consider a Weierstrass equation of the form

$$E : y^2 = 4x^3 + b_2x^2 + 2b_4x + b_6.$$

(See (III.1.3) and (A.1.2a).) The curve E is singular if and only if there is a point $(x_0, y_0) \in E$ satisfying

$$2y_0 = 12x_0^2 + 2b_2x_0 + 2b_4 = 0.$$

In other words, the singular points are exactly the points of the form $(x_0, 0)$ such that x_0 is a double root of the cubic polynomial $4x^3 + b_2 x^2 + 2b_4 x + b_6$. This polynomial has a double root if and only if its discriminant, which equals 16Δ, vanishes. This completes the proof of (i)–(iii). Further, since a cubic polynomial cannot have two double roots, E has at most one singular point.

(b) If two elliptic curves are isomorphic, then the transformation formulas (III.1.2) show that they have the same j-invariant. For the converse, we will assume that $\mathrm{char}(K) \geq 5$ (see (III.1.3) and (A.1.2b)). Let E and E' be elliptic curves with the same j-invariant, say with Weierstrass equations

$$E : y^2 = x^3 + Ax + B,$$
$$E' : y'^2 = x'^3 + A'x' + B'.$$

Then the assumption that $j(E) = j(E')$ means that

$$\frac{(4A)^3}{4A^3 + 27B^2} = \frac{(4A')^3}{4A'^3 + 27B'^2},$$

which yields

$$A^3 B'^2 = A'^3 B^2.$$

We look for an isomorphism of the form $(x, y) = (u^2 x', u^3 y')$ and consider three cases:

Case 1. $A = 0$ $(j = 0)$. Then $B \neq 0$, since $\Delta \neq 0$, so $A' = 0$, and we obtain an isomorphism using $u = (B/B')^{1/6}$.

Case 2. $B = 0$ $(j = 1728)$. Then $A \neq 0$, so $B' = 0$, and we take $u = (A/A')^{1/4}$.

Case 3. $AB \neq 0$ $(j \neq 0, 1728)$. Then $A'B' \neq 0$, since if one of them were 0, then both of them would be 0, contradicting $\Delta' \neq 0$. Taking $u = (A/A')^{1/4} = (B/B')^{1/6}$ gives the desired isomorphism.

(c) Assume that $j_0 \neq 0, 1728$ and consider the curve

$$E : y^2 + xy = x^3 - \frac{36}{j_0 - 1728} x - \frac{1}{j_0 - 1728}.$$

A simple calculations yields

$$\Delta = \frac{j_0^3}{(j_0 - 1728)^3} \qquad \text{and} \qquad j = j_0.$$

This gives the desired elliptic curve (in any characteristic) provided that $j_0 \neq 0, 1728$. To complete the list, we use the two curves

$$\begin{array}{llll} E : y^2 + y = x^3, & \Delta = -27, & j = 0, \\ E : y^2 = x^3 + x, & \Delta = -64, & j = 1728. \end{array}$$

(Notice that in characteristic 2 or 3 we have $1728 = 0$, so even in these cases one of the two curves will be nonsingular and fill in the missing value of j.) $\qquad\square$

Proposition 1.5. *Let E be an elliptic curve. Then the invariant differential ω associated to a Weierstrass equation for E is holomorphic and nonvanishing, i.e., $\mathrm{div}(\omega) = 0$.*

PROOF. Let $P = (x_0, y_0) \in E$ and

$$E : F(x, y) = y^2 + a_1 xy + a_3 y - x^3 - a_2 x^2 - a_4 x - a_6 = 0,$$

so

$$\omega = \frac{d(x - x_0)}{F_y(x, y)} = -\frac{d(y - y_0)}{F_x(x, y)}.$$

Thus P cannot be a pole of ω, since otherwise $F_y(P) = F_x(P) = 0$, which would say that P is a singular point of E. The map

$$E \longrightarrow \mathbb{P}^1, \qquad [x, y, 1] \longmapsto [x, 1],$$

is of degree 2, so $\mathrm{ord}_P(x - x_0) \leq 2$, and we have equality $\mathrm{ord}_P(x - x_0) = 2$ if and only if the quadratic polynomial $F(x_0, y)$ has a double root. In other words, either $\mathrm{ord}_P(x - x_0) = 1$, or else $\mathrm{ord}_P(x - x_0) = 2$ and $F_y(x_0, y_0) = 0$. Thus in both cases, we can use (II.4.3) to compute

$$\mathrm{ord}_P(\omega) = \mathrm{ord}_P(x - x_0) - \mathrm{ord}_P(F_y) - 1 = 0.$$

This shows that ω has no poles or zeros of the form (x_0, y_0), so it remains to check what happens at O.

Let t be a uniformizer at O. Since $\mathrm{ord}_O(x) = -2$ and $\mathrm{ord}_O(y) = -3$, we see that $x = t^{-2} f$ and $y = t^{-3} g$ for functions f and g satisfying $f(O) \neq 0, \infty$ and $g(O) \neq 0, \infty$. Now

$$\omega = \frac{dx}{F_y(x, y)} = \frac{-2t^{-3} f + t^{-2} f'}{2t^{-3} g + a_1 t^{-2} f + a_3} \, dt = \frac{-2f + t f'}{2g + a_1 t f + a_3 t^3} \, dt.$$

Here we are writing $f' = df/dt$; cf. (II.4.3). In particular, (II.4.3b) tells us that f' is regular at O. Hence assuming that $\mathrm{char}(K) \neq 2$, the function

$$\frac{-2f + t f'}{2g + a_1 t f + a_3 t^3}$$

is regular and nonvanishing at O, and thus

$$\mathrm{ord}_O(\omega) = 0.$$

Finally, if $\mathrm{char}(K) = 2$, then the same result follows from a similar calculation using $\omega = dy/F_x(x, y)$. We leave the details to the reader. $\square$

Next we look at what happens when a Weierstrass equation is singular.

Proposition 1.6. *If a curve E given by a Weierstrass equation is singular, then there exists a rational map $\phi : E \to \mathbb{P}^1$ of degree one, i.e., the curve E is birational to $\mathbb{P}^1$. (Note that since E is singular, we cannot use (II.2.4.1) to conclude that $E \cong \mathbb{P}^1$.)*

PROOF. Making a linear change of variables, we may assume that the singular point is $(x, y) = (0, 0)$. Checking partial derivatives, we see that the Weierstrass equation has the form

$$E : y^2 + a_1 xy = x^3 + a_2 x^2.$$

Then the rational map

$$E \longrightarrow \mathbb{P}^1, \qquad (x, y) \to [x, y],$$

has degree one, since it has an inverse given by

$$\mathbb{P}^1 \longrightarrow E, \qquad [1, t] \longmapsto (t^2 + a_1 t - a_2, t^3 + a_1 t^2 - a_2 t).$$

(To derive this formula, let $t = y/x$ and note that dividing the Weierstrass equation of E by x^2 yields $t^2 + a_1 t = x + a_2$. This shows that both x and $y = xt$ are in $\bar{K}(t)$.) $\square$

Legendre Form

There is another form of Weierstrass equation that is sometimes convenient.

Definition. A Weierstrass equation is in *Legendre form* if it can be written as

$$y^2 = x(x - 1)(x - \lambda).$$

Proposition 1.7. *Assume that* $\mathrm{char}(K) \neq 2$.
(a) *Every elliptic curve is isomorphic (over* $\bar{K}$*) to an elliptic curve in Legendre form*

$$E_\lambda : y^2 = x(x - 1)(x - \lambda)$$

for some $\lambda \in \bar{K}$ *with* $\lambda \neq 0, 1$.
(b) *The j-invariant of E_λ is*

$$j(E_\lambda) = 2^8 \frac{(\lambda^2 - \lambda + 1)^3}{\lambda^2(\lambda - 1)^2}.$$

(c) *The association*

$$\bar{K} \smallsetminus \{0, 1\} \longrightarrow \bar{K}, \qquad \lambda \longmapsto j(E_\lambda),$$

is surjective and exactly six-to-one except above $j = 0$ and $j = 1728$, where it is two-to-one and three-to-one, respectively (unless $\mathrm{char}(K) = 3$*, in which case it is one-to-one above $j = 0 = 1728$).*

PROOF. (a) Since $\mathrm{char}(K) \neq 2$, we know that E has a Weierstrass equation of the form

$$y^2 = 4x^3 + b_2 x^2 + 2b_4 x + b_6.$$

Replacing (x, y) by $(x, 2y)$ and factoring the cubic yields an equation of the form

$$y^2 = (x - e_1)(x - e_2)(x - e_3)$$

for some $e_1, e_2, e_3 \in \bar{K}$. Further, since

$$\Delta = 16(e_1 - e_2)^2(e_1 - e_3)^2(e_2 - e_3)^2 \neq 0,$$

we see that the e_i's are distinct. Now the substitution

$$x = (e_2 - e_1)x' + e_1, \qquad y = (e_2 - e_1)^{3/2}y'$$

gives an equation in Legendre form with

$$\lambda = \frac{e_3 - e_1}{e_2 - e_1} \in \bar{K}, \qquad \lambda \neq 0, 1.$$

(b) Calculation.

(c) One can work directly from the formula for $j(E_\lambda)$ in (b), an approach that we leave to the reader. Instead, we use the fact that the j-invariant classifies an elliptic curve up to isomorphism (III.1.4b). Thus suppose that $j(E_\lambda) = j(E_\mu)$. Then $E_\lambda \cong E_\mu$, so their Weierstrass equations (in Legendre form) are related by a change of variables

$$x = u^2 x' + r, \qquad y = u^3 y'.$$

Equating

$$x(x - 1)(x - \mu) = \left(x + \frac{r}{u^2}\right)\left(x + \frac{r-1}{u^2}\right)\left(x + \frac{r-\lambda}{u^2}\right),$$

there are six ways of assigning the linear terms to one another, and one easily checks that these lead to six possible values for μ in terms of λ,

$$\mu \in \left\{\lambda, \frac{1}{\lambda}, 1 - \lambda, \frac{1}{1-\lambda}, \frac{\lambda}{\lambda-1}, \frac{\lambda-1}{\lambda}\right\}.$$

Hence $\lambda \mapsto j(E_\lambda)$ is exactly six-to-one unless two or more of these values for μ coincide. Equating them by pairs shows that this occurs if and only if

$$\lambda \in \left\{-1, 2, \frac{1}{2}\right\} \implies \text{association is three-to-one}$$

or

$$\lambda^2 - \lambda + 1 = 0 \implies \text{association is two-to-one}.$$

These λ values correspond, respectively, to $j = 1728$ and $j = 0$. Finally, if K has characteristic 3, then these λ values coincide and the equation $j(\lambda) = 0 = 1728$ has the unique solution $\lambda = -1$. $\square$

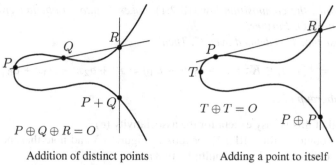

Addition of distinct points   Adding a point to itself

Figure 3.3: The composition law.

III.2 The Group Law

Let E be an elliptic curve given by a Weierstrass equation. Thus $E \subset \mathbb{P}^2$ consists of the points $P = (x, y)$ satisfying the Weierstrass equation, together with the point $O = [0, 1, 0]$ at infinity. Let $L \subset \mathbb{P}^2$ be a line. Then, since the equation has degree three, the line L intersects E at exactly three points, say P, Q, R. Of course, if L is tangent to E, then P, Q, R need not be distinct. The fact that $L \cap E$, taken with multiplicities, consists of exactly three points is a special case of Bézout's theorem [111, I.7.8]. However, since we give explicit formulas later in this section, there is no need to use a general theorem.

We define a composition law $\oplus$ on E by the following rule:

Composition Law 2.1. *Let $P, Q \in E$, let L be the line through P and Q (if $P = Q$, let L be the tangent line to E at P), and let R be the third point of intersection of L with E. Let L' be the line through R and O. Then L' intersects E at R, O, and a third point. We denote that third point by $P \oplus Q$.*

Various instances of the composition law (III.2.1) are illustrated in Figure 3.3. We now justify the use of the symbol $\oplus$.

Proposition 2.2. *The composition law (III.2.1) has the following properties*:
 (a) *If a line L intersects E at the (not necessarily distinct) points P, Q, R, then*

$$(P \oplus Q) \oplus R = O.$$

 (b) $P \oplus O = P$ *for all $P \in E$.*
 (c) $P \oplus Q = Q \oplus P$ *for all $P, Q \in E$.*
 (d) *Let $P \in E$. There is a point of E, denoted by $\ominus P$, satisfying*

$$P \oplus (\ominus P) = O.$$

 (e) *Let $P, Q, R \in E$. Then*

$$(P \oplus Q) \oplus R = P \oplus (Q \oplus R).$$

In other words, the composition law (III.2.1) *makes E into an abelian group with identity element O. Further:*

(f) *Suppose that E is defined over K. Then*

$$E(K) = \{(x, y) \in K^2 : y^2 + a_1 xy + a_3 y = x^3 + a_2 x^2 + a_4 x + a_6\} \cup \{O\}$$

is a subgroup of E.

PROOF. All of this is easy except for the associativity (e).

(a) This is obvious from (III.2.1), or look at Figure 3.3 and note that the tangent line to E at O intersects E with multiplicity 3 at O.

(b) Taking $Q = O$ in (III.2.1), we see that the lines L and L' coincide. The former intersects E at P, O, R and the latter at $R, O, P \oplus O$, so $P \oplus O = P$.

(c) This is also clear, since the construction of $P \oplus Q$ in (III.2.1) is symmetric in P and Q.

(d) Let the line through P and Q also intersect E at R. Then using (a) and (b), we find that

$$O = (P \oplus O) \oplus R = P \oplus R.$$

(e) Using the explicit formulas given later in this section (III.2.3), one can laboriously verify the associative law case by case. We leave this task to the reader. A more enlightening proof using the Riemann–Roch theorem is given in the next section (III.3.4e). For a geometric proof, see [95].

(f) If P and Q have coordinates in K, then the equation of the line connecting them has coefficients in K. If, further, E is defined over K, then the third point of intersection has coordinates given by a rational combination of the coordinates of coefficients of the line and of E, so will be in K. (If this is not clear, see (III.2.3) in this section for explicit formulas.)   □

Notation. From here on, we drop the special symbols $\oplus$ and $\ominus$ and simply write $+$ and $-$ for the group operation on an elliptic curve E. For $m \in \mathbb{Z}$ and $P \in E$, we let

$$[m]P = \overbrace{P + \cdots + P}^{m \text{ terms if } m > 0}, \qquad [m]P = \overbrace{-P - \cdots - P}^{|m| \text{ terms if } m < 0}, \qquad [0]P = O.$$

As promised, we now derive explicit formulas for the group operations on E. Let E be an elliptic curve given by a Weierstrass equation

$$F(x, y) = y^2 + a_1 xy + a_3 y - x^3 - a_2 x^2 - a_4 x - a_6 = 0,$$

and let $P_0 = (x_0, y_0) \in E$. Following the proof of (III.2.2d), in order to calculate $-P_0$, we take the line L through P_0 and O and find its third point of intersection with E. The line L is given by

$$L : x - x_0 = 0.$$

Substituting this into the equation for E, we see that the quadratic polynomial $F(x_0, y)$ has roots y_0 and y_0', where $-P = (x_0, y_0')$. Writing out

$$F(x_0, y) = c(y - y_0)(y - y_0')$$

and equating the coefficients of y^2 gives $c = 1$, and similarly equating the coefficients of y gives $y_0' = -y_0 - a_1 x_0 - a_3$. This yields

$$-P_0 = -(x_0, y_0) = (x_0, -y_0 - a_1 x_0 - a_3).$$

Next we derive a formula for the addition law. Let

$$P_1 = (x_1, y_1) \quad \text{and} \quad P_2 = (x_2, y_2)$$

be points of E. If $x_1 = x_2$ and $y_1 + y_2 + a_1 x_2 + a_3 = 0$, then we have already shown that $P_1 + P_2 = O$. Otherwise the line L through P_1 and P_2 (or the tangent line to E if $P_1 = P_2$) has an equation of the form

$$L : y = \lambda x + \nu;$$

formulas for λ and ν are given below. Substituting the equation of L into the equation of E, we see that $F(x, \lambda x + \nu)$ has roots x_1, x_2, x_3, where $P_3 = (x_3, y_3)$ is the third point of $L \cap E$. From (III.2.2a) we have

$$P_1 + P_2 + P_3 = O.$$

We write out

$$F(x, \lambda x + \nu) = x(x - x_1)(x - x_2)(x - x_3)$$

and equate coefficients. The coefficient of x^3 gives $c = -1$, and then the coefficient of x^2 yields

$$x_1 + x_2 + x_3 = \lambda^2 + a_a \lambda - a_2.$$

This gives a formula for x_3, and substituting into the equation of L gives the value of $y_3 = \lambda x_3 + \nu$. Finally, to find $P_1 + P_2 = -P_3$, we apply the negation formula to P_3. All of this is summarized in the following algorithm.

Group Law Algorithm 2.3. *Let E be an elliptic curve given by a Weierstrass equation*

$$E : y^2 + a_1 xy + a_3 y = x^3 + a_2 x^2 + a_4 x + a_6.$$

(a) *Let $P_0 = (x_0, y_0)$. Then*

$$-P_0 = (x_0, -y_0 - a_1 x_0 - a_3).$$

Next let

$$P_1 + P_2 = P_3 \quad \text{with} \quad P_i = (x_i, y_i) \in E \quad \text{for } i = 1, 2, 3.$$

(b) *If $x_1 = x_2$ and $y_1 + y_2 + a_1 x_2 + a_3 = 0$, then*

$$P_1 + P_2 = O.$$

Otherwise, define λ and ν by the following formulas:

	λ	ν
$x_1 \neq x_2$	$\dfrac{y_2 - y_1}{x_2 - x_1}$	$\dfrac{y_1 x_2 - y_2 x_1}{x_2 - x_1}$
$x_1 = x_2$	$\dfrac{3x_1^2 + 2a_2 x_1 + a_4 - a_1 y_1}{2y_1 + a_1 x_1 + a_3}$	$\dfrac{-x_1^3 + a_4 x_1 + 2a_6 - a_3 y_1}{2y_1 + a_1 x_1 + a_3}$

Then $y = \lambda x + \nu$ is the line through P_1 and P_2, or tangent to E if $P_1 = P_2$.

(c) *With notation as in (b), $P_3 = P_1 + P_2$ has coordinates*

$$x_3 = \lambda^2 + a_1 \lambda - a_2 - x_1 - x_2,$$
$$y_3 = -(\lambda + a_1)x_3 - \nu - a_3.$$

(d) *As special cases of (c), we have for $P_1 \neq \pm P_2$,*

$$x(P_1 + P_2) = \left(\frac{y_2 - y_1}{x_2 - x_1}\right)^2 + a_1 \left(\frac{y_2 - y_1}{x_2 - x_1}\right) - a_2 - x_1 - x_2,$$

and the duplication formula *for $P = (x, y) \in E$,*

$$x([2]P) = \frac{x^4 - b_4 x^2 - 2b_6 x - b_8}{4x^3 + b_2 x^2 + 2b_4 x + b_6},$$

where b_2, b_4, b_6, b_8 are the polynomials in the a_i's given in (III §1). (See also Exercise 3.25.)

Corollary 2.3.1. *With notation as in (III.2.3), a function $f \in \bar{K}(E) = \bar{K}(x, y)$ is said to be* even *if $f(P) = f(-P)$ for all $P \in E$. Then*

$$f \text{ is even} \qquad \text{if and only if} \qquad f \in \bar{K}(x).$$

PROOF. From (III.2.3), if $P = (x_0, y_0)$, then $-P = (x_0, -y_0 - a_1 x_0 - a_3)$. It follows immediately that every element of $\bar{K}(x)$ is even. Suppose now that $f \in \bar{K}(x, y)$ is even. Using the Weierstrass equation for E, we can write f in the form

$$f(x, y) = g(x) + h(x)y \qquad \text{for some } g, h \in \bar{K}(x).$$

Then the assumed evenness of f implies that

$$f(x, y) = f(x, -y - a_1 x - a_3),$$
$$g(x) + h(x)y = f(x) + h(x)(-y - a_1 x - a_3),$$
$$(2y + a_1 x + a_3)h(x) = 0.$$

This holds for all $(x, y) \in E$, so either h is identically 0, or else $2 = a_1 = a_3 = 0$. The latter implies that the discriminant satisfies $\Delta = 0$, contradicting our assumption that the Weierstrass equation is nonsingular (III.1.4a). Hence $h = 0$, and so $f(x, y) = g(x) \in \bar{K}(x)$. $\square$

Example 2.4. Let $E/\mathbb{Q}$ be the elliptic curve

$$E : y^2 = x^3 + 17.$$

A brief inspection reveals some points with integer coordinates,

$$P_1 = (-2, 3), \quad P_2 = (-1, 4), \quad P_3 = (2, 5), \quad P_4 = (4, 9), \quad P_5 = (8, 23),$$

and a short computer search gives some others,

$$P_6 = (43, 282), \quad P_7 = (52, 375), \quad P_8 = (5234, 378661).$$

Using the addition formula, one easily verifies relations such as

$$P_5 = [-2]P_1, \quad P_4 = P_1 - P_3, \quad [3]P_1 - P_3 = P_7.$$

Of course, there also are lots of points with nonintegral rational coordinates, for example

$$[2]P_2 = \left(\frac{127}{64}, -\frac{2651}{512} \right), \quad P_2 + P_3 = \left(-\frac{8}{9}, -\frac{109}{27} \right).$$

Now it is true, but not so easy to prove, that every rational point $P \in E(\mathbb{Q})$ can be written in the form

$$P = [m]P_1 + [n]P_3 \quad \text{for some } m, n \in \mathbb{Z},$$

and with this identification, the group $E(\mathbb{Q})$ is isomorphic to $\mathbb{Z} \times \mathbb{Z}$. Further, there are only 16 integral points $P = (x, y) \in E$, i.e., points with $x, y \in \mathbb{Z}$, namely $\{\pm P_1, \dots, \pm P_8\}$. (See [190].) These facts illustrate two fundamental theorems in the arithmetic of elliptic curves, namely that the group of rational points on an elliptic curve is finitely generated (the Mordell–Weil theorem, proven in Chapter VIII) and that the set of integral points on an elliptic curve is finite (Siegel's theorem, proven in Chapter IX).

Singular Weierstrass Equations

Suppose that a given Weierstrass equation has discriminant $\Delta = 0$, so (III.1.4a) tells us that it has a singular point. To what extent does our analysis of the composition law fail in this case? As we will see, everything is fine provided that we discard the singular point; and in fact, the resulting group has a particularly simple structure.

The reason that we will be interested in this situation is best illustrated by an example. Consider again the elliptic curve from (III.2.4),

$$E : y^2 = x^3 + 17.$$

This is an elliptic curve defined over $\mathbb{Q}$ with discriminant $\Delta = 2^4 3^3 17$. It is often useful to reduce the coefficients of E modulo p for various primes p and to consider E as a curve defined over the finite field $\mathbb{F}_p$. For almost all primes, namely those for which $\Delta \not\equiv 0 \pmod{p}$, the "reduced" curve is nonsingular, and hence is an elliptic curve defined over $\mathbb{F}_p$. However, for primes p that divide Δ, so in this example for $p \in \{2, 3, 17\}$, the "reduced" curve has a singular point, so it is no longer an elliptic curve. Thus even when dealing with nonsingular elliptic curves, say defined over $\mathbb{Q}$, we find singular curves naturally appearing. We will return to this reduction process in more detail in Chapter VII.

Definition. Let E be a (possibly singular) curve given by a Weierstrass equation. The *nonsingular part of E*, denoted by E_{ns}, is the set of nonsingular points of E. Similarly, if E is defined over K, then $E_{ns}(K)$ is the set of nonsingular points of $E(K)$.

We recall from (III.1.4a) that if E is singular, then there are two possibilities for the singularity, namely a node or a cusp, determined by whether $c_4 = 0$ or $c_4 \neq 0$, respectively.

Proposition 2.5. *Let E be a curve given by a Weierstrass equation with $\Delta = 0$, so E has a singular point S. Then the composition law (III.2.1) makes E_{ns} into an abelian group.*
  (a) *Suppose that E has a node, so $c_4 \neq 0$, and let*

$$y = \alpha_1 x + \beta_1 \quad and \quad y = \alpha_2 x + \beta_2$$

  be the distinct tangent lines to E at S. Then the map

$$E_{ns} \longrightarrow \bar{K}^*, \qquad (x, y) \longmapsto \frac{y - \alpha_1 x - \beta_1}{y - \alpha_2 x - \beta_2}$$

  is an isomorphism of abelian groups.
  (b) *Suppose that E has a cusp, so $c_4 = 0$, and let*

$$y = \alpha x + \beta$$

  be the tangent line to E at S. Then the map

$$E_{ns} \longrightarrow \bar{K}^+, \qquad (x, y) \longmapsto \frac{x - x(S)}{y - \alpha x - \beta}$$

  is an isomorphism of abelian groups.

Remark 2.6. For a group-theoretic description of $E_{ns}(K)$ when K is not algebraically closed, see Exercise 3.5.

PROOF. We first observe that E_{ns} is closed under the composition law (III.2.1), since if a line L intersects E_{ns} at two (not necessarily distinct) points, then L cannot contain the point S. This is true because S is a singular point of E, so S has multiplicity at least two in the intersection $E \cap L$; see Exercise 3.28. Thus if L also contains S, then $E \cap L$ would consist of four points (counted with multiplicity), contradicting Bézout's theorem [111, I.7.8].

We will verify that the maps in (a) and (b) are set bijections with the property that if a line L not hitting S intersects E_{ns} in three not necessarily distinct points, then the images of these three points in $\bar{K}^*$ (respectively $\bar{K}^+$) multiply to 1 (respectively sum to 0). Using this property, we will prove that the composition law (III.2.1) makes E_{ns} into an abelian group and that the maps in (a) and (b) are group isomorphisms.

Since the composition law (III.2.1) and the maps (a) and (b) are defined in terms of lines in $\mathbb{P}^2$, it suffices to prove the theorem after making a linear change of variables. We start by moving the singular point to $(0,0)$, yielding the Weierstrass equation

$$y^2 + a_1 xy = x^3 + a_2 x^2.$$

Let $s \in \bar{K}$ be a root of $s^2 + a_1 s - a_2 = 0$. Replacing y by $y + sx$ eliminates the x^2 term, giving the following equation for E, which we now write using homogeneous coordinates:

$$E : Y^2 Z + AXYZ - X^3 = 0.$$

Note that E has a node if $A \neq 0$ and a cusp if $A = 0$.

(a) The tangent lines to E at $S = [0,0,1]$ are $Y = 0$ and $Y + AX = 0$, so we are looking at the map

$$E_{ns} \longrightarrow \bar{K}^*, \qquad [X,Y,Z] \longmapsto 1 + \frac{AX}{Y}.$$

It is convenient to make one more variable change, so we let

$$X = A^2 X' - A^2 Y', \qquad Y = A^3 Y', \qquad Z = Z'.$$

Dropping the primes, this gives the equation

$$E : XYZ - (X - Y)^2 = 0.$$

We now dehomogenize by setting $Y = 1$, so $x = X/Y$ and $z = Z/Y$, which yields the equation

$$E : xz - (x - 1)^3 = 0$$

and the map

$$E_{ns} \longrightarrow \bar{K}^*, \qquad (x, z) \longmapsto x.$$

(Notice that in this new coordinate system, the singular point is now a point at infinity.) The inverse map is

$$\bar{K}^* \longrightarrow E_{ns}, \qquad t \longmapsto \left(t, \frac{(t-1)^3}{t} \right),$$

so we have a bijection of sets $\bar{K}^* \overset{\sim}{\longleftrightarrow} E_{ns}$. It remains to show that if a line, not going through $[0,0,1]$, intersects E at the three points (x_1, z_1), (x_2, z_2), and (x_3, z_3), then $x_1 x_2 x_3 = 1$. (See Figure 3.4.) Any such line has the form $z = ax + b$, so the three x-coordinates x_1, x_2, and x_3 are the roots of the cubic polynomial

$$x(ax + b) - (x - 1)^3 = -x^3 + (a + 3)x^2 + (b - 3)x + 1.$$

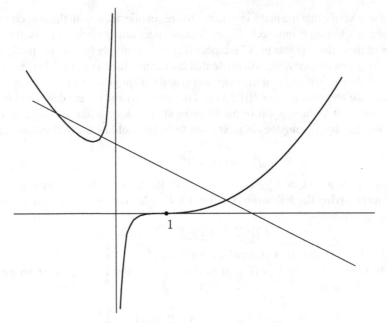

Figure 3.4: The curve $xz - (x - 1)^3 = 0$.

Looking at the constant term, we see that $x_1 x_2 x_3 = 1$, as desired.

(b) In this case $A = 0$ and the tangent line to E at $S = [0, 0, 1]$ is $Y = 0$, so we are looking at the map

$$E_{\mathrm{ns}} \longrightarrow \bar{K}^+, \qquad [X, Y, Z] \longmapsto X/Y.$$

Again dehomogenizing by setting $Y = 1$, we obtain

$$E : z - x^3 = 0,$$
$$E_{\mathrm{ns}} \longrightarrow \bar{K}^+, \qquad (x, z) \longmapsto x.$$

The inverse map is $t \mapsto (t, t^3)$. Finally, if the line $z = ax + b$ intersects E at the three points (x_1, z_1), (x_2, z_2), and (x_3, z_3), then the absence of an x^2-term in

$$(ax + b) - x^3$$

implies that $x_1 + x_2 + x_3 = 0$.   □

III.3 Elliptic Curves

Let E be a smooth curve of genus one. For example, the nonsingular Weierstrass equations studied in (III §1) and (III §2) define curves of this sort. As we have seen, such Weierstrass curves can be given the structure of an abelian group. In order to make a set into a group, clearly an initial requirement is to choose a distinguished (identity) element. This leads to the following definition.

Definition. An *elliptic curve* is a pair (E, O), where E is a nonsingular curve of genus one and $O \in E$. (We generally denote the elliptic curve by E, the point O being understood.) The elliptic curve E is *defined over* K, written E/K, if E is defined over K as a curve and $O \in E(K)$.

In order to connect this definition with the material in (III §1) and (III §2), we begin by using the Riemann–Roch theorem to show that every elliptic curve can be written as a plane cubic, and conversely, every smooth Weierstrass plane cubic curve is an elliptic curve.

Proposition 3.1. *Let E be an elliptic curve defined over K.*
(a) *There exist functions $x, y \in K(E)$ such that the map*

$$\phi : E \longrightarrow \mathbb{P}^2, \qquad \phi = [x, y, 1],$$

gives an isomorphism of E/K onto a curve given by a Weierstrass equation

$$C : Y^2 + a_1 XY + a_3 Y = X^3 + a_2 X^2 + a_4 X + a_6$$

with coefficients $a_1, \ldots, a_6 \in K$ and satisfying $\phi(O) = [0, 1, 0]$. The functions x and y are called Weierstrass coordinates *for the elliptic curve E.*
(b) *Any two Weierstrass equations for E as in* (a) *are related by a linear change of variables of the form*

$$X = u^2 X' + r, \qquad Y = u^3 Y' + su^2 X' + t,$$

with $u \in K^*$ and $r, s, t \in K$.
(c) *Conversely, every smooth cubic curve C given by a Weierstrass equation as in* (a) *is an elliptic curve defined over K with base point $O = [0, 1, 0]$.*

PROOF. (a) We look at the vector spaces $\mathcal{L}(n(O))$ for $n = 1, 2, \ldots$. By the Riemann–Roch theorem, more specifically from (II.5.5c) with $g = 1$, we have

$$\ell(n(O)) = \dim \mathcal{L}(n(O)) = n \qquad \text{for all } n \geq 1.$$

Thus we can choose functions $x, y \in K(E)$ as in (II.5.8) so that $\{1, x\}$ is a basis for $\mathcal{L}(2(O))$ and so that $\{1, x, y\}$ is a basis for $\mathcal{L}(3(O))$. Note that x must have a pole of exact order 2 at O, and similarly y must have a pole of exact order 3 at O.

Now we observe that $\mathcal{L}(6(O))$ has dimension 6, but it contains the seven functions

$$1, x, y, x^2, xy, y^2, x^3.$$

It follows that there is a linear relation

$$A_1 + A_2 x + A_3 y + A_4 x^2 + A_5 xy + A_6 y^2 + A_7 x^3 = 0,$$

where by (II.5.8) we may take $A_1, \ldots, A_7 \in K$. Note that $A_6 A_7 \neq 0$, since otherwise every term would have a pole at O of a different order, and so all of the A_j's would vanish. Replacing x and y by $-A_6 A_7 x$ and $A_6 A_7^2 y$, respectively, and dividing by $A_6^3 A_7^4$, we get a cubic equation in Weierstrass form. This gives a map

$$\phi : E \longrightarrow \mathbb{P}^2, \qquad \phi = [x, y, 1],$$

whose image C lies in the locus described by a Weierstrass equation. Note that $\phi : E \to C$ is a morphism from (II.2.1), and that it is surjective from (II.2.3). Further, we have $\phi(O) = [0, 1, 0]$, since y has a higher-order pole than x at the point O.

The next step is to show that the map $\phi : E \to C \subset \mathbb{P}^2$ has degree-one, or equivalently, to show that $K(E) = K(x, y)$. Consider the map $[x, 1] : E \to \mathbb{P}^1$. Since x has a double pole at O and no other poles, (II.2.6a) says that this map has degree 2. Thus $\big[K(E) : K(x) \big] = 2$. Similarly, the map $[y, 1] : E \to \mathbb{P}^1$ has degree 3, so $\big[K(E) : K(y) \big] = 3$. Therefore $\big[K(E) : K(x, y) \big]$ divides both 2 and 3, so it must equal 1.

Next we show that C is smooth. Suppose that C is singular. Then from (III.1.6), there is a rational map $\psi : C \to \mathbb{P}^1$ of degree one. It follows that the composition $\psi \circ \phi : E \to \mathbb{P}^1$ is a map of degree one between smooth curves, so from (II.2.4.1), it is an isomorphism. This contradicts the fact that E has genus one and $\mathbb{P}^1$ has genus zero (II.5.6). Therefore C is smooth, and now another application of (II.2.4.1) shows that the degree one map $\phi : E \to C$ is an isomorphism.

(b) Let $\{x, y\}$ and $\{x', y'\}$ be two sets of Weierstrass coordinate functions on E. Then x and x' have poles of order 2 at O, and y and y' have poles of order 3 at O. Hence $\{1, x\}$ and $\{1, x'\}$ are both bases for $\mathcal{L}\big(2(O) \big)$, and similarly $\{1, x, y\}$ and $\{1, x', y'\}$ are both bases for $\mathcal{L}\big(3(O) \big)$. Thus there are constants

$$u_1, u_2 \in K^* \quad \text{and} \quad r, s_2, t \in K$$

such that

$$x = u_1 x' + r \quad \text{and} \quad y = u_2 y' + s_2 x' + t.$$

Since both (x, y) and (x', y') satisfy Weierstrass equations in which the Y^2 and X^3 terms have coefficient 1, we have $u_1^3 = u_2^2$. Letting $u = u_2/u_1$ and $s = s_2/u^2$ puts the change of variables formula into the desired form.

(c) Let E be given by a nonsingular Weierstrass equation. We have seen (III.1.5) that the differential

$$\omega = \frac{dx}{2y + a_1 x + a_3} \in \Omega_E$$

has neither zeros nor poles, so $\operatorname{div}(\omega) = 0$. The Riemann–Roch theorem (II.5.5b) then tells us that

$$2 \operatorname{genus}(E) - 2 = \deg \operatorname{div}(\omega) = 0,$$

so E has genus one, and taking $[0, 1, 0]$ as the base point makes E into an elliptic curve. (For an alternative proof of (c) using the Hurwitz genus formula, see Exercise 2.7.) $\qquad \square$

Corollary 3.1.1. *Let E/K be an elliptic curve with Weierstrass coordinate functions x and y. Then*

$$K(E) = K(x, y) \qquad and \qquad \big[K(E) : K(x)\big] = 2.$$

PROOF. These two facts were proven during the course of proving (III.3.1a). $\square$

Remark 3.2. Note that (III.3.1b) does *not* imply that if two Weierstrass equations have coefficients in a given field K, then every change of variables mapping one to the other has coefficients in K. A simple example is the equation

$$y^2 = x^3 - x.$$

It has coefficients in $\mathbb{Q}$, yet it is mapped to itself by the substitution

$$x = -x', \qquad y = \sqrt{-1}\, y'.$$

We next use the Riemann–Roch theorem to describe a group law on the points of an elliptic curve E. Of course, this will turn out to be the group law described by (III.2.1) when E is given by a Weierstrass equation. We start with a simple lemma that serves to distinguish $\mathbb{P}^1$ from curves of genus one; see Exercise 2.5 for a generalization.

Lemma 3.3. *Let C be a curve of genus one and let $P, Q \in C$. Then*

$$(P) \sim (Q) \qquad if\ and\ only\ if \qquad P = Q.$$

PROOF. Suppose that $(P) \sim (Q)$ and choose $f \in \bar{K}(C)$ such that

$$\mathrm{div}(f) = (P) - (Q).$$

Then $f \in \mathcal{L}((Q))$. The Riemann–Roch theorem (II.5.5c) tells us that

$$\dim \mathcal{L}((Q)) = 1.$$

But $\mathcal{L}((Q))$ certainly contains the constant functions; hence $f \in \bar{K}$ and $P = Q$. $\square$

Proposition 3.4. *Let (E, O) be an elliptic curve.*
(a) *For every degree-0 divisor $D \in \mathrm{Div}^0(E)$ there exists a unique point $P \in E$ satisfying*

$$D \sim (P) - (O).$$

Define

$$\sigma : \mathrm{Div}^0(E) \longrightarrow E$$

to be the map that sends D to its associated P.
(b) *The map σ is surjective.*

(c) *Let $D_1, D_2 \in \mathrm{Div}^0(E)$. Then*

$$\sigma(D_1) = \sigma(D_2) \qquad \textit{if and only if} \qquad D_1 \sim D_2.$$

Thus σ induces a bijection of sets (which we also denote by σ),

$$\sigma : \mathrm{Pic}^0(E) \xrightarrow{\;\sim\;} E.$$

(d) *The inverse to σ is the map*

$$\kappa : E \xrightarrow{\;\sim\;} \mathrm{Pic}^0(E), \qquad P \longmapsto (\textit{divisor class of } (P) - (O)).$$

(e) *If E is given by a Weierstrass equation, then the "geometric group law" on E described by (III.2.1) and the "algebraic group law" induced from $\mathrm{Pic}^0(E)$ using σ are the same.*

PROOF. (a) Since E has genus one, the Riemann–Roch theorem (II.5.5c) says that

$$\dim \mathcal{L}(D + (O)) = 1.$$

Let $f \in \bar{K}(E)$ be a nonzero element of $\mathcal{L}(D + (O))$, so f is a basis for this one-dimensional vector space. Since

$$\mathrm{div}(f) \geq -D - (O) \qquad \text{and} \qquad \deg(\mathrm{div}(f)) = 0,$$

it follows that

$$\mathrm{div}(f) = -D - (O) + (P)$$

for some $P \in E$. Hence

$$D \sim (P) - (O),$$

which gives the existence of a point with the desired property.

Next suppose that $P' \in E$ has the same property. Then

$$(P) \sim D + (O) \sim (P'),$$

so (III.3.3) tells us that $P = P'$. Hence P is unique.

(b) For any $P \in E$, we have

$$\sigma((P) - (O)) = P.$$

(c) Let $D_1, D_2 \in \mathrm{Div}^0(E)$, and set $P_i = \sigma(D_i)$ for $i = 1, 2$. Then from the definition of σ we have

$$(P_1) - (P_2) \sim D_1 - D_2.$$

Thus if $P_1 = P_2$, then $D_1 \sim D_2$; and conversely, if $D_1 \sim D_2$, then $(P_1) \sim (P_2)$, so $P_1 = P_2$ from (III.3.3).

(d) Clear.

(e) Let E be given by a Weierstrass equation and let $P, Q, \in E$. It suffices to show that

$$\kappa(P + Q) = \kappa(P) + \kappa(Q).$$

(N.B. The first $+$ is addition on E using (III.2.1), while the second $+$ is addition of divisor classes in $\text{Pic}^0(E)$.)

Let

$$f(X, Y, Z) = \alpha X + \beta Y + \gamma Z = 0$$

give the line L in $\mathbb{P}^2$ going through P and Q, let R be the third point of intersection of L with E, and let

$$f'(X, Y, Z) = \alpha' X + \beta' Y + \gamma' Z = 0$$

be the line L' through R and O. Then from the definition of addition on E (III.2.1) and the fact that the line $Z = 0$ intersects E at O with multiplicity 3, we have

$$\text{div}(f/Z) = (P) + (Q) + (R) - 3(O),$$
$$\text{div}(f'/Z) = (R) + (P + Q) - 2(O).$$

Hence

$$(P + Q) - (P) - (Q) + (O) = \text{div}(f'/f) \sim 0,$$

so

$$\kappa(P + Q) - \kappa(P) - \kappa(Q) = 0.$$

This proves that κ is a group homomorphism.   □

Corollary 3.5. *Let E be an elliptic curve and let $D = \sum n_P(P) \in \text{Div}(E)$. Then D is a principal divisor if and only if*

$$\sum_{P \in E} n_P = 0 \qquad and \qquad \sum_{P \in E} [n_P]P = O.$$

(Note that the first sum is of integers, while the second is addition on E.)

PROOF. From (II.3.1b), every principal divisor has degree 0. Next let $D \in \text{Div}^0(E)$. We use (III.3.4a,e) to deduce that

$$D \sim 0 \iff \sigma(D) = O \iff \sum_{P \in E} [n_P]\sigma\big((P) - (O)\big) = O.$$

This is the desired result, since $\sigma\big((P) - (O)\big) = P$.   □

Remark 3.5.1. If we combine (III.3.4) and (II.3.4), we see that every elliptic curve E/K fits into an exact sequence

$$1 \longrightarrow \bar{K}^* \longrightarrow \bar{K}(E)^* \xrightarrow{\text{div}} \text{Div}^0(E) \xrightarrow{\sigma} E \longrightarrow 0,$$

where σ is the operation "sum the points in the divisor using the group law on E." Further, Exercise 2.13b implies that the sequence remains exact if we take $G_{\bar{K}/K}$-invariants,

$$1 \longrightarrow K^* \longrightarrow K(E)^* \xrightarrow{\text{div}} \text{Div}_K^0(E) \xrightarrow{\sigma} E(K) \longrightarrow 0.$$

(See also (X.3.8).)

We now prove the fundamental fact that the addition law on an elliptic curve is a *morphism*. Addition is a map $E \times E \to E$ and the variety $E \times E$ has dimension 2, so we cannot use (II.2.1) directly; but (II.2.1) will play a crucial role in the proof. One can also give a proof using explicit equations, but the algebra is somewhat lengthy; see (III.3.6.1).

Theorem 3.6. *Let E/K be an elliptic curve. Then the equations* (III.2.3) *giving the group law on E define morphisms*

$$\begin{array}{ll}
+ : E \times E \longrightarrow E, & \text{and} \quad - : E \longrightarrow E, \\
(P_1, P_2) \longmapsto P_1 + P_2, & \qquad P \longmapsto -P.
\end{array}$$

PROOF. First, the negation map

$$(x, y) \longmapsto (x, -y - a_1 x - a_3)$$

is clearly a rational map $E \to E$. Since E is smooth, it follows from (II.2.1) that negation is a morphism.

Next we fix a point $Q \neq O$ on E and consider the *translation-by-Q map*

$$\tau : E \longrightarrow E, \qquad \tau(P) = P + Q.$$

From the addition formula given in (III.2.3c), this is clearly a rational map; and thus, again using (II.2.1), it is a morphism. In fact, since τ has an inverse, namely $P \mapsto P - Q$, it is an isomorphism.

Finally, consider the general addition map $+ : E \times E \to E$. From (III.2.3c) we see that it is a morphism except possibly at pairs of points having one of the following forms,

$$(P, P), \quad (P, -P), \quad (P, O), \quad (O, P),$$

since for pairs of points not of this form, the rational functions

$$\lambda = \frac{y_2 - y_1}{x_2 - x_1} \quad \text{and} \quad \nu = \frac{y_1 x_2 - y_2 x_1}{x_2 - x_1}$$

on $E \times E$ are well-defined.

To deal with the four exceptional cases, we could work directly with the definition of morphism; see (III.3.6.1). However, we prefer to let the group law assist us. Thus let τ_1 and τ_2 be translation maps as above for points Q_1 and Q_2, respectively. Consider the composition of maps

$$\phi : E \times E \xrightarrow{\tau_1 \times \tau_2} E \times E \xrightarrow{+} E \xrightarrow{\tau_1^{-1}} E \xrightarrow{\tau_2^{-1}} E.$$

Since the group law on E is associative and commutative (III.2.2), the net effect of the above maps is as follows:

$$(P_1, P_2) \xrightarrow{\ \tau_1 \times \tau_2\ } (P_1 + Q_1, P_2 + Q_2)$$
$$\xrightarrow{\ +\ } P_1 + Q_1 + P_2 + Q_2$$
$$\xrightarrow{\ \tau_1^{-1}\ } P_1 + P_2 + Q_2$$
$$\xrightarrow{\ \tau_2^{-1}\ } P_1 + P_2.$$

Thus the rational map ϕ agrees with the addition map wherever they are both defined.

Further, since the τ_i's are isomorphisms, it follows from the above discussion that ϕ is a morphism except possibly at pairs of points of the form

$$(P - Q_1, P - Q_2), \quad (P - Q_1, -P - Q_2), \quad (P - Q_1, -Q_2), \quad (-Q_1, P - Q_2).$$

But Q_1 and Q_2 are arbitrary points. Hence by varying Q_1 and Q_2, we can find a finite set of rational maps

$$\phi_1, \phi_2, \ldots, \phi_n : E \times E \longrightarrow E$$

with the following properties:

(i) ϕ_1 is the addition map given in (III.2.3c).

(ii) For each $(P_1, P_2) \in E \times E$, some ϕ_i is defined at (P_1, P_2).

(iii) If ϕ_i and ϕ_j are both defined at (P_1, P_2), then $\phi_i(P_1, P_2) = \phi_j(P_1, P_2)$.

It follows that addition is defined on all of $E \times E$, so it is a morphism. $\qquad\square$

Remark 3.6.1. During the course of proving (III.3.6), we noted that the formulas in (III.2.3c) make it clear that the addition map $+ : E \times E \to E$ is a morphism except possibly at pairs of points of the form $(P, \pm P)$, (P, O), or (O, P). Rather than using translation maps to circumvent this difficulty, one can work directly with the definition of morphism using explicit equations. It turns out that this involves consideration of quite a few cases; we do one to illustrate the method.

Thus let $(x_1, y_1; x_2, y_2)$ be Weierstrass coordinates on $E \times E$. We will show explicitly that addition is a morphism at points of the form (P, P) with $P \neq O$ and $[2]P \neq O$. Note that addition is defined in general by the formulas given in (III.2.3c):

$$\lambda = \frac{y_2 - y_1}{x_2 - x_1}, \qquad\qquad \nu = \frac{y_1 x_2 - y_2 x_1}{x_2 - x_1} = y_1 - \lambda x_1,$$
$$x_3 = \lambda^2 + a_1 \lambda - a_2 - x_1 - x_2, \qquad y_3 = -(\lambda + a_1)x_3 - \nu - a_3.$$

Here we view λ, ν, x_3, y_3 as functions on $E \times E$, and addition is given by the map $[x_3, y_3, 1] : E \times E \to E$. Thus to show that addition is a morphism at (P, P), it suffices to show that λ is a morphism at (P, P). By assumption, both pairs of functions (x_1, y_1) and (x_2, y_2) satisfy the same Weierstrass equation. Subtracting one equation from the other and factoring yields

$$(y_1 - y_2)(y_1 + y_2 + a_1 x_1 + a_3)$$
$$= (x_1 - x_2)(x_1^2 + x_1 x_2 + x_2^2 + a_2 x_1 + a_2 x_2 + a_4 - a_1 y_2).$$

Thus λ, considered as a function on $E \times E$, may also be written as

$$\lambda(P_1, P_2) = \frac{x_1^2 + x_1 x_2 + x_2^2 + a_2 x_1 + a_2 x_2 + a_4 - a_1 y_2}{y_1 + y_2 + a_1 x_1 + a_3}.$$

Therefore, letting $P = (x, y)$, we have

$$\lambda(P, P) = \frac{3x^2 + 2a_2 x + a_4 - a_1 y}{2y + a_1 x + a_3}.$$

Hence λ is a morphism at (P, P) provided that $2y(P) + a_1 x(P) + a_3 \neq 0$, and we have excluded this case by our assumption that $[2]P \neq O$. We leave it as an exercise for the reader to deal similarly with the other cases.

III.4 Isogenies

Having examined in some detail the geometry of individual elliptic curves, we turn now to the study of the maps between curves. Since an elliptic curve has a distinguished zero point, it is natural to single out the maps that respect this property.

Definition. Let E_1 and E_2 be elliptic curves. An *isogeny* from E_1 to E_2 is a morphism

$$\phi : E_1 \longrightarrow E_2 \quad \text{satisfying} \quad \phi(O) = O.$$

Two elliptic curves E_1 and E_2 are *isogenous* if there is an isogeny from E_1 to E_2 with $\phi(E_1) \neq \{O\}$. We will see later (III.6.1) that this is an equivalence relation.

It follows from (II.2.3) that an isogeny satisfies either

$$\phi(E_1) = \{O\} \quad \text{or} \quad \phi(E_1) = E_2.$$

Thus except for the zero isogeny defined by $[0](P) = O$ for all $P \in E_1$, every other isogeny is a finite map of curves. Hence we obtain the usual injection of function fields (II §2),

$$\phi^* : \bar{K}(E_2) \longrightarrow \bar{K}(E_1).$$

The degree of ϕ, which is denoted by $\deg \phi$, is the degree of the finite extension $\bar{K}(E_1)/\phi^* \bar{K}(E_2)$, and similarly for the separable and inseparable degrees, denoted respectively by $\deg_s \phi$ and $\deg_i \phi$. We also refer to the map ϕ as being separable, inseparable, or purely inseparable according to the corresponding property of the field extension. Further, by convention we set

$$\deg[0] = 0.$$

This convention ensures that we have

$$\deg(\psi \circ \phi) = \deg(\psi) \deg(\phi) \quad \text{for all chains of isogenies} \quad E_1 \xrightarrow{\phi} E_2 \xrightarrow{\psi} E_3.$$

Elliptic curves are abelian groups, so the maps between them form groups. We denote the set of isogenies from E_1 to E_2 by

$$\mathrm{Hom}(E_1, E_2) = \{\text{isogenies } E_1 \to E_2\}.$$

The sum of two isogenies is defined by

$$(\phi + \psi)(P) = \phi(P) + \psi(P),$$

and (III.3.6) implies that $\phi + \psi$ is a morphism, so it is an isogeny. Hence $\mathrm{Hom}(E_1, E_2)$ is a group.

If $E_1 = E_2$, then we can also compose isogenies. Thus if E is an elliptic curve, we let

$$\mathrm{End}(E) = \mathrm{Hom}(E, E)$$

be the ring whose addition law is as given above and whose multiplication is composition,

$$(\phi\psi)(P) = \phi(\psi(P)).$$

(It is not obvious that the distributive law holds, but we will prove it later in this section; see (III.4.8).) The ring $\mathrm{End}(E)$ is called the *endomorphism ring of E*. The invertible elements of $\mathrm{End}(E)$ form the *automorphism group of E*, which is denoted by $\mathrm{Aut}(E)$. The endomorphism ring of an elliptic curve E is an important invariant of E that we will study in some detail throughout the rest of this chapter.

Of course, if E_1, E_2, and E are defined over a field K, then we can restrict attention to those isogenies that are defined over K. The corresponding groups of isogenies are denoted with the usual subscripts; thus

$$\mathrm{Hom}_K(E_1, E_2), \qquad \mathrm{End}_K(E), \qquad \mathrm{Aut}_K(E).$$

We have already seen an example (III.3.2) showing that $\mathrm{Aut}(E)$ may be strictly larger than $\mathrm{Aut}_K(E)$.

Example 4.1. For each $m \in \mathbb{Z}$ we define the *multiplication-by-m isogeny*

$$[m] : E \longrightarrow E$$

in the natural way. Thus if $m > 0$, then

$$[m](P) = \underbrace{P + P + \cdots + P}_{m \text{ terms}}.$$

For $m < 0$, we set $[m](P) = [-m](-P)$, and we have already defined $[0](P) = O$. Using (III.3.6), an easy induction shows that $[m]$ is a morphism, hence an isogeny, since it clearly sends O to O.

Notice that if E is defined over K, then $[m]$ is defined over K. We start our analysis of the group of isogenies by showing that if $m \neq 0$, then the multiplication-by-m map is nonconstant.

Proposition 4.2. (a) *Let E/K be an elliptic curve and let $m \in \mathbb{Z}$ with $m \neq 0$. Then the multiplication-by-m map*

$$[m] : E \longrightarrow E$$

is nonconstant.

(b) *Let E_1 and E_2 be elliptic curves. Then the group of isogenies*

$$\mathrm{Hom}(E_1, E_2)$$

is a torsion-free $\mathbb{Z}$-module.

(c) *Let E be an elliptic curve. Then the endomorphism ring $\mathrm{End}(E)$ is a (not necessarily commutative) ring of characteristic 0 with no zero divisors.*

PROOF. (a) We start by showing that $[2] \neq [0]$. The duplication formula (III.2.3d) says that if a point $P = (x, y) \in E$ has order 2, then it must satisfy

$$4x^3 + b_2 x^2 + 2b_4 x + b_6 = 0.$$

If $\mathrm{char}(K) \neq 2$, this shows immediately that there are only finitely many such points. Further, even for $\mathrm{char}(K) = 2$, the only way to have $[2] = [0]$ is for the cubic polynomial to be identically 0, which means that $b_2 = b_6 = 0$, which in turn implies that $\Delta = 0$. Hence in all cases we have $[2] \neq [0]$. Now, using the fact that $[mn] = [m] \circ [n]$, we are reduced to considering the case that m is odd.

Assume now that $\mathrm{char}(K) \neq 2$. Then, using long division, it is easy to verify that the polynomial

$$4x^3 + b_2 x^2 + 2b_4 x + b_6$$

does not divide the polynomial

$$x^4 - b_4 x^2 - 2b_6 x - b_8.$$

More precisely, if the first polynomial divides the second, then $\Delta = 0$; see Exercise 3.1. Hence we can find an $x_0 \in \bar{K}$ such that the first polynomial vanishes to higher order at x_0 than does the second. Choosing $y_0 \in \bar{K}$ so that $P_0 = (x_0, y_0) \in E$, the doubling formula implies that $[2]P_0 = O$. In other words, we have shown that E has a nontrivial point P_0 of order 2. Then for odd integers m we have

$$[m]P_0 = P_0 \neq O,$$

so clearly $[m] \neq [0]$.

Finally, if $\mathrm{char}(K) = 2$, then one can proceed as above using the "triplication formula" (Exercise 3.2) to produce a point of order 3. We leave this approach to the reader, since later in this chapter we prove a result (III.5.4) that includes the case of $\mathrm{char}(K) = 2$ and m odd.

(b) This follows immediately from (a). Suppose that $\phi \in \mathrm{Hom}(E_1, E_2)$ and $m \in \mathbb{Z}$ satisfy

$$[m] \circ \phi = [0].$$

Taking degrees gives

$$\big(\deg[m]\big)\big(\deg \phi\big) = 0,$$

so either $m = 0$, or else (a) implies that $\deg[m] \geq 1$, in which case we must have $\phi = [0]$.

(c) From (b), the endomorphism ring $\mathrm{End}(E)$ has characteristic 0. Suppose that $\phi, \psi \in \mathrm{End}(E)$ satisfy $\phi \circ \psi = [0]$. Then

$$(\deg \phi)(\deg \psi) = \deg(\phi \circ \psi) = 0.$$

It follows that either $\phi = [0]$ or $\psi = [0]$. Therefore $\mathrm{End}(E)$ is an integral domain.
□

Definition. Let E be an elliptic curve and let $m \in \mathbb{Z}$ with $m \geq 1$. The *m-torsion subgroup of E*, denoted by $E[m]$, is the set of points of E of order m,

$$E[m] = \{ P \in E : [m]P = O \}.$$

The *torsion subgroup of E*, denoted by E_{tors}, is the set of points of finite order,

$$E_{\mathrm{tors}} = \bigcup_{m=1}^{\infty} E[m].$$

If E is defined over K, then $E_{\mathrm{tors}}(K)$ denotes the points of finite order in $E(K)$.

The most important fact about the multiplication-by-m map is that it has degree m^2, from which one can deduce the structure of the finite group $E[m]$. We do not prove this result here, because it is an immediate corollary of the material on dual isogenies covered in (III §6). However, the reader should be aware that there are completely elementary, but rather messy, proofs that $\deg[m] = m^2$ using explicit formulas and induction. (See exercises 3.7, 3.8, and 3.9 for various approaches.)

Remark 4.3. Suppose that $\mathrm{char}(K) = 0$. Then the map

$$[\ ] : \mathbb{Z} \longrightarrow \mathrm{End}(E)$$

is usually the whole story, i.e., $\mathrm{End}(E) \cong \mathbb{Z}$. If E is strictly larger than $\mathbb{Z}$, then we say that E has *complex multiplication*, or CM for short. Elliptic curves with complex multiplication have many special properties; see (C §11) for a brief dicussion. On the other hand, if K is a finite field, then $\mathrm{End}(E)$ is always larger than $\mathbb{Z}$; see (V §3).

Example 4.4. Assume that $\mathrm{char}(K) \neq 2$ and let $i \in \bar{K}$ be a primitive fourth root of unity, i.e., $i^2 = -1$. Then, as noted in (III.3.2), the elliptic curve E/K given by the equation

$$E : y^2 = x^3 - x$$

has endomorphism ring $\mathrm{End}(E)$ strictly larger than $\mathbb{Z}$, since it contains a map, which we denote by $[i]$, given by

$$[i] : (x, y) \longmapsto (-x, iy).$$

Thus E has complex multiplication. Clearly $[i]$ is defined over K if and only if $i \in K$. Hence even if E is defined over K, it may happen that $\operatorname{End}_K(E)$ is strictly smaller than $\operatorname{End}(E)$.

Continuing with this example, we observe that

$$[i] \circ [i](x, y) = [i](-x, iy) = (x, -y) = -(x, y),$$

so $[i] \circ [i] = [-1]$. There is thus a ring homomorphism

$$\mathbb{Z}[i] \longrightarrow \operatorname{End}(E), \qquad m + ni \longmapsto [m] + [n] \circ [i].$$

If $\operatorname{char}(K) = 0$, this map is an isomorphism, $\mathbb{Z}[i] \cong \operatorname{End}(E)$, in which case

$$\operatorname{Aut}(E) \cong \mathbb{Z}[i]^* = \{\pm 1, \pm i\}$$

is a cyclic group of order 4.

Example 4.5. Again assume that $\operatorname{char}(K) \neq 2$ and let $a, b \in K$ satisfy $b \neq 0$ and $r = a^2 - 4b \neq 0$. Consider the two elliptic curves

$$E_1 : y^2 = x^3 + ax^2 + bx,$$
$$E_2 : Y^2 = X^3 - 2aX^2 + rX.$$

There are isogenies of degree 2 connecting these curves,

$$\phi : E_1 \longrightarrow E_2, \qquad\qquad \hat{\phi} : E_2 \longrightarrow E_1,$$
$$(x, y) \longmapsto \left(\frac{y^2}{x^2}, \frac{y(b - x^2)}{x^2} \right), \qquad (X, Y) \longmapsto \left(\frac{Y^2}{4X^2}, \frac{Y(r - X^2)}{8X^2} \right).$$

A direct computation shows that $\hat{\phi} \circ \phi = [2]$ on E_1 and $\phi \circ \hat{\phi} = [2]$ on E_2. The maps ϕ and $\hat{\phi}$ are examples of *dual isogenies*, which we discuss further in (III §6).

Example 4.6. Let K be a field of characteristic $p > 0$, let $q = p^r$, and let E/K be an elliptic curve given by a Weierstrass equation. We recall from (II §2) that the curve $E^{(q)}/K$ is defined by raising the coefficients of the equation for E to the qth power, and the Frobenius morphism ϕ_q is defined by

$$\phi_q : E \longrightarrow E^{(q)}, \qquad (x, y) \longmapsto (x^q, y^q).$$

Since $E^{(q)}$ is the zero locus of a Weierstrass equation, it will be an elliptic curve provided that its equation is nonsingular. Writing everything out in terms of Weierstrass coefficients and using the fact that the qth-power map $K \to K$ is a homomorphism, it is clear that

$$\Delta(E^{(q)}) = \Delta(E)^q \qquad \text{and} \qquad j(E^{(q)}) = j(E)^q.$$

In particular, the equation for $E^{(q)}$ is nonsingular.

Now suppose that $K = \mathbb{F}_q$ is a finite field with q elements. Then the qth-power map on K is the identity, so $E^{(q)} = E$ and ϕ_q is an endomorphism of E, called the *Frobenius endomorphism*. The set of points fixed by ϕ_q is exactly the finite group $E(\mathbb{F}_q)$. This fact lies at the heart of Hasse's proof of an estimate for $\#E(\mathbb{F}_q)$; see (V §1).

Example 4.7. Let E/K be an elliptic curve and let $Q \in E$. Then we can define a *translation-by-Q map*

$$\tau_Q : E \longrightarrow E, \qquad P \longmapsto P + Q.$$

The map τ_Q is clearly an isomorphism, since τ_{-Q} provides an inverse. Of course, it is not an isogeny unless $Q = O$.

Now consider an arbitrary morphism

$$F : E_1 \longrightarrow E_2$$

of elliptic curves. The composition

$$\phi = \tau_{-F(O)} \circ F$$

is an isogeny, since $\phi(O) = O$. This proves that any morphism F between elliptic curves can be written as

$$F = \tau_{F(O)} \circ \phi,$$

the composition of an isogeny and a translation.

An isogeny is a map between elliptic curves that sends O to O. Since an elliptic curve is a group, it might seem more natural to focus on those isogenies that are group homomorphisms. However, as we now show, it turns out that every isogeny is automatically a homomorphism.

Theorem 4.8. *Let*

$$\phi : E_1 \longrightarrow E_2$$

be an isogeny. Then

$$\phi(P + Q) = \phi(P) + \phi(Q) \qquad \text{for all } P, Q \in E_1.$$

PROOF. If $\phi(P) = O$ for all $P \in E$, there is nothing to prove. Otherwise, ϕ is a finite map, so by (II.3.7), it induces a homomorphism

$$\phi_* : \text{Pic}^0(E_1) \longrightarrow \text{Pic}^0(E_2)$$

defined by

$$\phi_*\left(\text{class of } \sum n_i(P_i)\right) = \text{class of } \sum n_i(\phi P_i).$$

On the other hand, from (III.3.4) we have *group isomorphisms*

$$\kappa_i : E_i \longrightarrow \text{Pic}^0(E_i), \qquad P \longmapsto \text{class of } (P) - (O).$$

Then, since $\phi(O) = O$, we obtain the following commutative diagram:

$$
\begin{array}{ccc}
E_1 & \xrightarrow{\ \cong\ } & \text{Pic}^0(E_1) \\
 & \scriptstyle \kappa_1 & \\
\phi \downarrow & & \downarrow \phi_* \\
 & & \\
E_2 & \xrightarrow{\ \cong\ } & \text{Pic}^0(E_2). \\
 & \scriptstyle \kappa_2 &
\end{array}
$$

Since κ_1, κ_2, and ϕ_* are all group homomorphisms and κ_2 is injective, it follows that ϕ is also a homomorphism. $\qquad\square$

Corollary 4.9. *Let* $\phi : E_1 \to E_2$ *be a nonzero isogeny. Then*

$$\ker \phi = \phi^{-1}(O)$$

is a finite group.

PROOF. It is a subgroup of E from (III.4.8), and it is finite (of order at most $\deg \phi$) from (II.2.6a). $\qquad\qquad\qquad\qquad\qquad\qquad\qquad\qquad\qquad\qquad\qquad\qquad\qquad\quad$ □

The next three results, (III.4.10), (III.4.11), and (III.4.12), encompass the basic Galois theory of elliptic function fields.

Theorem 4.10. *Let* $\phi : E_1 \to E_2$ *be a nonzero isogeny.*
(a) *For every* $Q \in E_2$,
$$\#\phi^{-1}(Q) = \deg_s \phi.$$

Further, for every $P \in E_1$,

$$e_\phi(P) = \deg_i \phi.$$

(b) *The map*

$$\ker \phi \longrightarrow \mathrm{Aut}\big(\bar{K}(E_1)/\phi^* \bar{K}(E_2)\big), \qquad T \longmapsto \tau_T^*,$$

is an isomorphism. (Here τ_T *is the translation-by-T map (III.4.7) and* τ_T^* *is the automorphism that* τ_T *induces on* $\bar{K}(E_Q)$.)
(c) *Suppose that* ϕ *is separable. Then* ϕ *is unramified,*

$$\# \ker \phi = \deg \phi,$$

and $\bar{K}(E_1)$ *is a Galois extension of* $\phi^* \bar{K}(E_2)$.

PROOF. (a) From (II.2.6b) we know that

$$\#\phi^{-1}(Q) = \deg_s \phi \qquad \text{for all but finitely many } Q \in E_2.$$

But for any $Q, Q' \in E_2$, if we choose some $R \in E_1$ with $\phi(R) = Q' - Q$, then the fact that ϕ is a homomorphism implies that there is a one-to-one correspondence

$$\phi^{-1}(Q) \longrightarrow \phi^{-1}(Q'), \qquad P \longmapsto P + R.$$

Hence
$$\#\phi^{-1}(Q) = \deg_s \phi \qquad \text{for all } Q \in E_2,$$

which proves the first assertion.

Now let $P, P' \in E_1$ with $\phi(P) = \phi(P') = Q$, and let $R = P' - P$. Then $\phi(R) = O$, so $\phi \circ \tau_R = \phi$. Therefore, using (II.2.6c) and the fact that τ_R is an iso-morphism,

$$e_\phi(P) = e_{\phi \circ \tau_R}(P) = e_\phi\big(\tau_R(P)\big)e_{\tau_R}(P) = e_\phi(P').$$

Hence every point in $\phi^{-1}(Q)$ has the same ramification index. We compute

$$(\deg_s \phi)(\deg_i \phi) = \deg \phi = \sum_{P \in \phi^{-1}(Q)} e_\phi(P) \qquad \text{from (II.2.6a)},$$
$$= (\#\phi^{-1}(Q))e_\phi(P) \qquad \text{for any } P \in \phi^{-1}(Q),$$
$$= (\deg_s \phi)e_\phi(P) \qquad \text{from above}.$$

Canceling $\deg_s \phi$ gives the second assertion.

(b) First, if $T \in \ker \phi$ and $f \in \bar{K}(E_2)$, then

$$\tau_T^*(\phi^* f) = (\phi \circ \tau_T)^* f = \phi^* f,$$

since $\phi \circ \tau_T = \phi$. Hence as an automorphism of $\bar{K}(E_1)$, the map τ_T^* fixes $\phi^* \bar{K}(E_2)$, so the map in (b) is well-defined. Next, since

$$\tau_S \circ \tau_T = \tau_{S+T} = \tau_T \circ \tau_S,$$

the map in (b) is a homomorphism. Finally, from (a) we have

$$\# \ker \phi = \deg_s \phi,$$

while a basic result from Galois theory says that

$$\# \operatorname{Aut}(\bar{K}(E_1)/\phi^* \bar{K}(E_2)) \le \deg_s \phi.$$

Hence to prove that the map $T \to \tau_T^*$ is an isomorphism, it suffices to show that it is injective. But if τ_T^* fixes $\bar{K}(E_1)$, then in particular every function on E_1 takes the same value at T and O. This clearly implies that $T = O$, since for example, the coordinate function x has a pole at O and no other poles.

(c) If ϕ is separable, then from (a) we see that

$$\#\phi^{-1}(Q) = \deg \phi \qquad \text{for all } Q \in E_2.$$

Hence ϕ is unramified (II.2.7), and putting $Q = O$ gives

$$\# \ker \phi = \deg \phi.$$

Then from (b) we find that

$$\# \operatorname{Aut}(\bar{K}(E_1)/\phi^* \bar{K}(E_2)) = [\bar{K}(E_1) : \phi^* \bar{K}(E_2)],$$

so $\bar{K}(E_1)/\phi^* \bar{K}(E_2)$ is a Galois extension. $\qquad \square$

Corollary 4.11. *Let*

$$\phi : E_1 \longrightarrow E_2 \qquad \text{and} \qquad \psi : E_1 \longrightarrow E_3$$

be nonconstant isogenies, and assume that ϕ is separable. If

$$\ker \phi \subset \ker \psi,$$

then there is a unique isogeny

$$\lambda : E_2 \longrightarrow E_3$$

satisfying $\psi = \lambda \circ \phi$.

PROOF. Since ϕ is separable, (III.4.10c) says that $\bar{K}(E_1)$ is a Galois extension of $\phi^* \bar{K}(E_2)$. Then the inclusion $\ker \phi \subset \ker \psi$ and the identification (III.4.10b) imply that every element of $\mathrm{Gal}\big(\bar{K}(E_1)/\phi^* \bar{K}(E_2)\big)$ fixes $\psi^* \bar{K}(E_3)$. Hence by Galois theory, there are field inclusions

$$\psi^* \bar{K}(E_3) \subset \phi^* \bar{K}(E_2) \subset \bar{K}(E_1).$$

Now (II.2.4b) gives a map

$$\lambda : E_2 \longrightarrow E_3$$

satisfying

$$\phi^* \big(\lambda^* \bar{K}(E_3)\big) = \psi^* \bar{K}(E_3),$$

and this in turn implies that

$$\lambda \circ \phi = \psi.$$

Finally, λ is an isogeny, since

$$\lambda(O) = \lambda\big(\phi(O)\big) = \psi(O) = O. \qquad \qquad \square$$

Proposition 4.12. *Let E be an elliptic curve and let Φ be a finite subgroup of E. There are a unique elliptic curve E' and a separable isogeny*

$$\phi : E \longrightarrow E' \qquad satisfying \qquad \ker \phi = \Phi.$$

Remark 4.13.1. The elliptic curve whose existence is asserted in this corollary is often denoted by the quotient E/Φ. This notation clearly indicates the group structure, but there is no a priori reason why this quotient group should correspond to the points of an elliptic curve, nor why the natural group homomorphism $E \to E/\Phi$ should be a morphism. In general, it turns out that the quotient of any variety by a finite group of automorphisms is again a variety (see [186, §7]). The case of curves is done in Exercise 3.13.

Remark 4.13.2. Suppose that E is defined over K and that Φ is $G_{\bar{K}/K}$-invariant. In other words, if $T \in \Phi$, then $T^\sigma \in \Phi$ for all $\sigma \in G_{\bar{K}/K}$. Then the curve E' and isogeny ϕ described in (III.4.12) can be defined over K; see Exercise 3.13e.

Remark 4.13.3. For a given curve E and subgroup Φ, Velu [297] describes how to explicitly write down equations for the curve $E' = E/\Phi$ and isogeny $\phi : E \to E'$.

PROOF OF (III.4.12). As in (III.4.10b), each point $T \in \Phi$ gives rise to an automorphism τ_T^* of $\bar{K}(E)$. Let $\bar{K}(E)^\Phi$ be the subfield of $\bar{K}(E)$ fixed by every element of Φ. Then Galois theory tells us that $\bar{K}(E)$ is a Galois extension of $\bar{K}(E)^\Phi$ with Galois group isomorphic to Φ.

The field $\bar{K}(E)^\Phi$ has transcendence degree one over $\bar{K}$, so from (II.2.4c) there are a unique smooth curve $C/\bar{K}$ and a finite morphism

$$\phi : E \longrightarrow C \qquad \text{satisfying} \qquad \phi^* \bar{K}(C) = \bar{K}(E)^\Phi.$$

We next show that ϕ is unramified. Let $P \in E$ and $T \in \Phi$. Then for *every* function $f \in \bar{K}(C)$,

$$f\big(\phi(P + T)\big) = \big((\tau_T^* \circ \phi^*)f\big)(P) = (\phi^* f)(P) = f\big(\phi(P)\big),$$

where the middle equality uses the fact that τ_T^* fixes every element of $\phi^* \bar{K}(C)$. It follows that $\phi(P + T) = \phi(P)$. Now let $Q \in C$ and choose any point $P \in E$ with $\phi(P) = Q$. Then

$$\phi^{-1}(Q) \supset \{P + T : T \in \Phi\}.$$

However, we also know from (II.2.7) that

$$\#\phi^{-1}(Q) \le \deg \phi = \#\Phi,$$

with equality if and only if ϕ is unramified. Since the points $P + T$ are distinct as T ranges over the elements of Φ, we conclude that ϕ is unramified at Q; and since Q was arbitrary, the map ϕ is unramified.

Finally, we apply the Hurwitz genus formula (II.2.7) to ϕ. Since ϕ is unramified, the formula reads

$$2\,\mathrm{genus}(E) - 2 = (\deg \phi)(\mathrm{genus}(C) - 2).$$

From this we conclude that C also has genus one, and hence C becomes an elliptic curve and ϕ becomes an isogeny if we take $\phi(O)$ to be the "zero point" on C. $\quad\square$

III.5 The Invariant Differential

Let E/K be an elliptic curve given by the usual Weierstrass equation

$$E : y^2 + a_1 xy + a_3 y = x^3 + a_2 x^2 + a_4 x + a_6.$$

We have seen (III.1.5) that the differential

$$\omega = \frac{dx}{2y + a_1 x + a_3} \in \Omega_E$$

has neither zeros nor poles. We now justify its name of *invariant differential* by proving that it is invariant under translation.

Proposition 5.1. *Let E and ω be as above, let $Q \in E$, and let $\tau_Q : E \to E$ be the translation-by-Q map (III.4.7). Then*

$$\tau_Q^* \omega = \omega.$$

PROOF. One can prove this proposition by a straightforward, but messy and un-enlightening, calculation as follows. Write $x(P + Q)$ and $y(P + Q)$ in terms of $x(P)$, $x(Q)$, $y(P)$, and $y(Q)$ using the addition formula (III.2.3c). Then use standard differentiation rules to calculate $dx(P+Q)$ as a rational function times $dx(P)$, treating $x(Q)$ and $y(Q)$ as constants. In this way one can directly verify that for a fixed value of Q,

$$\frac{dx(P + Q)}{2y(P + Q)) + a_1 x(P + Q) + a_3} = \frac{dx(P)}{2y(P)) + a_1 x(P) + a_3}.$$

We leave the details of this calculation to the reader and instead give a more illuminating proof.

Since Ω_E is a one-dimensional $\bar{K}(E)$-vector space (II.4.2), there is function $a_Q \in \bar{K}(E)^*$, depending a priori on Q, such that

$$\tau_Q^* \omega = a_Q \omega.$$

(Note that $a_Q \neq 0$, because τ_Q is an isomorphism.) We compute

$$\begin{aligned} \operatorname{div}(a_Q) &= \operatorname{div}(\tau_Q^* \omega) - \operatorname{div}(\omega) \\ &= \tau_Q^* \operatorname{div}(\omega) - \operatorname{div}(\omega) \\ &= 0 \qquad \text{since } \operatorname{div}(\omega) = 0 \text{ from (III.1.5).} \end{aligned}$$

Hence a_Q is a function on E having neither zeros nor poles, so (II.1.2) tells us that it is constant, i.e., $a_Q \in \bar{K}^*$.

Next consider the map

$$f : E \longrightarrow \mathbb{P}^1, \qquad Q \longmapsto [a_Q, 1].$$

From the calculation sketched earlier, even without doing it explicitly, it is clear that a_Q can be expressed as a rational function of $x(Q)$ and $y(Q)$. Hence f is a rational map from E to $\mathbb{P}^1$, and it is not surjective, since it misses both $[0, 1]$ and $[1, 0]$. We conclude from (II.2.1) and (II.2.3) that f is constant. Thus a_Q does not depend on Q, and we find its value by noting that

$$a_Q = a_O = 1 \qquad \text{for all } Q \in E.$$

This completes the proof that $\tau_Q^* \omega = \omega$. $\square$

Differential calculus is, in essence, a linearization tool. It will thus come as no surprise to learn that the enormous utility of the invariant differential on an elliptic curve lies in its ability to linearize the otherwise quite complicated addition law on the curve.

Theorem 5.2. *Let E and E' be elliptic curves, let ω be an invariant differential on E, and let*

$$\phi, \psi : E' \longrightarrow E$$

be isogenies. Then

$$(\phi + \psi)^*\omega = \phi^*\omega + \psi^*\omega.$$

N.B. *The two plus signs in this equation represent completely different operations. The first is addition in $\mathrm{Hom}(E', E)$, which is essentially addition using the group law on E. The second is the usual addition in the vector space of differentials Ω_E.*

PROOF. If $\phi = [0]$ or $\psi = [0]$, the result is clear. Next, if $\phi + \psi = [0]$, then using the fact that

$$\psi^* = (-\phi)^* = \phi^* \circ [-1]^*,$$

it suffices to check that

$$[-1]^*\omega = -\omega.$$

The negation formula

$$[-1](x, y) = (x, -y - a_1 x - a_3)$$

allows us to calculate

$$[-1]^* \left(\frac{dx}{2y + a_1 x + a_3} \right) = \frac{dx}{2(-y - a_1 x - a_3) + a_1 x + a_3}$$

$$= -\frac{dx}{2y + a_1 x + a_3},$$

which is the desired result. We now assume that ϕ, ψ, and $\phi + \psi$ are all nonzero.

Let (x_1, y_1) and (x_2, y_2) be "independent" Weierstrass coordinates on E. By this we mean that they satisfy the given Weierstrass equation for E, but satisfy no other algebraic relations. More formally,

$$([x_1, y_1, 1], [x_2, y_2, 1])$$

give coordinates for $E \times E$ sitting inside $\mathbb{P}^2 \times \mathbb{P}^2$. (Alternatively, (x_1, y_1) and (x_2, y_2) are "independent generic points of E" in the sense of Weil; see [41].)

Let

$$(x_3, y_3) = (x_1, y_1) + (x_2, y_2),$$

so x_3 and y_3 are rational combinations of x_1, x_2, y_1, y_2 given by the addition formula (III.2.3c) on E. Further, for any (x, y), let $\omega(x, y)$ denote the corresponding invariant differential,

$$\omega(x, y) = \frac{dx}{2y + a_1 x + a_3}.$$

Then, using the addition formula (III.2.3c) and standard rules for differentiation, we can express $\omega(x_3, y_3)$ in terms of $\omega(x_1, y_2)$ and $\omega(x_2, y_2)$. This yields

$$\omega(x_3, y_3) = f(x_1, y_1, x_2, y_2)\omega(x_1, y_1) + g(x_1, y_1, x_2, y_2)\omega(x_2, y_2),$$

where f and g are rational functions of the indicated variables. In doing this calculation, remember that since x_i and y_i satisfy the given Weierstrass equation, the differentials dx_i and dy_i are related by

$$(2y_i + a_1 x_i + a_3)\, dy_i = (3x_i^2 + 2a_2 x_i + a_4 - a_1 y_i)\, dx_i.$$

In this way, $\omega(x_3, y_3)$ can be expressed as a $\bar{K}(x_1, y_1, x_2, y_2)$-linear combination of dx_1 and dx_2.

We claim that both f and g are identically 1. Clearly this can be proven by an explicit calculation, a painful task that we leave for the reader. Instead, we use (III.5.1) to obtain the desired result. Suppose that we assign fixed values to x_2 and y_2, say by choosing some $Q \in E$ and setting

$$x_2 = x(Q) \qquad \text{and} \qquad y_2 = y(Q).$$

Then

$$dx_2 = dx(Q) = 0, \quad \text{so} \quad \omega(x_2, y_2) = 0,$$

while (III.5.1) tells us that

$$\omega(x_3, y_3) = \tau_Q^* \omega(x_1, y_1) = \omega(x_1, y_1).$$

Substituting these into the expression for $\omega(x_3, y_3)$, we find that

$$f(x_1, y_1, x(Q), y(Q)) = 1$$

as a rational function in $\bar{K}(x_1, y_1)$. Thus f does not depend on x_1 and y_1, so $f \in \bar{K}(x_2, y_2)$. But we also know that $f(x_2, y_2)$ satisfies $f(x(Q), y(Q)) = 1$ for *every* point $Q \in E$, so f must be identically 1. The same argument using x_2 and y_2 in place of x_1 and y_1 shows that g is also identically 1.

To recapitulate, we have shown that if

$$(x_3, y_3) = (x_1, y_1) + (x_2, y_2) \qquad (+ \text{ is addition on } E),$$

then

$$\omega(x_3, y_3) = \omega(x_1, y_1) + \omega(x_2, y_2) \qquad (+ \text{ is addition in } \Omega_E).$$

Now let (x', y') be Weierstrass coordinates on E' and set

$$(x_1, y_1) = \phi(x', y'), \qquad (x_2, y_2) = \psi(x', y'), \qquad (x_3, y_3) = (\phi + \psi)(x', y').$$

Substituting this into $\omega(x_3, y_3) = \omega(x_1, y_1) + \omega(x_2, y_2)$ yields

$$\big(\omega \circ (\phi + \psi)\big)(x', y') = (\omega \circ \phi)(x', y') + (\omega \circ \psi)(x', y'),$$

which says exactly that

$$(\phi + \psi)^* \omega = \phi^* \omega + \psi^* \omega. \qquad \qquad \square$$

Corollary 5.3. *Let ω be an invariant differential on an elliptic curve E. Let $m \in \mathbb{Z}$. Then*

$$[m]^* \omega = m\omega.$$

PROOF. The assertion is true for $m = 0$, since $[0]$ is the constant map, and it is true for $m = 1$, since $[1]$ is the identity map. We use (III.5.2) with $\phi = [m]$ and $\psi = [1]$ to obtain

$$[m+1]^* \omega = [m]^* \omega + \omega.$$

The desired result now follows by ascending and descending induction. □

As a first indication of the utility of the invariant differential, we give a new, less computational, proof of part of (III.4.2a).

Corollary 5.4. *Let E/K be an elliptic curve and let $m \in \mathbb{Z}$. Assume that $m \neq 0$ in K. Then the multiplication-by-m map on E is a finite separable endomorphism.*

PROOF. Let ω be an invariant differential on E. Then (III.5.3) and our assumption on m implies that

$$[m]^* \omega = m\omega \neq 0,$$

so certainly $[m] \neq [0]$. Hence $[m]$ is finite, and (II.4.2c) tells us that $[m]$ is separable. □

As a second application of (III.5.2) and (III.5.3), we examine when a linear combination involving the Frobenius morphism is separable.

Corollary 5.5. *Let E be an elliptic curve defined over a finite field $\mathbb{F}_q$ of characteristic p, let $\phi : E \to E$ be the qth-power Frobenius morphism (III.4.6), and let $m, n \in \mathbb{Z}$. Then the map*

$$m + n\phi : E \longrightarrow E$$

is separable if and only if $p \nmid m$. In particular, the map $1 - \phi$ is separable.

PROOF. Let ω be an invariant differential on E. From (II.4.2c) we know that a map $\psi : E \to E$ is inseparable if and only if $\psi^* \omega = 0$. We apply this criterion to the map $\psi = m + n\phi$. Using (III.5.2) and (III.5.3), we compute

$$(m + n\phi)^* \omega = m\omega + n\phi^* \omega.$$

Note that $\phi^* \omega = 0$, since ϕ is inseparable, or, by direct calculation,

$$\phi^* \left(\frac{dx}{2y + a_1 x + a_3} \right) = \frac{d(x^q)}{2y^q + a_1 x^q + a_3} = \frac{q x^{q-1} dx}{2y^q + a_1 x^q + a_3} = 0.$$

Hence

$$(m + n\phi)^* \omega = [m]^* \omega + [n]^* \circ \phi^* \omega = m\omega.$$

Since $m\omega = 0$ if and only if $p \mid m$, this gives the desired result. □

Corollary 5.6. *Let E/K be an elliptic curve and let ω be a nonzero invariant differential on E. We define a map from $\operatorname{End}(E)$ to $\bar{K}$ in the following way:*

$$\operatorname{End}(E) \longrightarrow \bar{K}, \qquad \phi \longmapsto a_\phi \quad \text{such that } \phi^*\omega = a_\phi\omega.$$

(a) *The map $\phi \mapsto a_\phi$ is a ring homomorphism.*
(b) *The kernel of $\phi \mapsto a_\phi$ is the set of inseparable endomorphisms of E.*
(c) *If $\operatorname{char}(K) = 0$, then $\operatorname{End}(E)$ is a commutative ring.*

PROOF. As in the proof of (III.5.1), the fact that Ω_E is a one-dimensional $\bar{K}(E)$-vector space (II.4.2) implies that $\phi^*\omega = a_\phi\omega$ for some function $a_\phi \in \bar{K}(E)$. We claim that $a_\phi \in \bar{K}$. This is clear if $a_\phi = 0$, while if $a_\phi \neq 0$, we use the fact that $\operatorname{div}(\omega) = 0$ to compute

$$\operatorname{div}(a_\phi) = \operatorname{div}(\phi^*\omega) - \operatorname{div}(\omega) = \phi^*\operatorname{div}(\omega) - \operatorname{div}(\omega) = 0.$$

Hence a_ϕ has no zeros or poles, so (II.1.2) says that $a_\phi \in \bar{K}$.
  (a) We use (III.5.2) to compute

$$a_{\phi+\psi}\omega = (\phi+\psi)^*\omega = \phi^*\omega + \psi^*\omega = a_\phi\omega + a_\psi\omega = (a_\phi + a_\psi)\omega.$$

This gives $a_{\phi+\psi} = a_\phi + a_\psi$. Similarly,

$$a_{\phi\circ\psi}\omega = (\phi\circ\psi)^*\omega = \psi^*(\phi^*\omega) = \psi^*(a_\phi\omega) = a_\phi\psi^*(\omega) = a_\phi a_\psi\omega,$$

which proves that $a_{\phi\circ\psi} = a_\phi a_\psi$.
  (b) We have

$$a_\phi = 0 \quad \Longleftrightarrow \quad \phi^*\omega = 0 \quad \Longleftrightarrow \quad \phi \text{ is inseparable (II.4.2c).}$$

  (c) If $\operatorname{char}(K) = 0$, then every endomorphism is separable, so (b) says that $\operatorname{End}(E)$ injects into $\bar{K}^*$. Hence $\operatorname{End}(E)$ is commutative. $\square$

III.6   The Dual Isogeny

Let $\phi : E_1 \to E_2$ be a nonconstant isogeny. We have seen (II.3.7) that ϕ induces a map

$$\phi^* : \operatorname{Pic}^0(E_1) \longrightarrow \operatorname{Pic}^0(E_1).$$

On the other hand, for $i = 1$ and 2 we have group isomorphisms (III.3.4)

$$\kappa_i : E_i \longrightarrow \operatorname{Pic}^0(E_i), \qquad P \longmapsto \text{class of } (P) - (O).$$

This gives a homomorphism going in the opposite direction to ϕ, namely the composition

$$E_2 \xrightarrow{\ \kappa_2\ } \operatorname{Pic}^0(E_2) \xrightarrow{\ \phi^*\ } \operatorname{Pic}^0(E_1) \xrightarrow{\ \kappa_1^{-1}\ } E_1.$$

Later in this section we will verify that this map may be computed as follows. Let $Q \in E_2$, and choose any $P \in E_1$ satisfying $\phi(P) = Q$. Then

$$\kappa_1^{-1} \circ \phi^* \circ \kappa_2(Q) = [\deg \phi](P).$$

It is by no means clear that the homomorphism $\kappa_1^{-1} \circ \phi^* \circ \kappa_2$ is an isogeny, i.e., that it is given by a rational map. The process of finding a point P satisfying $\phi(P) = Q$ involves taking roots of various polynomial equations. If ϕ is separable, one needs to check that applying $[\deg \phi]$ to P causes the conjugate roots to appear symmetrically. (It is actually reasonably clear that this is true if one explicitly writes out $\kappa_1^{-1} \circ \phi^* \circ \kappa_2$.) If ϕ is inseparable, this approach is more complicated. We now show that in all cases there is an actual isogeny that may be computed in the manner described above.

Theorem 6.1. *Let $E_1 \to E_2$ be a nonconstant isogeny of degree m.*
(a) *There exists a unique isogeny*

$$\hat{\phi} : E_2 \longrightarrow E_1 \qquad satisfying \qquad \hat{\phi} \circ \phi = [m].$$

(b) *As a group homomorphism, $\hat{\phi}$ equals the composition*

$$
\begin{array}{ccccc}
E_2 & \longrightarrow & \operatorname{Div}^0(E_2) & \xrightarrow{\phi^*} \operatorname{Div}^0(E_1) & \xrightarrow{\text{sum}} E_1, \\
Q & \longmapsto & (Q) - (O) & \sum n_P(P) & \longmapsto \sum [n_P]P.
\end{array}
$$

PROOF. (a) First we show uniqueness. Suppose that $\hat{\phi}$ and $\hat{\phi}'$ are two such isogenies. Then

$$(\hat{\phi} - \hat{\phi}') \circ \phi = [m] - [m] = [0].$$

Since ϕ is nonconstant, it follows from (II.2.3) that $\hat{\phi} - \hat{\phi}'$ must be constant, so $\hat{\phi} = \hat{\phi}'$.

Next suppose that $\psi : E_2 \to E_3$ is another nonconstant isogeny, say of degree n, and suppose that we know that $\hat{\phi}$ and $\hat{\psi}$ exist. Then

$$(\hat{\phi} \circ \hat{\psi}) \circ (\psi \circ \phi) = \hat{\phi} \circ [n] \circ \phi = [n] \circ \hat{\phi} \circ \phi = [nm].$$

Thus $\hat{\phi} \circ \hat{\psi}$ has the requisite property to be $\widehat{\psi \circ \phi}$. Hence using (II.2.12) to write an arbitrary isogeny ϕ as a composition, it suffices to prove the existence of $\hat{\phi}$ when ϕ is either separable or equal to the Frobenius morphism.

$\boxed{Case~1.~~\phi~\text{is separable}}$ Since ϕ has degree m, we have (III.4.10c)

$$\# \ker \phi = m,$$

so every element of $\ker \phi$ has order dividing m, i.e.,

$$\ker \phi \subset \ker[m].$$

It follows immediately from (III.4.11) that there is an isogeny

$$\hat{\phi} : E_2 \longrightarrow E_1 \qquad \text{satisfying} \qquad \hat{\phi} \circ \phi = [m].$$

$\boxed{Case\ 2.\ \phi\ \text{is a Frobenius morphism}}$ If ϕ is the qth-power Frobenius morphism with $q = p^e$, then ϕ is clearly the composition of the pth-power Frobenius morphism with itself e times. Hence it suffices to prove that $\hat{\phi}$ exists if ϕ is the pth-power Frobenius morphism, so in particular, $\deg \phi = p$ from (II.2.11).

We look at the multiplication-by-p map on E. Let ω be an invariant differential. Then from (III.5.3) and the fact that $\text{char}(K) = p$, we see that

$$[p]^* \omega = p\omega = 0.$$

We conclude from (III.4.2c) that $[p]$ is not separable, and thus when we decompose $[p]$ as a Frobenius morphism followed by a separable map (II.2.12), the Frobenius morphism does appear. In other words,

$$[p] = \psi \circ \phi^e$$

for some integer $e \geq 1$ and some separable isogeny ψ. Then we can take

$$\hat{\phi} = \psi \circ \phi^{e-1}.$$

(b) Let $Q \in E_2$. Then the image of Q under the indicated composition is

$$\text{sum}\big(\phi^*((Q) - (O))\big)$$

$$= \sum_{P \in \phi^{-1}(Q)} [e_\phi(P)]P - \sum_{T \in \phi^{-1}(O)} [e_\phi(T)]T \qquad \text{by definition of } \phi^*,$$

$$= [\deg_i \phi] \left(\sum_{P \in \phi^{-1}(Q)} P - \sum_{T \in \phi^{-1}(O)} T \right) \qquad \text{from (III.4.10a)},$$

$$= [\deg_i \phi] \circ [\#\phi^{-1}(Q)]P \qquad \text{for any } P \in \phi^{-1}(Q),$$

$$= [\deg \phi]P \qquad \text{from (III.4.10a)}.$$

But by construction,

$$\hat{\phi}(Q) = \hat{\phi} \circ \phi(P) = [\deg \phi]P,$$

so the two maps are the same.   □

Definition. Let $\phi : E_1 \to E_2$ be an isogeny. The *dual isogeny* to ϕ is the isogeny

$$\hat{\phi} : E_2 \longrightarrow E_1$$

given by (III.6.1a). (This assumes that $\phi \neq [0]$. If $\phi = [0]$, then we set $\hat{\phi} = [0]$.)

The next theorem gives the basic properties of the dual isogeny. From these basic facts we will be able to deduce a number of very important corollaries, including a good description of the kernel of the multiplication-by-m map.

Theorem 6.2. *Let*

$$\phi : E_1 \longrightarrow E_2$$

be an isogeny.
  (a) *Let* $m = \deg \phi$. *Then*

$$\hat{\phi} \circ \phi = [m] \quad on \ E_1 \qquad and \qquad \phi \circ \hat{\phi} = [m] \quad on \ E_2.$$

  (b) *Let* $\lambda : E_2 \to E_3$ *be another isogeny. Then*

$$\widehat{\lambda \circ \phi} = \hat{\phi} \circ \hat{\lambda}.$$

  (c) *Let* $\psi : E_1 \to E_2$ *be another isogeny. Then*

$$\widehat{\phi + \psi} = \hat{\phi} + \hat{\psi}.$$

  (d) *For all* $m \in \mathbb{Z}$,

$$\widehat{[m]} = [m] \qquad and \qquad \deg[m] = m^2.$$

  (e) $\deg \hat{\phi} = \deg \phi$.
  (f) $\hat{\hat{\phi}} = \phi$.

PROOF. If ϕ is constant, then the entire theorem is trivial, and similarly (b) and (c) are trivial if λ or ψ is constant. We may thus assume that all isogenies are nonconstant.
  (a) The first statement is the defining property of $\hat{\phi}$. For the second, consider

$$(\phi \circ \hat{\phi}) \circ \phi = \phi \circ [m] = [m] \circ \phi.$$

Hence $\phi \circ \hat{\phi} = [m]$, since ϕ is not constant.
  (b) Letting $n = \deg \lambda$, we have

$$(\hat{\phi} \circ \hat{\lambda}) \circ (\lambda \circ \phi) = \hat{\phi} \circ [n] \circ \phi = [n] \circ \hat{\phi} \circ \phi = [nm].$$

The uniqueness statement in (III.6.1a) implies that

$$\hat{\phi} \circ \hat{\lambda} = \widehat{\lambda \circ \phi}.$$

  (c) We give a proof in characteristic 0. See Exercise 3.31 for a proof in arbitrary characteristic.
    Let $x_1, y_1 \in K(E_1)$ and $x_2, y_2 \in K(E_2)$ be Weierstrass coordinates. We start by looking at E_2 considered as an elliptic curve defined over the field $K(E_1) = K(x_1, y_1)$.[1] Then another way of saying that $\phi : E_1 \to E_2$ is an isogeny is to note that $\phi(x_1, y_1) \in E_2(K(x_1, y_1))$, and similarly for $\psi(x_1, y_1)$ and $(\phi + \psi)(x_1, y_1)$. Now consider the divisor

---

[1] This is where we use the characteristic 0 assumption, since all of our results on elliptic curves have assumed that the base field is perfect.

$$D = \mathrm{div}\big((\phi + \psi)(x_1, y_1)\big) - \mathrm{div}\big(\phi(x_1, y_1)\big) + \mathrm{div}\big(\psi(x_1, y_1)\big) + (O)$$
$$\in \mathrm{Div}_{K(x_1, y_1)}(E_2).$$

The definition of $\phi + \psi$ implies that D sums to O, so (III.3.5) tells us that D is linearly equivalent to 0. Thus there is a function

$$f \in K(x_1, y_1)(E_2) = K(x_1, y_1, x_2, y_2)$$

that, *when considered as a function of x_2 and y_2*, has divisor D.

We now switch perspective and look at f as a function of x_1 and y_1. In other words, we treat f as a function on E_1 considered as an elliptic curve defined over $K(x_2, y_2)$. Suppose that $P_1 \in E_1\big(\overline{K(x_2, y_2)}\big)$ is a point satisfying $\phi(P_1) = (x_2, y_2)$. Then examining D, specifically the term $-\mathrm{div}\big(\phi(x_1, y_1)\big)$, we see that f has a pole at P_1, i.e., the function $f(x_1, y_1; x_2, y_2)$ has a pole if x_1, y_1, x_2, y_2 satisfy $(x_2, y_2) = \phi(x_1, y_1)$. Further,

$$\mathrm{ord}_{P_1}(f) = e_\phi(P_1).$$

Similarly, f has a pole at P_1 if $(x_2, y_2) = \psi(P_1)$, and it has a zero at P_1 if $(x_2, y_2) = (\phi + \psi)(P_1)$. It follows that as a function of x_1 and y_1, the divisor of f has the form

$$(\phi + \psi)^*\big((x_2, y_2)\big) - \phi^*\big((x_2, y_2)\big) - \psi^*\big((x_2, y_2)\big) + \sum n_i(P_i) \in \mathrm{Div}_{\overline{K(x_2, y_2)}}(E_1),$$

where the P_i's are in $E_1(\bar{K})$, i.e., $\sum n_i(P_i) \in \mathrm{Div}_{\bar{K}}(E_1)$. Since this is the divisor of a function, it sums to O, so using (III.6.1b), we conclude that the point

$$\widehat{(\phi + \psi)}(x_2, y_2) - \hat{\phi}(x_2, y_2) - \hat{\psi}(x_2, y_2)$$

does not depend on (x_2, y_2), i.e., it is in $E_1(\bar{K})$. Putting $(x_2, y_2) = O$ shows that it is equal to O, which completes the proof that

$$\widehat{\phi + \psi} = \hat{\phi} + \hat{\psi}.$$

(d) This is true for $m = 0$ by definition, and it is clearly true for $m = 1$. Using (c) with $\phi = [m]$ and $\psi = [1]$ yields

$$\widehat{[m + 1]} = \widehat{[m]} + \widehat{[1]},$$

and ascending and descending induction shows that $\widehat{[m]} = [m]$ holds for all m.

Now let $d = \deg[m]$ and consider the multiplication-by-d map. Thus

$$[d] = \widehat{[m]} \circ [m] \qquad\qquad \text{definition of dual isogeny,}$$
$$= [m^2] \qquad\qquad\qquad \text{since } \widehat{[m]} = [m].$$

Using the fact (III.4.2b) that the endomorphism ring of an elliptic curve is a torsion-free $\mathbb{Z}$-module, it follows that $d = m^2$.

(e) Let $m = \deg \phi$. Then using (d) and (a), we find that

$$m^2 = \deg[m] = \deg(\phi \circ \hat{\phi}) = (\deg \phi)(\deg \hat{\phi}) = m(\deg \hat{\phi}).$$

Hence $m = \deg \hat{\phi}$.

(f) Again let $m = \deg \phi$. Then using (a), (b), and (d) yields

$$\hat{\phi} \circ \phi = [m] = \widehat{[m]} = \widehat{\hat{\phi} \circ \phi} = \hat{\phi} \circ \hat{\hat{\phi}}.$$

Therefore

$$\phi = \hat{\hat{\phi}}.$$

$\square$

Definition. Let A be an abelian group. A function

$$d : A \longrightarrow \mathbb{R}$$

is a *quadratic form* if it satisfies the following conditions:

(i) $d(\alpha) = d(-\alpha)$ for all $\alpha \in A$.

(ii) The pairing

$$A \times A \longrightarrow \mathbb{R}, \qquad (\alpha, \beta) \longmapsto d(\alpha + \beta) - d(\alpha) - d(\beta),$$

is bilinear.

A quadratic form d is *positive definite* if it further satisfies:

(iii) $d(\alpha) \geq 0$ for all $\alpha \in A$.

(iv) $d(\alpha) = 0$ if and only if $\alpha = 0$.

Corollary 6.3. *Let E_1 and E_2 be elliptic curves. The degree map*

$$\deg : \operatorname{Hom}(E_1, E_2) \longrightarrow \mathbb{Z}$$

is a positive definite quadratic form.

PROOF. Everything is clear except for the fact that the pairing

$$\langle \phi, \psi \rangle = \deg(\phi + \psi) - \deg(\phi) - \deg(\psi)$$

is bilinear. To verify this, we use the injection

$$[\ ] : \mathbb{Z} \longrightarrow \operatorname{End}(E_1)$$

and compute

$$\begin{aligned}
[\langle \phi, \psi \rangle] &= [\deg(\phi + \psi)] - [\deg(\phi)] - [\deg(\psi)] \\
&= (\widehat{\phi + \psi}) \circ (\phi + \psi) - \hat{\phi} \circ \phi - \hat{\psi} \circ \psi \\
&= \hat{\phi} \circ \psi + \hat{\psi} \circ \phi \qquad \text{from (III.6.2c)}.
\end{aligned}$$

Using (III.6.2c) a second time, we see that this last expression is linear in both ϕ and ψ.

$\square$

Corollary 6.4. *Let E be an elliptic curve and let $m \in \mathbb{Z}$ with $m \neq 0$.*
(a) $\deg[m] = m^2$.
(b) *If $m \neq 0$ in K, i.e., if either $\mathrm{char}(K) = 0$ or $p = \mathrm{char}(K) > 0$ and $p \nmid m$, then*

$$E[m] = \frac{\mathbb{Z}}{m\mathbb{Z}} \times \frac{\mathbb{Z}}{m\mathbb{Z}}.$$

(c) *If $\mathrm{char}(K) = p > 0$, then one of the following is true:*

(i) $E[p^e] = \{O\}$ *for all $e = 1, 2, 3, \ldots$.*

(ii) $E[p^e] = \dfrac{\mathbb{Z}}{p^e\mathbb{Z}}$ *for all $e = 1, 2, 3, \ldots$.*

(Recall that $E[m]$ is another notation for $\ker[m]$, the set of points of order m on E.)

PROOF. (a) This was proven in (III.6.2d). We record it again here in order to point out that there are other ways of proving that $[m]$ has degree m^2; see for example exercises 3.7, 3.8, and 3.11. Then the fundamental description of $E[m]$ in (b) follows formally from (a).

(b) The assumption on m and the fact that $\deg[m] = m^2$ tells us that $[m]$ is a finite separable map. Hence from (III.4.10c),

$$\#E[m] = \deg[m] = m^2.$$

Further, for every integer d dividing m, we similarly have

$$\#E[d] = d^2.$$

Writing the finite group $E[m]$ as a product of cyclic groups, it is easy to see that the only possibility is

$$E[m] = \frac{\mathbb{Z}}{m\mathbb{Z}} \times \frac{\mathbb{Z}}{m\mathbb{Z}}.$$

(See Exercise 3.30.)

(c) Let ϕ be the pth-power Frobenius morphism. Then

$$
\begin{aligned}
\#E[p^e] &= \deg_s[p^e] && \text{from (III.4.10a),} \\
&= \left(\deg_s(\hat{\phi} \circ \phi)\right)^e && \text{from (III.6.2a),} \\
&= (\deg_s \hat{\phi})^e && \text{from (II.2.11b).}
\end{aligned}
$$

From (III.6.2e) and (II.2.11c) we have

$$\deg \hat{\phi} = \deg \phi = p,$$

so there are two cases. If $\hat{\phi}$ is inseparable, then $\deg_s \hat{\phi} = 1$, so

$$\#E[p^e] = 1 \qquad \text{for all } e.$$

Otherwise $\hat{\phi}$ is separable, so $\deg_s \hat{\phi} = p$ and

$$\#E[p^e] = p^e \qquad \text{for all } e.$$

Again writing $E[p^e]$ as a product of cyclic groups, it is easy to see that this implies that

$$E[p^e] = \frac{\mathbb{Z}}{p^e\mathbb{Z}}.$$

(For a more detailed analysis of $E[p^e]$ in characteristic p and its relationship to the endomorphism ring $\text{End}(E)$, see (V §3).) $\square$

III.7 The Tate Module

Let E/K be an elliptic curve and let $m \geq 2$ be an integer, prime to $\text{char}(K)$ if $\text{char}(K) > 0$. As we have seen,

$$E[m] \cong \frac{\mathbb{Z}}{m\mathbb{Z}} \times \frac{\mathbb{Z}}{m\mathbb{Z}},$$

the isomorphism being one between abstract groups. However, the group $E[m]$ comes equipped with considerably more structure than an abstract group. For example, each element σ of the Galois group $G_{\bar{K}/K}$ acts on $E[m]$, since if $[m]P = O$, then

$$[m](P^\sigma) = ([m]P)^\sigma = O^\sigma = O.$$

We thus obtain a representation

$$G_{\bar{K}/K} \longrightarrow \text{Aut}(E[m]) \cong \text{GL}_2(\mathbb{Z}/m\mathbb{Z}),$$

where the latter isomorphism involves choosing a basis for $E[m]$. Individually, for each m, these representations are not completely satisfactory, since it is generally easiesr to deal with representations whose matrices have coefficients in a ring of characteristic 0. We are going to fit together these mod m representations for varying m in order to create a characteristic 0 representation. To do this, we mimic the inverse limit construction of the ℓ-adic integers $\mathbb{Z}_\ell$ from the finite groups $\mathbb{Z}/\ell^n\mathbb{Z}$.

Definition. Let E be an elliptic curve and let $\ell \in \mathbb{Z}$ be a prime. The (ℓ-adic) Tate module of E is the group

$$T_\ell(E) = \varprojlim_n E[\ell^n],$$

the inverse limit being taken with respect to the natural maps

$$E[\ell^{n+1}] \xrightarrow{\quad [\ell] \quad} E[\ell^n].$$

Since each $E[\ell^n]$ is a $\mathbb{Z}/\ell^n\mathbb{Z}$-module, we see that the Tate module has a natural structure as a $\mathbb{Z}_\ell$-module. Further, since the multiplication-by-ℓ maps are surjective, the inverse limit topology on $T_\ell(E)$ is equivalent to the ℓ-adic topology that it gains by being a $\mathbb{Z}_\ell$-module.

Proposition 7.1. *As a $\mathbb{Z}_\ell$-module, the Tate module has the following structure*:

(a) $T_\ell(E) \cong \mathbb{Z}_\ell \times \mathbb{Z}_\ell$ *if $\ell \neq \mathrm{char}(K)$.*

(b) $T_p(E) \cong \{0\}$ *or* $\mathbb{Z}_p$ *if $p = \mathrm{char}(K) > 0$.*

PROOF. This follows immediately from (III.6.4b,c). $\square$

The action of $G_{\bar{K}/K}$ on each $E[\ell^n]$ commutes with the multiplication-by-ℓ map used to form the inverse limit, so $G_{\bar{K}/K}$ also acts on $T_\ell(E)$. Further, since the profinite group $G_{\bar{K}/K}$ acts continuously on each finite (discrete) group $E[\ell^n]$, the resulting action on $T_\ell(E)$ is also continuous.

Definition. The *ℓ-adic representation (of $G_{\bar{K}/K}$ associated to E)* is the homomorphism

$$\rho_\ell : G_{\bar{K}/K} \longrightarrow \mathrm{Aut}\big(T_\ell(E)\big)$$

induced by the action of $G_{\bar{K}/K}$ on the ℓ^n-torsion points of E.

Convention. From here on, the number ℓ always refers to a prime number that is different from the characteristic of K.

Remark 7.2. If we choose a $\mathbb{Z}_\ell$-basis for $T_\ell(E)$, we obtain a representation

$$G_{\bar{K}/K} \longrightarrow \mathrm{GL}_2(\mathbb{Z}_\ell),$$

and then the natural inclusion $\mathbb{Z}_\ell \subset \mathbb{Q}_\ell$ gives a representation

$$G_{\bar{K}/K} \longrightarrow \mathrm{GL}_2(\mathbb{Q}_\ell).$$

In this way we obtain a two-dimensional representation of $G_{\bar{K}/K}$ over a field of characteristic 0. More intrinsically, we can avoid choosing a basis by using the natural map

$$\rho_\ell : G_{\bar{K}/K} \longrightarrow \mathrm{Aut}\big(T_\ell(E)\big) \longhookrightarrow \mathrm{Aut}\big(T_\ell(E)\big) \otimes_{\mathbb{Z}_\ell} \mathbb{Q}_\ell.$$

Remark 7.3. The above construction is analogous to the following, which may be more familiar to the reader. Let

$$\boldsymbol{\mu}_{\ell^n} \subset \bar{K}^*$$

be the group of $(\ell^n)^{\mathrm{th}}$ roots of unity. Raising to the ℓth power gives maps

$$\boldsymbol{\mu}_{\ell^{n+1}} \xrightarrow{\ \zeta \mapsto \zeta^\ell\ } \boldsymbol{\mu}_{\ell^n},$$

and then taking the inverse limit yields the *Tate module of K*,

$$T_\ell(\boldsymbol{\mu}) = \varprojlim_n \boldsymbol{\mu}_{\ell^n}.$$

(More formally, $T_\ell(\boldsymbol{\mu})$ is the Tate module of the multiplicative group $\bar{K}^*$.) As abstract groups, we have

$$\boldsymbol{\mu}_{\ell^n} \cong \mathbb{Z}/\ell^n\mathbb{Z}   \quad \text{and} \quad   T_\ell(\boldsymbol{\mu}) \cong \mathbb{Z}_\ell.$$

Further, the natural action of $G_{\bar{K}/K}$ on each μ_{ℓ^n} induces an action on $T_\ell(\mu)$, so we obtain a 1-dimensional representation

$$G_{\bar{K}/K} \longrightarrow \text{Aut}\big(T_\ell(\mu)\big) \cong \mathbb{Z}_\ell^*.$$

For $K = \mathbb{Q}$, this cyclotomic representation is surjective, because the ℓ-power cyclotomic polynomials are irreducible over $\mathbb{Q}$.

Remark 7.3.1. In Chapter VI, when we study elliptic curves over the complex numbers, we will see (VI.5.6) that there is a natural way in which the m-torsion subgroup $E[m]$ may be identified with the homology group $H_1(E, \mathbb{Z}/m\mathbb{Z})$, and similarly $T_\ell(E)$ with $H_1(E, \mathbb{Z}_\ell)$. The utility of this identification is that while homology groups do not generally admit a Galois action, the torsion subgroup $E[m]$ and Tate module $T_\ell(E)$ do admit such an action. This idea has been vastly generalized by Grothendieck and others in the theory of étale cohomology.

The Tate module is a useful tool for studying isogenies. Let

$$\phi : E_1 \longrightarrow E_2$$

be an isogeny of elliptic curves. Then ϕ induces maps

$$\phi : E_1[\ell^n] \longrightarrow E_2[\ell^n],$$

and hence it induces a $\mathbb{Z}_\ell$-linear map

$$\phi_\ell : T_\ell(E_1) \longrightarrow T_\ell(E_2).$$

We thus obtain a natural homomorphism

$$\text{Hom}(E_1, E_2) \longrightarrow \text{Hom}\big(T_\ell(E_1), T_\ell(E_2)\big).$$

Further, if $E_1 = E_2 = E$, then the map

$$\text{End}(E) \longrightarrow \text{End}\big(T_\ell(E)\big)$$

is even a homomorphism of rings. It is not hard to show that these maps are injective (see Exercise 3.14), but the following result gives much stronger information about the structure of $\text{Hom}(E_1, E_2)$.

Theorem 7.4. *Let E_1 and E_2 be elliptic curves and let $\ell \neq \text{char}(K)$ be a prime. Then the natural map*

$$\text{Hom}(E_1, E_2) \otimes \mathbb{Z}_\ell \longrightarrow \text{Hom}\big(T_\ell(E_1), T_\ell(E_2)\big), \qquad \phi \longmapsto \phi_\ell,$$

is injective

PROOF. We start by proving the following statement:

$$\left[\begin{array}{l} \text{Let } M \subset \text{Hom}(E_1, E_2) \text{ be a finitely generated subgroup, and let} \\ M^{\text{div}} = \big\{\phi \in \text{Hom}(E_1, E_2) : [m] \circ \phi \in M \text{ for some integer } m \geq 1\big\}. \\ \text{Then } M^{\text{div}} \text{ is finitely generated.} \end{array} \right. \qquad (*)$$

To prove (∗), we extend the degree mapping to the finite-dimensional real vector space $M \otimes \mathbb{R}$, which we equip with the natural topology inherited from $\mathbb{R}$. Then the degree mapping is clearly continuous, so the set

$$U = \{\phi \in M \otimes \mathbb{R} : \deg \phi < 1\}$$

is an open neighborhood of 0. Further, since $\mathrm{Hom}(E_1, E_2)$ is a torsion-free $\mathbb{Z}$-module (III.4.2b), there is a natural inclusion

$$M^{\mathrm{div}} \subset M \otimes \mathbb{R}.$$

Further, it is clear that

$$M^{\mathrm{div}} \cap U = \{0\},$$

since every nonzero isogeny has degree at least one. Hence M^{div} is a discrete subgroup of the finite-dimensional vector space $M \otimes \mathbb{R}$, so it is finitely generated. This completes the proof of (∗).

We now turn to the proof of (III.7.4). Let $\phi \in \mathrm{Hom}(E_1, E_2) \otimes \mathbb{Z}_\ell$, and suppose that $\phi_\ell = 0$. Let

$$M \subset \mathrm{Hom}(E_1, E_2)$$

be some finitely generated subgroup with the property that $\phi \in M \otimes \mathbb{Z}_\ell$. Then, with notation as above, the group M^{div} is finitely generated, so it is also free, since (III.4.2b) tells us that it is torsion-free. Let

$$\psi_1, \ldots, \psi_t \in \mathrm{Hom}(E_1, E_2)$$

be a basis for M^{div}, and write

$$\phi = \alpha_1 \psi_1 + \cdots + \alpha_t \psi_t \quad \text{with} \quad \alpha_1, \ldots, \alpha_t \in \mathbb{Z}_\ell.$$

Now fix some $n \geq 1$ and choose $a_1, \ldots, a_t \in \mathbb{Z}$ with

$$a_i \equiv \alpha_i \pmod{\ell^n}.$$

Then the assumption that $\phi_\ell = 0$ implies that the isogeny

$$\psi = [a_1] \circ \psi_1 + \cdots + [a_t] \circ \psi_t \in \mathrm{Hom}(E_1, E_2)$$

annihilates $E_1[\ell^n]$. It follows from (III.4.11) that ψ factors through $[\ell^n]$, so there is an isogeny

$$\lambda \in \mathrm{Hom}(E_1, E_2) \quad \text{satisfying} \quad \psi = [\ell^n] \circ \lambda.$$

Further, λ is in M^{div}, so there are integers $b_i \in \mathbb{Z}$ such that

$$\lambda = [b_1] \circ \psi_1 + \cdots + [b_t] \circ \psi_t.$$

Then, since the ψ_i's form a $\mathbb{Z}$-basis for M^{div}, the fact that $\psi = [\ell^n] \circ \lambda$ implies that

$$a_i = \ell^n b_i,$$

and hence

$$\alpha_i \equiv 0 \pmod{\ell^n}.$$

This holds for all n, so we conclude that $\alpha_i = 0$, and hence that $\phi = 0$. (N.B. The reason that we need to use M^{div}, rather than working in M, is because it is essential that ϕ, ψ, and λ be written in terms of a $\mathbb{Z}$-basis that does not depend on the choice of ℓ^n.) $\qquad\square$

Corollary 7.5. *Let E_1 and E_2 be elliptic curves. Then*

$$\mathrm{Hom}(E_1, E_2)$$

is a free $\mathbb{Z}$-module of rank at most 4.

PROOF. We know from (III.4.2b) that $\mathrm{Hom}(E_1, E_2)$ is torsion-free. This implies that

$$\mathrm{rank}_{\mathbb{Z}} \mathrm{Hom}(E_1, E_2) = \mathrm{rank}_{\mathbb{Z}_\ell} \mathrm{Hom}(E_1, E_2) \otimes \mathbb{Z}_\ell,$$

in the sense that if one is finite, then the other is finite and they are equal. Next, from (III.7.4) we have the estimate

$$\mathrm{rank}_{\mathbb{Z}_\ell} \mathrm{Hom}(E_1, E_2) \otimes \mathbb{Z}_\ell \le \mathrm{rank}_{\mathbb{Z}_\ell} \mathrm{Hom}\big(T_\ell(E_1), T_\ell(E_2)\big).$$

Finally, choosing a $\mathbb{Z}_\ell$-basis for $T_\ell(E_1)$ and $T_\ell(E_2)$, we see from (III.7.1a) that

$$\mathrm{Hom}\big(T_\ell(E_1), T_\ell(E_2)\big) = M_2(\mathbb{Z}_\ell)$$

is the additive group of 2×2 matrices with $\mathbb{Z}_\ell$-coefficients. The $\mathbb{Z}_\ell$-rank of $M_2(\mathbb{Z}_\ell)$ is 4, which proves that $\mathrm{rank}_{\mathbb{Z}} \mathrm{Hom}(E_1, E_2)$ is at most 4. $\qquad\square$

Remark 7.6. By definition, an isogeny is defined over K if it commutes with the action of $G_{\bar{K}/K}$. Similarly, we can define

$$\mathrm{Hom}_K\big(T_\ell(E_1), T_\ell(E_2)\big)$$

to be the group of $\mathbb{Z}_\ell$-linear maps from $T_\ell(E_1)$ to $T_\ell(E_2)$ that commute with the action of $G_{\bar{K}/K}$ as given by the ℓ-adic representation. Then we have a homomorphism

$$\mathrm{Hom}_K(E_1, E_2) \otimes \mathbb{Z}_\ell \longrightarrow \mathrm{Hom}_K\big(T_\ell(E_1), T_\ell(E_2)\big),$$

and (III.7.4) tells us that this homomorphism is injective. It turns out that in many cases, it is an isomorphism.

Isogeny Theorem 7.7. *Let $\ell \ne \mathrm{char}(K)$ be a prime. The natural map*

$$\mathrm{Hom}_K(E_1, E_2) \otimes \mathbb{Z}_\ell \longrightarrow \mathrm{Hom}_K\big(T_\ell(E_1), T_\ell(E_2)\big)$$

is an isomorphism in the following two situations:

(a) K *is a finite field.* (Tate [282])

(b) K *is a number field.* (Faltings [82, 84])

The original proofs of both parts of (III.7.7) make heavy use of abelian varieties (higher-dimensional analogues of elliptic curves) and are thus unfortunately beyond the scope of this book. Indeed, the methods used to prove (III.7.7b) include virtually all of the tools needed for Faltings' proof of the Mordell conjecture. See also [237] for a proof of (III.7.7b) in the case that $j(E)$ is nonintegral, and [45, 160, 163] for alternative proofs of (III.7.7b).

One way to interpret (III.7.7) is to view the Tate modules as homology groups, specifically as the first homology with $\mathbb{Z}_\ell$-coefficients (III.7.3.1). Then (III.7.7) characterizes when a map between homology groups comes from an actual geometric map between the curves.

Remark 7.8. It is also natural to ask about the size of the image of $\rho_\ell(G_{\bar{K}/K})$ in $\mathrm{Aut}(T_\ell(E))$. The following theorem of Serre provides an answer for number fields. We do not include the proof. (But see (IX.6.3) and Exercise 9.7.)

Theorem 7.9. (Serre) *Let K be a number field and let E/K be an elliptic curve without complex multiplication.*

(a) $\rho_\ell(G_{\bar{K}/K})$ *is of finite index in* $\mathrm{Aut}(T_\ell(E))$ *for all primes $\ell \neq \mathrm{char}(K)$.*

(b) $\rho_\ell(G_{\bar{K}/K}) = \mathrm{Aut}(T_\ell(E))$ *for all but finitely many primes ℓ.*

PROOF. See [237] and [231]. $\qquad\qquad\qquad\qquad\qquad\qquad\qquad\qquad\square$

Remark 7.10. Let E/K be an elliptic curve. Then the elements of $\mathrm{End}_K(E)$ commute with the elements of $G_{\bar{K}/K}$ in their action on $T_\ell(E)$. If

$$\mathrm{End}_K(E) = \mathbb{Z},$$

this gives no additional information. However, if E has complex multiplication, then one can show (Exercise 3.24) that this forces the action of $\mathrm{Gal}_{\bar{K}/K}$ on $T_\ell(E)$ to be abelian, i.e., the image $\rho_\ell(\mathrm{Gal}_{\bar{K}/K})$ is an abelian subgroup of $\mathrm{Aut}(T_\ell(E)) \cong \mathrm{GL}_2(\mathbb{Z}_\ell)$. In particular, adjoining the coordinates of ℓ^n-torsion points to K leads to explicitly constructed abelian extensions of K, in much the same way that abelian extensions of $\mathbb{Q}$ are obtained by adjoining roots of unity. See (C §11) for a brief discussion, and [140, Part II], [249, Chapter 5], or [266, Chapter II] for further details.

III.8   The Weil Pairing

Let E/K be an elliptic curve. For this section we fix an integer $m \geq 2$, which we assume to be prime to $p = \mathrm{char}(K)$ if $p > 0$.

As an abstract group, the group of m-torsion points $E[m]$ has the form (III.6.4b)

$$E[m] \cong \mathbb{Z}/m\mathbb{Z} \times \mathbb{Z}/m\mathbb{Z}.$$

Thus $E[m]$ is a free $\mathbb{Z}/m\mathbb{Z}$-module of rank two. Every free module comes equipped with a natural nondegenerate alternating multilinear map, the determinant. Choosing a basis $\{T_1, T_2\}$ for $E[m]$, the determinant pairing on $E[m]$ is given by

$$\det : E[m] \longrightarrow \mathbb{Z}/m\mathbb{Z}, \qquad \det(aT_1 + bT_2, cT_1 + dT_2) = ad - bc,$$

where the value is, of course, independent of the choice of basis. However, a drawback of the determinant pairing on $E[m]$ is that it is not Galois invariant, i.e., if $P, Q \in E[m]$ and $\sigma \in G_{\bar{K}/K}$, then the values of $\det(P^\sigma, Q^\sigma)$ and $\det(P, Q)^\sigma$ need not be the same.

We can achieve Galois invariance by using instead a modified pairing of the form $\zeta^{\det(P,Q)}$, where ζ is a primitive m^{th} root of unity. In order to define this pairing intrinsically, we will make frequent use of (III.3.5), which says that a divisor $\sum n_i(P_i)$ is the divisor of a function if and only if both $\sum n_i = 0$ and $\sum [n_i] P_i = O$.

Let $T \in E[m]$. Then there is a function $f \in \bar{K}(E)$ satisfying

$$\text{div}(f) = m(T) - m(O).$$

Next take $T' \in E$ to be a point with $[m]T' = T$. Then there is similarly a function $g \in \bar{K}(E)$ satisfying

$$\text{div}(g) = [m]^*(T) - [m]^*(O) = \sum_{R \in E[m]} (T' + R) - (R).$$

(To see that this divisor sums to O, we observe that $\#E[m] = m^2$ from (III.6.4b) and that $[m^2]T' = O$.) It is easy to verify that the functions $f \circ [m]$ and g^m have the same divisor, so multiplying f by an appropriate constant from $\bar{K}^*$, we may assume that

$$f \circ [m] = g^m.$$

Now let $S \in E[m]$ be another m-torsion point, where we allow $S = T$. Then for any point $X \in E$, we have

$$g(X + S)^m = f([m]X + [m]S) = f([m]X) = g(X)^m.$$

Thus considered as a function of X, the function $g(X + S)/g(X)$ takes on only finitely many values, i.e., for every X, it is an mth root of unity. In particular, the morphism

$$E \longrightarrow \mathbb{P}^1, \qquad S \longmapsto g(X + S)/g(X)$$

is not surjective, so (II.2.3) says that it is constant. This allows us to define a pairing

$$e_m : E[m] \times E[m] \longrightarrow \boldsymbol{\mu}_m$$

by setting

$$e_m(S, T) = \frac{g(X + S)}{g(X)},$$

where $X \in E$ is any point such that $g(X + S)$ and $g(X)$ are both defined and nonzero. (As usual, μ_m denotes the group of mth roots of unity.) Note that although the function g is well-defined only up to multiplication by an element of $\bar{K}^*$, the value of $e_m(S, T)$ does not depend on this choice. The pairing that we have just defined is called the *Weil e_m-pairing*. We begin by proving some of its basic properties.

Proposition 8.1. *The Weil e_m-pairing has the following properties:*
  (a) *It is* bilinear:

$$e_m(S_1 + S_2, T) = e_m(S_1, T)e_m(S_2, T),$$
$$e_m(S, T_1 + T_2) = e_m(S, T_1)e_m(S, T_2).$$

  (b) *It is* alternating:

$$e_m(T, T) = 1.$$

   So in particular, $e_m(S, T) = e_m(T, S)^{-1}$.
  (c) *It is* nondegenerate:

$$\text{If } e_m(S, T) = 1 \text{ for all } S \in E[m], \text{ then } T = O.$$

  (d) *It is* Galois invariant:

$$e_m(S, T)^\sigma = e_m(S^\sigma, T^\sigma) \qquad \text{for all } \sigma \in G_{\bar{K}/K}.$$

  (e) *It is* compatible:

$$e_{mm'}(S, T) = e_m([m']S, T) \qquad \text{for all } S \in E[mm'] \text{ and } T \in E[m].$$

PROOF. (a) Linearity in the first factor is easy:

$$e_m(S_1 + S_2, T) = \frac{g(X + S_1 + S_2)}{g(X)} = \frac{g(X + S_1 + S_2)}{g(X + S_1)} \frac{g(X + S_1)}{g(X)}$$
$$= e_m(S_2, T)e_m(S_1, T).$$

Note how useful it is that in computing $e_m(S_2, T) = g(Y + S_2)/g(Y)$, we may choose any value for Y, for example we may take $Y = X + S_1$.

In order to prove linearity in the second factor, let $f_1, f_2, f_3, g_1, g_2, g_3$ be the appropriate functions for the points T_1, T_2, and $T_3 = T_1 + T_2$. Choose a function $h \in \bar{K}(E)$ with divisor

$$\text{div}(h) = (T_1 + T_2) - (T_1) - (T_2) + (O).$$

Then

$$\text{div}\left(\frac{f_3}{f_1 f_2}\right) = m \, \text{div}(h),$$

so

$$f_3 = c f_1 f_2 h^m \qquad \text{for some } c \in \bar{K}^*.$$

We compose with the multiplication-by-m map, use the fact that $f_i \circ [m] = g_i^m$, and take m^{th} roots to obtain

$$g_3 = c' \cdot g_1 \cdot g_2 \cdot (h \circ [m]) \qquad \text{for some } c' \in \bar{K}^*.$$

This allows us to compute

$$
e_m(S, T_1 + T_2) = \frac{g_3(X + S)}{g_3(X)} = \frac{g_1(X + S)g_2(X + S)h([m]X + [m]S)}{g_1(X)g_2(X)h([m]X)}
$$
$$
= e_m(S, T_1)e_m(S, T_2), \qquad \text{since } [m]S = O.
$$

(b) From (a) we have

$$e_m(S + T, S + T) = e_m(S, S)e_m(S, T)e_m(T, S)e_m(T, T),$$

so it suffices to show that $e_m(T, T) = 1$ for all $T \in E[m]$. For any $P \in E$, recall that $\tau_P : E \to E$ denotes the translation-by-P map (III.4.7). We compute

$$
\operatorname{div}\left(\prod_{i=0}^{m-1} f \circ \tau_{[i]T}\right) = m \sum_{i=0}^{m-1} ([1 - i]T) - ([-i]T) = 0.
$$

It follows that

$$\prod_{i=0}^{m-1} f \circ \tau_{[i]T}$$

is constant, and if we choose some $T' \in E$ satisfying $[m]T' = T$, then

$$\prod_{i=0}^{m-1} g \circ \tau_{[i]T'}$$

is also constant, because its mth power is the above product of f's. Therefore the product of the g's takes on the same value at X and at $X + T'$,

$$\prod_{i=0}^{m-1} g(X + [i]T') = \prod_{i=0}^{m-1} g(X + [i + 1]T').$$

Canceling like terms from each side gives

$$g(X) = g(X + [m]T') = g(X + T),$$

and hence

$$e_m(T, T) = \frac{g(X + T)}{g(X)} = 1.$$

(c) If $e_m(S, T) = 1$ for all $S \in E[m]$, then $g(X + S) = g(X)$ for all $S \in E[m]$, so (III.4.10b) tells us that $g = h \circ [m]$ for some function $h \in \bar{K}(E)$. But then

$$\left(h \circ [m]\right)^m = g^m = f \circ [m],$$

which implies that $f = h^m$. Hence

$$m \operatorname{div}(h) = \operatorname{div}(f) = m(T) - m(O),$$

so

$$\operatorname{div}(h) = (T) - (O).$$

It follows from (III.3.3) that $T = O$.

(d) Let $\sigma \in G_{\bar{K}/K}$. If f and g are the functions for T as above, then clearly f^σ and g^σ are the corresponding functions for T^σ. Then

$$e_m(S^\sigma, T^\sigma) = \frac{g^\sigma(X^\sigma + S^\sigma)}{g^\sigma(X^\sigma)} = \left(\frac{g(X + S)}{g(X)}\right)^\sigma = e_m(S, T)^\sigma.$$

(e) Taking f and g as usual, we have

$$\operatorname{div}(f^{m'}) = mm'(T) - mm'(O)$$

and

$$\left(g \circ [m']\right)^{mm'} = \left(f \circ [mm']\right)^{m'}.$$

Then directly from the definition of $e_{mm'}$ and e_m, we compute

$$e_{mm'}(S, T) = \frac{g \circ [m'](X + S)}{g \circ [m'](X)} = \frac{g(Y + [m']S)}{g(Y)} = e_m([m']S, T). \qquad \square$$

The basic properties of the Weil pairing imply its surjectivity, as in the next result.

Corollary 8.1.1. *There exist points $S, T \in E[m]$ such that $e_m(S, T)$ is a primitive m^{th} root of unity. In particular, if $E[m] \subset E(K)$, then $\mu_m \subset K^*$.*

PROOF. The image of $e_m(S, T)$ as S and T range over $E[m]$ is a subgroup of μ_m, say equal to μ_d. It follows that

$$1 = e_m(S, T)^d = e_m([d]S, T) \qquad \text{for all } S, T \in E[m].$$

The nondegeneracy of the e_m-pairing implies that $[d]S = O$, and since S is arbitrary, it follows from (III.6.4) that $d = m$. Finally, if $E[m] \subset E(K)$, then the Galois invariance of the e_m-pairing implies that $e_m(S, T) \in K^*$ for all $S, T \in E[m]$. Hence $\mu_m \subset K^*$. $\qquad \square$

Recall from (III §6) that associated to any isogeny $\phi : E_1 \to E_2$ is a dual isogeny $\hat{\phi} : E_2 \to E_1$ going in the opposite direction. The next proposition says that ϕ and $\hat{\phi}$ are dual (or adjoint) with respect to the Weil pairing.

Proposition 8.2. *Let $\phi : E_1 \to E_2$ be an isogeny of elliptic curves. Then for all m-torsion points $S \in E_1[m]$ and $T \in E_2[m]$,*

$$e_m(S, \hat{\phi}(T)) = e_m(\phi(S), T).$$

PROOF. Let

$$\mathrm{div}(f) = m(T) - m(O) \quad \text{and} \quad f \circ [m] = g^m$$

be as usual. Then

$$e_m(\phi S, T) = \frac{g(X + \phi S)}{g(X)}.$$

Choose a function $h \in \bar{K}(E_1)$ satisfying

$$\phi^*((T)) - \phi^*((O)) = (\hat{\phi}T) - (O) + \mathrm{div}(h).$$

Such an h exists because ((III.6.1ab) tells us that $\hat{\phi}T$ is precisely the sum of the points of the divisor on the left-hand side of this equality. Now we observe that

$$\mathrm{div}\left(\frac{f \circ \phi}{h^m}\right) = \phi^* \mathrm{div}(f) - m\, \mathrm{div}(h) = m(\hat{\phi}T) - m(O)$$

and

$$\left(\frac{g \circ \phi}{h \circ [m]}\right)^m = \frac{f \circ [m] \circ \phi}{(h \circ [m])^m} = \left(\frac{f \circ \phi}{h^m}\right) \circ [m].$$

Then directly from the definition of the e_m-pairing we obtain

$$\begin{aligned}
e_m(S, \hat{\phi}T) &= \frac{(g \circ \phi/h \circ [m])(X + S)}{(g \circ \phi/h \circ [m])(X)} \\
&= \frac{g(\phi X + \phi S)}{g(\phi X)} \cdot \frac{h([m]X)}{h([m]X + [m]S)} \\
&= e_m(\phi S, T). \qquad \qquad \square
\end{aligned}$$

Let ℓ be a prime number different from $\mathrm{char}(K)$. We are going to combine the pairings

$$e_{\ell^n} : E[\ell^n] \times E[\ell^n] \longrightarrow \boldsymbol{\mu}_{\ell^n}$$

for $n = 1, 2, \ldots$ in order to create an ℓ-adic Weil pairing on the Tate module,

$$e : T_\ell(E) \times T_\ell(E) \longrightarrow T_\ell(\boldsymbol{\mu}).$$

Recall that the inverse limits for $T_\ell(E)$ and $T_\ell(\boldsymbol{\mu})$ are formed using the maps

$$E[\ell^{n+1}] \xrightarrow{[\ell]} E[\ell^n] \quad \text{and} \quad \boldsymbol{\mu}_{\ell^{n+1}} \xrightarrow{\zeta \mapsto \zeta^\ell} \boldsymbol{\mu}_{\ell^n}.$$

Thus in order to show that the e_{ℓ^n}-pairings are compatible with taking the inverse limits, we must show that

$$e_{\ell^{n+1}}(S,T)^\ell = e_{\ell^n}\big([\ell]S,[\ell]T\big) \qquad \text{for all } S,T \in E[\ell^{n+1}].$$

We use linearity (III.8.1a) to observe that

$$e_{\ell^{n+1}}(S,T)^\ell = e_{\ell^{n+1}}(S,[\ell]T),$$

and then the desired compatibility relation follows by applying (III.8.1e) to the points S and $[\ell]T$ with $m = \ell^n$ and $m' = \ell$. This proves that the pairing $e : T_\ell(E) \times T_\ell(E) \to T_\ell(\mu)$ is well-defined. Further, it inherits all of the properties described in (III.8.1) and (III.8.2), which completes the proof of the following result.

Proposition 8.3. *There exists a bilinear, alternating, nondegenerate, Galois invariant pairing*

$$e : T_\ell(E) \times T_\ell(E) \longrightarrow T_\ell(\mu).$$

Further, if $\phi : E_1 \to E_2$ is an isogeny, then ϕ and its dual $\hat{\phi}$ are adjoints for the pairing, i.e., $e(\phi S, T) = e(S, \hat{\phi} T)$.

Remark 8.4. More generally, if $\phi : E_1 \to E_2$ is any nonconstant isogeny, then there is a Weil pairing

$$e_\phi : \ker \phi \times \ker \hat{\phi} \longrightarrow \mu_m.$$

See Exercise 3.15.

Remark 8.5. There is an alternative definition of the Weil pairing $e_m(S,T)$ that works as follows. Choose arbitrary points $X, Y \in E$ and functions $f_S, f_T \in \bar{K}(E)$ satisfying

$$\mathrm{div}(f_S) = m(X+S) - m(X) \qquad \text{and} \qquad \mathrm{div}(f_T) = m(Y+T) - m(Y).$$

Then

$$e_m(S,T) = \frac{f_S(Y+T)}{f_S(Y)} \bigg/ \frac{f_T(X+S)}{f_T(X)}.$$

We leave to the reader to prove that this quantity is well-defined and equal to the Weil pairing; see Exercise 3.16.

Recall that we have a representation (III §7)

$$\mathrm{End}(E) \longrightarrow \mathrm{End}\big(T_\ell(E)\big), \qquad \phi \longmapsto \phi_\ell.$$

Choosing a $\mathbb{Z}_\ell$-basis for $T_\ell(E)$, we can write ϕ_ℓ as a 2×2 matrix, and in particular we can compute

$$\det(\phi_\ell) \in \mathbb{Z}_\ell \quad \text{and} \quad \mathrm{tr}(\phi_\ell) \in \mathbb{Z}_\ell.$$

Of course, the value of the determinant and trace do not depend on the choice of basis.

The next result, whose proof uses the Weil pairing, shows how the determinant and trace values may be employed to compute the degree of an isogeny. These formulas are applied in Chapter V to count the number of points on an elliptic curve defined over a finite field (V.2.3.1). If we view the Tate module as a homology group (III.7.3.1), then (III.8.6) says that the degree of an isogeny can be computed topologically via its action on $H_1(E, \mathbb{Z}_\ell)$.

Proposition 8.6. *Let $\phi \in \mathrm{End}(E)$, and let $\phi_\ell : T_\ell(E) \to T_\ell(E)$ be the map that ϕ induces on the Tate module of E. Then*

$$\det(\phi_\ell) = \deg(\phi) \quad and \quad \mathrm{tr}(\phi_\ell) = 1 + \deg(\phi) - \deg(1 - \phi).$$

In particular, $\det(\phi_\ell)$ and $\mathrm{tr}(\phi_\ell)$ are in $\mathbb{Z}$ and are independent of ℓ.

PROOF. Let $\{v_1, v_2\}$ be a $\mathbb{Z}_\ell$-basis for $T_\ell(E)$ and write

$$\phi_\ell(v_1) = av_1 + bv_2, \qquad \phi_\ell(v_2) = cv_1 + dv_2,$$

so the matrix of ϕ_ℓ relative to this basis is

$$\phi_\ell = \begin{pmatrix} a & b \\ c & d \end{pmatrix}.$$

Using properties of the Weil pairing (III.8.3), we compute

$$
\begin{aligned}
e(v_1, v_2)^{\deg \phi} &= e([\deg \phi]v_1, v_2) & & \text{bilinearity of } e, \\
&= e(\hat{\phi}_\ell \phi_\ell v_1, v_2) & & \text{(III.6.1a)}, \\
&= e(\phi_\ell v_1, \phi_\ell v_2) & & \text{(III.8.3) and (III.6.2f)}, \\
&= e(av_1 + cv_2, bv_1 + dv_2) & & \\
&= e(v_1, v_2)^{ad-bc} & & \text{since } e \text{ is bilinear and alternating}, \\
&= e(v_1, v_2)^{\det \phi_\ell}.
\end{aligned}
$$

Since e is nondegenerate, we conclude that $\deg \phi = \det \phi_\ell$. Finally, for any 2×2 matrix A, a trivial calculation yields

$$\mathrm{tr}(A) = 1 + \det(A) - \det(1 - A). \qquad \square$$

III.9   The Endomorphism Ring

Let E be an elliptic curve. In this section we characterize which rings may occur as the endomorphism ring of E. So far we have accumulated the following information:

(i) $\mathrm{End}(E)$ has characteristic 0, no zero divisors, and rank at most four as a $\mathbb{Z}$-module (III.4.2c), (III.7.5).

(ii) $\text{End}(E)$ possesses an anti-involution $\phi \mapsto \hat{\phi}$ (III.6.2bcf).

(iii) For $\phi \in \text{End}(E)$, the product $\phi\hat{\phi}$ is a non-negative integer, and further, $\phi\hat{\phi} = 0$ if and only if $\phi = 0$ (III.6.2a), (III.6.3).

It turns out that any ring satisfying (i)–(iii) is of a very special sort. After giving the relevant definitions, we describe the general classification of rings satisfying (i)–(iii). This may then be applied to the particular case of $\text{End}(E)$.

Definition. Let $\mathcal{K}$ be a (not necessarily commutative) $\mathbb{Q}$-algebra that is finitely generated over $\mathbb{Q}$. An *order* $\mathcal{R}$ of $\mathcal{K}$ is a subring of $\mathcal{K}$ that is finitely generated as a $\mathbb{Z}$-module and satisfies $\mathcal{R} \otimes \mathbb{Q} = \mathcal{K}$.

Example 9.1. Let $\mathcal{K}$ be an imaginary quadratic field and let $\mathcal{O}$ be its ring of integers. Then for each integer $f \geq 1$, the ring $\mathbb{Z} + f\mathcal{O}$ is an order of $\mathcal{K}$. In fact, these are all of the orders of $\mathcal{K}$; see Exercise 3.20.

Definition. A *quaternion algebra* is an algebra of the form

$$\mathcal{K} = \mathbb{Q} + \mathbb{Q}\alpha + \mathbb{Q}\beta + \mathbb{Q}\alpha\beta$$

whose multiplication satisfies

$$\alpha^2, \beta^2 \in \mathbb{Q}, \qquad \alpha^2 < 0, \qquad \beta^2 < 0, \qquad \beta\alpha = -\alpha\beta.$$

Remark 9.2. These quaternion algebras are more properly called *definite quaternion algebras over* $\mathbb{Q}$, but since these are the only quaternion algebras that we use in this book, we generally drop the "definite" appellation.

Theorem 9.3. *Let* $\mathcal{R}$ *be a ring of characteristic* 0 *having no zero divisors, and assume that* $\mathcal{R}$ *has the following properties*:
(i) $\mathcal{R}$ *has rank at most four as a* $\mathbb{Z}$-*module.*
(ii) $\mathcal{R}$ *has an anti-involution* $\alpha \mapsto \hat{\alpha}$ *satisfying*

$$\widehat{\alpha + \beta} = \hat{\alpha} + \hat{\beta}, \qquad \widehat{\alpha\beta} = \hat{\beta}\hat{\alpha}, \qquad \hat{\hat{\alpha}} = \alpha, \qquad \hat{a} = a \quad \text{for } a \in \mathbb{Z} \subset \mathcal{R}.$$

(iii) *For* $\alpha \in \mathcal{R}$, *the product* $\alpha\hat{\alpha}$ *is a nonnegative integer, and* $\alpha\hat{\alpha} = 0$ *if and only if* $\alpha = 0$.
Then $\mathcal{R}$ *is one of the following types of rings*:
(a) $\mathcal{R} \cong \mathbb{Z}$.
(b) $\mathcal{R}$ *is an order in an imaginary quadratic extension of* $\mathbb{Q}$.
(c) $\mathcal{R}$ *is an order in a quaternion algebra over* $\mathbb{Q}$.

PROOF. Let $\mathcal{K} = \mathcal{R} \otimes \mathbb{Q}$. Since $\mathcal{R}$ is finitely generated as a $\mathbb{Z}$-module, it suffices to prove that $\mathcal{K}$ is either $\mathbb{Q}$, an imaginary quadratic field, or a quaternion algebra. We extend the anti-involution to $\mathcal{K}$ and define a (reduced) *norm* and *trace* from $\mathcal{K}$ to $\mathbb{Q}$ by

$$\text{N}\alpha = \alpha\hat{\alpha} \qquad \text{and} \qquad \text{T}\alpha = \alpha + \hat{\alpha}.$$

We make several observations about the trace. First, since

$$T\alpha = 1 + N\alpha - N(\alpha - 1),$$

we see that $T\alpha \in \mathbb{Q}$. Second, the trace is $\mathbb{Q}$-linear, since the involution fixes $\mathbb{Q}$. Third, if $\alpha \in \mathbb{Q}$, then $T\alpha = 2\alpha$. Finally, if $\alpha \in \mathcal{K}$ satisfies $T\alpha = 0$, then

$$0 = (\alpha - \alpha)(\alpha - \hat{\alpha}) = \alpha^2 - (T\alpha)\alpha + N\alpha = \alpha^2 + N\alpha,$$

so $\alpha^2 = -N\alpha$. Thus

$$\alpha \neq 0 \quad \text{and} \quad T\alpha = 0 \quad \Longrightarrow \quad \alpha^2 \in \mathbb{Q} \quad \text{and} \quad \alpha^2 < 0.$$

If $\mathcal{K} = \mathbb{Q}$, there is nothing to prove. Otherwise we can find some $\alpha \in \mathcal{K}$ with $\alpha \notin \mathbb{Q}$. Replacing α by $\alpha - \frac{1}{2}T\alpha$, we may assume that $T\alpha = 0$. Then $\alpha^2 \in \mathbb{Q}$ and $\alpha^2 < 0$, so $\mathbb{Q}(\alpha)$ is a quadratic imaginary field. If $\mathcal{K} = \mathbb{Q}(\alpha)$, we are again done.

Suppose now that $\mathcal{K} \neq \mathbb{Q}(\alpha)$ and choose some $\beta \in \mathcal{K}$ with $\beta \notin \mathbb{Q}(\alpha)$. We may replace β with

$$\beta - \frac{1}{2}T\beta - \frac{T(\alpha\beta)}{2\alpha^2}\alpha.$$

We know that $T\alpha = 0$ and $\alpha^2 \in \mathbb{Q}^*$, so an easy calculation shows that

$$T\beta = T(\alpha\beta) = 0.$$

In particular, $\beta^2 \in \mathbb{Q}$ and $\beta^2 < 0$. We next write

$$T\alpha = 0, \quad T\beta = 0, \quad T(\alpha\beta) = 0$$

as

$$\alpha = -\hat{\alpha}, \quad \beta = -\hat{\beta}, \quad \alpha\beta = -\hat{\beta}\hat{\alpha}$$

and substitute the first two equalities into the third to obtain

$$\alpha\beta = -\beta\alpha.$$

Hence

$$\mathbb{Q}[\alpha, \beta] = \mathbb{Q} + \mathbb{Q}\alpha + \mathbb{Q}\beta + \mathbb{Q}\alpha\beta$$

is a quaternion algebra. It remains to prove that $\mathbb{Q}[\alpha, \beta] = \mathcal{K}$, and to do this, it suffices to show that $1, \alpha, \beta, \alpha\beta$ are $\mathbb{Q}$-linearly independent, since then $\mathbb{Q}[\alpha, \beta]$ and $\mathcal{K}$ both have dimension 4 over $\mathbb{Q}$.

Suppose that

$$w + x\alpha + y\beta + z\alpha\beta = 0 \qquad \text{with } w, x, y, z \in \mathbb{Q}.$$

Taking the trace yields

$$2w = 0, \quad \text{so} \quad w = 0.$$

Next we multiply by α on the left and by β on the right to obtain

$$(x\alpha^2)\beta + (y\beta^2)\alpha + z\alpha^2\beta^2 = 0.$$

We know that 1, α, and β are $\mathbb{Q}$-linearly independent, since $\alpha \notin \mathbb{Q}$ and $\beta \notin \mathbb{Q}(\alpha)$. Hence this equation implies that

$$x\alpha^2 = y\beta^2 = z\alpha^2\beta^2 = 0,$$

and so $x = y = z = 0$, which completes the proof that 1, α, β, and $\alpha\beta$ are $\mathbb{Q}$-linearly independent. (We have used several times the fact that α^2 and β^2 are in $\mathbb{Q}^*$.) $\qquad\square$

Corollary 9.4. *The endomorphism ring of an elliptic curve E/K is either $\mathbb{Z}$, an order in an imaginary quadratic field, or an order in a quaternion algebra. If* $\operatorname{char}(K) = 0$, *then only the first two are possible.*

PROOF. We have proven in (III.4.2b), (III.6.2), and (III.6.3) all of the facts needed to apply (III.9.3) to the ring $\operatorname{End}(E)$. This proves the first part of the corollary. If $\operatorname{char}(K) = 0$, then (III.5.6c) says that $\operatorname{End}(E)$ is commutative, so in this case $\operatorname{End}(E)$ cannot be an order in a quaternion algebra. (See also Exercise 3.33 for a proof of this corollary that does not require knowing a priori that $\operatorname{End}(E)$ has rank at most four.) $\qquad\square$

Remark 9.4.1. If $\operatorname{char}(K) = 0$, then (III.5.6c) tells us that $\operatorname{End}(E) \otimes \mathbb{Q}$ is commutative, so it cannot be a quaternion algebra. (For alternative proofs of this important fact, see (VI.6.1b) and Exercise 3.18b.) On the other hand, if K is a finite field $\mathbb{F}_q$, then we will later see that $\operatorname{End}(E)$ is always larger than $\mathbb{Z}$ (V.3.1) and that there are always elliptic curves defined over $\mathbb{F}_{p^2}$ with $\operatorname{End}(E) \otimes \mathbb{Q}$ a quaternion algebra (V.4.1c). The complete description of $\operatorname{End}(E)$ is given in Deuring's comprehensive article [60].

The next definition and theorem are used in the exercises.

Definition. Let p be a prime or ∞, let $\mathbb{Q}_p$ be the p-adic rationals if p is finite, and let $\mathbb{Q}_\infty = \mathbb{R}$. A quaternion algebra $\mathcal{K}$ is said to *split at p* if

$$\mathcal{K} \otimes_\mathbb{Q} \mathbb{Q}_p \cong M_2(\mathbb{Q}_p),$$

where $M_2(K)$ is the algebra of 2×2 matrices with coefficients in K. Otherwise $\mathcal{K}$ is said to be *ramified at p*. The *invariant of $\mathcal{K}$ at p* is defined by

$$\operatorname{inv}_p \mathcal{K} = \begin{cases} 0 & \text{if } \mathcal{K} \text{ splits at } p, \\ \frac{1}{2} & \text{if } \mathcal{K} \text{ ramifies at } p. \end{cases}$$

Theorem 9.5. *Let $\mathcal{K}$ be a quaternion algebra.*
(a) *We have* $\operatorname{inv}_p(\mathcal{K}) = 0$ *for all but finitely many p, and*

$$\sum_p \operatorname{inv}_p(\mathcal{K}) \in \mathbb{Z}.$$

(Note that the sum includes $p = \infty$.)

(b) *Two quaternion algebras $\mathcal{K}$ and $\mathcal{K}'$ are isomorphic as $\mathbb{Q}$-algebras if and only if $\mathrm{inv}_p(\mathcal{K}) = \mathrm{inv}_p(\mathcal{K}')$ for all p.*

PROOF. This is a very special case of the fact that the central simple algebras over a field K are classified by the Brauer group $\mathrm{Br}(K) = H^2(G_{\bar{K}/K}, \bar{K}^*)$ [233, X §5], and the fundamental exact sequence from class field theory [288, §9.6]

$$0 \longrightarrow \mathrm{Br}(\mathbb{Q}) \longrightarrow \bigoplus_p \mathrm{Br}(\mathbb{Q}_p) \xrightarrow{\ \sum_p \mathrm{inv}_p\ } \frac{\mathbb{Q}}{\mathbb{Z}} \longrightarrow 0,$$

where

$$\mathrm{Br}(\mathbb{Q}_p) \xrightarrow[\mathrm{inv}_p]{\ \sim\ } \begin{cases} \mathbb{Q}/\mathbb{Z} & \text{if } p \neq \infty, \\ \{0, \tfrac{1}{2}\} & \text{if } p = \infty. \end{cases}$$

Quaternion algebras (definite and indefinite) correspond to elements of order 2 in $\mathrm{Br}(\mathbb{Q})$. $\qquad\square$

III.10 The Automorphism Group

If an elliptic curve is given by a Weierstrass equation, it is generally a nontrivial matter to determine the exact structure of its endomorphism ring. The situation is much simpler for the automorphism group.

Theorem 10.1. *Let E/K be an elliptic curve. Then its automorphism group $\mathrm{Aut}(E)$ is a finite group of order dividing 24. More precisely, the order of $\mathrm{Aut}(E)$ is given by the following table:*

$\#\,\mathrm{Aut}(E)$	$j(E)$	$\mathrm{char}(K)$
2	$j(E) \neq 0, 1728$	—
4	$j(E) = 1728$	$\mathrm{char}(K) \neq 2, 3$
6	$j(E) = 0$	$\mathrm{char}(K) \neq 2, 3$
12	$j(E) = 0 = 1728$	$\mathrm{char}(K) = 3$
24	$j(E) = 0 = 1728$	$\mathrm{char}(K) = 2$

PROOF. We restrict attention to $\mathrm{char}(K) \neq 2, 3$; see (III.1.3) and (A.1.2c). Then E is given by an equation

$$E : y^2 = x^3 + Ax + B,$$

and every automorphism of E has the form

$$x = u^2 x', \qquad y = u^3 y',$$

for some $u \in \bar{K}^*$. Such a substitution gives an automorphism of E if and only if

$$u^{-4}A = A \qquad \text{and} \qquad u^{-6}B = B.$$

If $AB \neq 0$, i.e., if $j(E) \neq 0, 1728$, then the only possibilities are $u = \pm 1$. Similarly, if $B = 0$, then $j(E) = 1728$ and $u^4 = 1$, and if $A = 0$, then $j(E) = 0$ and $u^6 = 1$. Hence $\mathrm{Aut}(E)$ is cyclic of order 2, 4, or 6, depending on whether $AB \neq 0$, $B = 0$, or $A = 0$. $\qquad\square$

It is worth remarking that the proof of (III.10.1) gives the structure of $\mathrm{Aut}(E)$ as a $G_{\bar{K}/K}$-module, at least for $\mathrm{char}(K) \neq 2, 3$. We record this as a corollary.

Corollary 10.2. *Let E/K be a curve over a field of characteristic not equal to 2 or 3, and let*

$$n = \begin{cases} 2 & \text{if } j(E) \neq 0, 1728, \\ 4 & \text{if } j(E) = 1728, \\ 6 & \text{if } j(E) = 0. \end{cases}$$

Then there is a natural isomorphism of $G_{\bar{K}/K}$-modules

$$\mathrm{Aut}(E) \cong \boldsymbol{\mu}_n.$$

PROOF. While proving (III.10.1), we showed that the map

$$[\ ] : \boldsymbol{\mu}_n \longrightarrow E, \qquad [\zeta](x, y) = (\zeta^2 x, \zeta^3 y),$$

is an isomorphism of abstract groups. It is clear that this map commutes with the action of $G_{\bar{K}/K}$, and hence it is an isomorphism of $G_{\bar{K}/K}$-modules. $\qquad\square$

Exercises

3.1. Show that the polynomials

$$x^4 - b_4 x^2 - 2b_6 x - b_8 \qquad \text{and} \qquad 4x^3 + b_2 x^2 + 2b_4 x + b_6$$

appearing in the duplication formula (III.2.3d) are relatively prime if and only if the discriminant of the associated Weierstrass equation is nonzero.

3.2. (a) Derive a *triplication formula*, analogous to to the duplication formula (III.2.3), i.e., express $x([3]P)$ as a rational function of $x(P)$ and $a_1, \ldots, a_6$.
   (b) Use the result from (a) to show that if $\mathrm{char}(K) \neq 3$, then E has a nontrivial point of order 3. Conclude that if $\gcd(m, 3) = 1$, then $[m] \neq [0]$. (*Warning.* You'll probably want to use a computer algebra package for this problem.)

3.3. Assume that $\mathrm{char}(K) \neq 3$ and let $A \in K^*$. Then Exercise 2.7 tells us that the curve

$$E : X^3 + Y^3 = AZ^3$$

is a curve of genus one, so together with the point $O = [1, -1, 0]$, it is an elliptic curve.
   (a) Prove that three points on E add to O if and only if they are collinear.
   (b) Let $P = [X, Y, Z] \in E$. Prove the formulas

$$-P = [Y, X, Z],$$
$$[2]P = \left[-Y(X^3 + AZ^3), X(Y^3 + AZ^3), X^3 Z - Y^3 Z \right].$$

   (c) Develop an analogous formula for the sum of two distinct points.
   (d) Prove that E has j-invariant 0.

3.4. Referring to (III.2.4), express each of the points $P_2, P_4, P_5, P_6, P_7, P_8$ in the form $[m]P_1 + [n]P_3$ with $m, n \in \mathbb{Z}$.

3.5. Let E/K be given by a singular Weierstrass equation.
  (a) Suppose that E has a node, and let the tangent lines at the node be

$$y = \alpha_1 x + \beta_1 \quad \text{and} \quad y = \alpha_2 x + \beta_2.$$

  (i) If $\alpha_1 \in K$, prove that $\alpha_2 \in K$ and

$$E_{\mathrm{ns}}(K) \cong K^*.$$

  (ii) If $\alpha_1 \notin K$, prove that $L = K(\alpha_1, \alpha_2)$ is a quadratic extension of K. Note that (i) tells us that $E_{\mathrm{ns}}(K) \subset E_{\mathrm{ns}}(L) \cong L^*$. Prove that

$$E_{\mathrm{ns}}(K) \cong \big\{ t \in L^* : \mathrm{N}_{L/K}(t) = 1 \big\}.$$

  (b) Suppose that E has a cusp. Prove that

$$E_{\mathrm{ns}}(K) \cong K^+.$$

3.6. Let C be a smooth curve of genus g, let $P_0 \in C$, and let $n \geq 2g + 1$ be an integer. Choose a basis $\{f_0, \ldots, f_m\}$ for $\mathcal{L}\big(n(P_0)\big)$ and define a map

$$\phi : [f_0, \ldots, f_m] : C \longrightarrow \mathbb{P}^m.$$

  (a) Prove that the image $C' = \phi(C)$ is a curve in $\mathbb{P}^m$.
  (b) Prove that the map $\phi : C \longrightarrow C'$ has degree one.
  (c) * Prove that C' is smooth and that $\phi : C \longrightarrow C'$ is an isomorphism.

3.7. This exercise gives an elementary, highly computational, proof that the multiplication-by-m map has degree m^2. Let E be given be the Weierstrass equation

$$E : y^2 + a_1 xy + a_3 y = x^3 + a_2 x^2 + a_4 x + a_6,$$

and let b_2, b_4, b_6, b_8 be the usual quantities. (If you're content to work with $\mathrm{char}(K) \neq 2, 3$, you may find it easier to use the short Weierstrass form $E : y^2 = x^3 + Ax + B$.)

  We define *division polynomials* $\psi_m \in \mathbb{Z}[a_1, \ldots, a_6, x, y]$ using initial values

$\psi_1 = 1,$

$\psi_2 = 2y + a_1 x + a_3,$

$\psi_3 = 3x^4 + b_2 x^3 + 3b_4 x^2 + 3b_6 x + b_8,$

$\psi_4 = \psi_2 \cdot \big(2x^6 + b_2 x^5 + 5b_4 x^4 + 10b_6 x^3 + 10b_8 x^2 + (b_2 b_8 - b_4 b_6)x + (b_4 b_8 - b_6^2) \big),$

and then inductively by the formulas

$$\psi_{2m+1} = \psi_{m+2}\psi_m^3 - \psi_{m-1}\psi_{m+1}^3 \qquad \text{for } m \geq 2,$$

$$\psi_2 \psi_{2m} = \psi_{m-1}^2 \psi_m \psi_{m+2} - \psi_{m-2}\psi_m \psi_{m+1}^2 \quad \text{for } m \geq 3.$$

Verify that ψ_{2m} is a polynomial for all $m \geq 1$, and then define further polynomials ϕ_m and ω_m by

$$\phi_m = x\psi_m^2 - \psi_{m+1}\psi_{m-1},$$

$$4y\omega_m = \psi_{m-1}^2 \psi_{m+2} + \psi_{m-2}\psi_{m+1}^2.$$

(a) Prove that if m is odd, then ψ_m, ϕ_m, and $y^{-1}\omega_m$ are polynomials in

$$\mathbb{Z}\big[a_1, \ldots, a_6, x, (2y + a_1 x + a_3)^2\big],$$

and similarly for $(2y)^{-1}\psi_m$, ϕ_m, and ω_m if m is even. So replacing $(2y + a_1 x + a_3)^2$ by $4x^3 + b_2 x^2 + 2b_4 x + b_6$, we may treat each of these quantities as a polynomial in $\mathbb{Z}[a_1, \ldots, a_6, x]$.

(b) As polynomials in x, show that

$$\phi_m(x) = x^{m^2} + \text{(lower order terms)},$$
$$\psi_m(x)^2 = m^2 x^{m^2 - 1} + \text{(lower order terms)}.$$

(c) If $\Delta \neq 0$, prove that $\phi_m(x)$ and $\psi_m(x)^2$ are relatively prime polynomials in $K[x]$.

(d) Continuing with the assumption that $\Delta \neq 0$, so E is an elliptic curve, prove that for any point $P = (x_0, y_0) \in E$ we have

$$[m]P = \left(\frac{\phi_m(P)}{\psi_m(P)^2}, \frac{\omega_m(P)}{\psi_m(P)^3} \right).$$

(e) Prove that the map $[m] : E \to E$ has degree m^2.

(f) Prove that the function $\psi_n \in K(E)$ has divisor

$$\operatorname{div}(\psi_n) = \sum_{T \in E[n]} (T) - n^2(O).$$

Thus ψ_n vanishes at precisely the nontrivial n-torsion points and has a corresponding pole at O.

(g) Prove that

$$\psi_{n+m}\psi_{n-m}\psi_r^2 = \psi_{n+r}\psi_{n-r}\psi_m^2 - \psi_{m+r}\psi_{m-r}\psi_n^2 \qquad \text{for all } n > m > r.$$

3.8. (a) Let $E/\mathbb{C}$ be an elliptic curve. We will prove later (VI.5.1.1) that there are a lattice $L \subset \mathbb{C}$ and a complex analytic isomorphism of groups $\mathbb{C}/L \cong E(\mathbb{C})$. (N.B. This isomorphism is given by convergent power series, not by rational functions.) Assuming this fact, prove that

$$\deg[m] = m^2 \qquad \text{and} \qquad E[m] = \frac{\mathbb{Z}}{m\mathbb{Z}} \times \frac{\mathbb{Z}}{m\mathbb{Z}}.$$

(b) Let K be a field with $\operatorname{char}(K) = 0$ and let E/K be an elliptic curve. Use (a) to prove that $\deg[m] = m^2$. (*Hint.* If K can be embedded into $\mathbb{C}$, then the result follows immediately from (a). Reduce to this case.)

3.9. Let E/K be an elliptic curve over a field K with $\operatorname{char}(K) \neq 2, 3$, and fix a a homogeneous Weierstrass equation for E,

$$F(X_0, X_1, X_2) = X_1^2 X_2 - X_0^3 - A X_0 X_2^2 - B X_2^3 = 0,$$

i.e., $x = X_0/X_2$ and $y = X_1/X_2$ are affine Weierstrass coordinates. Let $P \in E$.

(a) Prove that $[3]P = O$ if and only if the tangent line to E at P intersects E only at P.

(b) Prove that $[3]P = O$ if and only if the Hessian matrix

$$\left(\frac{\partial^2 F}{\partial X_i X_j}(P)\right)_{0 \le i, j \le 2}$$

has determinant 0.

(c) Prove that $E[3]$ consists of nine points.

3.10. Let E/K be an elliptic curve with Weierstrass coordinate functions x and y.
(a) Show that the map

$$\phi : E \longrightarrow \mathbb{P}^2, \qquad f = [1, x, y, x^2],$$

maps E isomorphically onto the intersection of two quadric surfaces in $\mathbb{P}^3$. (A quadric surface in $\mathbb{P}^3$ is the zero set of a homogeneous polynomial of degree two.) In particular, if $H \subset \mathbb{P}^3$ is a hyperplane, then $H \cap \phi(E)$ consists of exactly four points, counted with appropriate multiplicities.

(b) Show that $\phi(O) = [0, 0, 0, 1]$, and that the hyperplane $\{T_0 = 0\}$ intersects $\phi(E)$ at the single point $\phi(O)$ with multiplicity 4.

(c) Let $P, Q, R, S \in E$. Prove that $P + Q + R + S = O$ if and only if the four points $\phi(P), \phi(Q), \phi(R), \phi(S)$ are coplanar, i.e., if and only if there is a plane $H \subset \mathbb{P}^3$ such that the intersection $E \cap H$, counted with appropriate multiplicities, consists of the points $\phi(P), \phi(Q), \phi(R), \phi(S)$.

(d) Let $P \in E$. Prove that $[4]P = O$ if and only if there exists a hyperplane $H \subset \mathbb{P}^3$ satisfying $H \cap \phi(E) = \{P\}$. If $\operatorname{char}(K) \ne 2$, prove that there are exactly 16 such hyperplanes, and hence that $\#E[4] = 16$.

(e) Continuing with the assumption that $\operatorname{char}(K) \ne 2$, prove that there is a $\bar{K}$-linear change of coordinates such that $\phi(E)$ is given by equations of the form

$$T_0^2 + T_2^2 = T_0 T_3 \qquad \text{and} \qquad T_1^2 + \alpha T_2^2 = T_2 T_3.$$

For what value(s) of α do these equations define a nonsingular curve?

(f) Using the model in (e) and the addition law described in (c), find formulas for $-P$, for $P_1 + P_2$, and for $[2]P$, analogous to the formulas given in (III.2.3).

(g) What is the j-invariant of the elliptic curve described in (e)?

3.11. Generalize Exercise 3.10 as follows. Let E/K be an elliptic curve and choose a basis $f_1, \ldots, f_m$ for $\mathcal{L}(m(O))$. For $m \ge 3$, it follows from Exercise 3.6 that the map

$$\phi : E \longrightarrow \mathbb{P}^{m-1}, \qquad \phi = [f_1, \ldots, f_m],$$

is an isomorphism of E onto its image.
(a) Show that $\phi(E)$ is a curve of degree m, i.e., prove that the intersection of $\phi(E)$ and a hyperplane consists of m points, counted with appropriate multiplicities. (*Hint.* Find a hyperplane that intersects $\phi(E)$ at the single point $\phi(O)$ and show that it intersects with multiplicity m.)

(b) Let $P_1, \ldots, P_m \in E$. Prove that $P_1 + \cdots + P_m = O$ if and only if the points $\phi(P_1), \phi(P_2), \ldots, \phi(P_m)$ lie on a hyperplane. (Note that if some of the P_i's coincide, then the hyperplane is required to intersect $\phi(E)$ with correspondingly higher multiplicities at such points.)

(c) * Let $P \in E$. Prove that $[m]P = O$ if and only if there is a hyperplane $H \subset \mathbb{P}^{m-1}$ satisfying $H \cap \phi(E) = \{P\}$. If $\text{char}(K) = 0$ or $\text{char}(K) > m$, prove that there are exactly m^2 such points. Use this to deduce that $\deg[m] = m^2$.

3.12. Let $m \geq 2$ be an integer, prime to $\text{char}(K)$ if $\text{char}(K) > 0$. Prove that the natural map

$$\text{Aut}(E) \longrightarrow \text{Aut}\big(E[m]\big)$$

is injective except for $m = 2$, where the kernel is $[\pm 1]$. (You should be able to prove this directly, without using (III.10.1).)

3.13. Generalize (III.4.12) as follows. Let $C/\bar{K}$ be a smooth curve, and let Φ be a finite group of isomorphisms from C to itself. (For example, if E is an elliptic curve, then Φ might contain some translations by torsion points and $[\pm 1]$.) We observe that an element $\alpha \in \Phi$ acts on $\bar{K}(C)$ via the map

$$\alpha^* : \bar{K}(C) \longrightarrow \bar{K}(C), \qquad \alpha^*(f) = f \circ \alpha.$$

(a) Prove that there exist a unique smooth curve $C'/\bar{K}$ and a finite separable morphism $\phi : C \to C'$ such that $\phi^* \bar{K}(C') = \bar{K}(C)^\Phi$, where $\bar{K}(C)^\Phi$ denotes the subfield of $\bar{K}(C)$ fixed by every element of Φ.
(b) Let $P \in C$. Prove that

$$e_\phi(P) = \#\{\alpha \in \Phi : \alpha P = P\}.$$

(c) Prove that ϕ is unramified if and only if every nontrivial element of Φ has no fixed points.
(d) Express the genus of C' in terms of the genus of C, the number of elements in Φ, and the number of fixed points of elements of Φ.
(e) * Suppose that C is defined over K and that Φ is $G_{\bar{K}/K}$-invariant. The latter condition means that for all $\alpha \in \Phi$ and all $\sigma \in G_{\bar{K}/K}$ we have $\alpha^\sigma \in \Phi$. Prove that it is possible to find C' and ϕ as in (a) such that C' and ϕ are defined over K. Prove further that C' is unique up to isomorphism over K.

3.14. Prove directly that the natural map

$$\text{Hom}(E_1, E_2) \longrightarrow \text{Hom}\big(T_\ell(E_1), T_\ell(E_2)\big)$$

is injective. (*Hint.* If $\phi : E_1 \to E_2$ satisfies $\phi_\ell = 0$, then $E_1[\ell^n] \subset \ker \phi$ for all $n \geq 1$.) Note that this result is not as strong as (III.7.4).

3.15. Let E_1/K and E_2/K be elliptic curves, and let $\phi : E_1 \to E_2$ be an isogeny of degree m defined over K, where m is prime to $\text{char}(K)$ if $\text{char}(K) > 0$.
(a) Mimic the construction in (III §8) to construct a pairing

$$e_\phi : \ker \phi \times \ker \hat{\phi} \longrightarrow \mu_m.$$

(b) Prove that e_ϕ is bilinear, nondegenerate, and Galois invariant.
(c) Prove that e_ϕ is compatible in the sense that if $\psi : E_2 \to E_3$ is another isogeny, then

$$e_{\psi \circ \phi}(P, Q) = e_\psi(\phi P, Q) \qquad \text{for all } P \in \ker(\psi \circ \phi) \text{ and } Q \in \ker(\hat{\psi}).$$

3.16. *Alternative Definition of the Weil Pairing.* Let E be an elliptic curve. We define a pairing

$$\tilde{e}_m : E[m] \times E[m] \longrightarrow \mu_m$$

as follows: Let $P, Q \in E[m]$ and choose divisors D_P and D_Q in $\mathrm{Div}^0(E)$ that add to P and Q, respectively, i.e., such that $\sigma(D_P) = P$ and $\sigma(D_Q) = Q$, where σ is as in (III.3.4a). Assume further that D_P and D_Q are chosen with disjoint supports. Since P and Q have order m, there are functions $f_P, f_Q \in \bar{K}(E)$ satisfying

$$\mathrm{div}(f_P) = m D_P \qquad \text{and} \qquad \mathrm{div}(f_Q) = m D_Q.$$

We define

$$\tilde{e}_m = \frac{f_P(D_Q)}{f_Q(D_P)}.$$

(See Exercise 2.10 for the definition of the value of a function at a divisor.)
  (a) Prove that $\tilde{e}_m(P, Q)$ is well-defined, i.e., its value depends only on P and Q, independent of the various choices of D_P, D_Q, f_P, and f_Q. (*Hint.* Use Weil reciprocity, Exercise 2.11.)
  (b) Prove that $\tilde{e}_m(P, Q) \in \mu_m$.
  (c) * Prove that $\tilde{e}_m = e_m$, where e_m is the Weil pairing defined in (III §8).

3.17. Let $\mathcal{K}$ be a definite quaternion algebra. Prove that $\mathcal{K}$ is ramified at ∞. (*Hint.* The ring $M_2(\mathbb{R})$ contains zero divisors.)

3.18. Let E/K be an elliptic curve and suppose that $\mathcal{K} = \mathrm{End}(E) \otimes \mathbb{Q}$ is a quaternion algebra.
  (a) Prove that if $p \neq \infty$ and $p \neq \mathrm{char}(K)$, then $\mathcal{K}$ splits at p. (*Hint.* Use (III.7.4).)
  (b) Deduce that $\mathrm{char}(K) > 0$. (This gives an alternative proof of (III.5.6c).)
  (c) Prove that $\mathcal{K}$ is the unique quaternion algebra that is ramified at ∞ and $\mathrm{char}(K)$ and nowhere else.
  (d) * Prove that $\mathrm{End}(E)$ is a maximal order in $\mathcal{K}$. (Note that unlike number fields, a quaternion algebra may have more than one maximal order.)

3.19. Let $\mathcal{K}$ be a quaternion algebra.
  (a) Prove that $\mathcal{K} \otimes \bar{\mathbb{Q}} \cong M_2(\bar{\mathbb{Q}})$.
  (b) Prove that $\mathcal{K} \otimes \mathcal{K} \cong M_4(\mathbb{Q})$. This shows that $\mathcal{K}$ corresponds to an element of order 2 in the Brauer group $\mathrm{Br}(\mathbb{Q})$. (*Hint.* First show that $\mathcal{K} \otimes \mathcal{K}$ is simple, i.e., has no two-sided ideals. Then prove that the map

$$\mathcal{K} \otimes \mathcal{K} \longrightarrow \mathrm{End}(\mathcal{K}), \qquad a \otimes b \longmapsto (x \mapsto axb),$$

is an isomorphism.)

3.20. Let $\mathcal{K}$ be an imaginary quadratic field with ring of integers $\mathcal{O}$. Prove that the orders of $\mathcal{K}$ are precisely the rings $\mathbb{Z} + f\mathcal{O}$ for integers $f > 0$. The integer f is called the *conductor* of the order.

3.21. Let $C/\bar{K}$ be a curve of genus one. For any point $O \in C$, we can associate to the elliptic curve (C, O) its j-invariant $j(C, O)$. This exercise asks you to prove that the value of $j(C, O)$ is independent of the choice of the base point O. Thus we can assign a *j-invariant* to any curve C of genus one.

(a) Let (C, O) and (C', O') be curves of genus one with associated base points, and suppose that there is an isomorphism of curves $\phi : C \to C'$ satisfying $\phi(O) = O'$. Prove that $j(C, O) = j(C', O')$. (*Hint.* The j-invariant, which is defined in terms of the coefficients of a Weierstrass equation, is independent of the choice of the equation.)

(b) Prove that given any two points $O, O' \in C$, there is an automorphism of C taking O to O'.

(c) Use (a) and (b) to conclude that $j(C, O) = j(C, O')$.

3.22. Let C be a curve of genus one defined over K.

(a) Prove that $j(C) \in K$.

(b) Prove that C is an elliptic curve over K if and only if $C(K) \neq \emptyset$.

(c) Prove that C is always isomorphic, over $\bar{K}$, to an elliptic curve defined over K.

3.23. *Deuring Normal Form.* The following normal form for a Weierstrass equation is sometimes useful when dealing with elliptic curves over (algebraically closed) fields of arbitrary characteristic.

(a) Let E/K be an elliptic curve, and assume that either $\text{char}(K) \neq 3$ or $j(E) \neq 0$. Prove that E has a Weierstrass equation over $\bar{K}$ of the form

$$E : y^2 + \alpha xy + y = x^3 \qquad \text{with } \alpha \in \bar{K}.$$

(b) For the Weierstrass equation in (a), prove that $(0, 0) \in E[3]$.

(c) For what value(s) of α is the Weierstrass equation in (a) singular?

(d) Verify that

$$j(E) = \frac{\alpha^3 (\alpha^3 - 24)^2}{\alpha^3 - 27}.$$

3.24. Let E/K be an elliptic curve with complex multiplication over K, i.e., such that $\text{End}_K(E)$ is strictly larger than $\mathbb{Z}$. Prove that for all primes $\ell \neq \text{char}(K)$, the action of $G_{\bar{K}/K}$ on the Tate module $T_\ell(E)$ is abelian. (*Hint.* use the fact that the endomorphisms in $\text{End}_K(E)$ commute with the action of $G_{\bar{K}/K}$ on $T_\ell(E)$.)

3.25. Let E be an elliptic curve and let $P = (x, y) \in E$. As a supplement to the duplication formula (III.2.3d) for x, prove that the quantity $Y([2]P) = 2y([2]P) + a_1 x([2]P) + a_3$ is given by the formula

$$Y([2]P) = \frac{2x^6 + b_2 x^5 + 5b_4 x^4 + 10b_6 x^3 + 10b_8 x^2 + (b_2 b_8 - b_4 b_6)x + (b_4 b_8 - b_6^2)}{(2y + a_1 x + a_3)^3}.$$

3.26. Let E be the elliptic curve $y^2 = x^3 + x$ having complex multiplication by $\mathbb{Z}[i]$, let $m \geq 2$ be an integer, and let $T \in E[m]$ be a point of exact order m. In each of the following situations, prove that $\{T, [i]T\}$ is a basis for $E[m]$, and thus that $e_m(T, [i]T)$ is a primitive m^{th} root of unity.

(a) $m \equiv 3 \pmod 4$.

(b) m is prime, K is a field with $i \notin K$, and $E(K)[m]$ is nonzero.

The map ϕ is an example of a *distortion map*.

3.27. Let E/K be an elliptic curve and let $m \neq 0$ be an integer.

(a) Prove that $x \circ [m] \in K(x)$. In other words, prove that there is a rational function $F_m(x) \in K(x)$ satisfying $x([m]P) = F_m(x(P))$ for all $P \in E$.

(b) Prove that $F_m(F_n(x)) = F_{mn}(x)$.

(c) Compute $F_2(x)$ and $F_3(x)$ in terms of a given Weierstrass equation for E.

(d) A more intrinsic description of F_m is that it is the unique rational map $F_m : \mathbb{P}^1 \to \mathbb{P}^1$ fitting into the commutative diagram

$$
\begin{array}{ccccc}
E & \longrightarrow & E/\{\pm 1\} & \xrightarrow{\ x\ } & \mathbb{P}^1 \\
{\scriptstyle [m]}\big\downarrow & & & & \big\downarrow {\scriptstyle F_m} \\
E & \longrightarrow & E/\{\pm 1\} & \xrightarrow{\ x\ } & \mathbb{P}^1.
\end{array}
$$

Where is F_m ramified and what are the ramification indices at the ramification points?

(e) Find the fixed points of $F_m(x)$, i.e., the points $x \in \mathbb{P}^1(\bar{K})$ satisfying $F_m(x) = x$.

(f) For each fixed point $x \in \mathbb{P}^1(\bar{K})$ of $F_m(x)$, compute the value of the *multiplier* $F'_m(x)$. (*Hint.* The value should depend only on m, independent of the curve E.)

(g) A point $x \in \mathbb{P}^1(\bar{K})$ is called *preperiodic for F_m* if its forward orbit

$$
\big\{ x, F_m(x), F_m(F_m(x)), F_m(F_m(F_m(x))), \ldots \big\}
$$

is finite. Prove that the preperiodic points for F_m are exactly the points in $x\big(E(\bar{K})_{\text{tors}}\big)$. The rational map $F_m : \mathbb{P}^1 \to \mathbb{P}^1$ is an example of a *Lattès map*. Lattès maps are important in the theory of dynamical systems. In particular, Lattès proved that over $\mathbb{C}$, the map F_m is everywhere chaotic on $\mathbb{P}^1(\mathbb{C})$. For further information about elliptic curves and dynamical systems, see for example [14, §4.3], [179], or [267, §§1.6.3, 6.4–6.7].

3.28. Let $E \subset \mathbb{P}^2$ be a possibly singular curve given by a Weierstrass equation, and let $L \subset \mathbb{P}^2$ be a line.

(a) Prove directly from the equations that, counted with appropriate multiplicities, the intersection $E \cap L$ consists of exactly three points. (This is a special case of Bézout's theorem.)

(b) Let S be a singular point of E and suppose that $S \in L$. Prove that L intersects E at S with multiplicity at least two. Deduce that $E \cap L$ consists of S and at most one other point.

(c) More generally, let $C \subset \mathbb{P}^2$ be a curve, let $S \in C$ be a singular point of C, and let L be a line containing S. Prove that L intersects C at S with multiplicity at least two.

3.29. Let E be an elliptic curve.

(a) Fix a Weierstrass equation for E, fix a nonzero point $T \in E$, and write $x(P + T) = f\big(x(P), y(P)\big)$ for some function $f \in K(E) = K(x, y)$. Prove that f is a linear fractional transformation if and only if $T \in E[3]$, where a linear fractional transformation is a function of the form

$$
\frac{\alpha x + \beta y + \gamma}{\alpha' x + \beta' y + \gamma'}.
$$

(b) More generally, let $m \geq 3$, use a basis for $L\big(m(O)\big)$ to embed $E \hookrightarrow \mathbb{P}^{m-1}$, and let $T \in E$. Prove that the translation-by-T map $\tau_T : E \to E$ extends to an automorphism of $\mathbb{P}^{m-1}$ if and only if $T \in E[m]$.

3.30. Let A be a finite abelian group of order N^r. Suppose that for every $D \mid N$ we have $\#A[D] = D^r$, where $A[D]$ denotes the subgroup consisting of all elements of order D. Prove that

$$
A \cong \left(\frac{\mathbb{Z}}{N\mathbb{Z}}\right)^r.
$$

3.31. This exercise sketches an elementary proof of (III.6.2c) in arbitrary characteristic. We start with the case $\text{char}(K) \neq 2$. Let E/K be an elliptic curve.

(a) Use explicit formulas to prove that the doubling map $[2] : E \to E$ has degree 4.

(b) Use (a) to prove that $\deg[2^n] = 4^n$ for all $n \geq 1$.

(c) Use (b) and (III.4.10c) to deduce that $\#E[2^n] = 4^n$ for all $n \geq 1$. (This is where we use the assumption that $\text{char}(K) \neq 2$.)

(d) Use (c) and Exercise 3.30 to conclude that $E[2^n] \cong \mathbb{Z}/2^n\mathbb{Z} \times \mathbb{Z}/2^n\mathbb{Z}$ for all $n \geq 1$.

(e) Verify that the proof of the existence of dual isogenies (III.6.1) is valid in all characteristics.

(f) Suppose that $m \geq 1$ is an integer for which we know, a priori, that $\#E[m] = m^2$. Show that this suffices to prove the existence and basic properties of the Weil pairing $e_m : E[m] \times E[m] \to \mu_m$ as described in (III.8.1) and (III.8.2).

(g) Let $\phi : E_1 \to E_2$ and $\psi : E_1 \to E_2$ be isogenies of elliptic curves. Let $m = 2^n$, so (c) and (f) give the existence of the Weil pairing e_m on E_1 and E_2. Let $T_1 \in E_1[m]$ and $T_2 \in E_2[m]$ be m-torsion points. Use properties of the Weil pairing to prove that

$$e_m\big(T_1, \widehat{(\phi + \psi)}(T_2)\big) = e_m\big(T_1, \hat{\phi}(T_2) + \hat{\psi}(T_2)\big).$$

Since this holds for all $m = 2^n$, use the nondegeneracy of the Weil pairing to deduce that $\widehat{\phi + \psi} = \hat{\phi} + \hat{\psi}$.

(h) Use (g) to deduce that

$$[\hat{m}] = [m] \quad \text{and} \quad \deg[m] = m^2 \qquad \text{for all integers } m.$$

(Cf. (III.6.2d).)

(i) Let m be any integer such that $m \neq 0$ in K. Use (h) to prove that $\#E[m] = m^2$, and then observe that (f) gives the existence and standard properties of the Weil e_m-pairing.

(j) Finally, if $\text{char}(K) = 2$, replace (a) with a proof via explicit equations that $\deg[3] = 9$. Redo the rest of the exercise with 2^n replaced by 3^n.

3.32. Let $\phi \in \text{End}(E)$ be an endomorphism, and let

$$d = \deg\phi \qquad \text{and} \qquad a = 1 + \deg\phi - \deg(1 - \phi).$$

(a) Prove that $\phi^2 - [a] \circ \phi + [d] = [0]$ in $\text{End}(E)$.

(b) Let $\alpha, \beta \in \mathbb{C}$ be the complex roots of the polynomial $t^2 - at + d$. Prove that

$$|\alpha| = |\beta| = \sqrt{d}.$$

(c) Prove that $\deg(1 - \phi^n) = 1 + d^n - \alpha^n - \beta^n$ for all $n \geq 1$, and deduce that

$$\left|\deg(1 - \phi^n) - 1 - d^n\right| \leq 2d^{n/2}.$$

(d) Prove that

$$\exp\left(\sum_{n=1}^{\infty} \frac{\deg(1 - \phi^n)}{n} X^n\right) = \frac{1 - aX + dX^2}{(1 - X)(1 - dX)},$$

where the power series converges for $|X| < d^{-1}$.

(*Hint.* Use (III.8.6). For (b), use the fact that $\deg([m] + [n] \circ \phi) \geq 0$ for all $m, n \in \mathbb{Z}$.)

3.33. Let $\mathcal{K}$ be a $\mathbb{Q}$-division algebra, i.e., $\mathcal{K}$ is a (not necessarily commutative) $\mathbb{Q}$-algebra in which every nonzero element has a multiplicative inverse. This exercise sketches a proof of the following theorem, which can be used instead of (III.9.3) to prove (III.9.4). In particular, it is not necessary to know, a priori, that $\mathrm{End}(E)$ has rank at most four (III.7.4), (III.7.5).

Theorem. *Suppose that every element of $\mathcal{K}$ satisfies a quadratic equation with coefficients in $\mathbb{Q}$. Then either $\mathcal{K} = \mathbb{Q}$, $\mathcal{K}$ is a quadratic field, or $\mathcal{K}$ is a quaternion algebra.*

(a) Let $\alpha, \beta \in \mathcal{K}$. Prove that if $\beta \notin \mathbb{Q}(\alpha)$, then $\mathbb{Q}(\alpha) \cap \mathbb{Q}(\beta) = \mathbb{Q}$.
(b) Let $\alpha, \beta \in \mathcal{K}$. Prove that if $\alpha \notin \mathbb{Q}$ and $\alpha\beta = \beta\alpha$, then $\beta \in \mathbb{Q}(\alpha)$.
(c) Let $\alpha, \beta \in \mathcal{K}$. Prove that if $\alpha^2, \beta^2 \in \mathbb{Q}$, $\alpha \notin \mathbb{Q}$, and $\beta \notin \mathbb{Q}(\alpha)$, then $\alpha\beta + \beta\alpha \in \mathbb{Q}$.
(d) Let $\alpha \in \mathcal{K}$. Prove that there exists an $\alpha' \in \mathcal{K}$ such that $\mathbb{Q}(\alpha) = \mathbb{Q}(\alpha')$ and $\alpha'^2 \in \mathbb{Q}$.
(e) Let $\alpha, \beta \in \mathcal{K}^*$ satisfy $\alpha^2, \beta^2 \in \mathbb{Q}$. Prove that there exists a $\beta' \in \mathcal{K}$ such that $\mathbb{Q}(\alpha, \beta) = \mathbb{Q}(\alpha, \beta')$ and $\beta'^2, (\alpha\beta')^2 \in \mathbb{Q}$.
(f) Let $\alpha, \beta \in \mathcal{K}$ satisfy $\alpha \notin \mathbb{Q}$, $\beta \notin \mathbb{Q}(\alpha)$, and $\alpha^2, \beta^2, (\alpha\beta)^2 \in \mathbb{Q}$. Prove that $\alpha\beta = -\beta\alpha$.
(g) Prove the theorem.
(h) Use the theorem to prove (III.9.4).

3.34. Let K be a field. An *elliptic divisibility sequence* (EDS) over K is a sequence $(W_n)_{n\geq 1}$ defined by four initial conditions $W_1, W_2, W_3, W_4 \in K$ and satisfying the recurrence

$$W_{m+n}W_{m-n}W_1^2 = W_{m+1}W_{m-1}W_n^2 - W_{n+1}W_{n-1}W_m^2 \qquad \text{for all } m > n > 0.$$

An EDS in *nondegenerate* if $W_1 W_2 W_3 \neq 0$.
(a) Prove that a sequence $(W_n)_{n\geq 1}$ of elements of K with $W_1 W_2 W_3 \neq 0$ is an EDS if and only if it satisfies the two conditions

$$W_{2n+1}W_1^3 = W_{n+2}W_n^3 - W_{n-1}W_{n+1}^3 \qquad \text{for all } n \geq 2,$$
$$W_{2n}W_2W_1^2 = W_n(W_{n+1}W_{n-1}^2 - W_{n-2}W_{n+1}^2) \qquad \text{for all } n \geq 3.$$

(b) Prove that an EDS satisfies the more general recurrence

$$W_{m+n}W_{m-n}W_r^2 = W_{m+r}W_{m-r}W_n^2 - W_{n+r}W_{n-r}W_m^2 \qquad \text{for all } m > n > r > 0.$$

(c) Let (W_n) be an EDS and let $c \in K^*$. Prove that $(c^{n^2-1}W_n)$ is also an EDS.
(d) Let (W_n) be a nondegenerate EDS. Prove that (W_n/W_1) is an EDS. More generally, if $W_m \neq 0$, prove that $(W_{mn}/W_m)_{n\geq 1}$ is an EDS.

3.35. This exercise gives some examples of elliptic divisibility sequences (EDS).
(a) Prove that the sequence $1, 2, 3, \ldots$ is an EDS.
(b) Prove that the Fibonacci sequence is an EDS.
(c) More generally, let $(L_n)_{n\geq 1}$ be defined by a linear recurrence of the form

$$L_1 = 1, \qquad L_2 = A, \qquad L_{n+2} = AL_{n+1} - L_n \quad \text{for } n \geq 1.$$

Prove that (L_n) is an EDS.
(d) The most interesting EDS are associated to points on elliptic curves. Let E/K be an elliptic curve and let $P \in E(K)$ be a nonzero point. Define a sequence

$$W_n = \psi_n(P) \qquad \text{for } n \geq 1,$$

where ψ_n is the nth division polynomial for E as defined in Exercise 3.7. Prove that (W_n) is an EDS.

(e) Let (W_n) be an EDS associated to an elliptic curve E/K and nonzero point $P \in E(K)$ as in (d). Prove that P is a point of finite order at least 4 if and only if $W_n = 0$ for some $n \geq 4$.

(f) * Let (W_n) be an EDS associated to an elliptic curve E/K and a nonzero point $P \in E(K)$ of finite order. Let $r \geq 2$ be the smallest index such that $W_r = 0$. (The number r is called the *rank of apparition* of the sequence.) Assuming that $r \geq 4$, prove that there exist $A, B \in K^*$ such that

$$W_{ri+j} = W_j A^{ij} B^{i^2} \qquad \text{for all } i \geq 0 \text{ and all } j \geq 1.$$

(g) Suppose that K is a finite field and that the rank of apparition r of (W_n) is at least 4. Prove that the sequence (W_n) is periodic with period that is a multiple of r.

3.36. Let R be an integral domain, and let $(W_n)_{n \geq 1}$ be a nondegenerate elliptic divisibility sequence with $W_i \in R$ such that W_1 divides each of W_2, W_3, and W_4, and such that W_2 divides W_4.

(a) Prove that (W_n) is a *divisibility sequence*, in the sense that

$$m \mid n \quad \Longrightarrow \quad W_m \mid W_n.$$

(b) Suppose further that R is a principal ideal domain and that $\gcd(W_3, W_4) = 1$. Prove that (W_n) satisfies the stronger divisibility relation

$$W_{\gcd(m,n)} = \gcd(W_m, W_n) \qquad \text{for all } m, n \geq 1.$$

Chapter IV

The Formal Group
of an Elliptic Curve

Let E be an elliptic curve. In this chapter we study an "infinitesimal" neighborhood of E centered at the origin O. To do this, we start with the local ring $K[E]_O$ and take the completion of this ring at its maximal ideal. This leads to a power series ring in one variable, say $K[\![z]\!]$, for some uniformizer z at O. We then write the Weierstrass coordinate functions x and y as formal Laurent power series in z, and we construct a power series $F(z_1, z_2) \in K[\![z_1, z_2]\!]$ that formally gives the group law on E. Such a power series, which might be described as a "group law without any group elements," is an example of a *formal group*. In the remainder of this chapter we study in some detail the principal properties of arbitrary (one-parameter) formal groups. The advantage of suppressing all mention of the elliptic curve that motivated our study in the first place is that working with formal power series tends to be easier than working with quotients of polynomial rings. Then, of course, having obtained results for arbitrary formal groups, we can apply them in particular to the formal group associated to our original elliptic curve.

IV.1 Expansion Around O

In this section we investigate the structure of an elliptic curve and its addition law "close to the origin." To do this, it is convenient to make a change of variables, so we let

$$z = -\frac{x}{y} \quad \text{and} \quad w = -\frac{1}{y}, \qquad \text{so} \qquad x = \frac{z}{w} \quad \text{and} \quad y = -\frac{1}{w}.$$

The origin O on E is now the point $(z, w) = (0, 0)$, and z is a local uniformizer at O, i.e., the function z has a zero of order one at O. The usual Weierstrass equation for E becomes

$$w = z^3 + a_1 zw + a_2 z^2 w + a_3 w^2 + a_4 zw^2 + a_6 w^3 = f(z, w).$$

J.H. Silverman, *The Arithmetic of Elliptic Curves, Second Edition*, Graduate Texts in Mathematics 106, DOI 10.1007/978-0-387-09494-6_IV,
© Springer Science+Business Media, LLC 2009

The idea now is to substitute this equation into itself recursively so as to express w as a power series in z. Thus

$$
\begin{aligned}
w &= z^3 + (a_1 z + a_2 z^2)w + (a_3 + a_4 z)w^2 + a_6 w^3 \\
&= z^3 + (a_1 z + a_2 z^2)\big[z^3 + (a_1 z + a_2 z^2)w + (a_3 + a_4 z)w^2 + a_6 w^3\big] \\
&\quad + (a_3 + a_4 z)\big[z^3 + (a_1 z + a_2 z^2)w + (a_3 + a_4 z)w^2 + a_6 w^3\big]^2 \\
&\quad + a_6\big[z^3 + (a_1 z + a_2 z^2)w + (a_3 + a_4 z)w^2 + a_6 w^3\big] \\
&\quad\ \vdots \\
&= z^3 + a_1 z^4 + (a_1^2 + a_2)z^5 + (a_1^3 + 2a_1 a_2 + a_3)z^6 \\
&\quad + (a_1^4 + 3a_1^2 a_2 + 3a_1 a_3 + a_2^2 + a_4)z^7 + \cdots \\
&= z^3(1 + A_1 z + A_2 z^2 + \cdots),
\end{aligned}
$$

where each $A_n \in \mathbb{Z}[a_1, \ldots, a_6]$ is a polynomial in the coefficients of E. Of course, we need to prove that this procedure converges to a power series

$$
w(z) \in \mathbb{Z}[a_1, \ldots, a_6][\![z]\!],
$$

and we want the equality

$$
w(z) = f\big(z, w(z)\big)
$$

to be true in the ring $\mathbb{Z}[a_1, \ldots, a_6][\![z]\!]$.

To describe more precisely the algorithm for producing $w(z)$, we define a sequence of polynomials by

$$
f_1(z, w) = f(z, w) \qquad \text{and} \qquad f_{m+1}(z, w) = f_m\big(z, f(z, w)\big).
$$

Then we set

$$
w(z) = \lim_{m \to \infty} f_m(z, 0),
$$

provided that this limit makes sense in $\mathbb{Z}[a_1, \ldots, a_6][\![z]\!]$.

Proposition 1.1. (a) *The procedure described above gives a power series*

$$
w(z) = z^3(1 + A_1 z + A_2 z^2 + \cdots) \in \mathbb{Z}[a_1, \ldots, a_6][\![z]\!].
$$

(b) *The series $w(z)$ is the unique power series in $\mathbb{Z}[a_1, \ldots, a_6][\![z]\!]$ satisfying*

$$
w(z) = f\big(z, w(z)\big).
$$

(c) *If $\mathbb{Z}[a_1, \ldots, a_6]$ is made into a graded ring by assigning weights $\mathrm{wt}(a_i) = i$, then A_n is a homogeneous polynomial of weight n.*

PROOF. Parts (a) and (b) are special cases of Hensel's lemma, which we prove later in this section (IV.1.2). To prove the present proposition, use (IV.1.2) with

$$R = \mathbb{Z}[a_1, \ldots, a_6][[z]], \qquad I = (z),$$
$$F(w) = f(z, w) - w, \qquad a = 0, \qquad \alpha = -1.$$

Finally, to prove (c), we assign weights to z and w by setting

$$\mathrm{wt}(z) = -1 \qquad \text{and} \qquad \mathrm{wt}(w) = -3.$$

Then $f(z, w)$ is homogeneous of weight -3 in the graded ring $\mathbb{Z}[a_1, \ldots, a_6, z, w]$, and an easy induction on m shows that $f_m(z, w)$ is homogeneous of weight -3 for every $m \geq 1$. In particular,

$$f_m(z, 0) = z^3 (1 + B_1 z + B_2 z^2 + \cdots + B_N z^N)$$

is homogeneous of weight -3, so each $B_n \in \mathbb{Z}[a_1, \ldots, a_6]$ is homogeneous of weight n. Hence the A_n's have the same property, since $f_m(z, 0)$ converges to $w(z)$ as $m \to \infty$. $\qquad\square$

Lemma 1.2. (Hensel's Lemma) *Let R be a ring that is complete with respect to some ideal $I \subset R$, and let $F(w) \in R[w]$ be a polynomial. Suppose that there are an integer $n \geq 1$ and an element $a \in R$ satisfying*

$$F(a) \in I^n \qquad \text{and} \qquad F'(a) \in R^*.$$

Then for any $\alpha \in R$ satisfying $\alpha \equiv F'(a) \pmod{I}$, the sequence

$$w_0 = a, \qquad w_{m+1} = w_m - \frac{F(w_m)}{\alpha}$$

converges to an element $b \in R$ satisfying

$$F(b) = 0 \qquad \text{and} \qquad b \equiv a \pmod{I^n}.$$

If R is an integral domain, then these conditions determine b uniquely.

PROOF. To ease notation, we replace $F(w)$ by $F(w + a)/\alpha$, so we are now dealing with the recurrence

$$w_0 = 0, \qquad F(0) \in I^n, \qquad F'(0) \equiv 1 \pmod{I}, \qquad w_{m+1} = w_m - F(w_m).$$

Since $F(0) \in I^n$, it is clear that

$$w_m \in I^n \implies w_m - F(w_m) \in I^n,$$

from which it follows that

$$w_m \in I^n \quad \text{for all } m \geq 0.$$

We now show by induction that

$$w_m \equiv w_{m+1} \pmod{I^{m+n}} \qquad \text{for all } m \geq 0.$$

For $m = 0$, this just says that $F(0) \equiv 0 \pmod{I^n}$, which is one of our initial assumptions. Assume now that the desired congruence is true for all integers strictly smaller than m. Let X and Y be new variables and factor

$$F(X) - F(Y) = (X - Y)\big(F'(0) + XG(X,Y) + YH(X,Y)\big),$$

where G and H are polynomials in $R[X,Y]$. Then

$$
\begin{aligned}
w_{m+1} - w_m &= \big(w_m - F(w_m)\big) - \big(w_{m-1} - F(w_{m-1})\big) \\
&= (w_m - w_{m-1}) - \big(F(w_m) - F(w_{m-1})\big) \\
&= (w_m - w_{m-1})\big(1 - F'(0) - w_m G(w_m, w_{m-1}) \\
&\qquad - w_{m-1}H(w_m, w_{m-1})\big) \in I^{m+n}.
\end{aligned}
$$

Here the last line follows from the induction hypothesis and the assumptions that $F'(0) \equiv 1 \pmod{I}$ and $w_m, w_{m-1} \in I^n$. This proves that

$$w_m - w_{m+1} \in I^{m+n} \quad \text{for all } m \geq 0.$$

Since R is complete with respect to I, it follows that the sequence w_m converges to an element $b \in R$; and since every $w_m \in I^n$, we see that $b \in I^n$. Further, taking the limit of the relation $w_{m+1} = w_m - F(w_m)$ as $m \to \infty$ yields $b = b - F(b)$, so $F(b) = 0$.

Finally, to show uniqueness (under the assumption that R is an integral domain), suppose that $c \in I^n$ satisfies $F(c) = 0$. Then

$$0 = F(b) - F(c) = (b - c)\big(F'(0) + bG(b,c) + cH(b,c)\big).$$

If $b \neq c$, then $F'(0) + bG(b,c) + cH(b,c) = 0$, which would imply that

$$F'(0) = -bG(b,c) - cH(b,c) \in I.$$

This contradicts the assumption that $F'(0) \equiv 1 \pmod{I}$. Hence $b = c$. $\qquad\square$

Using the power series $w(z)$ from (IV.1.1), we derive *Laurent series* for x and y,

$$x(z) = \frac{z}{w(z)} = \frac{1}{z^2} - \frac{a_1}{z} - a_2 - a_3 z - (a_4 + a_1 a_3)z^2 - \cdots,$$

$$y(z) = -\frac{1}{w(z)} = -\frac{1}{z^3} + \frac{a_1}{z^2} + \frac{a_2}{z} + a_3 + (a_4 + a_1 a_3)z - \cdots.$$

Similarly, the invariant differential has an expansion

$$
\begin{aligned}
\omega(z) &= \frac{dx(z)}{2y(z) + a_1 x(z) + a_3} \\
&= \big(1 + a_1 z + (a_1^2 + a_2)z^2 + (a_1^3 + 2a_1 a_2 + 2a_3)z^3 \\
&\qquad + (a_1^4 + 3a_1^2 a_2 + 6a_1 a_3 + a_2^2 + 2a_4)z^4 + \cdots\big)\, dz.
\end{aligned}
$$

We note that the series for $x(z)$, $y(z)$, and $\omega(z)$ have coefficients in $\mathbb{Z}[a_1, \ldots, a_6]$. This is clear for $x(z)$ and $y(z)$, while for $\omega(z)$ it follows from the two expressions

$$\frac{\omega(z)}{dz} = \frac{dx(z)/dz}{2y + a_1 x + a_3} = \frac{-2z^{-3} + \cdots}{-2z^{-3} + \cdots} \in \mathbb{Z}\left[\frac{1}{2}, a_1, \ldots, a_6\right][\![z]\!],$$

$$\frac{\omega(z)}{dz} = \frac{dy(z)/dz}{3x^2 + 2a_2 x + a_4 - a_1 y} = \frac{-3z^{-4} + \cdots}{-3z^{-4} + \cdots} \in \mathbb{Z}\left[\frac{1}{3}, a_1, \ldots, a_6\right][\![z]\!],$$

which show that any denominator is simultaneously a power of 2 and a power of 3.

The pair $\big(x(z), y(z)\big)$ provides a formal solution to the Weierstrass equation

$$E : y^2 + a_1 xy + a_3 y = x^3 + a_2 x^2 + a_4 x + a_6,$$

i.e., a solution in the quotient field of the ring of formal power series. If E is defined over a field K, we might try to create points on E by evaluating these power series at $z \in K$. In general, there is no obvious way to assign a value to an infinite series such as $x(z)$ evaluated at some $z \in K$. However, suppose that K is a complete local field with ring of integers R and maximal ideal $\mathcal{M}$, and further suppose that $a_1, \ldots, a_6 \in R$. Then the power series $x(z)$ and $y(z)$ will converge for any $z \in \mathcal{M}$ and the result will be a point $\big(x(z), y(z)\big) \in E(K)$. This gives an injective map

$$\mathcal{M} \longrightarrow E(K), \qquad z \longmapsto \big(x(z), y(z)\big).$$

(The map is injective, since it has an inverse $(x, y) \mapsto -x/y$.) It is easy to characterize the image as consisting of those points (x, y) with $x^{-1} \in \mathcal{M}$. This map will be a key tool when we study elliptic curves over local fields in Chapter VII.

Returning now to formal power series, we look for the power series formally giving the addition law on E. Let z_1 and z_2 be independent indeterminates, and let $w_1 = w(z_1)$ and $w_2 = w(z_2)$. In the (z, w)-plane, the line connecting (z_1, w_1) to (z_2, w_2) has slope

$$\lambda = \lambda(z_1, z_2) = \frac{w_2 - w_1}{z_2 - z_1} = \sum_{n=3}^{\infty} A_{n-3} \frac{z_2^n - z_1^n}{z_2 - z_1} \in \mathbb{Z}[a_1, \ldots, a_6][\![z_1, z_2]\!].$$

Note that $\lambda(z_1, z_2)$ has no constant or linear terms, and that the A_n values come from (IV.1.1a). Letting

$$\nu = \nu(z_1, z_2) = w_1 - \lambda z_1 \in \mathbb{Z}[a_1, \ldots, a_6][\![z_1, z_2]\!],$$

the connecting line has equation $w = \lambda z - \nu$. Substituting this into the Weierstrass equation gives a cubic in z, two of whose roots are z_1 and z_2. Looking at the quadratic term, we see that the third root (say z_3) can be expressed as a power series in z_1 and z_2,

$$z_3 = z_3(z_1, z_2)$$

$$= -z_1 - z_2 + \frac{a_1 \lambda + a_3 \lambda^2 - a_2 \nu - 2a_4 \lambda \nu - 3a_6 \lambda^2 \nu}{1 + a_2 \lambda + a_4 \lambda^2 + a_6 \lambda^3}$$

$$\in \mathbb{Z}[a_1, \ldots, a_6][\![z_1, z_2]\!].$$

Letting

$$w_3 = \lambda(z_1, z_2) z_3(z_1, z_2) + \nu(z_1, z_2),$$

the three points (z_1, w_1), (z_2, w_2), and (z_3, w_3) are collinear on E, so they add to O using the group law. Further, the fact that (z_3, w_3) is on E means that $w_3 = f(z_3, w_3)$, while (IV.1.1b) says that the power series $w(z)$ described in (IV.1.1a) is the *unique* power series satisfying $w(z) = f(z, w(z))$. Hence $w_3 = w(z_3)$, i.e., we can compute the w-coordinate of $-(x_1, y_1) - (x_2, y_2)$ using the power series from (IV.1.1a).

In order to compute the sum of the first two points, we need the formula for the inverse. In the (x, y)-plane, the inverse of (x, y) is $(x, -y - a_1 x - a_3)$. Remembering that $z = -x/y$, we find that the inverse of (z, w) has z-coordinate

$$i(z) = \frac{x(z)}{y(z) + a_1 x(z) + a_3} = \frac{z^{-2} - a_1 z^{-1} - \cdots}{-z^{-3} + 2a_1 z^{-2} + \cdots} \in \mathbb{Z}[a_1, \ldots, a_6][[z]],$$

and an argument similar to that given above shows that the w-coordinate of the inverse $(x, -y - a_1 x - a_3)$ is equal to $w(i(z))$. This gives the formal addition law

$$
\begin{aligned}
F(z_1, z_2) &= i(z_3(z_1, z_2)) \\
&= z_1 + z + 2 - a_1 z_1 z_2 - a_2(z_1^2 z_2 + z_1 z_2^2) \\
&\quad + (2a_3 z_1^3 z_2 + (a_1 a_2 - 3a_3) z_1^2 z_2^2 + 2a_3 z_1 z_2^3) + \cdots \\
&\qquad\qquad\qquad\qquad\qquad\qquad \in \mathbb{Z}[a_1, \ldots, a_6][[z_1, z_2]].
\end{aligned}
$$

From properties of the addition law on E, we deduce that $F(z_1, z_2)$ has the corresponding properties:

$$
\begin{aligned}
F(z_1, z_2) &= F(z_2, z_1) && \text{(commutativity)}, \\
F(z_1, F(z_2, z)) &= F(F(z_1, z_2), z) && \text{(associativity)}, \\
F(z, i(z)) &= 0 && \text{(inverse)}.
\end{aligned}
$$

The power series $F(z_1, z_2)$ might be described as "a group law without any group elements." Such objects are called *formal groups*. We could now continue with the study of the particular formal group coming from our elliptic curve, but it is little more difficult to analyze arbitrary (one-parameter) formal groups, and in fact the abstraction tends to clarify the underlying structure, so we take the latter approach. The reader should, however, keep in mind the example of an elliptic curve while reading the remainder of this chapter.

IV.2 Formal Groups

In this section we define and prove some basic properties of formal groups.

Definition. Let R be a ring. A *(one-parameter commutative) formal group* $\mathcal{F}$ *over* R is a power series $F(X, Y) \in R[[X, Y]]$ with the following properties:

(a) $F(X, Y) = X + Y +$ (terms of degree ≥ 2).

(b) $F(X, F(Y, Z)) = F(F(X, Y), Z)$ (associativity).

(c) $F(X, Y) = F(Y, X)$ (commutativity).

(d) There is a unique power series $i(T) \in R[\![T]\!]$ such that $F(T, i(T)) = 0$ (inverse).

(e) $F(X, 0) = X$ and $F(0, Y) = Y$.

We call $F(X, Y)$ the *formal group law of $\mathcal{F}$*.

Remark 2.1. It is easy to prove that (a) and (b) imply (d) and (e); see Exercise 4.1. It is also true that (a) and (b) imply (c), provided that the ring R has no torsion nilpotent elements; see Exercise 4.2b. In this section we prove this last assertion when R has no torsion elements.

Definition. Let $(\mathcal{F}, F)$ and $(\mathcal{G}, G)$ be formal groups defined over R. A *homomorphism from $\mathcal{F}$ to $\mathcal{G}$ defined over R* is a power series $f \in R[\![T]\!]$ (with no constant term) that satisfies

$$f(F(X, Y)) = G(f(X), f(Y)).$$

The formal groups $\mathcal{F}$ and $\mathcal{G}$ are *isomorphic over R* if there are homomorphisms $f : \mathcal{F} \to \mathcal{G}$ and $g : \mathcal{G} \to \mathcal{F}$ defined over R such that

$$f(g(T)) = g(f(T)) = T.$$

Example 2.2.1. The *formal additive group*, denoted by $\hat{\mathbb{G}}_a$, is defined by

$$F(X, Y) = X + Y.$$

Example 2.2.2. The *formal multiplicative group*, denoted by $\hat{\mathbb{G}}_m$, is defined by

$$F(X, Y) = X + Y + XY = (1 + X)(1 + Y) - 1.$$

Example 2.2.3. Let E be an elliptic curve given by a Weierstrass equation with coefficients in R. The *formal group associated to E* is denoted by $\hat{E}$. It is defined by the power series $F(z_1, z_2)$ described in (IV §1).

Example 2.2.4. Let $(\mathcal{F}, F)$ be a formal group. We define homomorphisms

$$[m] : \mathcal{F} \longrightarrow \mathcal{F}$$

inductively for $m \in \mathbb{Z}$ by

$$[0](T) = 0, \quad [m + 1](T) = F([m](T), T), \quad [m - 1](T) = F([m](T), i(T)).$$

One may easily check by induction that $[m]$ is a homomorphism. We call $[m]$ the *multiplication-by-m map*. The following elementary proposition, which explains when $[m]$ is an isomorphism, will be of great importance. More precisely, the chain of implications

$$(\text{IV.2.3}) \Longrightarrow (\text{IV.3.2b}) \Longrightarrow (\text{VII.3.1})$$

proves a key fact required for the proof of the weak Mordell–Weil theorem (VIII.1.1).

Proposition 2.3. *Let $\mathcal{F}$ be a formal group over the ring R and let $m \in \mathbb{Z}$.*
(a) *$[m](T) = mT + (higher\text{-}order\ terms)$.*
(b) *If $m \in R^*$, then $[m] : \mathcal{F} \to \mathcal{F}$ is an isomorphism.*

PROOF. (a) For $m \geq 0$, the stated result is a trivial induction using the recursive definition of $[m]$ and the fact that $F(X, Y) = X + Y + \cdots$. Then, using

$$0 = F(T, i(T)) = T + i(T) + \cdots,$$

we see that $i(T) = -T + \cdots$, and now the downward induction for $m < 0$ is also clear.
(b) This follows from (a) and the following useful lemma.   □

Lemma 2.4. *Let $a \in R^*$ and let $f(T) \in R[\![T]\!]$ be a power series of the form*

$$f(T) = at + (higher\text{-}order\ terms).$$

Then there is a unique power series $g(T) \in R[\![T]\!]$ satisfying

$$f(g(T)) = T.$$

The series $g(T)$ also satisfies $g(f(T)) = T$.

PROOF. We construct a sequence of polynomials $g_n(T) \in R[T]$ satisfying

$$f(g_n(T)) \equiv T \pmod{T^{n+1}} \qquad \text{and} \qquad g_{n+1}(T) \equiv g_n(T) \pmod{T^{n+1}}.$$

Then the limit $g(T) = \lim g_n(T)$ exists in $R[\![T]\!]$ and clearly satisfies $f(g(T)) = T$.
    To start the induction, let $g_1(T) = a^{-1}T$. Now suppose that $g_{n-1}(T)$ has been constructed and has the desired properties. Then $g_n(T)$ must have the form

$$g_n(T) = g_{n-1}(T) + \lambda T^n$$

for some $\lambda \in R$, and we look for a value of λ that makes

$$f(g_n(T)) \equiv T \pmod{T^{n+1}}.$$

To do this, we use the induction hypothesis to compute

$$\begin{aligned}
f(g_n(T)) &= f(g_{n-1}(T) + \lambda T^n) \\
&\equiv f(g_{n-1}(T)) + a\lambda T^n \pmod{T^{n+1}} \\
&\equiv T + bT^n + a\lambda T^n \pmod{T^{n+1}} \qquad \text{for some } b \in R.
\end{aligned}$$

It thus suffices to take $\lambda = -a^{-1}b$, which we may do since $a \in R^*$. This completes the proof that $g(T)$ exists.
    Next we apply what we have proven, using the power series $g(T) = a^{-1}T + \cdots$ in place of $f(T)$. This gives us a power series $h(T)$ satisfying $g(h(T)) = T$. Then

$$g(f(T)) = g(f(g(h(T)))) = g(f \circ g(h(T))) = g(h(T)) = T.$$

Finally, suppose that $G(T) \in R[\![T]\!]$ is another power series satisfying $f(G(t)) = T$. Then

$$g(T) = g(f(G(T))) = (g \circ f)(G(T)) = G(T),$$

which shows that $g(T)$ is unique.   □

IV.3   Groups Associated to Formal Groups

A formal group is, in general, merely a group operation with no actual underlying group. However, if the ring R is local and complete and if the variables are assigned values in the maximal ideal $\mathcal{M}$ of R, then the power series defining the formal group converge and thus give $\mathcal{M}$ the structure of a group. This section is devoted to an initial analysis of such groups. We fix the following notation:

R a complete local ring.

$\mathcal{M}$ the maximal ideal of R.

k the residue field $R/\mathcal{M}$.

$\mathcal{F}$ a formal group defined over R, with formal group law $F(X, Y)$.

Definition. The *group associated to $\mathcal{F}/R$*, denoted by $\mathcal{F}(\mathcal{M})$, is the set $\mathcal{M}$ endowed with the group operations

$$x \oplus_{\mathcal{F}} y = F(x, y) \quad \text{(addition)} \quad \text{for } x, y \in \mathcal{M},$$
$$\ominus_{\mathcal{F}} x = i(x) \quad \text{(inversion)} \quad \text{for } x \in \mathcal{M}.$$

Similarly, for $n \geq 1$, we define $\mathcal{F}(\mathcal{M}^n)$ to be the subgroup of $\mathcal{F}(\mathcal{M})$ consisting of the set $\mathcal{M}^n$ together with the above group laws.

The assumption that R is complete ensures that the power series $F(x, y)$ and $i(x)$ converge in R for all $x, y \in \mathcal{M}$. The formal group axioms immediately imply that $\mathcal{F}(\mathcal{M})$ is a group and that $\mathcal{F}(\mathcal{M}^n)$ is a subgroup of $\mathcal{F}(\mathcal{M})$.

Example 3.1.1. The additive group $\hat{\mathbb{G}}_a(\mathcal{M})$ is just $\mathcal{M}$ with its usual addition law. Notice the exact sequence (of additive groups)

$$0 \longrightarrow \hat{\mathbb{G}}_a(\mathcal{M}) \longrightarrow R \longrightarrow k \longrightarrow 0.$$

Example 3.1.2. The multiplicative group $\hat{\mathbb{G}}_m(\mathcal{M})$ is the group of 1-units, i.e., the set $1 + \mathcal{M}$ with group law multiplication. Notice that we again have an exact sequence,

$$0 \longrightarrow \hat{\mathbb{G}}_m(\mathcal{M}) \xrightarrow{\ z \mapsto 1+z\ } R^* \longrightarrow k^* \longrightarrow 1.$$

Example 3.1.3. Let $\hat{E}$ be the formal group associated to an elliptic curve E/K as described in (IV.2.2.3), where K is the field of fractions of the complete local ring R. Then, as noted in (IV §1), the power series $x(z)$ and $y(z)$ give a well-defined map

$$\mathcal{M} \longrightarrow E(K), \quad z \longrightarrow P_z = \big(x(z), y(z) \big).$$

The construction of the power series for $\hat{E}$ imply that this map is a homomorphism of $\hat{E}(\mathcal{M})$ to $E(K)$.[1]

---

[1]More precisely, they imply that $P_{F(z, z')} = P_z + P_{z'}$ for *distinct* $z, z' \in \mathcal{M}$. For $z = z'$, we can let $z' \mapsto z$ and use the fact that the map $z \mapsto P_z$ and the addition law on $E(K)$ are continuous for the topology induced from K. Alternatively, we could do an explicit, albeit messy, calculation with power series and the duplication formula.

As we will see in Chapter VII, there is often an exact sequence

$$0 \longrightarrow \hat{E}(\mathcal{M}) \longrightarrow E(K) \longrightarrow \tilde{E}(k) \longrightarrow 0,$$

where $\tilde{E}$ is a certain elliptic curve defined over the residue field k. In this way, the study of $E(K)$ is reduced to the study of the formal group $\hat{E}(\mathcal{M})$ and the study of an elliptic curve over the smaller, and hopefully simpler, field k.

Proposition 3.2. *Let $\mathcal{F}/R$ be a formal group defined over a complete local ring.*
(a) *For each $n \geq 1$, the map*

$$\frac{\mathcal{F}(\mathcal{M}^n)}{\mathcal{F}(\mathcal{M}^{n+1})} \longrightarrow \frac{\mathcal{M}^n}{\mathcal{M}^{n+1}}$$

induced by the identity map on sets is an isomorphism of groups.
(b) *Let p be the characteristic of the residue field k, where p is allowed to equal 0. Then every element of finite order in $\mathcal{F}(\mathcal{M})$ has order that is a power of p. (See (IV §6) for a more precise description of the torsion subgroup of $\mathcal{F}(\mathcal{M})$.)*

PROOF. (a) Since the underlying sets are the same, it suffices to show that the map is a homomorphism. But this is clear, since for any $x, y \in \mathcal{M}^n$ we have

$$x \oplus_{\mathcal{F}} y = F(x, y)$$
$$= x + y + (\text{higher-order terms})$$
$$\equiv x + y \pmod{\mathcal{M}^{2n}}.$$

(b) We give two proofs of this important fact. Multiplying an arbitrary torsion element by an appropriate power of p, it suffices to prove that there are no nonzero torsion elements of order prime to p. So we let $m \geq 1$ with $p \nmid m$ (if $p = 0$, then m is arbitrary) and we suppose that $x \in \mathcal{F}(\mathcal{M})$ satisfies $[m](x) = 0$. We must show that $x = 0$.

For our first proof, we note that since m is prime to p, we have $m \notin \mathcal{M}$, so $m \in R^*$, since R is a local ring. It follows from (IV.2.3b) that $[m]$ is an automorphism of the formal group $\mathcal{F}/R$, so it induces an isomorphism

$$[m] : \mathcal{F}(\mathcal{M}) \xrightarrow{\ \sim\ } \mathcal{F}(\mathcal{M}).$$

In particular, multiplication-by-m has trivial kernel, so $x = 0$.

For the second proof, we assume that R is Noetherian and show inductively that $x \in \mathcal{M}^n$ for all $n \geq 1$. This will imply that $x = 0$ by Krull's theorem [8, Corollary 10.20], [73, Corollary 5.4]. We know that $x \in \mathcal{M}$. Suppose that $x \in \mathcal{M}^n$ and consider the image $\bar{x}$ of x in $\mathcal{F}(\mathcal{M}^n)/\mathcal{F}(\mathcal{M}^{n+1})$. On the one hand, $\bar{x}$ has order dividing m. On the other hand, from (a) we know that the quotient $\mathcal{F}(\mathcal{M}^n)/\mathcal{F}(\mathcal{M}^{n+1})$ is isomorphic to the k-vector space $\mathcal{M}^n/\mathcal{M}^{n+1}$, hence has only p-torsion. Therefore $\bar{x} = 0$, so $x \in \mathcal{M}^{n+1}$.   □

IV.4 The Invariant Differential

We return to the study of a formal group $\mathcal{F}$ defined over an arbitrary ring R. In a formal setting of this sort, a differential form is simply an expression $P(T)\,dT$ with $P(T) \in R[[T]]$. Of particular interest are those differential forms that respect the group structure of $\mathcal{F}$.

Definition. An *invariant differential* on a formal group $\mathcal{F}/R$ is a differential form

$$\omega(T) = P(T)\,dT \in R[[T]]\,dt$$

satisfying

$$\omega \circ F(T,S) = \omega(T).$$

Writing this out, $\omega(T) = P(T)\,dT$ is an invariant differential if it satisfies

$$P\big(F(T,S)\big)F_X(T,S) = P(T),$$

where $F_X(X,Y)$ is the partial derivative of F with respect to its first variable. An invariant differential is said to be *normalized* if $P(0) = 1$.

Example 4.1.1. On the additive group $\hat{\mathbb{G}}_a$, the differential $\omega = dT$ is invariant.

Example 4.1.2. On the multiplicative group $\hat{\mathbb{G}}_m$, the following is an invariant differential:

$$\omega = \frac{dT}{1+T} = (1 - T + T^2 - T^3 + \cdots)\,dT.$$

Proposition 4.2. *Let $\mathcal{F}/R$ be a formal group. There exists a unique normalized invariant differential on $\mathcal{F}/R$. It is given by the formula*

$$\omega = F_X(0,T)^{-1}\,dT.$$

Every invariant differential on $\mathcal{F}/R$ is of the form $a\omega$ for some $a \in R$.

PROOF. Suppose that $P(T)\,dT$ is an invariant differential on $\mathcal{F}/R$, so it satisfies

$$P\big(F(T,S)\big)F_X(T,S) = P(T),$$

Putting $T = 0$ and remembering that $F(0,S) = S$ gives

$$P(S)F_X(0,S) = P(0).$$

Since $F_X(0,S) = 1 + \cdots$, we see that $P(T)$ is completely determined by the value $P(0)$, and further that every invariant differential is of the form $a\omega$ with $a \in R$ and

$$\omega = F_X(0,T)^{-1}\,dT.$$

Since this differential ω is normalized, it remains only to show that it is invariant.
   We must prove that

$$F_X\big(0, F(T,S)\big)^{-1} F_X(T,S) = F_X(0,T)^{-1}.$$

To do this, we differentiate the associative law

$$F\big(U, F(T,S)\big) = F\big(F(U,T), S\big)$$

with respect to U to obtain (using the chain rule!)

$$F_X\big(U, F(T,S)\big) = F_X\big(F(U,T), S\big) F_X(U,T).$$

Putting $U = 0$ and using the fact that $F(0,T) = T$ yields

$$F_X\big(0, F(T,S)\big) = F_X(T,S) F_X(0,T),$$

which is the desired result.   □

Before stating the first corollary, we set the notation $f'(T)$ for the *formal derivative* of a power series $f(T) \in R[\![T]\!]$, i.e., $f'(T)$ is obtained by formally differentiating $f(T)$ term by term.

Corollary 4.3. *Let $\mathcal{F}/R$ and $\mathcal{G}/R$ be formal groups with normalized differentials $\omega_{\mathcal{F}}$
and $\omega_{\mathcal{G}}$. Let $f : \mathcal{F} \to \mathcal{G}$ be a homomorphism. Then*

$$\omega_{\mathcal{G}} \circ f = f'(0)\omega_{\mathcal{F}}.$$

PROOF. Let $F(X,Y)$ and $G(X,Y)$ be the formal group laws for $\mathcal{F}$ and $\mathcal{G}$. We claim
that $\omega_{\mathcal{G}} \circ f$ is an invariant differential for $\mathcal{F}$. To prove this, we compute

$$\begin{aligned}
(\omega_{\mathcal{G}} \circ f)\big(F(T,S)\big) &= \omega_{\mathcal{G}}\big(G(f(T), f(S))\big) &&\text{since } f \text{ is a homomorphism,}\\
&= (\omega_{\mathcal{G}} \circ f)(T) &&\text{since } \omega_{\mathcal{G}} \text{ is invariant for } \mathcal{G}.
\end{aligned}$$

It follows from (IV.4.2) that $\omega_{\mathcal{G}} \circ f$ is equal to $a\omega_{\mathcal{F}}$ for some $a \in R$. Comparing
coefficients of T on each side gives $a = f'(0)$.   □

Corollary 4.4. *Let $\mathcal{F}/R$ be a formal group and let $p \in \mathbb{Z}$ be a prime. There there
are power series $f(T), g(T) \in R[\![T]\!]$ with $f(0) = g(0) = 0$ such that*

$$[p](T) = pf(T) + g(T^p).$$

PROOF. Let $\omega(T)$ be the normalized invariant differential on $\mathcal{F}$. From (IV.2.3a) we
have $[p]'(0) = p$, so (IV.4.3) implies that

$$p\omega(T) = \big(\omega \circ [p]\big)(T) = (1 + \cdots)[p]'(T)\, dT.$$

The series $(1 + \cdots)$ is invertible in $R[\![T]\!]$, from which it follows that

$$[p]'(T) \in pR[\![T]\!].$$

Therefore every term aT^n in the series $[p](T)$ satisfies either $a \in pR$ or $p \mid n$.   □

IV.5   The Formal Logarithm

Integrating an invariant differential might, one hopes, yield a homomorphism to the additive group. Unfortunately, integration tends to introduce denominators, but at least in characteristic 0 things work fairly well.

Definition. Let R be a torsion-free[2] ring, let $K = R \otimes \mathbb{Q}$, let $\mathcal{F}/R$ be a formal group, and let

$$\omega(T) = (1 + c_1 T + c_2 T^2 + c_3 T^3 + \cdots) \, dT$$

be the normalized invariant differential on $\mathcal{F}/R$. The *formal logarithm of $\mathcal{F}/R$* is the power series

$$\log_{\mathcal{F}}(T) = \int \omega(T) = T + \frac{c_1}{2} T^2 + \frac{c_2}{3} T^3 + \cdots \in K[\![T]\!].$$

The *formal exponential of $\mathcal{F}/R$* is the unique power series $\exp_{\mathcal{F}}(T) \in K[\![T]\!]$ satisfying

$$\log_{\mathcal{F}} \circ \exp_{\mathcal{F}}(T) = \exp_{\mathcal{F}}(T) \circ \log_{\mathcal{F}}(T) = T.$$

The existence and uniqueness of $\exp_{\mathcal{F}}$ are ensured by (IV.2.4).

Example 5.1. The formal group law and invariant differential of the formal multiplicative group $\mathcal{F} = \hat{\mathbb{G}}_m$ are

$$F_{\mathcal{F}}(X, Y) = X + Y + XY \qquad \text{and} \qquad \omega_{\mathcal{F}}(T) = (1 + T)^{-1} dT.$$

Then the formal logarithm and exponential are given by

$$\log_{\mathcal{F}}(T) = \int (1 + T)^{-1} dT = \sum_{n=1}^{\infty} \frac{(-1)^{n-1} T^n}{n},$$

$$\exp_{\mathcal{F}}(T) = \sum_{n=1}^{\infty} \frac{T^n}{n!}.$$

(We recall that the "identity element" is at $T = 0$, so $\log_{\mathcal{F}}(T)$ and $\exp_{\mathcal{F}}(T)$ are the standard Taylor series expansions of $\log(1 + T)$ and $e^T - 1$.)

Proposition 5.2. *Let R be a torsion-free ring and let $\mathcal{F}/R$ be a formal group. Then*

$$\log_{\mathcal{F}} : \mathcal{F} \longrightarrow \hat{\mathbb{G}}_a$$

is an isomorphism of formal groups over $K = R \otimes \mathbb{Q}$. (N.B. The presence of denominators in the coefficients of the power series $\log_{\mathcal{F}}(T)$ means that $\log_{\mathcal{F}}$ generally does not give an isomorphism of formal groups over R.)

---

[2] The assumption that R has no torsion elements means that if $n \in \mathbb{Z}$ and $\alpha \in R$ satisfy $n\alpha = 0$, then either $n = 0$ or $\alpha = 0$. Equivalently, the natural map $R \to K = R \otimes \mathbb{Q}$ is an injection.

PROOF. Let $\omega(T)$ be the normalized invariant differential on $\mathcal{F}/R$, so

$$\omega\big(F(T,S)\big) = \omega(T).$$

Integrating with respect to T gives

$$\log_{\mathcal{F}} F(T,S) = \log \mathcal{F}(T) + C(S)$$

for some "constant of integration" $C(S) \in K[\![S]\!]$. Taking $T = 0$ shows that $C(S) = \log_{\mathcal{F}}(S)$, which proves that $\log_{\mathcal{F}}$ is a homomorphism. Further, it has $\exp_{\mathcal{F}}$ as its inverse, so $\log_{\mathcal{F}}$ is an isomorphism. $\qquad\square$

Application 5.3. Let R be a torsion-free ring and suppose that $F(X,Y) \in R[\![X,Y]\!]$ is a power series satisfying

$$F\big(X, F(Y,Z)\big) = F\big(F(X,Y), Z\big), \qquad F(X,0) = 0, \qquad F(0,Y) = Y.$$

We observe that the construction of the invariant differential, formal logarithm, and formal exponential, and the proofs of their basic properties used only these three properties of $F(X,Y)$. Thus letting $K = R \otimes \mathbb{Q}$, this proves the existence of power series $\log(T), \exp(T) \in K[\![T]\!]$ satisfying

$$F(X,Y) = \exp\big(\log(X) + \log(Y)\big).$$

In particular, we see that $F(X,Y) = F(Y,X)$. In other words, every one-parameter formal group over a torsion-free ring is automatically commutative. (See Exercise 4.2b for a more precise statement.)

For certain applications it is useful to have a bound on the denominators appearing in $\log_{\mathcal{F}}$ and $\exp_{\mathcal{F}}$. The answer for $\log_{\mathcal{F}}$ is clear from the definition, while for $\exp_{\mathcal{F}}$ we use the following calculation.

Lemma 5.4. *Let R be a torsion-free ring, let $K = R \otimes \mathbb{Q}$, and let*

$$f(T) = \sum_{n=1}^{\infty} \frac{a_n}{n!} T^n \in K[\![T]\!]$$

be a power series with $a_n \in R$ and $a_1 \in R^*$. Then there is a unique power series $g(T) \in K[\![T]\!]$ satisfying $f\big(g(T)\big) = T$; cf. (IV.2.4). The series $g(T)$ has the form

$$g(T) = \sum_{n=1}^{\infty} \frac{b_n}{n!} T^n$$

with $b_n \in R$.

PROOF. Differentiating $f\big(g(T)\big) = T$ gives

$$f'\big(g(T)\big)g'(T) = 1.$$

and evaluating at $T = 0$ shows that

$$b_1 = g'(0) = \frac{1}{f'(0)} = \frac{1}{a_1} \in R^*.$$

Differentiating again yields

$$f'\big(g(T)\big)g''(T) + f''\big(g(T)\big)g'(T)^2 = 0.$$

Repeated differentiation shows that for every $n \geq 2$, the quantity $f'\big(g(T)\big)g^{(n)}(T)$ can be expressed as a polynomial (with integer coefficients) in the variables

$$f^{(i)}\big(g(T)\big) \quad \text{with } 1 \leq i \leq n \quad \text{and} \quad g^{(j)}(T) \quad \text{with } 1 \leq j \leq n-1.$$

Evaluating at $T = 0$ expresses $a_1 b_n$ as a polynomial in $a_1, \ldots, a_n, b_1, \ldots, b_{n-1}$. Since $a_1, b_1 \in R^*$, an easy induction shows that every $b_n \in R$. $\qquad\square$

Proposition 5.5. *Let R be a torsion-free ring and let $\mathcal{F}/R$ be a formal group. Then*

$$\log_{\mathcal{F}}(T) = \sum_{n=1}^{\infty} \frac{a_n}{n} T^n \quad \text{and} \quad \exp_{\mathcal{F}}(T) = \sum_{n=1}^{\infty} \frac{b_n}{n!} T^n$$

with $a_n, b_n \in R$ and $a_1 = b_1 = 1$.

PROOF. The expression for $\log_{\mathcal{F}}$ follows directly from the definition of the formal logarithm, and then (IV.5.4) implies that $\exp_{\mathcal{F}}$ has the specified form. $\qquad\square$

IV.6   Formal Groups over Discrete Valuation Rings

Let R be a complete local ring with maximal ideal $\mathcal{M}$, and let $\mathcal{F}/R$ be a formal group. As we have seen (IV.3.2b), the associated group $\mathcal{F}(\mathcal{M})$ has no torsion of order prime to $p = \operatorname{char}(R/\mathcal{M})$. We analyze more closely the p-primary torsion when R is a discrete valuation ring.

Theorem 6.1. *Let R be a discrete valuation ring that is complete with respect to its maximal ideal $\mathcal{M}$, let $p = \operatorname{char}(R/\mathcal{M})$, and let v be the valuation on R. Let $\mathcal{F}/R$ be a formal group, and suppose that $x \in \mathcal{F}(\mathcal{M})$ has exact order p^n for some $n \geq 1$, i.e.,*

$$[p^n](x) = 0 \quad \text{and} \quad [p^{n-1}](x) \neq 0.$$

Then

$$v(x) \leq \frac{v(p)}{p^n - p^{n-1}}.$$

PROOF. The statement is trivial (and uninteresting) if $\operatorname{char}(R) \neq 0$ or if $p = 0$, since then $v(p) = \infty$, so we may assume that $\operatorname{char}(R) = 0$ and that $p > 0$. From (IV.4.4) we know that there are power series $f(T), g(T) \in R[\![T]\!]$ such that

$$[p](T) = pf(T) + g(T^p),$$

and (IV.2.3a) tells us that $f(T) = T + \cdots$. We are going to prove the theorem by induction on n.

Suppose first that $x \neq 0$ and $[p](x) = 0$. Thus

$$0 = pf(x) + g(x^p).$$

Since R is a discrete valuation ring and the linear term of $f(T)$ is T, the only way that the leading term of $pf(x)$ can be eliminated is to have

$$v(px) \geq v(x^p).$$

Hence

$$v(p) \geq (p-1)v(x).$$

Now assume that the theorem is true for n, and let $x \in \mathcal{F}(\mathcal{M})$ have exact order p^{n+1}. Then

$$v\big([p](x)\big) = v\big(pf(x) + g(x^p)\big) \geq \min\{v(px), v(x^p)\}.$$

The point $[p](x)$ has exact order p^n, so the induction hypothesis tells us that

$$\frac{v(p)}{p^n - p^{n-1}} \geq v\big([p](x)\big),$$

and therefore

$$\frac{v(p)}{p^n - p^{n-1}} \geq \min\{v(px), v(x^p)\}.$$

Since $v(x) > 0$ and $n \geq 1$, it is not possible to have

$$\frac{v(p)}{p^n - p^{n-1}} \geq v(px).$$

We conclude that

$$\frac{v(p)}{p^n - p^{n-1}} \geq v(x^p) = pv(x),$$

which is the desired result.   □

Example 6.1.1. Let $\mathcal{F}$ be a formal group defined over $\mathbb{Z}_p$, the ring of p-adic integers. If $p \geq 2$, then (IV.6.1) says that $\mathcal{F}(p\mathbb{Z}_p)$ has no torsion at all, and even for $p = 2$ it has at most elements of order 2. The same holds for the ring of integers in any finite *unramified* extension of $\mathbb{Q}_p$. For a general finite extension, the determining factor for possible p-primary torsion is the ramification degree of the extension, i.e., the value of $v(p)$ if one takes v to be a normalized valuation.

Next we show that a large piece of $\mathcal{F}(\mathcal{M})$ looks like the additive group. The idea is to use the formal logarithm to define a map, but the presence of denominators means that convergence is no longer automatic. The following two lemmas will thus be useful.

Lemma 6.2. *Let v be a valuation and let $p \in \mathbb{Z}$ be a prime such that $0 < v(p) < \infty$. Then for all integers $n \geq 1$,*

$$v(n!) \leq \frac{(n-1)v(p)}{p-1}.$$

PROOF. We compute

$$v(n!) = \sum_{i=1}^{\infty} \left[\frac{n}{p^i}\right] v(p) \leq \sum_{i=1}^{[\log_p n]} \frac{nv(p)}{p^i} = \frac{nv(p)}{p-1}\left(1 - p^{-[\log_p n]}\right) \leq \frac{(n-1)v(p)}{p-1}.$$

$\square$

Lemma 6.3. *Let R be a ring of characteristic 0 that is complete with respect to a discrete valuation v, and let $p \in \mathbb{Z}$ be a prime with $v(p) > 0$.*
(a) *Let $f(T)$ be a power series of the form*

$$f(T) = \sum_{n=1}^{\infty} \frac{a_n}{n} T^n \quad \text{with} \quad a_n \in R.$$

If $x \in R$ satisfies $v(x) > 0$, then $f(x)$ converges in R.
(b) *Let $g(T)$ be a power series of the form*

$$g(T) = \sum_{n=1}^{\infty} \frac{b_n}{n!} T^n \quad \text{with} \quad b_n \in R.$$

If $x \in R$ satisfies $v(x) > v(p)/(p-1)$, then $g(x)$ converges in R. If further $b_1 \in R^*$, then

$$v\bigl(g(x)\bigr) = v(x).$$

PROOF. (a) For a general term of $f(x)$ we have

$$v(a_n x^n / n) \geq nv(x) - v(n) \qquad \text{since } a_n \in R,$$
$$\geq nv(x) - (\log_p n)v(p).$$

This last expression goes to ∞ as n goes to infinity. Since v is nonarchimedean and R is complete, the series $f(x)$ converges.
(b) For a general term of the series $g(x)$, we have

$$v(b_n x^n / n!) \geq nv(x) - v(n!) \qquad\qquad \text{since } b_n \in R,$$
$$\geq nv(x) - (n-1)\frac{v(p)}{p-1} \qquad\qquad \text{from (IV.6.2)},$$
$$= v(x) + (n-1)\left(v(x) - \frac{v(p)}{p-1}\right).$$

We are assuming that $v(x) > v(p)/(p-1)$, so

$$v(b_n x^n/n!) \to \infty \qquad \text{as } n \to \infty,$$

and

$$v(b_n x^n/n!) > v(x) \qquad \text{for all } n \geq 2.$$

Since v is nonarchimedean, the former implies that $g(x)$ converges; and if $b_1 \in R^*$, so $v(b_1 x) = v(x)$, then the latter shows that the leading term dominates. $\qquad\square$

Theorem 6.4. *Let K be a field of characteristic 0 that is complete with respect to a normalized discrete valuation v, i.e., $v(K^*) = \mathbb{Z}$, let R be the valuation ring of K, let $\mathcal{M}$ be the maximal ideal of R, and let p be a prime with $v(p) > 0$. Consider a formal group $\mathcal{F}/R$.*

(a) *The formal logarithm induces a homomorphism*

$$\log_{\mathcal{F}} : \mathcal{F}(\mathcal{M}) \longrightarrow K,$$

where the group law on K is addition.

(b) *Let $r > v(p)/(p-1)$ be an integer. Then the formal logarithm induces an isomorphism*

$$\log_{\mathcal{F}} : \mathcal{F}(\mathcal{M}^r) \xrightarrow{\ \sim\ } \hat{\mathbb{G}}_a(\mathcal{M}^r).$$

PROOF. (a) From (IV.5.2) we have an identity of power series

$$\log_{\mathcal{F}}\big(F(X,Y)\big) = \log_{\mathcal{F}}(X) + \log_{\mathcal{F}}(Y).$$

Hence $\log_{\mathcal{F}}$ will be a homomorphism on $\mathcal{M}$ provided that $\log_{\mathcal{F}}(x)$ converges for $x \in \mathcal{M}$. The convergence follows from (IV.5.5) and (IV.6.3a).

(b) Similarly, since (IV.5.2) says that $\log_{\mathcal{F}}$ and $\exp_{\mathcal{F}}$ are inverse maps as formal power series, it suffices to show that for all $x \in \mathcal{M}^r$, the power series $\log_{\mathcal{F}}(x)$ and $\exp_{\mathcal{F}}(x)$ converge to values in $\mathcal{M}^r$. This follows directly from the estimates given in (IV.5.5) and (IV.6.3b). (Note that since v is normalized, the conditions $x \in \mathcal{M}^r$ and $v(x) \geq r$ are equivalent.) $\qquad\square$

Remark 6.5. If $r > v(p)/(p-1)$, then (IV.6.4) implies that $\mathcal{F}(\mathcal{M}^r)$ is torsion-free, since $\hat{\mathbb{G}}_a(\mathcal{M}^r)$ certainly has no torsion. We thus recover the $n = 1$ case of (IV.6.1).

IV.7   Formal Groups in Characteristic p

For this section we let R be a ring of characteristic $p > 0$.

Definition. Let $\mathcal{F}/R$ and $\mathcal{G}/R$ be formal groups, and let $f : \mathcal{F} \to \mathcal{G}$ be a homomorphism defined over R. The *height of f*, denoted by $\mathrm{ht}(f)$, is the largest integer h such that

$$f(T) = g(T^{p^h})$$

for some power series $g(T) \in R[[T]]$. (If $f = 0$, we set $\mathrm{ht}(f) = \infty$.) The *height of the formal group $\mathcal{F}$*, denoted by $\mathrm{ht}(\mathcal{F})$, is the height of the multiplication-by-p map $[p] : \mathcal{F} \to \mathcal{F}$.

Remark 7.1. If $m \geq 1$ is prime to p, then $\text{ht}([m]) = 0$, since (IV.2.3a) says that $[m](T) = mT + \cdots$. On the other hand, (IV.4.4) implies that $\text{ht}([p]) \geq 1$, so the height of a formal group over a ring of positive characteristic is always positive.

Proposition 7.2. Let $\mathcal{F}/R$ and $\mathcal{G}/R$ be formal groups, and let $f : \mathcal{F} \to \mathcal{G}$ be a homomorphism defined over R.
  (a) If $f'(0) = 0$, then $f(T) = f_1(T^p)$ for some $f_1 \in R[\![T]\!]$.
  (b) Write $f(T) = g(T^{p^h})$ with $h = \text{ht}(f)$. Then $g'(0) \neq 0$.

PROOF. (a) Let $\omega_{\mathcal{F}}$ and $\omega_{\mathcal{G}}$ be the normalized invariant differentials on $\mathcal{F}$ and $\mathcal{G}$. Then

$$
\begin{aligned}
0 = f'(0)\omega_{\mathcal{F}}(T) &\qquad \text{since } f'(0) = 0, \\
= \omega_{\mathcal{G}}(f(T)) &\qquad \text{from (IV.4.3),} \\
= (1 + \cdots)f'(T)\,dT.
\end{aligned}
$$

Hence $f'(T) = 0$, so $f(T) = f_1(T^p)$.
  (b) Let $q = p^h$, and if $F(X,Y) = \sum a_{ij}X^iY^j$ is the power series defining the formal group $\mathcal{F}$, let $\mathcal{F}^{(q)}$ denote the formal group defined by the power series $F^{(q)}(X,Y) = \sum a_{ij}^q X^iY^j$. Using the fact that $\text{char}(R) = p$, it is easy to check that $\mathcal{F}^{(q)}$ is a formal group. We claim that g is a homomorphism from $\mathcal{F}^{(q)}$ to $\mathcal{G}$. To prove this, we compute:

$$
\begin{aligned}
g(F^{(q)}(X,Y)) = g(F(S,T)^q) &\qquad \text{writing } S^q = X \text{ and } T^q = Y, \\
= f(F(S,T)) \\
= G(f(S), f(T)) &\qquad \text{since } f \text{ is a homomorphism,} \\
= G(g(S^q), g(T^q)) \\
= G(g(X), g(Y)).
\end{aligned}
$$

Suppose that $g'(0)$ equals 0. Then from (a) we have $g(T) = g_1(T^p)$, which implies that

$$
f(T) = g(T^{p^h}) = g_1(T^{p^{h+1}}).
$$

This contradicts the assumption that $h = \text{ht}(f)$. Therefore $g'(0) \neq 0$.   □

Next we show that the height behaves well under composition.

Proposition 7.3. Let $\mathcal{F}/R$, $\mathcal{G}/R$, and $\mathcal{H}/R$ be formal groups, and let

$$
\mathcal{F} \xrightarrow{\ f\ } \mathcal{G} \xrightarrow{\ g\ } \mathcal{H}
$$

be a chain of homomorphisms defined over R. Then

$$
\text{ht}(g \circ f) = \text{ht}(f) + \text{ht}(g).
$$
   □

PROOF. Write

$$f(T) = f_1(T^{p^{\mathrm{ht}(f)}}) \qquad \text{and} \qquad g(T) = g_1(T^{p^{\mathrm{ht}(g)}}).$$

Then

$$(g \circ f)(T) = g_1\big(f_1(T^{p^{\mathrm{ht}(f)}})^{\mathrm{ht}(g)}\big) = g_1\big(\tilde{f}_1(T^{p^{\mathrm{ht}(f)+\mathrm{ht}(g)}})\big),$$

where $\tilde{f}_1$ is obtained from f_1 by raising each coefficient of f_1 to the $p^{\mathrm{ht}(g)}$ power. We know from (IV.7.2b) that g_1 and f_1 have nonzero linear terms, so it follows that

$$\mathrm{ht}(g \circ f) = \mathrm{ht}(f) + \mathrm{ht}(g). \qquad \square$$

We now resume our study of elliptic curves by giving a relationship between the inseparable degree of an isogeny and the height of the associated map of formal groups.

Theorem 7.4. *Let K be a field of characteristic $p > 0$, let E_1/K and E_2/K be elliptic curves, and let $\phi : E_1 \to E_2$ be a nonzero isogeny defined over K. Further, let $f : \hat{E}_1 \to \hat{E}_2$ be the homomorphism of formal groups induced by ϕ. Then*

$$\deg_i(\phi) = p^{\mathrm{ht}(f)}.$$

Corollary 7.5. *Let E/K be an elliptic curve defined over a field of positive characteristic. Then*

$$\mathrm{ht}(\hat{E}) = 1 \text{ or } 2.$$

PROOF. We start with two special cases.

Case 1. ϕ is the p^r-power Frobenius map.
Then (II.2.11) says that $\deg_i \phi = p^r$, while $f(T) = T^{p^r}$, so clearly $\mathrm{ht}(f) = r$.

Case 2. ϕ is separable.
Let ω be an invariant differential on E_2/K, and let $\omega(T)$ be the corresponding differential on the formal group $\hat{E}_2$. Since ϕ is separable by assumption, (II.4.2c) tells us that $\phi^*\omega \neq 0$, so using (IV.4.3) we conclude that

$$(\omega \circ f)(T) = f'(0)\omega(T) \neq 0.$$

It follows that $f'(0) \neq 0$, and hence $\mathrm{ht}(f) = 0$.

We now use the fact (II.2.12) that every isogeny is the composition of a Frobenius map and a separable map. The theorem then follows from the two cases already considered, since inseparable degrees multiply under composition, while (IV.7.3) tells us that heights add under composition.

The corollary is immediate on applying the theorem with $\phi = [p]$, since the map $[p]$ has degree p^2 from (III.6.4a). $\qquad \square$

Exercises

4.1. Let $F(X,Y) \in R[\![X,Y]\!]$ be a power series satisfying

$$F(X,Y) = X + Y + \cdots \qquad \text{and} \qquad F\big(X, F(Y,Z)\big) = F\big(F(X,Y), Z\big).$$

(a) Show that there is a unique power series $i(T) \in R[\![T]\!]$ satisfying $F\big(T, i(T)\big) = 0$. Prove that $i(T)$ also satisfies $F\big(i(T), T\big) = 0$.

(b) Prove that $F(X,0) = X$ and $F(0,Y) = Y$.

4.2. (a) Let $R = \mathbb{F}_p[\epsilon]/(\epsilon^2)$. Prove that

$$F(X,Y) = X + Y + \epsilon XY^p$$

defines a noncommutative formal group, i.e., F has all of the properties to be a formal group law except that $F(X,Y) \neq F(Y,X)$.

(b) * Let R be a ring. Prove that there exists a noncommutative formal group defined over R if and only if there are a nonzero element $\epsilon \in R$ and positive integers m and n such that $m\epsilon = \epsilon^n = 0$.

4.3. Let R be the ring of integers in a finite extension of $\mathbb{Q}_p$, and let $\mathcal{F}/R$ be a formal group.

(a) Prove that for every $x \in \mathcal{F}(\mathcal{M})$,

$$\lim_{n \to \infty} [p^n](x) = 0.$$

(b) Prove that for every $\alpha \in \mathbb{Z}_p$ there exists a unique homomorphism $[\alpha] : \mathcal{F} \to \mathcal{F}$ satisfying

$$[\alpha](T) = \alpha T + \cdots \in R[\![T]\!].$$

4.4. Let R and $\mathcal{F}/R$ be as in Exercise 4.3, and let h be the height of the formal group over the residue field $R/\mathcal{M}$ obtained by reducing modulo $\mathcal{M}$ the coefficients of the formal group law for $\mathcal{F}$. Prove that there is a finite extension R' of R with maximal ideal $\mathcal{M}'$ such that the p-torsion in $\mathcal{F}(\mathcal{M}')$ is isomorphic to $(\mathbb{Z}/p\mathbb{Z})^h$. (*Hint.* Use the p-adic version of the Weierstrass preparation theorem [143, Chapter 5, Theorem 11.2].) This provides an alternative proof of (IV.7.5).

4.5. Let E be the elliptic curve $y^2 = x^3 + Ax$.

(a) Let $w(z) = \sum A_n z^n$ be the power series for E described in (IV §1). Prove that

$$A_n = 0 \quad \text{for all} \quad n \not\equiv 3 \pmod{4}.$$

(b) Let $F(X,Y) = \sum F_n(X,Y)$ be the formal group law for E, where $F_n(X,Y)$ is a homogeneous polynomial of degree n. Prove that

$$F_n = 0 \quad \text{for all} \quad n \not\equiv 1 \pmod{4}.$$

(c) Prove results analogous to (a) and (b) for the curve $y^2 = x^3 + B$.

4.6. Using the notation from (IV.6.1), let $k = R/\mathcal{M}$, and let h be the height of the formal group $\tilde{\mathcal{F}}/k$ obtained by reducing the coefficients of the formal group law $F(X,Y)$ modulo $\mathcal{M}$. Suppose that $x \in \mathcal{F}(\mathcal{M})$ has exact order p^{n+1}. Prove that

$$v(x) \leq \left[\frac{v(p)}{p^{hn}(p^h - 1)} \right].$$

Since every formal group has height $h \geq 1$, this strengthens (IV.6.1).

Chapter V

Elliptic Curves over Finite Fields

In this chapter we study elliptic curves defined over a finite field $\mathbb{F}_q$. The most important arithmetic quantity associated to such a curve is its number of rational points. We start by a proving a theorem of Hasse that says that if $E/\mathbb{F}_q$ is an elliptic curve, then $E(\mathbb{F}_q)$ has approximately q points, with an error of no more than $2\sqrt{q}$. Following Weil, we then reinterpret and extend this result in terms of a certain generating function, the zeta function of the curve. In the final two sections we study in some detail the endomorphism ring of an elliptic curve defined over a finite field, and in particular we give a relationship between $\mathrm{End}(E)$ and the existence of nontrivial p-torsion points. We fix the following notation for Chapter V:

q	a power of a prime p.
$\mathbb{F}_q$	a finite field with q elements.
$\bar{\mathbb{F}}_q$	an algebraic closure of $\mathbb{F}_q$.

V.1 Number of Rational Points

Let $E/\mathbb{F}_q$ be an elliptic curve defined over a finite field. We wish to estimate the number of points in $E(\mathbb{F}_q)$, or equivalently, one more than the number of solutions to the equation

$$E : y^2 + a_1 xy + a_3 y = x^3 + a_2 x^2 + a_4 x + a_6 \qquad \text{with } (x, y) \in \mathbb{F}_q^2.$$

Since each value of x yields at most two values for y, a trivial upper bound is

$$\#E(\mathbb{F}_q) \le 2q + 1.$$

However, since a "randomly chosen" quadratic equation has a 50% chance of being solvable in $\mathbb{F}_q$, we expect that the right order of magnitude should be q, rather than $2q$.

J.H. Silverman, *The Arithmetic of Elliptic Curves, Second Edition*, Graduate Texts in Mathematics 106, DOI 10.1007/978-0-387-09494-6_V,
© Springer Science+Business Media, LLC 2009

The next result, which was conjectured by E. Artin in his thesis and proven by Hasse in the 1930s, shows that this heuristic reasoning is correct.

Theorem 1.1. (Hasse) *Let $E/\mathbb{F}_q$ be an elliptic curve defined over a finite field. Then*

$$\left| \#E(\mathbb{F}_q) - q - 1 \right| \leq 2\sqrt{q}.$$

PROOF. Choose a Weierstrass equation for E with coefficients in $\mathbb{F}_q$, and let

$$\phi : E \longrightarrow E, \qquad (x, y) \longmapsto (x^q, y^q),$$

be the qth-power Frobenius morphism (III.4.6). Since the Galois group $G_{\bar{\mathbb{F}}_q/\mathbb{F}_q}$ is (topologically) generated by the qth-power map on $\bar{\mathbb{F}}_q$, we see that for any point $P \in E(\bar{\mathbb{F}}_q)$,

$$P \in E(\mathbb{F}_q) \qquad \text{if and only if} \qquad \phi(P) = P.$$

Thus

$$E(\mathbb{F}_q) = \ker(1 - \phi),$$

so using (III.5.5) and (III.4.10c), we find that

$$\#E(\mathbb{F}_q) = \#\ker(1 - \phi) = \deg(1 - \phi).$$

(Note the importance of knowing that the map $1 - \phi$ is separable.) Since the degree map on $\text{End}(E)$ is a positive definite quadratic form (III.6.3) and since $\deg \phi = q$, the following version of the Cauchy–Schwarz inequality gives the desired result.   □

Lemma 1.2. *Let A be an abelian group, and let*

$$d : A \longrightarrow \mathbb{Z}$$

be a positive definite quadratic form. Then

$$\left| d(\psi - \phi) - d(\phi) - d(\psi) \right| \leq 2\sqrt{d(\phi)d(\psi)} \qquad \text{for all } \psi, \phi \in A.$$

PROOF. For $\psi, \phi \in A$, let

$$L(\psi, \phi) = d(\psi - \phi) - d(\phi) - d(\psi)$$

be the bilinear form associated to the quadratic form d. Since d is positive definite, we have for all $m, n \in \mathbb{Z}$,

$$0 \leq d(m\psi - n\phi) = m^2 d(\psi) + mnL(\psi, \phi) + n^2 d(\phi).$$

In particular, taking

$$m = -L(\psi, \phi) \qquad \text{and} \qquad n = 2d(\psi)$$

yields

$$0 \leq d(\psi)\left(4d(\psi)d(\phi) - L(\psi, \phi)^2\right).$$

This gives the desired inequality, provided that $\psi \neq 0$, while for $\psi = 0$ the original inequality is trivial.   □

Application 1.3. Let $\mathbb{F}_q$ be a finite field with q odd. We can use Hasse's result to estimate the value of certain character sums on $\mathbb{F}_q$. Thus let

$$f(x) = ax^3 + bx^2 + cx + d \in K[x]$$

be a cubic polynomial with distinct roots in $\bar{\mathbb{F}}_q$, and let

$$\chi : \mathbb{F}_q^* \longrightarrow \{\pm 1\}$$

be the unique nontrivial character of order 2, i.e., $\chi(t) = 1$ if and only if t is a square in $\mathbb{F}_q^*$. Extend χ to $\mathbb{F}_q$ by setting $\chi(0) = 0$. We can use χ to count the $\mathbb{F}_q$-rational points on the elliptic curve

$$E : y^2 = f(x).$$

Each $x \in \mathbb{F}_q$ yields zero, one, or two points $(x, y) \in E(\mathbb{F}_q)$ according to whether the value $f(x)$ is, respectively, a nonsquare, equal to zero, or a square in $\mathbb{F}_q$. Thus in terms of χ we obtain (remember the extra point at infinity)

$$\#E(\mathbb{F}_q) = 1 + \sum_{x \in \mathbb{F}_q} \big(1 + \chi(f(x))\big)$$

$$= 1 + q + \sum_{x \in \mathbb{F}_q} \chi(f(x)).$$

Comparing this with (V.1.1) yields the following result.

Corollary 1.4. *With notation as above,*

$$\left| \sum_{x \in \mathbb{F}_q} \chi(f(x)) \right| \le 2\sqrt{q}.$$

We note that the sum in (V.1.4) consists of q terms, each of which is ± 1, so (V.1.4) says that as x runs through $\mathbb{F}_q$, the values of the cubic polynomial $f(x)$ tend to be equally distributed between squares and nonsquares. Indeed, if one takes a random sequence $(\epsilon_1, \ldots, \epsilon_q)$ of ones and negative ones, then the expected value of $|\epsilon_1 + \cdots + \epsilon_q|^2$ is q, so (V.1.4) says that the set of values of $\big(\chi(f(x))\big)_{x \in \mathbb{F}_q}$ looks like a random sequence.

Remark 1.5. Hasse's theorem (V.1.1) gives a bound for the number of points in $E(\mathbb{F}_q)$, but it does not provide a practical algorithm for computing $\#E(\mathbb{F}_q)$ when q is large. See (XI §3).

Remark 1.6. Let $E/\mathbb{F}_q$ be an elliptic curve, and let $P, Q \in E(\mathbb{F}_q)$ be points such that Q is in the subgroup generated by P. The *elliptic curve discrete logarithm problem* (ECDLP) asks for an integer m satisfying $Q = [m]P$. If q is small, we can compute $P, [2]P, [3]P, \ldots$ until we find Q, but for large values of q it is quite difficult to find m. This has led people to create public key cryptosystems based on the difficulty of solving the ECDLP. See (XI §§4–7) for a discussion of elliptic curve cryptography and the ECDLP.

V.2   The Weil Conjectures

In 1949, André Weil made a series of remarkable conjectures concerning the number of points on varieties defined over finite fields. In this section we state Weil's conjectures and prove them for elliptic curves.

For each integer $n \geq 1$, let $\mathbb{F}_{q^n}$ be the extension of $\mathbb{F}_q$ of degree n, so $\#\mathbb{F}_{q^n} = q^n$. Let $V/\mathbb{F}_q$ be a projective variety, say V is the set of solutions to

$$f_1(x_0, \ldots, x_N) = \cdots = f_m(x_0, \ldots, x_N) = 0,$$

where $f_1, \ldots, f_m$ are homogeneous polynomials with coefficients in $\mathbb{F}_q$. Then $V(\mathbb{F}_{q^n})$ is the set of points of V with coordinates in $\mathbb{F}_{q^n}$. We encode the number of points in $V(\mathbb{F}_{q^n})$ for all $n \geq 1$ into a generating function.

Definition. The *zeta function of $V/\mathbb{F}_q$* is the power series

$$Z(V/\mathbb{F}_q; T) = \exp\left(\sum_{n=1}^{\infty} (\#V(\mathbb{F}_{q^n})) \frac{T^n}{n}\right).$$

Here, for any power series $F(T) \in \mathbb{Q}[[T]]$ with no constant term, we define the power series $\exp(F(T))$ to be the series $\sum_{k \geq 0} F(T)^k/k!$. Note that if we know the series $Z(V/\mathbb{F}_q; T)$, then we can recover the numbers $\#V(\mathbb{F}_{q^n})$ by the formula

$$\#V(\mathbb{F}_{q^n}) = \frac{1}{(n-1)!} \frac{d^n}{dT^n} \log Z(V/\mathbb{F}_q; T)\Big|_{T=0}.$$

The reason for defining $Z(V/\mathbb{F}_q; T)$ in this way, rather than using the more natural series $\sum \#V(\mathbb{F}_{q^n})T^n$, will soon be apparent.

Example 2.1. Let $V = \mathbb{P}^N$. Then a point of $V(\mathbb{F}_{q^n})$ is given by homogeneous coordinates $[x_0, \ldots, x_N]$ with $x_i \in \mathbb{F}_{q^n}$ not all zero. Two sets of coordinates give the same point if they differ by multiplication by an element of $\mathbb{F}_{q^n}^*$. Hence

$$\#\mathbb{P}^N(\mathbb{F}_{q^n}) = \frac{q^{n(N+1)} - 1}{q^n - 1} = \sum_{i=0}^{N} q^{ni},$$

so

$$\log Z(\mathbb{P}^n/\mathbb{F}_q; T) = \sum_{n=1}^{\infty} \left(\sum_{i=0}^{N} q^{ni}\right) \frac{T^n}{n} = \sum_{i=0}^{N} -\log(1 - q^i T).$$

Thus

$$Z(\mathbb{P}^n/\mathbb{F}_q; T) = \frac{1}{(1-T)(1-qT)\cdots(1-q^N T)}.$$

Notice that in this case the zeta function is actually in $\mathbb{Q}(T)$. In general, if there are numbers $\alpha_1, \ldots, \alpha_r \in \mathbb{C}$ such that

$$\#V(\mathbb{F}_{q^n}) = \pm\alpha_1^n \pm \cdots \pm \alpha_r^n \qquad \text{for all } n = 1, 2, \ldots,$$

then $Z(V/\mathbb{F}_q; T)$ is a rational function.

Theorem 2.2. (Weil Conjectures) *Let $V/\mathbb{F}_q$ be a smooth projective variety of dimension N.*

(a) Rationality
$$Z(V/\mathbb{F}_q; T) \in \mathbb{Q}(T).$$

(b) Functional Equation
There is an integer ϵ, called the Euler characteristic *of V, such that*
$$Z(V/\mathbb{F}_q; 1/q^N T) = \pm q^{N\epsilon/2} T^\epsilon Z(V/\mathbb{F}_q; T).$$

(c) Riemann Hypothesis
The zeta function factors as
$$Z(V/\mathbb{F}_q; T) = \frac{P_1(T) \cdots P_{2N-1}(T)}{P_0(T) P_2(T) \cdots P_{2N}(T)}$$

with each $P_i(T) \in \mathbb{Z}[T]$, with
$$P_0(T) = 1 - T \qquad and \qquad P_{2N}(T) = 1 - q^N T,$$

and such that for every $0 \le i \le 2N$, the polynomial $P_i(T)$ factors over $\mathbb{C}$ as
$$P_i(T) = \prod_{j=1}^{b_i} (1 - \alpha_{ij} T) \quad with \quad |\alpha_{ij}| = q^{1/2}.$$

The quantity b_i, i.e., the degree of $P_i(T)$, is called the ith Betti number *of V.*

This conjecture was proposed by Weil in 1949 [305] and proven by him for curves and for abelian varieties. The rationality of the zeta function in general was established by Dwork [70] in 1960 using techniques of p-adic functional analysis. Soon thereafter the ℓ-adic cohomology theory developed by M. Artin, Grothendieck, and others was used to give another proof of rationality and to establish the functional equation. Then, in 1973, Deligne [60] proved the Riemann hypothesis. For a nice overview of Deligne's proof, see [123].

We are going to prove the Weil conjectures for elliptic curves. Let ℓ be a prime different from $p = \text{char}(\mathbb{F}_q)$. Recall that there is a representation (III §7)
$$\text{End}(E) \longrightarrow \text{End}(T_\ell(E)), \qquad \psi \longmapsto \psi_\ell,$$

and choosing a $\mathbb{Z}_\ell$-basis for $T_\ell(E)$, we can write ψ_ℓ as a 2×2 matrix and compute its determinant and trace, $\det(\psi_\ell), \text{tr}(\psi_\ell) \in \mathbb{Z}_\ell$.

Proposition 2.3. *Let $\psi \in \text{End}(E)$. Then*
$$\det(\psi_\ell) = \deg(\psi) \qquad and \qquad \text{tr}(\psi_\ell) = 1 + \deg(\psi) - \deg(1 - \psi).$$

In particular, $\det(\psi_\ell)$ and $\text{tr}(\psi_\ell)$ are in $\mathbb{Z}$ and are independent of ℓ.

PROOF. We already proved this result (III.8.6). $\qquad\qquad\qquad\qquad\qquad\qquad \square$

We apply (V.2.3) to an elliptic curve over a finite field. This enables us to compute the number of points and to deduce an important property of the Frobenius endomorphism.

Theorem 2.3.1. *Let $E/\mathbb{F}_q$ be an elliptic curve, let*

$$\phi : E \longrightarrow E, \qquad (x, y) \longmapsto (x^q, y^q),$$

be the qth-power Frobenius endomorphism, and let

$$a = q + 1 - \#E(\mathbb{F}_q).$$

(a) *Let $\alpha, \beta \in \mathbb{C}$ be the roots of the polynomial $T^2 - aT + q$. Then α and β are complex conjugates satisfying $|\alpha| = |\beta| = \sqrt{q}$, and for every $n \geq 1$,*

$$\#E(\mathbb{F}_{q^n}) = q^n + 1 - \alpha^n - \beta^n.$$

(b) *The Frobenius endomorphism satisfies*

$$\phi^2 - a\phi + q = 0 \qquad in \ \text{End}(E).$$

PROOF. We observed in (V §1) that (III.5.5) and (III.4.10c) imply that

$$\#E(\mathbb{F}_q) = \deg(1 - \phi).$$

We use (V.2.3) to compute

$$\det(\phi_\ell) = \deg(\phi) = q,$$
$$\text{tr}(\phi_\ell) = 1 + \deg(\phi) - \deg(1 - \phi) = 1 + q - \#E(\mathbb{F}_q) = a.$$

Hence the characteristic polynomial of ϕ_ℓ is

$$\det(T - \phi_\ell) = T^2 - \text{tr}(\phi_\ell)T + \det(\phi_\ell) = T^2 - aT + q.$$

(a) Since the characteristic polynomial of ϕ_ℓ has coefficients in $\mathbb{Z}$, we can factor it over $\mathbb{C}$ as

$$\det(T - \phi_\ell) = T^2 - aT + q = (T - \alpha)(T - \beta).$$

For every rational number $m/n \in \mathbb{Q}$ we have

$$\det\left(\frac{m}{n} - \phi_\ell\right) = \frac{\det(m - n\phi_\ell)}{n^2} = \frac{\deg(m - n\phi)}{n^2} \geq 0.$$

Thus the quadratic polynomial $\det(T - \phi_\ell) = T^2 - aT + q \in \mathbb{Z}[T]$ is nonnegative for all $T \in \mathbb{R}$, so either it has complex conjugate roots or it has a double root. In either case we have $|\alpha| = |\beta|$, and then from

$$\alpha\beta = \det\phi_\ell = \deg\phi = q,$$

we deduce that

$$|\alpha| = |\beta| = \sqrt{q}.$$

This gives the first part of (a).

Similarly, for each integer $n \geq 1$, the $(q^n)^{\text{th}}$-power Frobenius endomorphism satisfies

$$\#E(\mathbb{F}_{q^n}) = \deg(1 - \phi^n).$$

It follows that the characteristic polynomial of ϕ_ℓ^n is given by

$$\det(T - \phi_\ell^n) = (T - \alpha^n)(T - \beta^n).$$

(To see this, put ϕ_ℓ into Jordan normal form, so it is upper triangular with α and β on the diagonal.) In particular,

$$
\begin{aligned}
\#E(\mathbb{F}_{q^n}) &= \deg(1 - \phi^n) \\
&= \det(1 - \phi_\ell^n) && \text{from (V.2.3),} \\
&= 1 - \alpha^n - \beta^n + q^n.
\end{aligned}
$$

(b) The Cayley–Hamilton theorem tells us that ϕ_ℓ satisfies its characteristic polynomial, so $\phi_\ell^2 - a\phi_\ell + q = 0$. Applying (V.2.3) gives

$$\deg(\phi^2 - a\phi + q) = \det(\phi_\ell^2 - a\phi_\ell + q) = \det(0) = 0,$$

so $\phi^2 - a\phi + q$ is the zero map in $\mathrm{End}(E)$. $\qquad\square$

Using (V.2.3.1a), it is easy to verify the Weil conjectures for elliptic curves.

Theorem 2.4. *Let $E/\mathbb{F}_q$ be an elliptic curve. Then there is an $a \in \mathbb{Z}$ such that*

$$Z(E/\mathbb{F}_q; T) = \frac{1 - aT + qT^2}{(1 - T)(1 - qT)}.$$

Further,

$$Z(E/\mathbb{F}_q; 1/qT) = Z(E/\mathbb{F}_q; T),$$

and

$$1 - aT + qT^2 = (1 - \alpha T)(1 - \beta T) \quad \text{with} \quad |\alpha| = |\beta| = \sqrt{q}.$$

PROOF. We compute

$$
\begin{aligned}
\log Z(E/\mathbb{F}_q; T) &= \sum_{n=1}^{\infty} \frac{\#E(\mathbb{F}_{q^n}) T^n}{n} && \text{by definition,} \\
&= \sum_{n=1}^{\infty} \frac{(1 - \alpha^n - \beta^n + q^n) T^n}{n} && \text{from (V.2.3.1a),} \\
&= -\log(1 - T) + \log(1 - \alpha T) + \log(1 - \beta T) - \log(1 - qT).
\end{aligned}
$$

Hence

$$Z(E/\mathbb{F}_q; T) = \frac{(1 - \alpha T)(1 - \beta T)}{(1 - T)(1 - qT)}.$$

This is the desired result, since (V.2.3.1a) says that α and β are complex conjugates of absolute value $\sqrt{q}$ and that they satisfy

$$a = \alpha + \beta = \mathrm{tr}(\phi_\ell) = 1 + q - \deg(1 - \phi) \in \mathbb{Z}.$$

Finally, the functional equation is immediate (with $\epsilon = 0$). $\qquad\square$

Remark 2.5. To see why (V.2.2c) is called the Riemann hypothesis, we make a change of variables by setting $T = q^{-s}$. This gives a function of s,

$$\zeta_{E/\mathbb{F}_q}(s) = Z(E/\mathbb{F}_q; q^{-s}) = \frac{1 - aq^{-s} + q^{1-2s}}{(1 - q^{-s})(1 - q^{1-s})}.$$

The functional equation reads

$$\zeta_{E/\mathbb{F}_q}(s) = \zeta_{E/\mathbb{F}_q}(1 - s),$$

which certainly looks familiar. Further, the Riemann hypothesis for $Z(E/\mathbb{F}_q; T)$ says that if $\zeta_{E/\mathbb{F}_q}(s) = 0$, then $|q^s| = \sqrt{q}$, which is equivalent to $\mathrm{Re}(s) = \frac{1}{2}$.

Remark 2.6. Let $E/\mathbb{F}_q$ be an elliptic curve. The quantity

$$a = q + 1 - \#E(\mathbb{F}_q)$$

is called the *trace of Frobenius*, because, as we saw during the proof of (V.2.3.1), it is equal to the trace of the q-power Frobenius map considered as a linear transformation of $T_\ell(E)$. Thus if ϕ denotes the q-power Frobenius map, then (V.2.3) gives

$$\mathrm{tr}(\phi_\ell) = 1 + \deg(\phi) - \deg(1 - \phi) = 1 + q - \#E(\mathbb{F}_q) = a.$$

V.3   The Endomorphism Ring

Let K be a (not necessarily finite) field of characteristic p, and let E/K be an elliptic curve. We have seen (III.6.4) that there are two possibilities for the group of p-torsion points $E[p]$, namely 0 and $\mathbb{Z}/p\mathbb{Z}$. Similarly, as described in (III §9), there are several possibilities for the endomorphism ring $\mathrm{End}(E)$. We now show that the seemingly unrelated values of $E[p]$ and $\mathrm{End}(E)$ are in fact far from independent.

Theorem 3.1. ([60]) *Let K be a field of characteristic p, and let E/K be an elliptic curve. For each integer $r \geq 1$, let*

$$\phi_r : E \longrightarrow E^{(p^r)} \qquad and \qquad \hat{\phi}_r : E^{(p^r)} \longrightarrow E$$

be the p^r-power Frobenius map and its dual.

(a) *The following are equivalent.*

  (i) $E[p^r] = 0$ *for one (all)* $r \geq 1$.

  (ii) $\hat{\phi}_r$ *is (purely) inseparable for one (all)* $r \geq 1$.

  (iii) *The map* $[p] : E \to E$ *is purely inseparable and* $j(E) \in \mathbb{F}_{p^2}$.

  (iv) $\text{End}(E)$ *is an order in a quaternion algebra.*

  (v) *The formal group* $\hat{E}/K$ *associated to* E *has height* 2. (*See* (IV §7).)

(b) *If the equivalent conditions in* (a) *do not hold, then*

$$E[p^r] = \mathbb{Z}/p^r\mathbb{Z} \quad \text{for all } r \geq 1,$$

and the formal group $\hat{E}/K$ *has height* 1. *If further* $j(E) \in \bar{\mathbb{F}}_p$, *then* $\text{End}(E)$ *is an order of a quadratic imaginary field.* (*For the case that* $j(E)$ *is transcendental over* $\mathbb{F}_p$, *see Exercise* 5.8.)

Definition. If E has the properties given in (V.3.1a), then we say that E is *supersingular*, or that E has *Hasse invariant* 0. Otherwise we say that E is *ordinary*, or that E has *Hasse invariant* 1.

Remark 3.2.1. There are other characterizations of supersingular elliptic curves that are important in various applications. See [111, IV §4] for a description in terms of sheaf cohomology and [140, Appendix 2 §5] for a description involving residues of differentials.

Remark 3.2.2. Do not confuse the notions of singularity and supersingularity. A supersingular elliptic curve is, by definition, an elliptic curve, so it is nonsingular. The origin of this potentially confusing terminology is as follows. Historically, elliptic curves defined over $\mathbb{C}$ whose endomorphism rings are larger than $\mathbb{Z}$ were called singular, where "singular" was used in the sense of "unusual" or "rare." However, in this sense, all elliptic curves defined over $\bar{\mathbb{F}}_p$ are singular! The endomorphism rings of most elliptic curves over $\bar{\mathbb{F}}_p$ are orders in imaginary quadratic fields. It is only the rare and unusual curve whose endomorphism ring is an order in a quaternion algebra, whence the term "supersingular."

PROOF OF V.3.1. Conditions (i)–(v) are invariant under field extension, so we may assume that K is algebraically closed, and in particular, a perfect field. For notational convenience, we let $\phi = \phi_1$.

  (a) Since the Frobenius map is purely inseparable (II.2.11b), we have

$$\deg_s(\hat{\phi}_r) = \deg_s[p^r] = \left(\deg_s[p]\right)^r = \left(\deg_s \hat{\phi}\right)^r.$$

Combining this with (III.4.10a) yields

$$\#E[p^r] = \deg_s(\hat{\phi}_r) = \deg(\hat{\phi})^r,$$

from which the equivalence of (i) and (ii) follows immediately.

  Next, from (IV.7.4) and the fact that ϕ is purely inseparable, we have

$$\deg_i \hat{\phi} = \frac{\deg_i[p]}{p} = p^{\text{ht}(\hat{E})-1}.$$

Since $\hat{\phi}$ has degree p, this shows that (ii) and (v) are equivalent.

We now prove that (ii) $\Rightarrow$ (iii) $\Rightarrow$ (iv) $\Rightarrow$ (ii).

(ii) $\Rightarrow$ (iii). From (ii) it is immediate that $[p] = \hat{\phi} \circ \phi$ is purely inseparable, so we must show that $j(E) \in \mathbb{F}_{p^2}$. We apply (II.2.12) to the map $\hat{\phi} : E^{(p)} \to E$. Since $\hat{\phi}$ is purely inseparable by assumption, it follows from (II.2.12) and comparison of degrees that $\hat{\phi}$ factors as

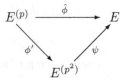

where ϕ' is the pth-power Frobenius map on $E^{(p)}$ and where ψ has degree one. It follows from (II.2.4.1) that ψ is an isomorphism, so

$$j(E) = j\big(E^{(p^2)}\big) = j(E)^{p^2}.$$

(For the second equality, see (III.4.6).) Hence $j(E) \in \mathbb{F}_{p^2}$.

(iii) $\Rightarrow$ (iv). Suppose that $\text{End}(E)$ is not an order in a quaternion algebra. We proceed to derive a contradiction. From (III.9.4) we find that

$$\mathcal{K} = \text{End}(E) \otimes \mathbb{Q}$$

is a number field, since it is either $\mathbb{Q}$ or an imaginary quadratic extension of $\mathbb{Q}$.

Let E' be any elliptic curve that is isogenous to E, say $\psi : E \to E'$. Since $\psi \circ [p] = [p] \circ \psi$, and since $[p] : E \to E$ is purely inseparable by assumption, comparing inseparability degrees shows that $[p] : E' \to E'$ is also purely inseparable. Hence

$$\#E'[p] = \deg_s[p] = 1,$$

so from the already proven implications (i) $\Rightarrow$ (ii) $\Rightarrow$ (iii), we conclude that $j(E') \in \mathbb{F}_{p^2}$. This shows that up to isomorphism, there are only finitely many elliptic curves that are isogenous to E.

Now choose a prime $\ell \in \mathbb{Z}$ with $\ell \neq p$ such that ℓ remains prime in $\text{End}(E')$ for every elliptic curve E' that is isogenous to E. Since there are only finitely many such $\text{End}(E')$ and each is a subring of $\mathcal{K}$, it is easy to find such an ℓ; see Exercise 5.5. From (III.6.4b) we know that

$$E[\ell^i] \cong \mathbb{Z}/\ell^i\mathbb{Z} \times \mathbb{Z}/\ell^i\mathbb{Z},$$

so we can choose a sequence of subgroups

$$\Phi_1 \subset \Phi_2 \subset \cdots \subset E \qquad \text{with } \Phi_i \cong \mathbb{Z}/\ell^i\mathbb{Z}.$$

Let $E_i = E/\Phi_i$ be the quotient of E by Φ_i, so from (III.4.12) there is an isogeny $E \to E_i$ with kernel Φ_i. We know from above that up to isomorphism, there are only finitely many distinct E_i, so we can choose integers $m, n > 0$ such that E_{m+n} and E_m are isomorphic. Composing this isomorphism with the natural projection from E_m to E_{m+n} yields an endomorphism of E_m,

$$\lambda : E_m \xrightarrow{\text{proj}} E_{m+n} \cong E_m.$$

Note that the kernel of λ is *cyclic* of order ℓ^n, since $\ker(\lambda) = \Phi_{m+n}/\Phi_m$. But ℓ is prime in the ring $\operatorname{End}(E_m)$, so by comparing degrees we must have $\lambda = u \circ [\ell^{n/2}]$ for some $u \in \operatorname{Aut}(E_m)$. (Also n must be even.) However, the kernel of $[\ell^{n/2}]$ is not cyclic for any $n > 0$. This contradiction proves that $\mathcal{K}$ is not a number field, and hence that $\operatorname{End}(E)$ is an order in a quaternion algebra.

(iv) $\Rightarrow$ (ii). Suppose that (ii) is false, so $\hat{\phi}_r$ is separable for all $r \geq 1$. We will prove that $\operatorname{End}(E)$ is commutative, which contradicts (iv).

First we show that the natural map

$$\operatorname{End}(E) \longrightarrow \operatorname{End}(T_p(E))$$

is injective. Suppose that $\psi \in \operatorname{End}(E)$ goes to 0. Then from the definition of $T_p(E)$ we have $\psi(E[p^r])$ for all $r \geq 1$. Since $[p^r] = \phi_r \circ \hat{\phi}_r$ and since ϕ_r is surjective (II.2.3), it follows that

$$\phi_r(\ker \psi) \supset \ker \hat{\phi}_r,$$

and thus

$$\# \ker \psi \geq \# \ker \hat{\phi}_r \quad \text{for all } r \geq 1.$$

On the other hand, we know that

$$\# \ker \hat{\phi}_r = \deg \hat{\phi}_r \qquad \text{from (III.4.10c), since } \hat{\phi}_r \text{ is separable,}$$

$$\deg \hat{\phi}_r = \deg \phi_r \qquad \text{from (III.6.2e),}$$

$$\deg \phi_r = p^r \qquad \text{from (II.2.11c).}$$

Therefore $\# \ker \psi \geq p^r$ for all $r \geq 1$, which implies that $\psi = 0$.

Next, from (III.7.1b) we see that $T_p(E)$ is either 0 or $\mathbb{Z}_p$. Further, we have $T_p(E)/pT_p(E) \cong E[p]$, and by assumption $E[p] \neq 0$, so we have $T_p(E) = \mathbb{Z}_p$. Combining this fact with the injection proven earlier, we have

$$\operatorname{End}(E) \longhookrightarrow \operatorname{End}(T_p(E)) \cong \operatorname{End}(\mathbb{Z}_p) \cong \mathbb{Z}_p.$$

Therefore $\operatorname{End}(E)$ is commutative.

(b) From (III.6.4c) we know that $E[p^r]$ is equal to either 0 or $\mathbb{Z}/p^r\mathbb{Z}$ for every $r \geq 1$. Hence if condition (i) of (a) is false, then we must have

$$E[p^r] \cong \mathbb{Z}/p^r\mathbb{Z} \qquad \text{for all } r \geq 1.$$

Further, since (v) is assumed to be false, we can use (IV.7.5) to conclude that $\hat{E}/K$ has height 1.

Next we suppose that $j(E) \in \bar{\mathbb{F}}_p$ and that E does not satisfy the conditions in (a). We use (III.1.4b,c) to find an elliptic curve E' defined over a finite field $\mathbb{F}_{p^r}$ such that E' is isomorphic to E. Then ϕ_r is an endomorphism of E'. Suppose that

$$\phi_r \in \mathbb{Z} \subset \operatorname{End}(E').$$

Comparing degrees yields

$$\phi_r = [\pm p^{r/2}]$$

for some (even) integer r, and then (III.4.10) and (II.2.11b) tell us that

$$\#E[p^{r/2}] = \deg_s \phi_r = 1.$$

This contradicts the assumption that (i) is false. Therefore $\phi_r \notin \mathbb{Z}$, so $\operatorname{End}(E')$ is strictly larger than $\mathbb{Z}$. By assumption, it is not an order in a quaternion algebra, so from (III.9.4), the only remaining possibility is that $\operatorname{End}(E')$ is an order in an imaginary quadratic field. Since $\operatorname{End}(E') = \operatorname{End}(E)$, this completes the proof of (b). $\quad\square$

V.4   Calculating the Hasse Invariant

From (V.3.1a) we see that up to isomorphism, there are only finitely many elliptic curves with Hasse invariant 0, since each such curve has j-invariant in $\mathbb{F}_{p^2}$. For $p = 2$, one easily checks that the only supersingular curve (over $\bar{\mathbb{F}}_2$) is

$$E : y^2 + y = x^3.$$

(See also Exercise 5.7.) For $p > 2$, the next theorem gives a criterion for determining whether an elliptic curve is supersingular.

Theorem 4.1. *Let $\mathbb{F}_q$ be a finite field of characteristic $p \geq 3$.*
(a) *Let $E/\mathbb{F}_q$ be an elliptic curve given by a Weierstrass equation*

$$E : y^2 = f(x),$$

     where $f(x) \in \mathbb{F}_q[x]$ is a cubic polynomial with distinct roots in $\bar{\mathbb{F}}_q$. Then E is supersingular if and only if the coefficient of x^{p-1} in $f(x)^{(p-1)/2}$ is zero.
(b) *Let $m = (p-1)/2$, and define a polynomial*

$$H_p(t) = \sum_{i=0}^{m} \binom{m}{i}^2 t^i.$$

Let $\lambda \in \bar{\mathbb{F}}_q$ with $\lambda \neq 0, 1$. Then the elliptic curve

$$E : y^2 = x(x-1)(x-\lambda)$$

is supersingular if and only if $H_p(\lambda) = 0$.

(c) *The polynomial $H_p(t)$ has distinct roots in $\bar{\mathbb{F}}_q$. There is one supersingular curve in characteristic 3, and for $p \geq 5$, the number of supersingular elliptic curves (up to $\bar{\mathbb{F}}_q$-isomorphism) is*

$$\left[\frac{p}{12}\right] + \begin{cases} 0 & \text{if } p \equiv 1 \pmod{12}, \\ 1 & \text{if } p \equiv 5 \pmod{12}, \\ 1 & \text{if } p \equiv 7 \pmod{12}, \\ 2 & \text{if } p \equiv 11 \pmod{12}. \end{cases}$$

Remark 4.1.1. The results of (V.4.1) (and more) are mostly in [60]. Our proof of (a) follows [154], and the proof of (c) is from [119]. For a beautiful generalization to curves of higher genus, see [154].

PROOF. Let

$$\chi : \mathbb{F}_q^* \longrightarrow \{\pm 1\}$$

be the unique nontrivial character of order 2, and extend χ to $\mathbb{F}_q$ by setting $\chi(0) = 0$. As we have seen in (V.1.3), the character χ can be used to count the number of points of E,

$$\#E(\mathbb{F}_q) = 1 + q + \sum_{x \in \mathbb{F}_q} \chi(f(x)).$$

Since $\mathbb{F}_q^*$ is cyclic of order $q - 1$, for any $z \in \mathbb{F}_q$ we have

$$\chi(z) = z^{(q-1)/2} \quad \text{in } \mathbb{F}_q.$$

Hence

$$\#E(\mathbb{F}_q) = 1 + \sum_{x \in \mathbb{F}_q} f(x)^{(q-1)/2} \qquad \text{as an equality in } \mathbb{F}_q.$$

Again using the cyclic nature of $\mathbb{F}_q^*$, we have the easy result

$$\sum_{x \in \mathbb{F}_q} x^i = \begin{cases} -1 & \text{if } q - 1 \mid i, \\ 0 & \text{if } q - 1 \nmid i. \end{cases}$$

Since $f(x)$ is a polynomial of degree 3, if we expand $f(x)^{(q-1)/2}$, we see that the expansion has terms of the form x^n for $0 \leq n \leq \frac{3}{2}(q - 1)$. Hence when we sum over $x \in \mathbb{F}_q$, the only nonzero term comes from x^{q-1}. Thus if we let

$$A_q = \text{coefficient of } x^{q-1} \text{ in } f(x)^{(q-1)/2},$$

then

$$\#E(\mathbb{F}_q) = 1 - A_q.$$

However, note that this equality is taking place in $\mathbb{F}_q$, so it is actually only a formula for $\#E(\mathbb{F}_q)$ modulo p.

On the other hand, letting $\phi : E \to E$ be the q-power Frobenius endomorphism, we have (V §2)

$$\#E(\mathbb{F}_q) = \deg(1 - \phi) = 1 - a + q,$$

where

$$a = 1 - \deg(1 - \phi) + \deg(\phi).$$

(Thus $[a] = \phi + \hat{\phi}$.) Equating these two expressions for $\#E(\mathbb{F}_q)$, we find that

$$a = A_q \quad \text{as an equality in } \mathbb{F}_q.$$

Since a is an integer, this proves that

$$A_q = 0 \quad \Longleftrightarrow \quad a \equiv 0 \pmod{p}.$$

But $\hat{\phi} = [a] - \phi$, so we find that

$$a \equiv 0 \pmod{p} \Longleftrightarrow \hat{\phi} \text{ is inseparable} \qquad \text{(III.5.5)},$$
$$\Longleftrightarrow E \text{ is supersingular} \qquad \text{(V.3.1a(ii))}.$$

This proves that

$$A_q = 0 \quad \Longleftrightarrow \quad E \text{ is supersingular}.$$

It remains to show that $A_q = 0$ if and only if $A_p = 0$. Writing

$$f(x)^{(p^{r+1}-1)/2} = f(x)^{(p^2-1)/2}\left(f(x)^{(p-1)/2}\right)^{p^r}$$

and equating coefficients (remembering that f is a cubic) yields

$$A_{p^{r+1}} = A_{p^r} A_p^{p^r}.$$

An easy induction on r gives the desired result.

(b) This is a special case of (a). We need the coefficient of x^{p-1} in the expression $\left(x(x-1)(x-\lambda)\right)^m$, so the coefficient of x^m in $(x-1)^m(x-\lambda)^m$. That coefficient is

$$\sum_{i=0}^{m} \binom{m}{i}(-\lambda)^i \binom{m}{m-i}(-1)^{m-i},$$

which differs from $H_p(\lambda)$ by a factor of $(-1)^m$.

(b) Let $\mathcal{D}$ be the differential operator

$$\mathcal{D} = 4t(1-t)\frac{d^2}{dt^2} + 4(1-2t)\frac{d}{dt} - 1.$$

Then by a direct calculation and using the fact that $m = (p-1)/2$, we find that

$$\mathcal{D}H_p(t) = p\sum_{i=0}^{m}(p-2-4i)\binom{m}{i}^2 t^i.$$

In particular, since $\operatorname{char}(\mathbb{F}_q) = p$, we see that

$$DH_p(t) = 0 \quad \text{in } \mathbb{F}_q[t].$$

Hence the only possible multiple roots of $H_p(t)$ in $\bar{\mathbb{F}}_q$ are $t = 0$ and $t = 1$. We compute directly

$$H_p(0) = 1 \quad \text{and} \quad H_p(1) = \binom{p-1}{m} \equiv (-1)^m \pmod{p}.$$

Thus the roots of $H_p(t)$ are distinct, and each root λ gives a supersingular elliptic curve

$$E_\lambda : y^2 = x(x-1)(x-\lambda).$$

It remains to determine to what extent the resulting E_λ are isomorphic to one another.

For $p = 3$ we have $H_p(t) = 1 + t$, so there is exactly one supersingular elliptic curve in characteristic 3. It has j-invariant $j(-1) = 1728 = 0$.

Assume now that $p \geq 5$. We recall from (III.1.7) that the association

$$\lambda \rightarrow j(\lambda) = j(E_\lambda)$$

is six-to-one except over $j = 0$ and $j = 1728$, where it is, respectively, two-to-one and three-to-one. Further, if $H_p(\lambda) = 0$, then for every λ' satisfying $j(\lambda) = j(\lambda')$ we must have $H_p(\lambda') = 0$, since $E_\lambda \cong E_{\lambda'}$ and the roots of $H_p(t)$ are precisely those values of λ for which E_λ is supersingular.

For convenience, let $\epsilon_p(j) = 1$ if the elliptic curve with the indicated j-invariant is supersingular, and let $\epsilon_p(j) = 0$ if it is ordinary. Then, using the fact that $H_p(t)$ has distinct roots, the above discussion shows that the number of supersingular elliptic curves in characteristic $p \geq 5$ is

$$\frac{1}{6}\left(\frac{p-1}{2} - 2\epsilon_p(0) - 3\epsilon_p(1728)\right) + \epsilon_p(0) + \epsilon_p(1728)$$

$$= \frac{p-1}{12} + \frac{1}{2}\epsilon_p(0) + \frac{1}{2}\epsilon_p(1728).$$

We will compute below in (V.4.4) and (V.4.5) that

$$\epsilon_p(0) = \begin{cases} 0 & \text{if } p \equiv 1 \pmod 3, \\ 1 & \text{if } p \equiv 2 \pmod 3, \end{cases} \quad \text{and} \quad \epsilon_p(1728) = \begin{cases} 0 & \text{if } p \equiv 1 \pmod 4, \\ 1 & \text{if } p \equiv 3 \pmod 4. \end{cases}$$

Taking the four possibilities for $p \pmod{12}$ gives the stated result. $\qquad\square$

Remark 4.2. The differential operator $\mathcal{D}$ that we used to prove (V.4.1c) may seem mysterious. It is called a *Picard–Fuchs differential operator* for the Legendre equation

$$y^2 = x(x-1)(x-t).$$

It arises quite naturally when one views the Legendre equation as defining a family of elliptic curves parametrized by the complex variable t, i.e., when E is viewed as an elliptic surface over $\mathbb{P}^1$. For an instructive discussion of this connection, see [46, §2.10].

Example 4.3. For $p = 11$ we have

$$H_{11}(t) = t^5 + 3t^4 + t^3 + t^2 + 3t + 1$$
$$\equiv (t^2 - t + 1)(t + 1)(t - 2)(t + 5) \quad (\text{mod } 11).$$

The supersingular j-invariants in characteristic 11 are $j = 0$ and $j = 1728 = 1$.

Example 4.4. We compute for which primes $p \geq 5$ the elliptic curve

$$E : y^2 = x^3 + 1$$

with $j = 0$ is supersingular. The criterion (V.4.1a) says that we need to compute the coefficient of x^{p-1} in the polynomial $(x^3 + 1)^{(p-1)/2}$. If $p \equiv 2 \pmod{3}$, then there is no x^{p-1} term, so E is supersingular. On the other hand, if $p \equiv 1 \pmod 3$, then the coefficient of x^{p-1} is $\binom{(p-1)/2}{(p-1)/3}$, which is nonzero modulo p, so in this case E is ordinary.

Example 4.5. Similarly, we compute for which primes $p \geq 3$ the elliptic curve

$$E : y^2 = x^3 + x$$

with $j = 1728$ is supersingular. This is determined by the coefficient of $x^{(p-1)/2}$ in the polynomial $(x^2 + 1)^{(p-1)/2}$. This coefficient is equal to 0 if $p \equiv 3 \pmod 4$ and to $\binom{(p-1)/2}{(p-1)/4}$ if $p \equiv 1 \pmod 4$. Hence E is supersingular if $p \equiv 3 \pmod 4$ and ordinary if $p \equiv 1 \pmod 4$.

These examples might suggest that for a given Weierstrass equation with coefficients in $\mathbb{Z}$, the resulting elliptic curve is supersingular in characteristic p for half of the primes. This is in fact true, *provided* that the elliptic curve has complex multiplication over $\bar{\mathbb{Q}}$, as do the $j = 0$ and $j = 1728$ curves. (There is a more precise result due to Deuring that we do not give, but see for example [266, exercise 2.30].) The situation for elliptic curves not having complex multiplication is quite different. For such curves, supersingular primes seem to be very rare.

Example 4.6. Let E be the elliptic curve given by the equation

$$E : y^2 + y = x^3 - x^2 - 10x - 20,$$

so $j(E) = -2^{12}31^3/11^5$. Then either by using the criterion (V.4.1a) directly, or else using Exercise 5.1 and [19, Table 3], one finds that the only primes $p < 100$ for which E is supersingular in characteristic p are $p \in \{2, 19, 29\}$. More generally, D.H. Lehmer calculated that there are exactly 27 primes $p < 31500$ for which E is supersingular.

It is not hard to prove that for any elliptic curve $E/\mathbb{Q}$, there are infinitely many primes p such that E is ordinary; see Exercise 5.11. We conclude by stating two theorems and one conjecture; the proofs of the theorems are unfortunately beyond the scope of this book. For simplicity we state everything over $\mathbb{Q}$, but suitable versions apply over any number field.

Theorem 4.7. (Serre [234], Elkies [78]) *Let $E/\mathbb{Q}$ be an elliptic curve without complex multiplication. Then the set of supersingular primes has density 0. More precisely, for every $\epsilon > 0$ we have*

$$\#\{p < x : E/\mathbb{F}_p \text{ is supersingular}\} \ll x^{3/4+\epsilon}.$$

Conjecture 4.8. (Lang–Trotter [145]) *Let $E/\mathbb{Q}$ be an elliptic curve without complex multiplication. Then*

$$\#\{p < x : E/\mathbb{F}_p \text{ is supersingular}\} \sim \frac{c\sqrt{x}}{\log x}$$

as $x \to \infty$, where $c > 0$ is a constant depending on E.

Although (V.4.8) is still an open question, a weaker result due to Elkies says that there are infinitely many supersingular primes.

Theorem 4.9. (Elkies [77]; see also [30]) *Let $E/\mathbb{Q}$ be an elliptic curve without complex multiplication. Then there are infinitely many primes p for which $E/\mathbb{F}_p$ is supersingular.*

Exercises

5.1. Verify the Weil conjectures for $V = \mathbb{P}^N$.

5.2. Let $V/\mathbb{F}_q$ be a smooth projective variety of dimension N defined over a finite field, and let ϵ be the Euler characteristic of V as described in (V.2.2b). Prove that up to ± 1, the function

$$q^{\epsilon s/2} Z(V/\mathbb{F}_q; q^{-s})$$

is invariant under the substitution $s \mapsto N - s$.

5.3. Let A be a square matrix with coefficients in a field. Prove that

$$\exp\left(\sum_{n=1}^{\infty} \frac{(\operatorname{tr} A^n) T^n}{n}\right) = \frac{1}{\det(1 - AT)}.$$

5.4. Let $E/\mathbb{F}_q$ and $E'/\mathbb{F}_q$ be elliptic curves defined over a finite field.
   (a) If E and E' are isogenous over $\mathbb{F}_q$, prove that

$$\#E(\mathbb{F}_q) = \#E'(\mathbb{F}_q).$$

   Deduce that $Z(E/\mathbb{F}_q, T) = Z(E'/\mathbb{F}_q, T)$.
   (b) Prove the converse, i.e., if $\#E(\mathbb{F}_q) = \#E'(\mathbb{F}_q)$, then E and E' are isogenous. (*Hint*. Use (III.7.7a).)

5.5. Let $\mathcal{K}/\mathbb{Q}$ be an imaginary quadratic field, and let $\mathcal{R}_1, \ldots, \mathcal{R}_n$ be orders in $\mathcal{K}$. Prove that there is a prime $\ell \in \mathbb{Z}$ such that $\ell\mathcal{R}_i$ is a prime ideal of $\mathcal{R}_i$ for all $i = 1, 2, \ldots, n$.

5.6. Let $E/\mathbb{F}_q$ be an elliptic curve.

(a) Prove that there are integers $m \geq 1$ and $n \geq 1$ with $\gcd(m, q) = 1$ such that $E(\mathbb{F}_q) \cong \mathbb{Z}/m\mathbb{Z} \times \mathbb{Z}/mn\mathbb{Z}$.

(b) With notation as in (a), prove that $q \equiv 1 \pmod{m}$.

(c) Suppose that $q = p \geq 5$ is prime and that E is supersingular. Prove that either $m = 1$ or $m = 2$. If $p \equiv 1 \pmod 4$, prove that $m = 1$.

5.7. Let K be a field of characteristic 2 and let E/K be an elliptic curve defined over K. Prove that E is supersingular if and only if $j(E) = 0$.

5.8. * Let $\mathrm{char}(K) = p > 0$, and let E/K be an elliptic curve with $j(E) \notin \bar{\mathbb{F}}_p$. Prove that $\mathrm{End}(E) = \mathbb{Z}$. (*Hint.* From (III.9.4), it suffices to show that $\mathrm{End}(E)$ is not an order in an imaginary quadratic field.)

5.9. Prove the following *mass formula* of Eichler and Deuring:

$$\sum_{\substack{E/\bar{\mathbb{F}}_p \\ \text{supersingular}}} \frac{1}{\# \mathrm{Aut}(E)} = \frac{p - 1}{24}.$$

5.10. Let $E/\mathbb{F}_q$ be an elliptic curve, let $\phi : E \to E$ be the qth-power Frobenius endomorphism, and let $p = \mathrm{char}(\mathbb{F}_q)$.

(a) Prove that E is supersingular if and only if

$$\mathrm{tr}(\phi) \equiv 0 \pmod p.$$

(The trace of ϕ is computed in $\mathrm{End}\big(T_\ell(E)\big)$ for any prime $\ell \neq p$.)

(b) Suppose that $q = p \geq 5$ is prime. Prove that E is supersingular if and only if

$$\#E(\mathbb{F}_p) = p + 1.$$

(c) Write down all elliptic curves $E/\mathbb{F}_3$, determine which ones are supersingular by explicitly calculating $\#E(\mathbb{F}_3)$ and using (a), and show that (b) is false when $p = 3$.

(d) Repeat (c) for $p = 2$.

(e) Let p^i be the largest power of p such that $p^{2i} \mid q$. Prove that

$$\mathrm{tr}(\phi) \equiv 0 \pmod p \iff \mathrm{tr}(\phi) \equiv 0 \pmod{p^i}.$$

(f) Prove that there do not exist any elliptic curves $E/\mathbb{F}_8$ satisfying either $\#E(\mathbb{F}_8) = 7$ or $\#E(\mathbb{F}_8) = 11$. (*Hint.* Use (e).)

5.11. Let E be an elliptic curve defined over $\mathbb{Q}$, and fix a Weierstrass equation for E having coefficients in $\mathbb{Z}$. Prove that there are infinitely many primes $p \in \mathbb{Z}$ such that the reduced curve $E/\mathbb{F}_p$ has Hasse invariant 1. (*Hint.* Fix a prime ℓ and consider those primes p that split completely in the field $\mathbb{Q}\big(E[\ell]\big)$ obtained by adjoining to $\mathbb{Q}$ the coordinates of all ℓ-torsion points of E. Then use Exercise 5.10.)

5.12. Prove that for every prime $p \geq 3$, the elliptic curve

$$E : y^2 = x^3 + x$$

satisfies

$$\#E(\mathbb{F}_p) \equiv 0 \pmod 4.$$

5.13. Let $E/\mathbb{F}_q$ be an elliptic curve, and for each $n \geq 1$, let

$$a_n = q^n + 1 - \#E(\mathbb{F}_{q^n}).$$

(By convention, we set $a_0 = 2$.) Prove that

$$a_{n+2} = a_1 a_{n+1} - q a_n \qquad \text{for all } n \geq 0.$$

(This linear recurrence gives a way to compute a_n from the initial values $a_0 = 2$ and $a_1 = q + 1 - \#E(\mathbb{F}_q)$.)

5.14. Let $E/\mathbb{F}_q$ be an elliptic curve, let $m \geq 1$ be an integer satisfying $\gcd(q - 1, m) = 1$, let $P \in E(\mathbb{F}_q)$ be a point of exact order m, and let d be an integer such that $q^d \equiv 1 \pmod{m}$. Prove that $E[m] \subset E(\mathbb{F}_{q^d})$. (*Hint.* Note that $\mu_m \subset \mathbb{F}_{q^d}$ and use the Weil e_m-pairing to study the action of the Frobenius map on a basis for $E[m]$.)

5.15. Let $E/\mathbb{F}_p$ be a supersingular elliptic curve with $p \geq 5$ prime, and let $n \geq 1$ be an integer. Prove that

$$\#E(\mathbb{F}_{p^n}) = \begin{cases} p^n + 1 & \text{if } n \text{ is odd,} \\ \left(p^{n/2} - (-1)^{n/2}\right)^2 & \text{if } n \text{ is even.} \end{cases}$$

5.16. Let $E/\mathbb{F}_{p^2}$ be a supersingular elliptic curve.
  (a) Prove that the multiplication-by-p map may be written in the form

$$[p](x,y) = \left(g(x^{p^2}, y^{p^2}), h(x^{p^2}, y^{p^2})\right)$$

  with rational functions $g, h \in \mathbb{F}_{p^2}(X, Y)$.
  (b) Prove that g and h are polynomials, i.e., $g, h \in \mathbb{F}_{p^2}[X, Y]$.
  (c) Assume that $p \geq 3$ and take a Weierstrass equation for E with $a_1 = a_3 = 0$. Prove that $g = X$ and $h = \pm Y$.
  (d) Assume that $p \geq 5$ and that E is defined over $\mathbb{F}_p$. Prove that $h = -Y$. Let $\phi : E \to E$ be the pth-power Frobenius map on E. Prove that $\phi^2 = [-p]$ and that $\hat{\phi} = -\phi$.

5.17. Let $E/\mathbb{F}_q$ be an elliptic curve and suppose that we know, a priori, that the zeta function of E has the form

$$Z(E/K; T) = \frac{1 - aT + qT^2}{(1 - T)(1 - qT)} = \frac{(1 - \alpha T)(1 - \beta T)}{(1 - T)(1 - qT)}$$

with $a \in \mathbb{Z}$ and $\alpha, \beta \in \mathbb{C}$. Use this formula to prove that (cf. (V.2.3.1))

$$\#E(\mathbb{F}_{q^n}) = q^n + 1 - \alpha^n - \beta^n.$$

(*Hint.* Take the logarithmic derivative, i.e., take the logarithm of both sides and then differentiate with respect to T.)

5.18. Let $E/\mathbb{F}_q$ be an elliptic curve, let $P, Q \in E(\mathbb{F}_q)$ be points such that Q is in the subgroup generated by P, and let n be the order of P in the group $E(\mathbb{F}_q)$. Suppose that we want to solve the ECDLP, i.e., find an integer m satisfying $Q = [m]P$.
  (a) If we naively compute $P, [2]P, [3]P, \ldots$ until we find Q, approximately how many multiples of P would we expect to compute?

(b) Let $N \geq 1$ be an integer, let $R = [-N]P$, and consider the following two lists:

List 1: P, $[2]P$, $[3]P$, $\ldots$, $[N]P$.

List 2: $Q + R$, $Q + [2]R$, $Q + [3]R$, $\ldots$, $Q + [N]R$.

How large should we choose N (in terms of n) to guarantee that the two lists contain a common element?

(c) Show how to use a match between the two lists in order to solve the ECDLP.

Chapter VI

Elliptic Curves over $\mathbb{C}$

Evaluation of the integral giving arc length on a circle, namely $\int dx/\sqrt{1-x^2}$, leads to an inverse trigonometric function. The analogous problem for the arc length of an ellipse yields an integral that is not computable in terms of so-called elementary functions. The indeterminacy of the sign of the square root means that such integrals are not well-defined on $\mathbb{C}$; instead, they are more naturally studied on an associated Riemann surface. For the arc length integral of an ellipse, this Riemann surface turns out to be the set of complex points on an elliptic curve E. We thus begin our study of elliptic curves over $\mathbb{C}$ by studying certain *elliptic integrals*, which are line integrals on $E(\mathbb{C})$. Indeed, the reason that elliptic curves are so named is because they are the Riemann surfaces associated to arc length integrals of ellipses. In terms of their geometry, ellipses and elliptic curves actually have little in common, the former having genus zero and the latter genus one.

The study of elliptic integrals leads to questions that are fairly difficult to answer if one restricts attention to integrals. However, just as for the more familiar circular (trigonometric) functions, it is much easier to develop a theory of the inverse function to the integral. Thus trigonometry is not generally built up around the function $\int dx/\sqrt{1-x^2}$, but rather around its inverse $\sin(x)$. In (VI §§2, 3) we give the rudiments of the theory of *elliptic functions*, which are meromorphic functions having two $\mathbb{R}$-linearly independent periods. We then relate this theory back to our original study of elliptic integrals and use the relationship to make various deductions about elliptic curves over $\mathbb{C}$. In the final section of this chapter we amplify on the remark that the study of elliptic curves over $\mathbb{C}$ essentially encompasses the theory of elliptic curves over arbitrary algebraically closed fields of characteristic 0.

The analytic theory of elliptic functions and integrals is a beautiful, but vast, body of knowledge. The contents of this chapter represent a very modest beginning in the study of that theory. Further, we have restricted ourselves to the function theory of a single elliptic curve. There is another sort of function theory that is quite important, namely the theory of *modular functions*, in which one studies functions whose domain is the set of all elliptic curves over $\mathbb{C}$. We do not discuss modular functions in the body of this book, but see (C §12) for a brief discussion and a list of references for further reading.

J.H. Silverman, *The Arithmetic of Elliptic Curves, Second Edition*, Graduate Texts in Mathematics 106, DOI 10.1007/978-0-387-09494-6_VI,
© Springer Science+Business Media, LLC 2009

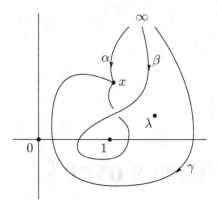

Figure 6.1: Three paths for a line integral.

VI.1   Elliptic Integrals

Let E be an elliptic curve defined over $\mathbb{C}$. Since $\mathrm{char}(\mathbb{C}) = 0$ and $\mathbb{C}$ is algebraically closed, there is a Weierstrass equation for E in Legendre form (III.1.7),

$$E : Y^2 = x(x-1)(x-\lambda).$$

The natural map

$$E(\mathbb{C}) \longrightarrow \mathbb{P}^1(\mathbb{C}), \qquad (x,y) \longmapsto x,$$

is a double cover ramified over precisely the four points $0, 1, \lambda, \infty \in \mathbb{P}^1(\mathbb{C})$.

We know from (III.1.5) that $\omega = dx/y$ is a holomorphic differential form on E. Suppose that we try to define a map by the rule

$$E(\mathbb{C}) \xrightarrow{\ ?\ } \mathbb{C}, \qquad P \longmapsto \int_O^P \omega \,,$$

where the integral is along some path connecting O to P. Unfortunately, this map is not well-defined, since it depends on the choice of path. We let $P = (x, y) \in E(\mathbb{C})$ and look more closely at what is happening in $\mathbb{P}^1(\mathbb{C})$.

We are attempting to compute the complex line integral

$$\int_\infty^x \frac{dt}{\sqrt{t(t-1)(t-\lambda)}} \,.$$

This line integral is not path-independent, because the square root is not single-valued. Thus in Figure 6.1, the three integral $\int_\alpha \omega$, $\int_\beta \omega$, and $\int_\gamma \omega$ need not be equal.

In order to make the integral well-defined, it is necessary to make branch cuts. For example, the integral will be path-independent on the complement of the branch cuts illustrated in Figure 6.2, because in this region it is possible to define a single-valued branch of $\sqrt{t(t-1)(t-\lambda)}$. More generally, since the square root is double-valued,

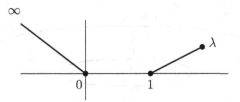

Figure 6.2: Branch cuts that make the integral single-valued.

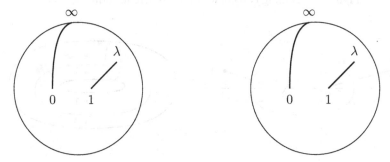

Figure 6.3: Branch cuts on the sphere.

we should take two copies of $\mathbb{P}^1(\mathbb{C})$, make branch cuts as indicated in Figure 6.3, and glue them together along the branch cuts to form the Riemann surface illustrated in Figure 6.4. (Note that $\mathbb{P}^1(\mathbb{C}) = \mathbb{C} \cup \{\infty\}$ is topologically a 2-sphere.) It is readily seen that the resulting Riemann surface is a torus, and it is on this surface that we should study the integral $\int dt/\sqrt{t(t-1)(t-\lambda)}$. In fact, elliptic curves arose when people began to study such integrals, and the reason that elliptic curves acquired their name is because such "elliptic integrals" arise when one attempts to calculate the arc length of an ellipse. (See Exercise 6.13b.)

Returning now to our hypothetical map

$$E(\mathbb{C}) \longrightarrow \mathbb{C}, \qquad P \longmapsto \int_O^P \omega,$$

we see that the indeterminacy comes from integrating across branch cuts in $\mathbb{P}^1(\mathbb{C})$, or equivalently around noncontractible loops on the torus. Figure 6.5 illustrates two closed paths α and β for which the integrals $\int_\alpha \omega$ and $\int_\beta \omega$ may be nonzero. We thus obtain two complex numbers, which are called *periods of E*,

$$\omega_1 = \int_\alpha \omega \qquad \text{and} \qquad \omega_2 = \int_\beta \omega.$$

Notice that the paths α and β generate the first homology group of the torus. Thus any two paths from O to P differ by a path that is homologous to $n_1\alpha + n_2\beta$ for

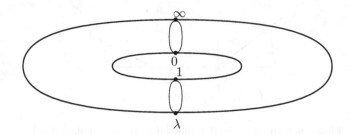

Figure 6.4: Joining two copies of the sphere to form a torus.

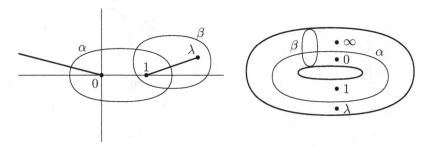

Figure 6.5: Paths on $\mathbb{P}^1(\mathbb{C})$ and on the torus.

some integers $n_1, n_2 \in \mathbb{Z}$. Thus the integral $\int_O^P \omega$ is well-defined up to addition of a number of the form $n_1\omega_1 + n_2\omega_2$, which suggests that we should look at the set

$$\Lambda = \{n_1\omega_1 + n_2\omega_2 : n_1, n_2 \in \mathbb{Z}\}.$$

The preceding discussion shows that there is a well-defined map

$$F : E(\mathbb{C}) \longrightarrow \mathbb{C}/\Lambda, \qquad P \longmapsto \int_O^P \omega \pmod{\Lambda}.$$

The set Λ is clearly a subgroup of $\mathbb{C}$, so the quotient $\mathbb{C}/\Lambda$ is a group. Using the translation invariance of ω that we proved in (III.5.1), we easily verify that F is a homomorphism:

$$\int_O^{P+Q} \omega \equiv \int_O^P \omega + \int_P^{P+Q} \omega \equiv \int_O^P \omega + \int_O^Q \tau_P^*\omega \equiv \int_O^P \omega + \int_O^Q \omega \pmod{\Lambda}.$$

The quotient space $\mathbb{C}/\Lambda$ will be a Riemann surface, i.e., a one-dimensional complex manifold, if and only if Λ is a lattice, or equivalently, if and only if the periods ω_1 and ω_2 that generate Λ are linearly independent over $\mathbb{R}$. This turns out to be the case, and further, the map F is a complex analytic isomorphism from $E(\mathbb{C})$

to $\mathbb{C}/\Lambda$. However, rather than proving these statements here, we instead turn to the study of the space $\mathbb{C}/\Lambda$ for a given lattice Λ. In (VI §3) we construct the inverse to the map F and prove that $\mathbb{C}/\Lambda$ is analytically isomorphic to $E_\Lambda(\mathbb{C})$ for a certain elliptic curve $E_\Lambda/\mathbb{C}$. We then apply the uniformization theorem (VI.5.1), which says that every elliptic curve $E/\mathbb{C}$ is isomorphic to some E_Λ, to deduce (VI.5.2) that the periods of $E/\mathbb{C}$ are $\mathbb{R}$-linearly independent and that F is a complex analytic isomorphism. (For a direct proof of the $\mathbb{R}$-linear independence of ω_1 and ω_2 using only Stokes's theorem in $\mathbb{R}^2$, see [46, §2.9].)

VI.2 Elliptic Functions

Let $\Lambda \subset \mathbb{C}$ be a lattice, that is, Λ is a discrete subgroup of $\mathbb{C}$ that contains an $\mathbb{R}$-basis for $\mathbb{C}$. In this section we study meromorphic functions on the quotient space $\mathbb{C}/\Lambda$, or equivalently, meromorphic functions on $\mathbb{C}$ that are periodic with respect to the lattice Λ.

Definition. An *elliptic function* (*relative to the lattice* Λ) is a meromorphic function $f(z)$ on $\mathbb{C}$ that satisfies

$$f(z + \omega) = f(z) \qquad \text{for all } z \in \mathbb{C} \text{ and all } \omega \in \Lambda.$$

The set of all such functions is denoted by $\mathbb{C}(\Lambda)$. It is clear that $\mathbb{C}(\Lambda)$ is a field.

Definition. A *fundamental parallelogram* for Λ is a set of the form

$$D = \{a + t_1\omega_1 + t_2\omega_2 : 0 \le t_1, t_2 < 1\},$$

where $a \in \mathbb{C}$ and $\{\omega_1, \omega_2\}$ is a basis for Λ. Note that the definition of D implies that the natural map $D \to \mathbb{C}/\Lambda$ is bijective. We denote the closure of D in $\mathbb{C}$ by $\bar{D}$. A lattice and three different fundamental parallelograms are illustrated in Figure 6.6.

Proposition 2.1. *A holomorphic elliptic function, i.e., an elliptic function with no poles, is constant. Similarly, an elliptic function with no zeros is constant.*

PROOF. Suppose that $f(z) \in \mathbb{C}(\Lambda)$ is holomorphic. Let D be a fundamental parallelogram for Λ. The periodicity of f implies that

$$\sup_{z \in \mathbb{C}} |f(z)| = \sup_{z \in \bar{D}} |f(z)|.$$

The function f is continuous and the set $\bar{D}$ is compact, so $|f(z)|$ is bounded on $\bar{D}$. Hence f is bounded on all of $\mathbb{C}$, so Liouville's theorem [3, Chapter 4, §2.3] tells us that f is constant. This proves the first statement. Finally, if f has no zeros, then $1/f$ is holomorphic, hence constant. $\qquad\square$

Let f be an elliptic function and let $w \in \mathbb{C}$. Then, just as for any meromorphic function, we can look at its order of vanishing and its residue, which we denote by

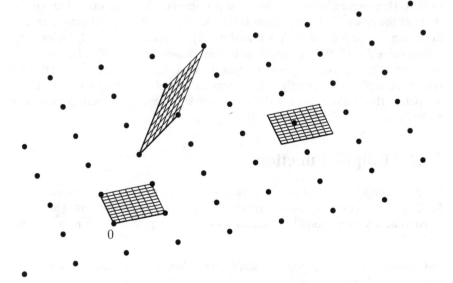

Figure 6.6: A lattice and three fundamental parallelograms.

$$\mathrm{ord}_w(f) = \text{order of vanishing of } f \text{ at } w,$$
$$\mathrm{res}_w(f) = \text{residue of } f \text{ at } w.$$

(See [3, Chapter 4, §§3.2, 5.1].) The fact that f is elliptic implies that the order and the residue of f do not change if we replace w by $w+\omega$ for any $\omega \in \Lambda$. This prompts the following convention.

Notation. The notation $\sum_{w \in \mathbb{C}/\Lambda}$ denotes a sum over $w \in D$, where D is a fundamental parallelogram for Λ. By implication, the value of the sum is independent of the choice of D and only finitely many terms of the sum are nonzero.

Notice that (VI.2.1) is the complex analogue of (II.1.2), which says that an algebraic function that has no poles is constant. The next theorem and corollary continue this theme by proving for $\mathbb{C}/\Lambda$ results that are analogous to (II.3.1) and (III.3.5)

Theorem 2.2. *Let $f \in \mathbb{C}(\Lambda)$ be an elliptic function relative to Λ.*

(a) $\displaystyle\sum_{w \in \mathbb{C}/\Lambda} \mathrm{res}_w(f) = 0.$

(b) $\displaystyle\sum_{w \in \mathbb{C}/\Lambda} \mathrm{ord}_w(f) = 0.$

(c) $\displaystyle\sum_{w \in \mathbb{C}/\Lambda} \mathrm{ord}_w(f)w \in \Lambda.$

PROOF. Let D be a fundamental parallelogram for Λ such that $f(z)$ has no zeros or poles on the boundary ∂D of D. All three parts of the theorem are simple applications of the residue theorem [3, Chapter 4, Theorem 19] applied to appropriately chosen functions on D.

(a) The residue theorem tells us that

$$\sum_{w \in \mathbb{C}/\Lambda} \operatorname{res}_w(f) = \frac{1}{2\pi i} \int_{\partial D} f(z)\, dz.$$

The periodicity of f implies that the integrals along the opposite sides of the parallelogram cancel, so the total integral around the boundary of D is zero.

(b) The periodicity of $f(z)$ implies that $f'(z)$ is also periodic, so applying (a) to the elliptic function $f'(z)/f(z)$ gives

$$\sum_{w \in \mathbb{C}/\Lambda} \operatorname{res}_w(f'/f) = 0.$$

Since $\operatorname{res}_w(f'/f) = \operatorname{ord}_w(f)$, this is the desired result.

(c) We apply the residue theorem to the function $z f'(z)/f(z)$ to obtain

$$\sum_{w \in \mathbb{C}/\Lambda} \operatorname{ord}_w(f) w = \frac{1}{2\pi i} \int_{\partial D} \frac{z f'(z)}{f(z)}\, dz$$

$$= \frac{1}{2\pi i} \left(\int_a^{a+\omega_1} + \int_{a+\omega_1}^{a+\omega_1+\omega_2} + \int_{a+\omega_1+\omega_2}^{a+\omega_2} + \int_{a+\omega_2}^{a} \right) \frac{z f'(z)}{f(z)}\, dz.$$

In the second (respectively third) integral we make the change of variable $z \mapsto z + \omega_1$ (respectively $z \mapsto z + \omega_2$). Then the periodicity of $f'(z)/f(z)$ yields

$$\sum_{w \in \mathbb{C}/\Lambda} \operatorname{ord}_w(f) w = -\frac{\omega_2}{2\pi i} \int_a^{a+\omega_1} \frac{f'(z)}{f(z)}\, dz + \frac{\omega_1}{2\pi i} \int_a^{a+\omega_2} \frac{f'(z)}{f(z)}\, dz.$$

If $g(z)$ is any meromorphic function, then the integral

$$\frac{1}{2\pi i} \int_a^b \frac{g'(z)}{g(z)}\, dz$$

is the winding number around 0 of the path

$$[0,1] \longrightarrow \mathbb{C}, \qquad t \longmapsto g\big((1-t)a + tb\big).$$

In particular, if $g(a) = g(b)$, then the integral is an integer. Thus the periodicity of $f'(z)/f(z)$ implies that $\sum \operatorname{ord}_w(f) w$ has the form $-\omega_2 n_2 + \omega_1 n_1$ for integers n_1 and n_2, so it is in Λ. $\qquad\square$

Definition. The *order* of an elliptic function is its number of poles (counted with multiplicity) in a fundamental parallelogram. Equivalently, (VI.2.2b) says that the order is the number of zeros.

Corollary 2.3. *A nonconstant elliptic function has order at least* 2.

PROOF. If $f(z)$ has a single simple pole, then (VI.2.2a) tells us that the residue at that pole is 0, so $f(z)$ is actually holomorphic. Now apply (VI.2.1). $\square$

We now define the *divisor group of* $\mathbb{C}/\Lambda$, denoted by $\mathrm{Div}(\mathbb{C}/\Lambda)$, to be the group of formal linear combinations

$$\sum_{w \in \mathbb{C}/\Lambda} n_w(w) \quad \text{with } n_w \in \mathbb{Z} \text{ and } n_w = 0 \text{ for all but finitely many } w.$$

Then for $D = \sum n_w(w) \in \mathrm{Div}(\mathbb{C}/\Lambda)$, we define

$$\deg D = \text{degree of } D = \sum n_w$$

and

$$\mathrm{Div}^0(\mathbb{C}/\Lambda) = \{ D \in \mathrm{Div}(\mathbb{C}/\Lambda) : \deg D = 0 \}.$$

Further, for any $f \in \mathbb{C}(\Lambda)^*$ we define the *divisor of* f to be

$$\mathrm{div}(f) = \sum_{w \in \mathbb{C}/\Lambda} \mathrm{ord}_w(f)(w).$$

We see from (VI.2.2b) that $\mathrm{div}(f) \in \mathrm{Div}^0(\mathbb{C}/\Lambda)$. The map

$$\mathrm{div} : \mathbb{C}(\Lambda)^* \to \mathrm{Div}^0(\mathbb{C}/\Lambda)$$

is clearly a homomorphism, since each ord_w is a valuation. Finally, we define a *summation map*

$$\mathrm{sum} : \mathrm{Div}^0(\mathbb{C}/\Lambda) \longrightarrow \mathbb{C}/\Lambda, \qquad \mathrm{sum}\left(\sum n_w(w)\right) = \sum n_w w \pmod{\Lambda}.$$

The next result gives an exact sequence that encompasses our main results so far for $\mathbb{C}/\Lambda$, plus one fact (VI.3.4) that will be proven in the next section.

Theorem 2.4. *The following is an exact sequence*:

$$1 \longrightarrow \mathbb{C}^* \longrightarrow \mathbb{C}(\Lambda)^* \xrightarrow{\ \mathrm{div}\ } \mathrm{Div}^0(\mathbb{C}/\Lambda) \xrightarrow{\ \mathrm{sum}\ } \mathbb{C}/\Lambda \longrightarrow 0.$$

PROOF. Exactness on the left is clear, and exactness on the right follows from $\mathrm{sum}\big((w) - (0)\big) = w$. Exactness at $\mathbb{C}(\Lambda)^*$ is (VI.2.1), and exactness at $\mathrm{Div}^0(\mathbb{C}/\Lambda)$ is (VI.2.2c) and (VI.3.4). $\square$

VI.3 Construction of Elliptic Functions

In order to show that the results of (VI §2) are not vacuous, we must construct some nonconstant elliptic functions. We know from (VI.2.3) that any such function has order at least 2. Following Weierstrass, we look for a function with a pole of order 2 at $z = 0$.

Definition. Let $\Lambda \subset \mathbb{C}$ be a lattice. The *Weierstrass $\wp$-function (relative to Λ)* is defined by the series

$$\wp(z; \Lambda) = \frac{1}{z^2} + \sum_{\substack{\omega \in \Lambda \\ \omega \neq 0}} \left(\frac{1}{(z - \omega)^2} - \frac{1}{\omega^2} \right).$$

The *Eisenstein series of weight $2k$ (for Λ)* is the series

$$G_{2k}(\Lambda) = \sum_{\substack{\omega \in \Lambda \\ \omega \neq 0}} \omega^{-2k}.$$

(For notational convenience, we write $\wp(z)$ and G_{2k} if the lattice Λ has been fixed.)

Theorem 3.1. *Let $\Lambda \subset \mathbb{C}$ be a lattice.*
(a) *The Eisenstein series $G_{2k}(\Lambda)$ is absolutely convergent for all $k > 1$.*
(b) *The series defining the Weierstrass $\wp$-function converges absolutely and uniformly on every compact subset of $\mathbb{C} \smallsetminus \Lambda$. The series defines a meromorphic function on $\mathbb{C}$ having a double pole with residue 0 at each lattice point and no other poles*
(c) *The Weierstrass $\wp$-function is an even elliptic function.*

PROOF. Since Λ is discrete in $\mathbb{C}$, it is not hard to see that there is a constant $c = c(\Lambda)$ such that for all $N \geq 1$, the number of points in an annulus satisfies

$$\#\{\omega \in \Lambda : N \leq |\omega| < N + 1\} < cN.$$

(See Exercise 6.2.) This allows us to estimate

$$\sum_{\substack{\omega \in \Lambda \\ |\omega| \geq 1}} \frac{1}{|\omega|^{2k}} \leq \sum_{N=1}^{\infty} \frac{\#\{\omega \in \Lambda : N \leq |\omega| < N + 1\}}{N^{2k}} < \sum_{N=1}^{\infty} \frac{c}{N^{2k-1}} < \infty.$$

(b) If $|\omega| > 2|z|$, then

$$\left| \frac{1}{(z - \omega)^2} - \frac{1}{\omega^2} \right| = \left| \frac{z(2\omega - z)}{\omega^2(z - \omega)^2} \right| \leq \frac{|z|(2|\omega| + |z|)}{|\omega|^2(|\omega| - |z|)^2} \leq \frac{10|z|}{|\omega|^2}.$$

It follows from (a) that the series for $\wp(z)$ is absolutely convergent for all $z \in \mathbb{C} \smallsetminus \Lambda$, and that it is uniformly convergent on every compact subset of $\mathbb{C} \smallsetminus \Lambda$. Therefore the series defines a holomorphic function on $\mathbb{C} \smallsetminus \Lambda$, and it is clear from the series expansion that $\wp(z)$ has a double pole with residue 0 at each point in Λ.

(c) Replacing ω by $-\omega$ in the series for $\wp$, it is clear that $\wp(z) = \wp(-z)$, so $\wp$ is an even function. We know from (b) that the series for $\wp$ is uniformly convergent, so we can compute its derivative by differentiating term by term,

$$\wp'(z) = -2 \sum_{\omega \in \Lambda} \frac{1}{(z-\omega)^2}.$$

It is clear from this expression that $\wp'$ is an elliptic function, so $\wp'(z+\omega) = \wp'(z)$ for all $\omega \in \Lambda$. Integrating this equality with respect to z yields

$$\wp(z+\omega) = \wp(z) + c(\omega) \qquad \text{for all } z \in \mathbb{C},$$

where $c(\omega) \in \mathbb{C}$ is independent of z. Setting $z = -\frac{1}{2}\omega$ and using the evenness of $\wp(z)$ shows that $c(\omega) = 0$, so $\wp$ is an elliptic function. $\qquad\qquad\square$

Next we show that every elliptic function is a rational function of the Weierstrsss $\wp$-function and its derivative. This result is the analytic analogue of (III.3.1.1).

Theorem 3.2. *Let $\Lambda \subset \mathbb{C}$ be a lattice. Then*

$$\mathbb{C}(\Lambda) = \mathbb{C}\big(\wp(z), \wp'(z)\big),$$

i.e., every elliptic function is a rational combination of $\wp$ and $\wp'$.

PROOF. Let $f(z) \in \mathbb{C}(\Lambda)$. Writing

$$f(z) = \frac{f(z) + f(-z)}{2} + \frac{f(z) - f(-z)}{2},$$

we see that it suffices to prove the theorem for functions that are either odd or even. Further, if $f(z)$ is odd, then $f(z)\wp'(z)$ is even, so we are reduced to the case that f is an even elliptic function.

The assumption that f is even implies that

$$\operatorname{ord}_w f = \operatorname{ord}_{-w} f \quad \text{for every } w \in \mathbb{C}.$$

Further, we claim that if $2w \in \Lambda$, then $\operatorname{ord}_w f$ is even. To see this, we differentiate $f(z) = f(-z)$ repeatedly to obtain

$$f^{(i)}(z) = (-1)^{i-1} f^{(i)}(-z).$$

If $2w \in \Lambda$, then $f^{(i)}(z)$ has the same value at w and $-w$, so

$$f^{(i)}(w) = f^{(i)}(-w) = (-1)^{i-1} f^{(i)}(w).$$

Thus $f^{(i)}(w) = 0$ for odd values of i, so $\operatorname{ord}_w f$ is even.

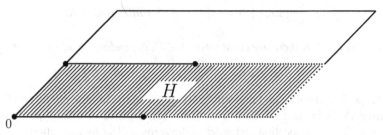

Figure 6.7: Half a fundamental parallelogram.

Let D be a fundamental parallelogram for Λ, and let H be "half" of D. In other words, H is a fundamental domain for $(\mathbb{C}/\Lambda)/\{\pm 1\}$, or equivalently, $\mathbb{C}$ is a disjoint union

$$\mathbb{C} = (H + \Lambda) \cup (-H + \Lambda),$$

as illustrated in Figure 6.7. The above discussion implies that the divisor of f has the form

$$\sum_{w \in H} n_w \big((w) + (-w)\big)$$

for certain integers $n_w \in \mathbb{Z}$. Note that for $2w \in \Lambda$, we are using the fact that $\mathrm{ord}_w f$ is even.

Consider the function

$$g(z) = \prod_{w \in H \smallsetminus \{0\}} \big(\wp(z) - \wp(w)\big)^{n_w}.$$

The divisor of $\wp(z) - \wp(w)$ is $(w) + (-w) - 2(0)$, so we see that f and g have exactly the same zeros and poles except possibly at $w = 0$. But then (VI.2.2b) implies that they have the same order at 0, too. Thus $f(z)/g(z)$ is a holomorphic elliptic function; hence it is constant from (VI.2.1). Therefore there is a constant c such that $f(z) = cg(z) \in \mathbb{C}\big(\wp(z), \wp'(z)\big)$. $\qquad\square$

In order to prove a converse to (VI.2.2), it is convenient to introduce a "theta function" for Λ.

Definition. The *Weierstrass σ-function (relative to Λ)* is the function defined by the product

$$\sigma(z) = \sigma(z; \Lambda) = z \prod_{\substack{w \in \Lambda \\ w \neq 0}} \left(1 - \frac{z}{\omega}\right) e^{(z/\omega) + \frac{1}{2}(z/\omega)^2}.$$

The next lemma describes the basic facts about $\sigma(z)$ that are needed for our applications. For further material about σ, see exercises 6.3 and 6.4 and [266, I §5].

Lemma 3.3. (a) *The infinite product for $\sigma(z)$ defines a holomorphic function on all of $\mathbb{C}$. It has simple zeros at each $z \in \Lambda$ and no other zeros.*

(b) $$\frac{d^2}{dz^2}\log\sigma(z) = -\wp(z) \qquad \text{for all } z \in \mathbb{C} \smallsetminus \Lambda.$$

(c) *For every $\omega \in \Lambda$ there are constants $a, b \in \mathbb{C}$, depending on ω, such that*

$$\sigma(z+\omega) = e^{az+b}\sigma(z) \qquad \text{for all } z \in \mathbb{C}.$$

PROOF. (a) The absolute and uniform convergence of the infinite product on $\mathbb{C}$ follows from (VI.3.1a) and standard facts about convergence of infinite products [3, Chapter 5, §2.3]. The location and order of the zeros is clear by inspection.
(b) The logarithm of $\sigma(z)$ is

$$\log\sigma(z) = \log z + \sum_{\substack{\omega \in \Lambda \\ \omega \neq 0}} \left\{ \log\left(1 - \frac{z}{\omega}\right) - \frac{z}{\omega} - \frac{1}{2}\left(\frac{z}{\omega}\right)^2 \right\},$$

and (a) tells us that we may differentiate term by term. The second derivative, up to sign, is exactly the series defining $\wp(z)$.
(c) The Weierstrass $\wp$-function is elliptic (VI.3.1c), so $\wp(z+\omega) = \wp(z)$. Integrating twice with respect to z and using (b) yields

$$\log\sigma(z+\omega) = \log\sigma(z) + az + b$$

for constants of integration $a, b \in \mathbb{C}$. $\square$

Proposition 3.4. *Let $n_1, \dots, n_r \in \mathbb{Z}$ and $z_1, \dots, z_r \in \mathbb{C}$ satisfy*

$$\sum n_i = 0 \qquad \text{and} \qquad \sum n_i z_i \in \Lambda.$$

Then there exists an elliptic function $f(z) \in \mathbb{C}(\Lambda)$ satisfying

$$\mathrm{div}(f) = \sum n_i(z_i).$$

More precisely, if we choose the n_i and z_i to satisfy $\sum n_i z_i = 0$, then we may take

$$f(z) = \prod \sigma(z - z_i)^{n_i}.$$

PROOF. Let $\lambda = \sum n_i z_i \in \Lambda$. Replacing

$$n_1(z_1) + \cdots + n_r(z_r) \qquad \text{by} \qquad n_1(z_1) + \cdots + n_r(z_r) + (0) - (\lambda),$$

we may assume that $\sum n_i z_i = 0$. Then (VI.3.3a) implies that

$$f(z) = \prod \sigma(z - z_i)^{n_i}$$

has the correct zeros and poles, while (VI.3.3c) allows us to compute (for any $\omega \in \Lambda$)

$$\frac{f(z+\omega)}{f(z)} = \prod e^{(a(z-z_i)+b)n_i} = e^{(az+b)\sum n_i} \cdot e^{-a\sum n_i z_i} = 1.$$

Therefore $f(z) \in \mathbb{C}(\Lambda)$. $\square$

We next derive the Laurent series expansions for $\wp(z)$ around $z = 0$, from which we will deduce the fundamental *algebraic* relation satisfied by $\wp(z)$ and $\wp'(z)$.

Theorem 3.5. (a) *The Laurent series for $\wp(z)$ around $z = 0$ is given by*

$$\wp(z) = \frac{1}{z^2} + \sum_{k=1}^{\infty} (2k + 1)G_{2k+2}z^{2k}.$$

(b) *For all $z \in \mathbb{C} \setminus \Lambda$, the Weierstrass $\wp$-function and its derivative satisfy the relation*

$$\wp'(z)^2 = 4\wp(z)^3 - 60G_4\wp(z) - 140G_6.$$

PROOF. (a) For all z with $|z| < |\omega|$ we have

$$\frac{1}{(z - \omega)^2} - \frac{1}{\omega^2} = \frac{1}{\omega^2}\left(\frac{1}{(1 - z/\omega)^2} - 1\right) = \sum_{n=1}^{\infty} (n + 1)\frac{z^n}{\omega^{n+2}}.$$

Substituting this formula into the series for $\wp(z)$ and reversing the order of summation gives the desired result.

(b) We write out the first few terms of various Laurent expansions:

$$\wp'(z)^2 = 4z^{-6} - 24G_4z^{-2} - 80G_6 + \cdots,$$
$$\wp(z)^3 = z^{-6} + 9G_4z^{-2} + 15G_6 + \cdots,$$
$$\wp(z) = z^{-2} + 3G_4z^2 + \cdots.$$

Comparing these expansions, we see that the function

$$f(z) = \wp'(z)^2 - 4\wp(z)^3 + 60G_4\wp(z) + 140G_6$$

is holomorphic at $z = 0$ and satisfies $f(0) = 0$. But $f(z)$ is an elliptic function relative to Λ, and from (VI.3.1b) it is holomorphic away from Λ, so $f(z)$ is a holomorphic elliptic function. Then (VI.2.1) says that $f(z)$ is constant, and the fact that $f(0) = 0$ implies that f is identically zero. □

Remark 3.5.1. It is standard notation to set

$$g_2 = g_2(\Lambda) = 60G_4(\Lambda) \quad \text{and} \quad g_3 = g_3(\Lambda) = 140G_6(\Lambda).$$

Then the algebraic relation satisfied by $\wp(z)$ and $\wp'(z)$ reads

$$\wp'(z)^2 = 4\wp(z)^3 - g_2\wp(z) - g_3.$$

Let $E/\mathbb{C}$ be an elliptic curve. The group law $E \times E \to E$ is given by everywhere locally defined rational functions (III.3.6), so we see in particular that $E = E(\mathbb{C})$ is a complex Lie group, i.e., it is a complex manifold with a group law given locally by complex analytic functions. Similarly, if $\Lambda \subset \mathbb{C}$ is a lattice, then $\mathbb{C}/\Lambda$ with its natural addition is a complex Lie group. The next result says that $\mathbb{C}/\Lambda$ is always complex analytically isomorphic to an elliptic curve.

Proposition 3.6. *Let* $g_2 = g_2(\Lambda)$ *and* $g_3 = g_3(\Lambda)$ *be the quantities associated to a lattice* $\Lambda \subset \mathbb{C}$.

(a) *The polynomial*

$$f(x) = 4x^2 - g_2 x - g_3$$

    has distinct roots, so its discriminant

$$\Delta(\Lambda) = g_2^3 - 27 g_3^3$$

    is nonzero.

(b) *Let* $E/\mathbb{C}$ *be the curve*

$$E : y^2 = 4x^2 - g_2 x - g_3,$$

    which from (a) *is an elliptic curve. Then the map*

$$\phi : \mathbb{C}/\Lambda \longrightarrow E(\mathbb{C}) \subset \mathbb{P}^2(\mathbb{C}), \qquad z \longmapsto \big[\wp(z), \wp'(z), 1\big],$$

    is a complex analytic isomorphism of complex Lie groups, i.e., it is an isomorphism of Riemann surfaces that is also a group homomorphism.

PROOF. (a) Let $\{\omega_1, \omega_2\}$ be a basis for Λ and let $\omega_3 = \omega_1 + \omega_2$. Then, since $\wp'(z)$ is an odd elliptic function, we see that

$$\wp'\left(\frac{\omega_i}{2}\right) = -\wp'\left(\frac{-\omega_i}{2}\right) = -\wp'\left(\frac{\omega_i}{2}\right),$$

so $\wp'(\omega_i/2) = 0$. It follows from (VI.3.5b) that $f(x)$ vanishes at each of the values $x = \wp(\omega_i/2)$, so it suffices to show that these three values are distinct.

The function $\wp(z) - \wp(\omega_i/2)$ is even, so it has at least a double zero at $z = \omega_i/2$. However, it is an elliptic function of order 2, so it has only these zeros in an appropriate fundamental parallelogram. Hence $\wp(\omega_j/2) \neq \wp(\omega_i/2)$ for $j \neq i$.

(b) The image of ϕ is contained in $E(\mathbb{C})$ from (VI.3.5b). To see that ϕ is surjective, let $(x, y) \in E(\mathbb{C})$. Then $\wp(z) - x$ is a nonconstant elliptic function, so from (VI.2.1) it has a zero, say $z = a$. It follows that $\wp'(a)^2 = y^2$, so replacing a by $-a$ if necessary, we obtain $\wp'(a) = y$. Then $\phi(a) = (x, y)$.

Next suppose that $\phi(z_1) = \phi(z_2)$. Assume first that $2z_1 \notin \Lambda$. Then the function $\wp(z) - \wp(z_1)$ is an elliptic function of order 2 that vanishes at z_1, $-z_1$, and z_2. It follows that two of these values are congruent modulo Λ, so the assumption that $2z_1 \notin \Lambda$ tells us that $z_2 \equiv \pm z_1 \pmod{\Lambda}$ for some choice of sign. Then

$$\wp'(z_1) = \wp'(z_2) = \wp'(\pm z_1) = \pm \wp'(z_1)$$

implies that $z_2 \equiv z_1 \pmod{\Lambda}$. (Note that $\wp'(z_1) \neq 0$ from the proof of (a).) Similarly, if $2z_1 \in \Lambda$, then $\wp(z) - \wp(z_1)$ has a double zero at z_1 and vanishes at z_2, so we again conclude that $z_2 \equiv z_1 \pmod{\Lambda}$. This proves that ϕ is injective.

Next we show that ϕ is an analytic isomorphism by computing its effect on the cotangent spaces of $\mathbb{C}/\Lambda$ and $E(\mathbb{C})$. At every point of $E(\mathbb{C})$, the differential form dx/y is holomorphic and nonvanishing. Further, we see that

$$\phi^*\left(\frac{dx}{dy}\right) = \frac{d\wp(z)}{\wp'(z)} = dz$$

is also holomorphic and nonvanishing at every point of $\mathbb{C}/\Lambda$. Hence ϕ is a local analytic isomorphism, and the bijectivity of ϕ then implies that it is a global isomorphism.

Finally, we must check that ϕ is a homomorphism. Let $z_1, z_2 \in \mathbb{C}$. Using (VI.3.4), we can find a function $f(z) \in \mathbb{C}(\Lambda)$ with divisor

$$\text{div}(f) = (z_1 + z_2) - (z_1) - (z_2) + (0).$$

Then (VI.3.2) allows us to write $f(z) = F(\wp(z), \wp'(z))$ for a rational function $F(X, Y) \in \mathbb{C}(X, Y)$. Treating $F(x, y)$ as an element of $\mathbb{C}(x, y) = \mathbb{C}(E)$, we have

$$\text{div}(F) = (\phi(z_1 + z_2)) + (\phi(z_1)) + (\phi(z_2)) + (\phi(0)).$$

It follows from (III.3.5) that

$$\phi(z_1 + z_2) = \phi(z_1) + \phi(z_2),$$

which completes the proof of the proposition. $\qquad\qquad\square$

VI.4 Maps Analytic and Maps Algebraic

In this section we investigate complex analytic maps between complex tori. It turns out that they all have a particularly simple form, and, somewhat more surprisingly, the maps that they induce on the corresponding elliptic curves via (VI.3.6b) turn out to be isogenies, i.e., they are given by rational functions.

Let Λ_1 and Λ_2 be lattices in $\mathbb{C}$, and suppose that $\alpha \in \mathbb{C}$ has the property that $\alpha\Lambda_1 \subset \Lambda_2$. Then scalar multiplication by α induces a well-defined holomorphic homomorphism

$$\phi_\alpha : \mathbb{C}/\Lambda_1 \longrightarrow \mathbb{C}/\Lambda_2, \qquad \phi_\alpha(z) = \alpha z \ (\text{mod } \Lambda_2).$$

We now show that these are essentially the only holomorphic maps from $\mathbb{C}/\Lambda_1$ to $\mathbb{C}/\Lambda_2$.

Theorem 4.1. (a) *With notation as above, the association*

$$\{\alpha \in \mathbb{C} : \alpha\Lambda_1 \subset \Lambda_2\} \longrightarrow \left\{\begin{array}{l} \text{holomorphic maps} \\ \phi : \mathbb{C}/\Lambda_1 \to \mathbb{C}/\Lambda_2 \\ \text{with } \phi(0) = 0 \end{array}\right\}$$

$$\alpha \qquad\qquad \longmapsto \qquad\qquad \phi_\alpha$$

is a bijection.

(b) *Let E_1 and E_2 be elliptic curves corresponding to lattices Λ_1 and Λ_2, respectively, as in (VI.3.6b). Then the natural inclusion*

$$\{\text{isogenies } \phi : E_1 \to E_2\} \longrightarrow \left\{ \begin{array}{l} \text{holomorphic maps} \\ \phi : \mathbb{C}/\Lambda_1 \to \mathbb{C}/\Lambda_2 \\ \text{with } \phi(0) = 0 \end{array} \right\}$$

is a bijection.

PROOF. (a) If $\phi_\alpha = \phi_\beta$, then

$$\alpha z \equiv \beta z \pmod{\Lambda_2} \qquad \text{for all } z \in \mathbb{C}.$$

Hence the map $z \mapsto (\alpha - \beta)z$ sends $\mathbb{C}$ to Λ_2. But Λ_2 is discrete, so the map must be constant, which implies that $\alpha = \beta$.

Next let $\phi : \mathbb{C}/\Lambda_1 \to \mathbb{C}/\Lambda_2$ be a holomorphic map with $\phi(0) = 0$. Then, since $\mathbb{C}$ is simply connected, we can lift ϕ to a holomorphic map $f : \mathbb{C} \to \mathbb{C}$ with $f(0) = 0$ so that the following diagram commutes:

$$\begin{array}{ccc} \mathbb{C} & \xrightarrow{\ f\ } & \mathbb{C} \\ \downarrow & & \downarrow \\ \mathbb{C}/\Lambda_1 & \xrightarrow{\ \phi\ } & \mathbb{C}/\Lambda_2. \end{array}$$

Thus

$$f(z + \omega) \equiv f(z) \pmod{\Lambda_2} \qquad \text{for all } \omega \in \Lambda_1 \text{ and all } z \in \mathbb{C}.$$

Again using the discreteness of Λ_2, we see that the difference $f(z + \omega) - f(z)$ must be independent of z. Differentiating, we find that

$$f'(z + \omega) = f'(z) \qquad \text{for all } \omega \in \Lambda_1 \text{ and all } z \in \mathbb{C},$$

so $f'(z)$ is a holomorphic elliptic function. It follows from (VI.2.1) that $f'(z)$ is constant, so $f(z) = \alpha z + \gamma$ for some $\alpha, \gamma \in \mathbb{C}$. The assumption that $f(0) = 0$ implies that $\gamma = 0$, and now $f(\Lambda_1) \subset \Lambda_2$ tells us that $\alpha\Lambda_1 \subset \Lambda_2$. Hence $\phi = \phi_\alpha$.

(b) First note that since an isogeny is given locally by *everywhere defined* rational functions, i.e., an isogeny is a morphism, the map induced between the corresponding complex tori is holomorphic. Thus our association

$$\text{Hom}(E_1, E_2) \longrightarrow \text{Holomorphic Maps}(\mathbb{C}/\Lambda_1, \mathbb{C}/\Lambda_2)$$

is well-defined, and it is clearly injective.

It remains to prove surjectivity. From (a) it suffices to consider a map of the form ϕ_α, where $\alpha \in \mathbb{C}^*$ satisfies $\alpha\Lambda_1 \subset \Lambda_2$. The induced map on Weierstrass equations is given by

$$\begin{array}{ccc} E_1 & \longrightarrow & E_2, \\ [\wp(z, \Lambda_1), \wp'(z, \Lambda_1), 1] & \longmapsto & [\wp(\alpha z, \Lambda_2), \wp'(\alpha z, \Lambda_2), 1], \end{array}$$

so we must show that $\wp(\alpha z, \Lambda_2)$ and $\wp'(\alpha z, \Lambda_2)$ can be expressed as rational expressions in $\wp(z, \Lambda_1)$ and $\wp'(z, \Lambda_1)$. Using the fact that $\alpha\Lambda_1 \subset \Lambda_2$, we see that for any $\omega_1 \in \Lambda_1$,

$$\wp\big(\alpha(z + \omega), \Lambda_2\big) = \wp(\alpha z + \alpha\omega, \Lambda_2) = \wp(\alpha z, \Lambda_2),$$

and similarly for $\wp'(\alpha z, \Lambda_2)$. Thus $\wp(\alpha z, \Lambda_2)$ and $\wp'(\alpha z, \Lambda_2)$ are in the field $\mathbb{C}(\Lambda_1)$. The desired result now follows immediately from (VI.3.2), which tells us that $\mathbb{C}(\Lambda_1) = \mathbb{C}\big(\wp(z, \Lambda_1), \wp'(z, \Lambda_1)\big)$. $\qquad\square$

Corollary 4.1.1. *Let $E_1/\mathbb{C}$ and $E_2/\mathbb{C}$ be elliptic curves corresponding to lattices Λ_1 and Λ_2 as in (VI.3.6b). Then E_1 and E_2 are isomorphic over $\mathbb{C}$ if and only if Λ_1 and Λ_2 are* homothetic, *i.e., there exists some $\alpha \in \mathbb{C}^*$ such that $\Lambda_1 = \alpha\Lambda_2$.*

Remark 4.2. Since the maps ϕ_α are clearly homomorphisms, (VI.4.1.1) implies that every complex analytic map from $E_1(\mathbb{C})$ to $E_2(\mathbb{C})$ taking O to O is necessarily a homomorphism. This is the analytic analogue of (III.4.8), which says that every isogeny of elliptic curves is a homomorphism.

VI.5 Uniformization

The uniformization theorem for elliptic curves says that every elliptic curve over $\mathbb{C}$ is parametrized by elliptic functions. The most natural proof of this fact uses the theory of modular functions, that is, functions whose domain is the set of lattices in $\mathbb{C}$. For example, $g_2(\Lambda)$ and $g_3(\Lambda)$ are modular functions. The proof is not difficult, but it would take us rather far afield, so we are content to state the result and use it to make various deductions.

Theorem 5.1. (Uniformization Theorem) *Let $A, B \in \mathbb{C}$ be complex numbers satisfying $4A^3 - 27B^2 \neq 0$. Then there exists a unique lattice $\Lambda \subset \mathbb{C}$ satisfying*

$$g_2(\Lambda) = A \quad and \quad g_3(\Lambda) = B.$$

PROOF. The proof may be found in many textbooks; see for example [5, Theorem 2.9], [210, I.3.13], [249, §4.2], [266, I.4.3], or [232, VII Proposition 5]. $\qquad\square$

Corollary 5.1.1. *Let $E/\mathbb{C}$ be an elliptic curve. There exist a lattice $\Lambda \subset \mathbb{C}$, unique up to homothety, and a complex analytic isomorphism*

$$\phi : \mathbb{C}/\Lambda \longrightarrow E(\mathbb{C}), \qquad \phi(z) = \big[\wp(z, \Lambda), \wp'(z, \Lambda), 1\big],$$

of complex Lie groups.

PROOF. The existence is immediate from (VI.3.6b) and (VI.5.1), and the uniqueness is (VI.4.1.1). $\qquad\square$

We are now in a position to prove the results left undone in (VI §1).

Proposition 5.2. *Let $E/\mathbb{C}$ be an elliptic curve with Weierstrass coordinate functions x and y.*

(a) *Let α and β be closed paths on $E(\mathbb{C})$ giving a basis for $H_1(E, \mathbb{Z})$. Then the periods*

$$\omega_1 = \int_\alpha \frac{dx}{y} \qquad \text{and} \qquad \omega_2 = \int_\beta \frac{dx}{y}$$

are $\mathbb{R}$-linearly independent.

(b) *Let Λ be the lattice generated by ω_1 and ω_2. Then the map*

$$F : E(\mathbb{C}) \longrightarrow \mathbb{C}/\Lambda, \qquad F(P) = \int_O^P \frac{dx}{y} \pmod{\Lambda},$$

is a complex analytic isomorphism of Lie groups. Its inverse is the map described in (VI.5.1.1).

PROOF. (a) From (VI.5.1.1), there exists some lattice Λ_1 such that the map

$$\phi_1 : \mathbb{C}/\Lambda_1 \longrightarrow E(\mathbb{C}), \qquad \phi_1(z) = \big[\wp(z, \Lambda_1), \wp'(z, \Lambda_1), 1 \big],$$

is a complex analytic isomorphism. It follows that $\phi_1^{-1} \circ \alpha$ and $\phi_1^{-1} \circ \beta$ are a basis for $H_1(\mathbb{C}/\Lambda_1, \mathbb{Z})$. (Here we are viewing α and β as continuous maps from the unit circle to $E(\mathbb{C})$.) We observe that $H_1(\mathbb{C}/\Lambda, \mathbb{Z})$ is naturally isomorphic to the lattice Λ_1 via the map $\gamma \mapsto \int_\gamma dz$, while the differential dx/y on E pulls back to

$$\phi_1^* \left(\frac{dx}{y} \right) = \frac{d\wp(z)}{\wp'(z)} = dz \qquad \text{on } \mathbb{C}/\Lambda_1.$$

Therefore the periods

$$\omega_1 = \int_\alpha \frac{dx}{y} = \int_{\phi_1^{-1} \circ \alpha} dz \qquad \text{and} \qquad \omega_2 = \int_\beta \frac{dx}{y} = \int_{\phi_1^{-1} \circ \beta} dz$$

are a basis for Λ_1, so in particular, they are linearly independent.

(b) We have just shown that the lattice Λ_1 corresponding to E in (VI.5.1.1) is precisely the lattice Λ generated by the periods of E. The composition $F \circ \phi$ thus gives an analytic map

$$F \circ \phi : \mathbb{C}/\Lambda \longrightarrow \mathbb{C}/\Lambda, \qquad (F \circ \phi)(z) = \int_O^{(\wp(z), \wp'(z))} \frac{dx}{y}.$$

Since

$$F^*(dz) = \frac{dx}{y} \qquad \text{and} \qquad \phi^* \left(\frac{dx}{y} \right) = \frac{d\wp(z)}{\wp'(z)} = dz,$$

we see that

$$(F \circ \phi)^* dz = dz.$$

On the other hand, (VI.4.1a) says that any analytic map $\mathbb{C}/\Lambda \to \mathbb{C}/\Lambda$ has the form $\psi_a(z) = az$ for some number $a \in \mathbb{C}^*$. Since $\psi_a^*(dz) = a\,dz$, we see that $(F \circ \phi)(z) = z$, i.e., the composition $F \circ \phi$ is the identity map. But we already know from (VI.3.6b) that ϕ is an analytic isomorphism, so $F = \phi^{-1}$ is, too. $\qquad\square$

Much of the preceding material may be summarized as an equivalence of categories.

Theorem 5.3. *The following categories are equivalent:*
(a) *Objects: Elliptic curves over* $\mathbb{C}$.
   Maps: Isogenies.
(b) *Objects: Elliptic curves over* $\mathbb{C}$.
   Maps: Complex analytic maps taking O to O.
(c) *Objects: Lattices* $\Lambda \subset \mathbb{C}$, *up to homothety.*
   Maps: $\mathrm{Map}(\Lambda_1, \Lambda_2) = \{\alpha \in \mathbb{C} : \alpha\Lambda_1 \subset \Lambda_2\}$.

PROOF. The one-to-one correspondence between elliptic curves over $\mathbb{C}$ and lattices modulo homothety follows from (VI.3.6b), (VI.5.1.1), and (VI.5.2). The matchup of the maps in (a), (b), and (c) is precisely the content of (VI.4.1). □

Remark 5.3.1. The equivalence of (a) and (b) in (VI.5.3) is a very special case of a general principle called GAGA (*Géométrie Algébrique et Géométrie Analytique*; see [229]). GAGA says (among other things) that any complex analytic map between projective varieties over $\mathbb{C}$ is necessarily given by rational functions. For an introductory discussion, see [111, Appendix B].

We now use the uniformization theorem (really (VI.5.1.1)) to make some general deductions about elliptic curves over $\mathbb{C}$. It is worth remarking that even without knowing (VI.5.1.1), everything that we are about to prove would at least apply to those elliptic curves that occur in (VI.3.6b). The uniformization theorem merely says that this class of curves includes every elliptic curve over $\mathbb{C}$.

Proposition 5.4. *Let* $E/\mathbb{C}$ *be an elliptic curve and let* $m \geq 1$ *be an integer.*
(a) *There is an isomorphism of abstract groups*
$$E[m] \cong \mathbb{Z}/m\mathbb{Z} \times \mathbb{Z}/m\mathbb{Z}.$$
(b) *The multiplication-by-m map* $[m] : E \to E$ *has degree* m^2.

PROOF. (a) From (VI.5.1.1), we know that $E(\mathbb{C})$ is isomorphic to $\mathbb{C}/\Lambda$ for some lattice $\Lambda \subset \mathbb{C}$. Hence
$$E[m] \cong \left(\frac{\mathbb{C}}{\Lambda}\right)[m] \cong \frac{\frac{1}{m}\Lambda}{\Lambda} \cong \left(\frac{\mathbb{Z}}{m\mathbb{Z}}\right)^2.$$

(b) Since $\mathrm{char}(\mathbb{C}) = 0$ and the map $[m]$ is unramified, the degree of $[m]$ is equal to the number of points in $E[m] = [m]^{-1}\{O\}$. □

Let $E/\mathbb{C}$ be an elliptic curve. Note that (VI.4.1) allows us to identify $\mathrm{End}(E)$ with a certain subring of $\mathbb{C}$. Thus if $E(\mathbb{C}) \cong \mathbb{C}/\Lambda$ as in (VI.5.1.1), then
$$\mathrm{End}(E) \cong \{\alpha \in \mathbb{C} : \alpha\Lambda \subset \Lambda\}.$$

Since Λ is unique up to homothety (VI.4.1.1), this ring is independent of the choice of Λ. We use this description of $\mathrm{End}(E)$ to completely characterize the endomorphism rings that may occur. We recall the following definition from (III §9).

Definition. Let $\mathcal{K}$ be a number field. An *order* $\mathcal{R}$ of $\mathcal{K}$ is a subring of $\mathcal{K}$ that is finitely generated as a $\mathbb{Z}$-module and satisfies $\mathcal{R} \otimes \mathbb{Q} = \mathcal{K}$.

Theorem 5.5. *Let $E/\mathbb{C}$ be an elliptic curve, and let ω_1 and ω_2 be generators for the lattice Λ associated to E by (VI.5.1.1). Then one of the following is true:*
  (i) $\operatorname{End}(E) = \mathbb{Z}$.
  (ii) *The field $\mathbb{Q}(\omega_2/\omega_1)$ is an imaginary quadratic extension of $\mathbb{Q}$, and $\operatorname{End}(E)$ is isomorphic to an order in $\mathbb{Q}(\omega_1/\omega_2)$.*

PROOF. Let $\tau = \omega_1/\omega_2$. Multiplying Λ by ω_1/ω_2 shows that Λ is homothetic to $\mathbb{Z} + \mathbb{Z}\tau$, so we may replace Λ by $\mathbb{Z} + \mathbb{Z}\tau$. Let

$$\mathcal{R} = \{\alpha \in \mathbb{C} : \alpha\Lambda \subset \Lambda\},$$

so $\mathcal{R} \cong \operatorname{End}(E)$ from (VI.4.1). Then, for any $\alpha \in \mathcal{R}$, there are integers a, b, c, d such that
$$\alpha = a + b\tau \qquad \text{and} \qquad \alpha\tau = c + d\tau.$$

Eliminating τ from these equations yields

$$\alpha^2 - (a + d)\alpha + ad - bc = 0.$$

This proves that $\mathcal{R}$ is an integral extension of $\mathbb{Z}$.

Now suppose that $\mathcal{R} \neq \mathbb{Z}$ and choose some $\alpha \in \mathcal{R} \smallsetminus \mathbb{Z}$. Then, with notation as above, we have $b \neq 0$, so eliminating α gives a nontrivial equation

$$b\tau^2 - (a - d)\tau - c = 0.$$

It follows that $\mathbb{Q}(\tau)$ is an imaginary quadratic extension of $\mathbb{Q}$ (note that $\tau \notin \mathbb{R}$). Finally, since $\mathcal{R} \subset \mathbb{Q}(\tau)$ and $\mathcal{R}$ is integral over $\mathbb{Z}$, it follows that $\mathcal{R}$ is an order in $\mathbb{Q}(\tau)$. □

Proposition 5.6. *Let $E/\mathbb{C}$ be an elliptic curve, and fix a lattice Λ and an isomorphism $E(\mathbb{C}) \cong \mathbb{C}/\Lambda$.*
  (a) *There is a natural isomorphism*

$$H_1\big(E(\mathbb{C}), \mathbb{Z}\big) \xrightarrow{\ \sim\ } \Lambda, \qquad \gamma \longmapsto \int_\gamma dz.$$

  (b) *There is a natural isomorphism*

$$H_1\big(E(\mathbb{C}), \mathbb{Z}/m\mathbb{Z}\big) \xrightarrow{\ \sim\ } E[m].$$

PROOF. (a) We proved this during the course of proving (VI.5.2a).
  (b) From (a) we have

$$H_1\big(E(\mathbb{C}), \mathbb{Z}/m\mathbb{Z}\big) \cong H_1\big(E(\mathbb{C}), \mathbb{Z}\big) \otimes \mathbb{Z}/m\mathbb{Z} \cong \Lambda \otimes \mathbb{Z}/m\mathbb{Z} \cong \Lambda/m\Lambda.$$

On the other hand, using the identification $E(\mathbb{C}) \cong \mathbb{C}/\Lambda$, we obtain an isomorphism

$$E(\mathbb{C})[m] \cong (\mathbb{C}/\Lambda)[m] = \{z \in \mathbb{C} : mz \in \Lambda\}/\Lambda \xrightarrow[z \to mz]{\ \sim\ } \Lambda/m\Lambda. \qquad □$$

VI.6   The Lefschetz Principle

The *Lefschetz principle* says, roughly, that algebraic geometry over an arbitrary algebraically closed field of characteristic 0 is "the same" as algebraic geometry over $\mathbb{C}$.
One can, of course, make this precise by formulating an equivalence of suitably defined categories, but we will be content here to give an informal presentation.

Our first observation is that if the given field K can be embedded as a subfield
of $\mathbb{C}$, then everything proceeds smoothly. For example, if $K \subset \mathbb{C}$ is any field and
if E/K is an elliptic curve, then the fact that $[m] : E \to E$ is a finite algebraic map
implies that $E[m] \subset E(\bar{K}) \subset E(\mathbb{C})$. (To see this, note that for any $P \in E(\bar{K})$,
the set $[m]^{-1}(P)$ is finite and invariant under $G_{\bar{K}/K}$, so every point in $[m]^{-1}(P)$ is
defined over $\bar{K}$.) Hence, using (VI.5.4), we obtain a proof that

$$E[m] = E(\bar{K})[m] = E(\mathbb{C})[m] \cong (\mathbb{Z}/m\mathbb{Z})^2.$$

Note that the embedding $K \subset \mathbb{C}$ need not be topological (assuming that K has
a topology in the first place). It does not matter that we may have used the topology
of $\mathbb{C}$ to reach our conclusions, e.g., using the analytic isomorphism $E(\mathbb{C}) \cong \mathbb{C}/\Lambda$,
as long as our hypotheses and conclusions are purely algebraic

Our second observation is that theorems in algebraic geometry generally deal
with finite (or sometimes countable) sets. For example, any variety is defined by a
finite set of polynomial equations (Hilbert basis theorem), and each equation has only
finitely many coefficients. Similarly, an algebraic map between varieties is given by
a finite set of polynomials, each having a finite number of coefficients. Now suppose
that $\{V_1, V_2, \ldots\}$ is a finite (or countable) set of varieties defined over some field K
of characteristic 0, and suppose that $\{\phi_1, \phi_2, \ldots\}$ is a finite (or countable) set of
rational maps defined over K that map the various V_i to one another. Let $K_0 \subset K$
be the field generated over $\mathbb{Q}$ by all of the coefficients of all of the polynomials
defining all of the V_i and all of the ϕ_j. It is clear that the transcendence degree of K_0
over $\mathbb{Q}$ has cardinality at most that of the natural numbers, so we can use Zorn's
lemma to embed K_0 in $\mathbb{C}$. Then using the above discussion concerning subfields
of $\mathbb{C}$, we are able to reduce most algebro-geometric questions concerning the V_i and
the ϕ_j to the corresponding questions over $\mathbb{C}$, where we may be able to profitably
employ techniques from complex analysis and differential geometry.

To illustrate the Lefschetz principle, we prove two results.

Theorem 6.1. *Let K be a field of characteristic 0 and let E/K be an elliptic
curve.*

(a) *Let $m \geq 1$ be an integer. Then*

$$E[m] \cong \mathbb{Z}/m\mathbb{Z} \times \mathbb{Z}/m\mathbb{Z}.$$

(b) *The endomorphism ring of E is either $\mathbb{Z}$ or an order in a quadratic imaginary
extension of $\mathbb{Q}$, cf. (III.5.6c) and (III.9.4).*

PROOF. (a) This is immediate from (VI.5.4) and the Lefschetz principle.

(b) Here we can apply the Lefschetz principle to (VI.5.5), once we note that $\text{End}(E)$ is countably (in fact, finitely) generated from (III.7.5). Alternatively, even without (III.7.5), we can argue as follows. If $\text{End}(E)$ is neither $\mathbb{Z}$ nor quadratic imaginary, then it contains a finitely generated subring that is neither $\mathbb{Z}$ nor imaginary quadratic. Applying the Lefschetz principle to the maps in this subring contradicts (VI.5.5). $\qquad\square$

Exercises

6.1. Let $\Lambda = \mathbb{Z}\omega_1 + \mathbb{Z}\omega_2$ be a lattice. Suppose that $\theta(z)$ is an entire function, i.e., holomorphic on all of $\mathbb{C}$, with the property that there are constants $a_1, a_2 \in \mathbb{C}$ such that

$$\theta(z + \omega_1) = a_1\theta(z) \quad \text{and} \quad \theta(z + \omega_2) = a_2\theta(z) \quad \text{for all } z \in \mathbb{C}.$$

Prove that

$$\theta(z) = be^{cz} \quad \text{for some } b, c \in \mathbb{C}.$$

6.2. Let $\Lambda \subset \mathbb{C}$ be a lattice.
(a) Prove that every fundamental parallelogram for Λ has the same area. Denote this area by $A(\Lambda)$.
(b) Prove that as $R \to \infty$,

$$\#\{\omega \in \Lambda : |\omega| \leq R\} = \frac{\pi R^2}{A(\Lambda)} + O(R).$$

(The big-O constant depends on Λ, of course.)
(c) Prove that there is a constant $c(\Lambda)$ such that for all $R > 0$,

$$\#\{\omega \in \Lambda : R \leq |\omega| < R+1\} < cR.$$

6.3. (a) Prove that for all $z, a \in \mathbb{C} \smallsetminus \Lambda$,

$$\wp(z) - \wp(a) = -\frac{\sigma(z+a)\sigma(z-a)}{\sigma(z)^2\sigma(a)^2}.$$

(*Hint.* Compare zeros and poles.)
(b) Prove that

$$\wp'(z) = -\frac{\sigma(2z)}{\sigma(z)^4}.$$

(c) Prove that for every integer n, the function $\sigma(nz)/\sigma(z)^{n^2}$ is in $\mathbb{C}(\Lambda)$.
(d) More precisely, prove that

$$(-1)^{n-1}\left(1!2!\cdots(n-1)!\right)^2 \frac{\sigma(nz)}{\sigma(z)^{n^2}} = \det\left(\wp^{(i+j-1)}(z)\right)_{1\leq i,j\leq n-1}.$$

(See also exercises 6.15 and 6.16.)

6.4. Define the *Weierstrass ζ-function* $\zeta(z)$ (not to be confused with the Riemann ζ-function) by the series

$$\zeta(z) = \frac{1}{z} + \sum_{\substack{w \in \Lambda \\ w \neq 0}} \left(\frac{1}{z-w} + \frac{1}{w} + \frac{z}{w^2}\right).$$

(a) Prove that

$$\frac{d}{dz}\log\sigma(z) = \zeta(z) \qquad \text{and} \qquad \frac{d}{dz}\zeta(z) = -\wp(z).$$

(b) Prove that

$$\zeta(-z) = -\zeta(z),$$

and that for all $\omega \in \Lambda$ there exists a constant $\eta(\omega) \in \mathbb{C}$ satisfying

$$\zeta(z + \omega) = \zeta(z) + \eta(\omega).$$

If $\omega \notin 2\Lambda$, prove that $\eta(\omega) = 2\zeta(\omega/2)$.

(c) Prove that the map $\eta : \Lambda \to \mathbb{C}$ given in (b) is bilinear.

(d) Write $\Lambda = \mathbb{Z}\omega_1 + \mathbb{Z}\omega_2$ with $\text{Im}(\omega_1/\omega_2) > 0$. Prove *Legendre's relation*

$$\omega_1\eta(\omega_2) - \omega_2\eta(\omega_1) = 2\pi i.$$

(*Hint*. Integrate $\zeta(z)$ around a fundamental parallelogram.) The two numbers $\eta(\omega_1)$ and $\eta(\omega_2)$ are called *quasiperiods*.

(e) Prove that

$$\sigma(z + \omega) = \pm e^{\eta(\omega)(z+\omega/2)}\sigma(z),$$

where the sign is positive if $\omega \in 2\Lambda$ and negative otherwise.

(f) Extend $\eta : \Lambda \to \mathbb{C}$ to an $\mathbb{R}$-linear map $\eta : \mathbb{C} \to \mathbb{C}$ by identifying $\Lambda \otimes_{\mathbb{Z}} \mathbb{R}$ with $\mathbb{C}$. Let

$$G(z) = e^{-z\eta(z)/2}\sigma(z).$$

Prove that

$$\left|G(z + \omega)\right| = \left|G(z)\right| \qquad \text{for all } \omega \in \Lambda \text{ and all } z \in \mathbb{C}.$$

Thus $\left|G(z)\right|$ defines a real analytic function from $(\mathbb{C}/\Lambda) \smallsetminus \{0\}$ to $\mathbb{R}$.

6.5. Verify the values of the following indefinite integrals.

(a) $\displaystyle\int \wp(z)^2 \, dz = \frac{1}{6}\wp'(z) + \frac{1}{12}g_2 z + C.$

(b) $\displaystyle\int \wp(z)^3 \, dz = \frac{1}{120}\wp'''(z) - \frac{3}{20}g_2\zeta(z) + \frac{1}{10}g_3 z + C.$

6.6. For a lattice $\Lambda \subset \mathbb{C}$, let $g_2(\Lambda)$ and $g_3(\Lambda)$ be as in (VI.3.5.1), and define

$$\Delta(\Lambda) = g_2(\Lambda)^3 - 27g_3(\Lambda)^2 \qquad \text{and} \qquad j(\Lambda) = 1728\frac{g_2(\Lambda)^3}{\Delta(\Lambda)}.$$

(a) Let $\alpha \in \mathbb{C}^*$. Prove that

$$g_2(\alpha\Lambda) = \alpha^{-4}g_2(\Lambda) \qquad \text{and} \qquad g_3(\alpha\Lambda) = \alpha^{-6}g_3(\Lambda),$$

and deduce that

$$\Delta(\alpha\Lambda) = \alpha^{-12}\Delta(\Lambda) \qquad \text{and} \qquad j(\alpha\Lambda) = j(\Lambda).$$

(b) Prove that $j(\Lambda_1) = j(\Lambda_2)$ if and only if there is an $\alpha \in \mathbb{C}^*$ such that $\alpha\Lambda_1 = \Lambda_2$, i.e., if and only if Λ_1 and Λ_2 are homothetic.

(c) Prove that
$$j(\mathbb{Z} + \mathbb{Z}i) = 1728 \qquad \text{and} \qquad j(\mathbb{Z} + \mathbb{Z}e^{2\pi i/3}) = 0.$$

6.7. *Elliptic curves over* $\mathbb{R}$. Let $E/\mathbb{C}$ be an elliptic curve corresponding to a lattice $\Lambda \subset \mathbb{C}$.
  (a) Prove that E is isomorphic to a curve defined over $\mathbb{R}$ if and only if there is an $\alpha \in \mathbb{C}^*$ such that $\alpha\Lambda$ is mapped to itself by complex conjugation. (*Hint.* First prove that $j(\Lambda) = j(\overline{\Lambda})$.)
  (b) Suppose that E is defined over $\mathbb{R}$ and that we have chosen a lattice Λ for E as in (a), so Λ is invariant under complex conjugation. Prove that $\Delta(\Lambda) \in \mathbb{R}$, and that $E(\mathbb{R})$ is connected if and only if $\Delta(\Lambda) < 0$.
  (c) Let $E/\mathbb{C}$ be given by a Legendre equation
$$E : y^2 = x(x - 1)(x - \lambda).$$

  Prove that $\lambda \in \mathbb{R}$ if and only if E can be defined over $\mathbb{R}$ and $E[2] \subset E(\mathbb{R})$.
  (d) If E is defined over $\mathbb{R}$ and $E[2] \subset E(\mathbb{R})$, prove that there is a lattice for E that is rectangular, i.e., of the form $\mathbb{Z}\omega_1 + \mathbb{Z}\omega_2 i$ with $\omega_1, \omega_2 \in \mathbb{R}$.

6.8. Let $\mathcal{K}/\mathbb{Q}$ be an imaginary quadratic field, let $\mathcal{R}$ be the ring of integers of $\mathcal{K}$, and let $h_\mathcal{R}$ denote the class number of $\mathcal{R}$.
  (a) Prove that up to isomorphism, there are exactly $h_\mathcal{R}$ elliptic curves $E/\mathbb{C}$ with endomorphism ring $\text{End}(E) \cong \mathcal{R}$.
  (b) If E is a curve as in (a), prove that $j(E)$ is an algebraic number and that its degree satisfies
$$\big[\mathcal{K}\big(j(E)\big) : \mathcal{K}\big] \leq h_\mathcal{R}.$$

In fact, $\mathcal{K}\big(j(E)\big)$ is the Hilbert class field of $\mathcal{K}$, so the inequality in (b) is an equality. See (C §11) and the references listed there.

6.9. Let $E_1/\mathbb{C}$ and $E_2/\mathbb{C}$ be elliptic curves, and assume that E_1 has complex multiplication. Prove that E_1 is isogenous to E_2 if and only if
$$\text{End}(E_1) \otimes \mathbb{Q} \cong \text{End}(E_2) \otimes \mathbb{Q}.$$

6.10. Let $\phi : E_1 \to E_2$ be an isogeny of elliptic curves over $\mathbb{C}$, and let $\phi_\alpha : \mathbb{C}/\Lambda_1 \to \mathbb{C}/\Lambda_2$ be the corresponding analytic map induced by $z \mapsto \alpha z$ as in (VI.4.1), so in particular we have $\alpha\Lambda_1 \subset \Lambda_2$.
  (a) Prove that $\deg \phi$ equals the index $(\Lambda_2 : \alpha\Lambda_1)$.
  (b) Let $m = \deg \phi$. Prove that the dual isogeny $\hat{\phi} : E_2 \to E_1$ corresponds to the analytic map induced by $z \mapsto m\alpha^{-1}z$.
  (c) Assume that $\Lambda_1 = \Lambda_2$. Prove that $\deg \phi = \text{N}_{\mathbb{Q}(\alpha)/\mathbb{Q}}(\alpha)$. Deduce that $\hat{\phi}$ corresponds to the analytic map induced by $z \mapsto \bar{\alpha}z$, where $\bar{\alpha}$ is the complex conjugate of α.

Elliptic Integrals. Exercises 6.11–6.13 develop a minute portion of the classical theory of elliptic integrals.

6.11. Let $E/\mathbb{C}$ be an elliptic curve given by a Legendre equation
$$E : Y^2 = X(X - 1)(X - \lambda).$$

(a) Prove that there is a $k \in \mathbb{C} \smallsetminus \{0, \pm 1\}$ such that E has an equation of the form

$$E; y^2 = (1 - x^2)(1 - k^2 x^2).$$

(*Hint.* Let $X = (ax + b)/(cx + d)$ and $Y = ey/(cx + d)^2$ for an appropriate choice of $a, b, c, d, e \in \mathbb{C}$.)

(b) For a given value of λ, find all possible values of k. Conversely, given k, find all values of λ.

(c) Express the j-invariant $j(E)$ in terms of k.

(d) Suppose that $\lambda \in \mathbb{R}$. (See Exercise 6.7.) Show that k may be chosen to be real and to satisfy $0 < k < 1$.

6.12. *Complete Elliptic Integrals.* Let E be an elliptic curve given by an equation

$$E : y^2 = (1 - x^2)(1 - k^2 x^2).$$

To simplify matters, assume that $0 < k < 1$ (cf. Exercise 6.11d). Define *complete elliptic integrals to the modulus k* by

$$K(k) = \int_0^1 \frac{dx}{y} = \int_0^1 \frac{1}{\sqrt{(1 - x^2)(1 - k^2 x^2)}} \, dx,$$

$$T(k) = \int_0^1 \frac{y}{1 - x^2} \, dx = \int_0^1 \sqrt{\frac{1 - k^2 x^2}{1 - x^2}} \, dx.$$

(a) Make appropriate branch cuts and prove that the lattice for E is generated by the periods

$$4 \int_0^1 \frac{1}{\sqrt{(1 - x^2)(1 - k^2 x^2)}} \, dx \quad \text{and} \quad 2i \int_1^{1/k} \frac{1}{\sqrt{(1 - x^2)(1 - k^2 x^2)}} \, dx.$$

(b) The *complementary modulus* to k is the quantity k' defined by

$$k^2 + k'^2 = 1 \quad \text{and} \quad 0 < k' < 1.$$

Prove that

$$\int_1^{1/k} \frac{1}{\sqrt{(1 - x^2)(1 - k^2 x^2)}} \, dx = \int_0^1 \frac{1}{\sqrt{(1 - X^2)(1 - k'^2 X^2)}} \, dX.$$

(*Hint.* Let $x = (1 - k'^2 X^2)^{-1/2}$.) Conclude that the period lattice of the elliptic curve $E/\mathbb{C}$ is generated by $4K(k)$ and $2iK(k')$.

(c) Prove the transformation formulas

$$K\left(\frac{2\sqrt{k}}{1 + k}\right) = (1 + k)K(k) \quad \text{and} \quad K\left(\frac{1 - k}{1 + k}\right) = \frac{1 + k}{2}K(k').$$

6.13. (a) Show that the complete elliptic integrals defined in Exercise 6.12 may also be written as

$$K(k) = \int_0^{\pi/2} \frac{d\theta}{\sqrt{1 - k^2 \sin^2 \theta}} \quad \text{and} \quad T(k) = \int_0^{\pi/2} \sqrt{1 - k^2 \sin^2 \theta} \, d\theta.$$

(b) Prove that the arc length of the ellipse

$$x^2/a^2 + y^2/b^2 = 1 \qquad \text{with } a \geq b > 0$$

is given by the complete elliptic integral

$$4aT\left(\sqrt{1 - \left(\frac{b}{a}\right)^2}\right).$$

(c) Prove that the arc length of the lemniscate

$$r^2 = \cos(2\theta)$$

is given by the complete elliptic integral $2\sqrt{2}K\left(1/\sqrt{2}\right)$. Prove that it also equals

$$4\int_0^1 \frac{dx}{\sqrt{1 - x^4}}.$$

Thus the integral giving the arc length of the lemniscate resembles the integral giving the arc length of the unit circle, i.e., $2\pi = 4\int_0^1 dx/\sqrt{1 - x^2}$.

6.14. *The Arithmetic–Geometric Mean.* For initial values $a, b \in \mathbb{R}$ with $a \geq b > 0$, we define sequences $\{a_n\}$ and $\{b_n\}$ recursively by

$$a_0 = a, \qquad b_0 = b, \qquad a_{n+1} = \frac{a_n + b_n}{2}, \qquad b_n = \sqrt{a_n b_n}.$$

(a) Prove that

$$0 \leq a_{n+1} - b_{n+1} \leq \frac{1}{2}(a_n - b_n).$$

Deduce that the limit

$$M(a, b) = \lim_{n \to \infty} a_n = \lim_{n \to \infty} b_n$$

exists. The quantity $M(a, b)$ is called the *arithmetic–geometric mean of a and b*.

(b) Prove that

$$M(a, b) = M(a_1, b_1) = M(a_2, b_2) = \cdots$$

and

$$M(ca, cb) = cM(a, b) \qquad \text{for } c > 0.$$

(c) Define an integral $I(a, b)$ by

$$I(a, b) = \int_0^{\pi/2} \frac{d\theta}{\sqrt{a^2 \cos^2 \theta + b^2 \sin^2 \theta}}.$$

Prove that $I(a, b)$ is related to the complete elliptic integrals described in exercises 6.12 and 6.13 by showing that

$$I(a, b) = a^{-1}K\left(\frac{2\sqrt{k}}{1 + k}\right) \qquad \text{and} \qquad I(a_1, b_1) = a_1^{-2}K(k)$$

for $k = (a - b)/(a + b)$.

(d) Prove that
$$M(a,b)I(a,b) = \pi/2.$$

(*Hint.* Use (c) and Exercise 6.12c to prove that $I(a,b) = I(a_1,b_1)$. Then calculate the limit of $I(a_n,b_n)$ as $n \to \infty$.)

Combining (c) and (d), observe that the complete elliptic integral $K(k)$ for $0 < k < 1$ may be computed in terms of the arithmetic–geometric mean.

(e) Prove that the rate of convergence of $M(a,b)$ predicted by (a), namely
$$a_n - b_n \le 2^{-n}(a - b),$$

is far slower than in reality. More precisely, use (b) to show that it suffices to compute $M(a,b)$ in the case that $b \ge 1$, and under this assumption, prove that
$$a_{n+m} - b_{n+m} \le 8 \left(\frac{a_n - b_n}{8} \right)^{2^m} \qquad \text{for all } m, n \ge 0.$$

In particular, since eventually $a_n - b_n < 8$, the sequences $\{a_n\}$ and $\{b_n\}$ converge doubly exponentially.

(f) Prove that
$$\int_0^1 \frac{dz}{\sqrt{1 - z^4}} = \frac{\pi}{2} M(\sqrt{2}, 1),$$

and use this equality to numerically calculate the value of the complete elliptic integral on the left-hand side. It was the observation that these two numbers, calculated independently, agree to eleven decimal places that led Gauss to initiate an extensive study of the arithmetic–geometric mean. For a fascinating account of this subject, see [52].

6.15. Let $E/\mathbb{C}$ be an elliptic curve and let ψ_n be the division polynomial defined in Exercise 3.7. Considered as a function on $\mathbb{C}/\Lambda$, prove that $\psi_n(z)$ is given by
$$\psi_n(z) = (-1)^{n+1} \frac{\sigma(nz)}{\sigma(z)^{n^2}}.$$

(*Hint.* Use the description of $\operatorname{div}(\psi_n)$ in Exercise 3.7f. Then evaluate $z^{n^2-1}\psi_n(z)$ as $z \to 0$ to find the constant.)

6.16. Let $(W_n)_{n\ge 1}$ be an elliptic divisibility sequence over $\mathbb{C}$. (See Exercise 3.34 for the definition of elliptic divisibility sequence.) Assume that $W_1 = 1$ and $W_2 W_3 W_4 \ne 0$. Prove that there are a lattice $\Lambda \subset \mathbb{C}$ and a complex number $u \in \mathbb{C}$ such that
$$W_n = \frac{\sigma(nu)}{\sigma(u)^{n^2}} \qquad \text{for all } n \ge 1.$$

More precisely, prove that Λ and u exist, provided that a certain polynomial in W_2, W_3, and W_4 does not vanish.

Chapter VII

Elliptic Curves over Local Fields

In this chapter we study the group of rational points on an elliptic curve defined over a field that is complete with respect to a discrete valuation. We start with some basic facts concerning Weierstrass equations and "reduction modulo π." This enables us to break our problem into several pieces, and then, by examining each piece individually, to deduce a great deal about the group of rational points as a whole. Unless explicitly stated otherwise, we use the following notation:

K a local field, complete with respect to a discrete valuation v.

R $= \{x \in K : v(x) \geq 0\}$, the ring of integers of K.

R^* $= \{x \in K : v(x) = 0\}$, the unit group of R.

$\mathcal{M}$ $= \{x \in K : v(x) > 0\}$, the maximal ideal of R.

π a uniformizer for R, i.e., $\mathcal{M} = \pi R$.

k $= R/\mathcal{M}$, the residue field of R.

We further assume that v is normalized so that $v(\pi) = 1$. Note that by convention, $v(0) = \infty$ is assigned a value larger than every real number. Finally, in keeping with our general policy, we assume that both K and k are perfect fields.

VII.1 Minimal Weierstrass Equations

Let E/K be an elliptic curve, and let

$$E : y^2 + a_1 xy + a_3 y = x^3 + a_2 x^2 + a_4 x + a_6$$

be a Weierstrass equation for E/K. The substitution $(x, y) \mapsto (u^{-2}x, u^{-3}y)$ leads to a new equation in which a_i is replaced by $u^i a_i$, so if we choose u to be divisible

J.H. Silverman, *The Arithmetic of Elliptic Curves, Second Edition*, Graduate Texts in Mathematics 106, DOI 10.1007/978-0-387-09494-6_VII,
© Springer Science+Business Media, LLC 2009

by a sufficiently large power of π, then we obtain a Weierstrass equation all of whose coefficients are in R. Having done this, the discriminant Δ satisfies $v(\Delta) \geq 0$. Finally, since v is discrete, among all such Weierstrass equations with coefficients in R, we can choose one that minimizes the value of $v(\Delta)$.

Definition. Let E/K be an elliptic curve. A Weierstrass equation for E is called a *minimal (Weierstrass) equation for E at v* if $v(\Delta)$ is minimized subject to the condition that $a_1, a_2, a_3, a_4, a_6 \in R$. This minimal value of $v(\Delta)$ is called the *valuation of the minimal discriminant of E at v*.

Remark 1.1. How can we tell whether a given Weierstrass equation is minimal? First, by definition, all of the a_i must be in R, so in particular, the discriminant Δ is in R. If the equation is not minimal, then (III.1.2) says that there is a coordinate change giving a new equation with discriminant $\Delta' = u^{-12}\Delta \in R$. Thus $v(\Delta)$ can be changed only by multiples of 12, so we conclude that

$$a_i \in R \text{ and } v(\Delta) < 12 \quad \Longrightarrow \quad \text{the equation is minimal.}$$

Similarly, since $c_4' = u^{-4}c_4$ and $c_6' = u^{-6}c_6$, we have

$$a_i \in R \text{ and } v(c_4) < 4 \quad \Longrightarrow \quad \text{the equation is minimal,}$$
$$a_i \in R \text{ and } v(c_6) < 6 \quad \Longrightarrow \quad \text{the equation is minimal.}$$

If $\operatorname{char}(k) \neq 2, 3$, then a converse holds. More precisely, if the equation is minimal, then $v(\Delta) < 12$ or $v(c_4) < 4$; see Exercise 7.1. For arbitrary K, there is an algorithm of Tate that determines whether a given equation is minimal; see [266, IV §9] or [283].

Example 1.2. Let p be a prime and consider the Weierstrass equation

$$E : y^2 + xy + y = x^3 + x^2 + 22x - 9$$

over the field $\mathbb{Q}_p$. This equation has discriminant $\Delta = -2^{15}5^2$ and $c_4 = -5 \cdot 211$. From (VII.1.1), this is a minimal Weierstrass equation at p for every prime $p \in \mathbb{Z}$.

Proposition 1.3. (a) *Every elliptic curve E/K has a minimal Weierstrass equation.*

(b) *A minimal Weierstrass equation is unique up to a change of coordinates*

$$x = u^2 x' + r, \qquad y = u^3 y' + u^2 s x' + t,$$

with $u \in R^*$ and $r, s, t \in R$.

(c) *The invariant differential*

$$\omega = \frac{dx}{2y + a_1 x + a_3}$$

associated to a minimal Weierstrass equation is unique up to multiplication by an element of R^*.

(d) *Conversely, if one starts with any Weierstrass equation whose coefficients are in R, then any change of coordinates*

$$x = u^2 x' + r, \qquad y = u^3 y' + u^2 s x' + t,$$

used to produce a minimal Weierstrass equation satisfies $u, r, s, t \in R$.

PROOF. (a) One can easily find some Weierstrass equation with all $a_i \in R$, and among such equations, there exists (at least) one that minimizes $v(\Delta)$, since v is discrete.

(b) We know from (III.3.1b) that any Weierstrass equation for E/K is unique up to the indicated change of coordinates with $u \in K^*$ and $r, s, t \in K$. Now suppose that the given equation and the new equation are both minimal. From the definition of minimality, we have $v(\Delta') = v(\Delta)$. We now apply the transformation formulas described in (III §1, Table 3.1). The transformation formula for Δ says that $u^{12} \Delta' = \Delta$, so we see that $u \in R^*$. Similarly, the transformation formula for b_6 (respectively for b_8) shows that $4r^3$ (respectively $3r^4$) is in R, hence $r \in R$. Finally, the transformation formula for a_2 gives $s \in R$, and the transformation formula for a_6 gives $t \in R$.

(c) Clear from (b), since $\omega' = u\omega$.

(d) Since the new equation is to be minimal, we know that $v(\Delta') \le v(\Delta)$, and we also have $u^{12} \Delta' = \Delta$. Hence $v(u) \ge 0$, so $u \in R$. Now the proof of (b) can be repeated to show that $r, s, t \in R$. $\qquad\square$

VII.2 Reduction Modulo π

We next look at the operation of "reduction modulo π," which we denote by a tilde. Thus, for example, the natural reduction map $R \to k = R/\pi R$ is denoted by $t \mapsto \tilde{t}$. Having chosen a minimal Weierstrass equation for E/K, we can reduce its coefficients modulo π to obtain a (possibly singular) curve over k, namely

$$\tilde{E} : y^2 + \tilde{a}_1 xy + \tilde{a}_3 y = x^3 + \tilde{a}_2 x^2 + \tilde{a}_4 x + \tilde{a}_6.$$

The curve $\tilde{E}/k$ is called the *reduction of E modulo π*. Since we started with a minimal equation for E, (VII.1.3b) tells us that the equation for $\tilde{E}$ is unique up to the standard change of coordinates (III.3.1b) for Weierstrass equations over the residue field k.

Next let $P \in E(K)$. We can find homogeneous coordinates $P = [x_0, y_0, z_0]$ with $x_0, y_0, z_0 \in R$ and at least one of x_0, y_0, z_0 in R^*. Then the reduced point

$$\tilde{P} = [\tilde{x}_0, \tilde{y}_0, \tilde{z}_0]$$

is in $\tilde{E}(k)$. This defines a *reduction map*

$$E(K) \longrightarrow \tilde{E}(k), \qquad P \longmapsto \tilde{P}.$$

More generally, in a similar fashion we can define a *reduction map*

$$\mathbb{P}^n(K) \longrightarrow \mathbb{P}^n(k).$$

Then the reduction map for $E(K) \subset \mathbb{P}^2(K)$ is just the restriction of the reduction map on $\mathbb{P}^2(K)$.

The curve $\tilde{E}/k$ may be singular (more on this later), but in any case we recall (III.2.5) that the set of nonsingular points $\tilde{E}_{ns}(k)$ forms a group. We define two subsets of $E(K)$ as follows:

$$E_0(K) = \{P \in E(K) : \tilde{P} \in \tilde{E}_{ns}(k)\},$$
$$E_1(K) = \{P \in E(K) : \tilde{P} = \tilde{O}\}.$$

In words, $E_0(K)$ is the set of points with *nonsingular reduction* and $E_1(K)$ is the *kernel of reduction*. From (VII.1.3b), these two sets do not depend on which minimal Weierstrass equation we choose.

Proposition 2.1. *There is an exact sequence of abelian groups*

$$0 \longrightarrow E_1(K) \longrightarrow E_0(K) \longrightarrow \tilde{E}_{ns}(k) \longrightarrow 0,$$

where the right-hand map is reduction modulo π.

PROOF. We begin by showing that the reduction map is surjective. To do this, we use Hensel's lemma and the completeness of K. Thus let

$$f(x,y) = y^2 + a_1 xy + a_3 y - x^3 - a_2 x^2 - a_4 x - a_6 = 0$$

be a minimal Weierstrass equation for E, let $\tilde{f}(x,y)$ be the corresponding polynomial with coefficients reduced modulo π, and let $\tilde{P} = (\tilde{\alpha}, \tilde{\beta}) \in \tilde{E}_{ns}(k)$ be a point. Since $\tilde{P}$ is a nonsingular point of $\tilde{E}$, we know that either

$$\frac{\partial \tilde{f}}{\partial x}(\tilde{P}) \neq 0 \qquad \text{or} \qquad \frac{\partial \tilde{f}}{\partial y}(\tilde{P}) \neq 0,$$

say the latter. (The other case is done similarly.) Choose any $x_0 \in R$ with $\tilde{x}_0 = \tilde{\alpha}$ and look at the equation

$$f(x_0, y) = 0.$$

When reduced modulo π, this equation has $\tilde{\beta}$ as a *simple* root, since by assumption $(\partial \tilde{f}/\partial y)(\tilde{x}_0, \tilde{\beta}) \neq 0$. Thus Hensel's lemma [142, Chapter II, Proposition 2] tells us that the mod π root $\tilde{\beta}$ can be lifted to a $y_0 \in R$ such that $\tilde{y}_0 = \tilde{\beta}$ and $f(x_0, y_0) = 0$. Then the point $P = (x_0, y_0) \in E_0(K)$ reduces to $\tilde{P}$, which completes the proof that the reduction map $E_0(K) \to \tilde{E}_{ns}(k)$ is surjective.

Our next task is to prove that $E_0(K)$ is a subgroup of $E(K)$ and that the reduction map $E_0(K) \to \tilde{E}_{ns}(k)$ is a homomorphism. Note that once we have proven these two facts, the exactness of

$$0 \longrightarrow E_1(K) \longrightarrow E_0(K) \longrightarrow \tilde{E}_{ns}(k) \longrightarrow 0$$

at the left and center follows directly from the definition of $E_1(K)$, so the proof of (VII.2.1) will be complete.

The group laws on $E(K)$ and $E_{ns}(k)$ are defined by taking intersections with lines in $\mathbb{P}^2$. For any line L defined over K, we can find an equation for L of the form

$$L : Ax + By + Cz = 0$$

such that $A, B, C \in R$ and at least one of A, B, C is in R^*. Then the reduction of L is given by the equation

$$\tilde{L} : \tilde{A}x + \tilde{B}y + \tilde{C}z = 0,$$

and it is clear that if $P \in \mathbb{P}^2(K)$ is a point on the line L, then the reduced point $\tilde{P}$ is on the reduced line $\tilde{L}$.

Let $P_1, P_2 \in E_0(K)$ and $P_3 \in E(K)$ be points satisfying $P_1 + P_2 + P_3 = O$. Thus there is a line L that intersects E at the three points P_1, P_2, P_3, counted with appropriate multiplicities. We are going to prove that $\tilde{L}$ intersects $\tilde{E}$ at $\tilde{P}_1, \tilde{P}_2, \tilde{P}_3$ with the correct multiplicities, from which it follows that $P_3 \in E_0(K)$ and that $\tilde{P}_1 + \tilde{P}_2 + \tilde{P}_3 = \tilde{O}$. However, since there are many cases to consider, we will be content to prove two cases and leave the others to the reader; see Exercise 7.15.

Suppose first that the reduced points $\tilde{P}_1, \tilde{P}_2, \tilde{P}_3$ are distinct. Then

$$\tilde{L} \cap \tilde{E} = \{\tilde{P}_1, \tilde{P}_2, \tilde{P}_3\}$$

consists of three distinct points, the first two of which are in $E_{ns}(k)$ by assumption. It follows from (III.2.5) that $\tilde{P}_3$ is also in $E_{ns}(k)$; see also Exercise 3.28(b). Hence $P_3 \in E_0(K)$ and $\tilde{P}_1 + \tilde{P}_2 + \tilde{P}_3 = \tilde{O}$, which is the desired result in this case.

To handle the second case, we use the following general result.

Lemma 2.1.1. *Let $P, Q \in E_0(K)$ be distinct points whose reductions satisfy $\tilde{P} = \tilde{Q}$, and let L be the line through P and Q. Then the line $\tilde{L}$ is tangent to $\tilde{E}$ at $\tilde{P}$.*

PROOF. We assume that $\tilde{P} \neq \tilde{O}$ and leave the reader to handle the case $\tilde{P} = \tilde{O}$. As above, we choose a minimal Weierstrass equation

$$E : f(x, y) = y^2 + a_1 xy + a_3 y - x^3 - a_2 x^2 - a_4 x - a_6 = 0,$$

and we let $\tilde{f}(x, y)$ be the corresponding polynomial with coefficients reduced modulo π. Write

$$P = (\alpha, \beta) \in E(K) \qquad \text{and} \qquad Q = (\alpha + \mu, \beta + \lambda) \in E(K).$$

The assumption that $\tilde{P} = \tilde{Q} \neq \tilde{O}$ implies that $\alpha, \beta \in R$ and $\mu, \lambda \in \mathcal{M}$. Further, the assumption that $P \in E_0(K)$ means that $\tilde{P}$ is a nonsingular point of $\tilde{E}$, so either

$$\frac{\partial \tilde{f}}{\partial x}(\tilde{P}) \neq 0 \qquad \text{or} \qquad \frac{\partial \tilde{f}}{\partial y}(\tilde{P}) \neq 0.$$

We do the case that $(\partial \tilde{f}/\partial y)(\tilde{P}) \neq 0$ and leave the other case to the reader.

The fact that $f(P) = f(Q) = 0$ allows us to compute the first few terms of the Taylor expansion of $f(x, y)$ around Q. Thus

$$
\begin{aligned}
0 &= f(\alpha + \mu, \beta + \lambda) \\
&= f(\alpha, \beta) + \frac{\partial f}{\partial x}(\alpha, \beta)\mu + \frac{\partial f}{\partial y}(\alpha, \beta)\lambda + a\mu^2 + b\mu\lambda + c\lambda^2 \\
&\qquad\qquad\qquad\qquad\qquad\qquad\qquad\qquad \text{for some } a, b, c \in R, \\
&= \frac{\partial f}{\partial x}(\alpha, \beta)\mu + \frac{\partial f}{\partial y}(\alpha, \beta)\lambda + a\mu^2 + b\mu\lambda + c\lambda^2.
\end{aligned}
$$

The assumption that $(\partial \tilde{f}/\partial y)(\tilde{P}) \neq 0$ is equivalent to $(\partial f/\partial y)(\alpha, \beta) \in R^*$, so

$$
v(\lambda) = v\left(\frac{\partial f}{\partial y}(\alpha, \beta)\lambda\right) = v\left(\frac{\partial f}{\partial x}(\alpha, \beta)\mu + a\mu^2 + b\mu\lambda + c\lambda^2\right) \geq v(\mu).
$$

Thus $\lambda/\mu \in R$, so dividing the Taylor expansion by μ and reducing modulo π gives the congruence

$$
\frac{\partial f}{\partial x}(P) + \frac{\partial f}{\partial y}(P) \cdot \frac{\lambda}{\mu} \equiv 0 \pmod{\mathcal{M}}.
$$

This tells us that the slope of the tangent line to $\tilde{E}$ at the point $\tilde{P}$ is

$$
\frac{dy}{dx}(\tilde{P}) = -\frac{(\partial \tilde{f}/\partial x)(\tilde{P})}{(\partial \tilde{f}/\partial y)(\tilde{P})} = \widetilde{\lambda/\mu}.
$$

The line L through P and Q is given by the equation

$$
L : y - \beta = \frac{\lambda}{\mu}(x - \alpha).
$$

We have shown that $\lambda/\mu \in R$, so the reduction of L is the line through $\tilde{P}$ having slope $\widetilde{\lambda/\mu}$. This proves that $\tilde{L}$ is tangent to $\tilde{E}$ at $\tilde{P}$, which completes the proof of the lemma when $\tilde{P} \neq \tilde{O}$ and $(\partial \tilde{f})(\partial y)(\tilde{P}) \neq 0$. The other cases are proven similarly. $\square$

Returning now to the proof of (VII.2.1), let $P_1, P_2 \in E_0(K)$ and $P_3 \in E(K)$ be distinct points satisfying $P_1 + P_2 + P_3 = O$, and suppose that their reductions satisfy

$$
\tilde{P}_1 = \tilde{P}_2 \neq \tilde{P}_3.
$$

Let L be the line through P_1, P_2, P_3. We apply (VII.2.1.1) with $P = P_1$ and $Q = P_2$. This tells us that $\tilde{L}$ is tangent to $\tilde{E}$ at $\tilde{P}_1$, and we also have $\tilde{P}_3 \in \tilde{L}$, so we find that $2\tilde{P}_1 + \tilde{P}_3 = \tilde{O}$. Since we are assuming that $\tilde{P}_1 = \tilde{P}_2$, we conclude that $P_3 \in E_{\text{ns}}(k)$ and that $\tilde{P}_1 + \tilde{P}_2 + \tilde{P}_3 = \tilde{O}$. $\square$

Note that if $v(\Delta) = 0$, so $\tilde{\Delta} \neq 0$, then $\tilde{E}$ is nonsingular, so $\tilde{E}_{\text{ns}} = \tilde{E}$ and $E_0(K) = E(K)$. In this case, (VII.2.1) says that $E(K)$ is built from two pieces, namely $E_1(K)$ and $\tilde{E}(k)$. The group $\tilde{E}(k)$ is the set of points on an elliptic curve defined over a field that is smaller than K, and indeed we often consider the situation in which k is a finite field, in which case we analyzed $E(k)$ is some detail in Chapter V.

The next proposition shows that the other part, $E_1(K)$, is also an object with which we are already familiar.

Proposition 2.2. *Let E/K be given by a minimal Weierstrass equation, let $\hat{E}/R$ be the formal group associated to E as in (IV.2.2.3), and let $w(z) \in R[\![x]\!]$ be the power series from (IV.1.1). Then the map*

$$\hat{E}(\mathcal{M}) \longrightarrow E_1(K), \qquad z \longmapsto \left(\frac{z}{w(z)}, -\frac{1}{w(z)} \right),$$

is an isomorphism of groups. (We understand that $z = 0$ goes to $O \in E_1(K)$. For the definition of $\hat{E}(\mathcal{M})$, see (IV §3).)

PROOF. From (IV.1.1b), the point $\bigl(z/w(z), -1/w(z) \bigr)$, when considered as a pair of power series, satisfies the Weierstrass equation for E. Since

$$w(z) = z^3(1 + \cdots) \in R[\![z]\!],$$

we see that $w(z)$ converges for every $z \in \mathcal{M}$. It follows that $\bigl(z/w(z), -1/w(z) \bigr)$ is in $E(K)$ for $z \in \mathcal{M}$, and since $v\bigl(-1/w(z)\bigr) = -3v(z) < 0$, it is even in $E_1(K)$. Thus we have a well-defined map of sets

$$\hat{E}(\mathcal{M}) \longrightarrow E_1(K), \qquad z \longmapsto \left(\frac{z}{w(z)}, -\frac{1}{w(z)} \right).$$

Further, in deriving the power series giving the group law on $\tilde{E}$, we simply used the group law on E in the (z, w)-plane and replaced w with $w(z)$. Therefore the map is a homomorphism. Further, since $w(z) = 0$ only for $z = 0$, the map is injective, so it remains to show that the image is all of $E_1(K)$.

Let $(x, y) \in E_1(K)$. Since (x, y) reduces modulo π to the point at infinity on $\tilde{E}(k)$, we see that $v(x) < 0$ and $v(y) < 0$. But then from the Weierstrass equation $y^2 + \cdots = x^3 + \cdots$, we must have

$$3v(x) = 2v(y) = -6r$$

for some integer $r \geq 1$. Hence $x/y \in \mathcal{M}$, so the map

$$E_1(K) \longrightarrow \hat{E}(\mathcal{M}), \qquad (x, y) \longmapsto -\frac{x}{y},$$

is well-defined. Again, since the group law on $\hat{E}(\mathcal{M})$ is defined using the group law on E, this map is a homomorphism, and it is clearly injective. Hence we have two injections

$$\hat{E}(\mathcal{M}) \lhook\joinrel\longrightarrow E_1(K) \lhook\joinrel\longrightarrow \hat{E}(\mathcal{M})$$

whose composition is the identity map, so they are isomorphisms. $\qquad\square$

VII.3   Points of Finite Order

In this section we analyze the points of finite order in the group $E(K)$. Although we later prove a stronger result (VII.3.4), we start with an easy proposition that provides a crucial ingredient in the proof of the weak Mordell–Weil theorem (VIII.1.1).

Proposition 3.1. *Let E/K be an elliptic curve and let $m \geq 1$ be an integer that is relatively prime to* char(k).
(a) *The subgroup $E_1(K)$ has no nontrivial points of order m.*
(b) *Assume further that the reduced curve $\tilde{E}/k$ is nonsingular. Then the reduction map*
$$E(K)[m] \longrightarrow \tilde{E}(k)$$
is injective, where $E(K)[m]$ denotes the set of points of order m in $E(K)$.

PROOF. From (VII.2.1) we have an exact sequence
$$0 \longrightarrow E_1(K) \longrightarrow E_0(K) \longrightarrow \tilde{E}_{\mathrm{ns}}(k) \longrightarrow 0.$$

We know from (VII.2.2) that $E_1(K) \cong \hat{E}(\mathcal{M})$, where $\hat{E}$ is the formal group associated to E, and our general result on formal groups (IV.3.2b) says that $\hat{E}(\mathcal{M})$ has no nontrivial elements of order m. This proves (a). If we further assume that $\tilde{E}$ is nonsingular, then $E_0(K) = E(K)$ and $\tilde{E}_{\mathrm{ns}}(k) = \tilde{E}(k)$, so the m-torsion of $E(K)$ injects into $\tilde{E}(k)$, which proves (b). ☐

Application 3.2. Repeated use of (VII.3.1) generally provides the quickest method for finding the torsion subgroup of an elliptic curve defined over a number field. Thus let K be a number field and let K_v be its completion at the discrete valuation v. It is clear that $E(K)$ injects into $E(K_v)$, so by applying (VII.3.1) for several different v, we can obtain information about the torsion in $E(K)$. We illustrate with several examples over $\mathbb{Q}$.

Example 3.3.1. Let $E/\mathbb{Q}$ be the elliptic curve
$$E : y^2 + y = x^3 - x + 1.$$

The discriminant of E is $\Delta = -611 = -13 \cdot 47$, so $\tilde{E}$ is nonsingular modulo 2. It is easy to check that $\tilde{E}(\mathbb{F}_2) = \{O\}$ and $E(\mathbb{Q})[2] = \{O\}$; hence (VII.3.1) implies that $E(\mathbb{Q})$ has no nonzero torsion points.

Example 3.3.2. Let $E/\mathbb{Q}$ be the elliptic curve
$$E : y^2 = x^3 + 3.$$

It has discriminant $\Delta = -2^4 \cdot 3^5$, so $\tilde{E}$ is nonsingular modulo p for every prime $p \geq 5$. One easily checks that
$$\#\tilde{E}(\mathbb{F}_5) = 6 \qquad \text{and} \qquad \#\tilde{E}(\mathbb{F}_7) = 13.$$

Hence $E(\mathbb{Q})$ has no nontrivial torsion. In particular, the point $(1, 2) \in E(\mathbb{Q})$ has infinite order, so $E(\mathbb{Q})$ is an infinite set, two facts that are by no means obvious. For a complete analysis of $E(\mathbb{Q})_{\mathrm{tors}}$ for curves of the form $y^2 = x^3 + D$, see [94] or Exercise 10.19.

Example 3.3.3. Let $E/\mathbb{Q}$ be the elliptic curve

$$E : y^2 = x^3 + x$$

having discriminant $\Delta = -64$. The point $(0,0) \in E(\mathbb{Q})$ is a point of order 2. We compute

$$\#\tilde{E}(\mathbb{F}_3) = 4, \qquad \#\tilde{E}(\mathbb{F}_5) = 4, \qquad \#\tilde{E}(\mathbb{F}_7) = 8.$$

It is not hard to check (Exercise 5.12) that $\#E(\mathbb{F}_p)$ is divisible by 4 for every prime $p \geq 3$. However, we gain additional information by looking at the group structure modulo different primes. Thus

$$\tilde{E}(\mathbb{F}_3) = \{O, (0,0), (2,1), (2,2)\} \cong \mathbb{Z}/4\mathbb{Z},$$
$$\tilde{E}(\mathbb{F}_5) = \{O, (0,0), (2,0), (2,0)\} \cong (\mathbb{Z}/2\mathbb{Z})^2.$$

Since $E(\mathbb{Q})_{\text{tors}}$ injects into both of these groups, we see that $(0,0)$ is the only nonzero torsion point in $E(\mathbb{Q})$.

The next result, which is due to Cassels, gives a precise bound on the denominator of a torsion point. Following Katz–Lang [135, Theorem III.3.7], we give a proof based on general facts concerning formal groups. For an exposition of Cassel's original proof, which involves a careful analysis of division polynomials, see [36, Theorem 17.2] or [135, Theorem III.1.5].

Theorem 3.4. *Assume that* $\text{char}(K) = 0$ *and that* $p = \text{char}(k) > 0$. *Let* E/K *be an elliptic curve given by a Weierstrass equation*

$$E : y^2 + a_1 xy + a_3 y = x^3 + a_2 x^2 + a_4 x + a_6$$

with all $a_i \in R$. *(Note that the equation need not be minimal.) Let* $P \in E(K)$ *be a point of exact order* $m \geq 2$.
(a) *If* m *is not a power of* p, *then* $x(P), y(P) \in R$.
(b) *If* $m = p^n$, *then*

$$\pi^{2r} x(P), \pi^{3r} y(P) \in R \qquad \text{with} \qquad r = \left[\frac{v(p)}{p^n - p^{n-1}}\right],$$

where $[t]$ *denotes the greatest integer in* t.

PROOF. If $x(P) \in R$, there is nothing to prove, so we assume that $v(x(P)) < 0$. If the equation for E is not minimal and if (x', y') are coordinates for a minimal equation, then we see from (VII.1.3d) that

$$v(x(P)) \geq v(x'(P)) \qquad \text{and} \qquad v(y(P)) \geq v(y'(P)).$$

It thus suffices to prove the theorem for a minimal Weierstrass equation.

Since $v(x(P)) < 0$, we see from the Weierstrass equation (and the nonarchimedean nature of v) that

$$3v\big(x(P)\big) = 2v\big(y(P)\big) = -6s \qquad \text{for some integer } s \geq 1.$$

Further, the point P is in $E_1(K)$, the kernel of the reduction map, so under the isomorphism in (VII.2.2), the point P corresponds to the element $-x(P)/y(P)$ in the formal group $\hat{E}(\mathcal{M})$. But (IV.3.2b) tells us that $\hat{E}(\mathcal{M})$ contains no torsion of order prime to p, which proves (a).

To prove (b), we use (IV.6.1). The assumption that $-x(P)/y(P)$ has exact order p^n in $\hat{E}(\mathcal{M})$ implies that

$$s = v\left(-\frac{x(P)}{y(P)}\right) \leq \frac{v(p)}{p^n - p^{n-1}}.$$

Since $\pi^{2s}x(P)$ and $\pi^{3s}y(P)$ are in R, this gives the desired result. $\square$

Application 3.5. Let $E/\mathbb{Q}$ be an elliptic curve given by a Weierstrass equation having coefficients in $\mathbb{Z}$, and let $P \in E(\mathbb{Q})$ be a point of exact order m. Embedding $E(\mathbb{Q})$ into $E(\mathbb{Q}_p)$ for various primes, we deduce integrality conditions on the coordinates of P. Thus if m is not a prime power, then (VII.3.4a) implies that $x(P), y(P) \in \mathbb{Z}$. And if $m = p^n$ is a prime power, letting v be the normalized valuation associated to p, we have

$$\left[\frac{v(p)}{p^n - p^{n-1}}\right] = \left[\frac{1}{p^n - p^{n-1}}\right] = 0$$

unless $p = 2$ and $n = 1$. We conclude that $x(P), y(P) \in \mathbb{Z}$ for every torsion point $P \in E(\mathbb{Q})$ whose exact order is at least 3. This is best possible, as shown by the example

$$E : y^2 + xy = x^3 + 4x + 1, \qquad \left(-\frac{1}{4}, \frac{1}{8}\right) \in E(\mathbb{Q})[2].$$

For a further discussion of torsion points over number fields, see (VIII §7).

VII.4 The Action of Inertia

In this section we reinterpret the injectivity of torsion (VII.3.1b) in terms of the action of the Galois group on torsion points. We set the following notation:

K^{nr} the maximal unramified extension of K.

I_v the inertia subgroup of $G_{\bar{K}/K}$.

Unramified extensions of K correspond to extensions of the residue field k, so the absolute Galois group of K decomposes as

$$1 \longrightarrow G_{\bar{K}/K^{\mathrm{nr}}} \longrightarrow G_{\bar{K}/K} \longrightarrow G_{K^{\mathrm{nr}}/K} \longrightarrow 1.$$
$$\qquad\qquad \Big\| \qquad\qquad\qquad\qquad\qquad \Big\|$$
$$\qquad\qquad I_v \qquad\qquad\qquad\qquad\qquad G_{\bar{k}/k}$$

In other words, the inertia group I_v is the set of elements of $G_{\bar{K}/K}$ that act trivially on the residue field $\bar{k}$. (For basic properties of local fields, see [92, §7], [142, Chapters I and II], or [233, Chapters I–IV]. Remember that both K and k are assumed to be perfect.)

Definition. Let Σ be a set on which $G_{\bar{K}/K}$ acts. We say that Σ is *unramified at v* if the action of I_v on Σ is trivial.

Let E/K be an elliptic curve. We have seen (III §7) that $G_{\bar{K}/K}$ acts on the torsion subgroups $E[m]$ and on the Tate modules $T_\ell(E)$ of E.

Proposition 4.1. *Let E/K be an elliptic curve such that the reduced curve $\tilde{E}/k$ is nonsingular.*
(a) *Let $m \geq 1$ be an integer that is relatively prime to $\mathrm{char}(k)$, i.e., satisfying $v(m) = 0$. Then $E[m]$ is unramified at v.*
(b) *Let ℓ be a prime with $\ell \neq \mathrm{char}(k)$. Then $T_\ell(E)$ is unramified at v.*

PROOF. (a) Let K'/K be a finite extension satisfying $E[m] \subset E(K')$, and let

$$R' = \text{the ring of integers of } K',$$

$$\mathcal{M}' = \text{the maximal ideal of } R',$$

$$k' = \text{the residue field of } R', \text{ i.e., } k' = R'/\mathcal{M}',$$

$$v' = \text{the valuation of } K'.$$

Our assumption that E has nonsingular reduction means that if we take a minimal Weierstrass equation for E at v, then its discriminant satisfies $v(\Delta) = 0$. Since the restriction of v' to K is a multiple of v, we see that $v'(\Delta) = 0$, so the Weierstrass equation is also minimal at v' and the reduced curve $\tilde{E}/k'$ is nonsingular. Now (VII.3.1b) implies that the reduction map

$$E[m] \longrightarrow \tilde{E}(k')$$

is injective.

Let $\sigma \in I_v$ and $P \in E[m]$. We need to show that $P^\sigma = P$. From the definition of the inertia group, the element σ acts trivially on $\tilde{E}(k')$, so

$$\widetilde{P^\sigma - P} = \tilde{P}^\sigma - \tilde{P} = \tilde{O}.$$

But $P^\sigma - P$ is clearly in $E[m]$, so the injectivity of the map $E[m] \hookrightarrow \tilde{E}(k')$ tells us that $P^\sigma - P = O$.
(b) This follows immediately from (a) and the definition of $T_\ell(E)$ as the inverse limit of $E[\ell^n]$. □

There is a converse to (VII.4.1) that is known as the criterion of Néron–Ogg–Shafarevich. It characterizes nonsingularity of $\tilde{E}/k$ in terms of the action of the inertia group on torsion points. We return to this topic in (VII §7), after first studying the reduced curve $\tilde{E}$ more closely.

VII.5   Good and Bad Reduction

Let E/K be an elliptic curve. From our general knowledge of Weierstrass equations (III.1.4), the reduced curve $\tilde{E}$ is of one of three types. We classify E according to these possibilities.

Definition. Let E/K be an elliptic curve, and let $\tilde{E}$ be the reduction modulo $\mathcal{M}$ of a minimal Weierstrass equation for E.
(a) E has *good* (or *stable*) *reduction* if $\tilde{E}$ is nonsingular.
(b) E has *multiplicative* (or *semistable*) *reduction* if $\tilde{E}$ has a node.
(c) E has *additive* (or *unstable*) *reduction* if $\tilde{E}$ has a cusp.
In cases (b) and (c) we say that E has *bad reduction*. If E has multiplicative reduction, then the reduction is said to be *split* if the slopes of the tangent lines at the node are in k, and otherwise it is said to be *nonsplit*.

It is quite easy to read off the reduction type of an elliptic curve from a minimal Weierstrass equation.

Proposition 5.1. *Let E/K be an elliptic curve given by a minimal Weierstrass equation*

$$E : y^2 + a_1 xy + a_3 y = x^3 + a_2 x^2 + a_4 x + a_6.$$

Let Δ be the discriminant of this equation, and let c_4 be the usual expression involving $a_1, \ldots, a_6$ as described in (III §1).
(a) *E has good reduction if and only if $v(\Delta) = 0$, i.e., $\Delta \in R^*$. In this case $\tilde{E}/k$ is an elliptic curve.*
(b) *E has multiplicative reduction if and only if $v(\Delta) > 0$ and $v(c_4) = 0$, i.e., $\Delta \in \mathcal{M}$ and $c_4 \in R^*$. In this case $\tilde{E}_{\text{ns}}$ is the multiplicative group,*

$$\tilde{E}_{\text{ns}}(\bar{k}) \cong \bar{k}^*.$$

(c) *E has additive reduction if and only if $v(\Delta) > 0$ and $v(c_4) > 0$, i.e., $\Delta, c_4 \in \mathcal{M}$. In this case $\tilde{E}_{\text{ns}}$ is the additive group,*

$$\tilde{E}_{\text{ns}}(\bar{k}) \cong \bar{k}^+.$$

PROOF. The reduction type of E follows from (III.1.4) applied to the reduced Weierstrass equation over the field k. Then the group $\tilde{E}_{\text{ns}}(\bar{k})$ is given by (III.2.5). $\square$

Example 5.2. Let $p \geq 5$ be a prime. Then the elliptic curve

$$E_1 : y^2 = x^3 + px^2 + 1$$

has good reduction over $\mathbb{Q}_p$, while

$$E_2 : y^2 = x^3 + x^2 + p$$

has (split) multiplicative reduction over $\mathbb{Q}_p$, and

$$E_3 : y^2 = x^3 + p$$

has additive reduction over $\mathbb{Q}_p$. If we go to the extension field $\mathbb{Q}(\sqrt[6]{p})$, then E_3 attains good reduction, since the substitution

$$x \longmapsto \sqrt[3]{p}\, x', \qquad y \longmapsto \sqrt{p}\, y',$$

yields a minimal Weierstrass equation having good reduction. On the other hand, the curve E_3 has multiplicative reduction over every extension of $\mathbb{Q}_p$. This is true in general; after extending the ground field, additive reduction turns into either multiplicative or good reduction, while the latter two do not change; see (VII.5.4). This suggests the origin of the terms stable, semistable, and unstable, although they also have quite precise definitions in terms of the stability of points in moduli space. For a high-powered account of the general theory, see [187].

When an elliptic curve E/K has bad reduction, it is often useful to know whether it attains good reduction over some extension of K. We give this property a name.

Definition. Let E/K be an elliptic curve. We say that E/K has *potential good reduction* if there is a finite extension K'/K such that E has good reduction over K'.

Example 5.3. If K is a finite extension of $\mathbb{Q}_p$ and if E/K has complex multiplication, then one can show that E has potential good reduction; see Exercise 7.10.

The next proposition explains how reduction type behaves under field extension, and the proposition immediately following provides a useful characterization of when an elliptic curve has potential good reduction.

Proposition 5.4. (Semistable reduction theorem) *Let E/K be an elliptic curve.*
 (a) *Let K'/K be an unramified extension. Then the reduction type of E over K (good, multiplicative, or additive) is the same as the reduction type of E over K'.*
 (b) *Let K'/K be a finite extension. If E has either good or multiplicative reduction over K, then it has the same reduction type over K'.*
 (c) *There exists a finite extension K'/K such that E has either good or (split) multiplicative reduction over K'.*

Proposition 5.5. *Let E/K be an elliptic curve. Then E has potential good reduction if and only if its j-invariant is integral, i.e., if and only if $j(E) \in R$.*

PROOF OF (VII.5.4). (a) For arbitrary characteristic this follows from Tate's algorithm; see [266, IV §9] or [283]. We prove the result under the assumption that $\mathrm{char}(k) \geq 5$, so E has a minimal Weierstrass equation over K of the form

$$E : y^2 = x^3 + Ax + B.$$

Let R' be the ring of integers of K', let v' be the valuation on K' extending v, and let

$$x = (u')^2 x', \qquad y = (u')^3 y',$$

be a change of coordinates that produces a minimal equation for E over K'. Since K'/K is unramified, we can find a $u \in K$ with $u/u' \in R'^*$. Then the substitution

$$x = u^2 x', \qquad y = u^3 y',$$

also gives a minimal equation for E/K', since

$$v'(u^{-12}\Delta) = v'((u')^{-12}\Delta).$$

But this new equation has coefficients in R, so by the minimality of the original equation over K, we have $v(u) = 0$. Hence the original equation is also minimal over K'. Further, since $v(\Delta) = v'(\Delta)$ and $v(c_4) = v'(c_4)$, we see from (VII.5.1) that the reduction type of E over K is the same as its reduction type over K'.

(b) Take a minimal Weierstrass equation for E over K with corresponding quantities Δ and c_4, and let R' and v' be as in the proof of (a). Further, let

$$x = u^2 x' + r, \qquad y = u^3 y' + su^2 x' + t,$$

be a change of coordinates giving a minimal Weierstrass equation for E over K. The quantities Δ' and c_4' associated to this new equation satisfy

$$0 \le v'(\Delta') = v'(u^{-12}\Delta) \qquad \text{and} \qquad 0 \le v'(c_4') = v'(u^{-4}c_4).$$

From (VII.1.3d) we have $u \in R'$, and hence

$$0 \le v'(u) \le \min\left\{\frac{1}{12}v'(\Delta), \frac{1}{4}v'(c_4)\right\}.$$

However, for good (respectively multiplicative) reduction, (VII.5.1a,b) tells us that $v(\Delta) = 0$ (respectively $v(c_4) = 0$), so in both cases we have $v'(u) = 0$. Hence

$$v'(\Delta') = v'(\Delta) \qquad \text{and} \qquad v'(c_4') = v'(c_4),$$

and another application of (VII.5.1) shows that E has good (respectively multiplicative) reduction over K'.

(c) We assume that $\mathrm{char}(k) \ne 2$ and take a finite extension of K such that E/K has a Weierstrass equation in Legendre normal form (III.1.7),

$$E : y^2 = x(x-1)(x-\lambda), \qquad \lambda \ne 0, 1.$$

(The case $\mathrm{char}(k) = 2$ is covered in (A.1.4a).) For the Legendre equation we have

$$c_4 = 16(\lambda^2 - \lambda + 1) \qquad \text{and} \qquad \Delta = 16\lambda^2(\lambda - 1)^2.$$

We consider three cases.

Case 1. $\lambda \in R$ and $\lambda \not\equiv 0$ or $1 \pmod{\mathcal{M}}$. Then $\Delta \in R^*$, so the given equation has good reduction.

Case 2. $\lambda \in R$ **and** $\lambda \equiv 0$ **or** 1 **(mod** $\mathcal{M}$**).** Then $\Delta \in \mathcal{M}$ and $c_4 \in R^*$, so the given equation has multiplicative reduction.

Case 3. $\lambda \notin R$. Let $r \geq 1$ be the integer such that $\pi^r \lambda \in R^*$. Then, replacing K by $K(\sqrt{\pi})$ if necessary, the substitutions $x = \pi^{-r} x'$ and $y = \pi^{-3r/2} y'$ give the Weierstrass equation

$$(y')^2 = x'(x' - \pi^r)(x' - \pi^r \lambda).$$

This equation for E has integral coefficients and $\Delta' \in \mathcal{M}$ and $c_4' \in R^*$, so E has multiplicative reduction.

Finally, we note that in Cases (2) and (3), if the multiplicative reduction is not already split, then it becomes split over a quadratic extension. $\qquad\square$

PROOF OF (VII.5.5). As in the proof of (VII.5.4c), we make the assumption that $\mathrm{char}(k) \neq 2$ and we take a finite extension of K such that E has a Weierstrass equation in Legendre form (III.1.7),

$$E : y^2 = x(x - 1)(x - \lambda), \qquad \lambda \neq 0, 1.$$

(For $\mathrm{char}(k) = 2$, see (A.1.4b).) By assumption, we have $j = j(E) \in R$, and an easy computation (III.1.7b) shows that j and λ are related by the equation

$$256\big(1 - \lambda(1 - \lambda)\big)^3 - j\lambda^2 (1 - \lambda)^2 = 0.$$

This equation and the integrality of j imply that

$$\lambda \in R \qquad \text{and} \qquad \lambda \not\equiv 0 \text{ or } 1 \ (\mathrm{mod}\ \mathcal{M}).$$

Thus the given Legendre equation has integral coefficients and good reduction.

Conversely, suppose that E has potential good reduction. Let K'/K be a finite extension such that E has good reduction over K', let R' be the ring of integers of K', and let Δ' and c_4' be the quantities associated to a minimal Weierstrass equation for E over K'. Since E has good reduction over K', we have $\Delta' \in (R')^*$, and hence

$$j(E) = \frac{(c_4')^3}{\Delta'} \in R'.$$

But $j(E) \in K$, since E is defined over K, so $j(E) \in R$. $\qquad\square$

VII.6   The Group E/E_0

Recall that the group $E_0(K)$ consists of the points of $E(K)$ that do not reduce to a singular point of $\tilde{E}(k)$. Further, from (VII.2.1) we know that $E_0(K)$ is made up of two pieces that we have analyzed fairly closely, namely $\tilde{E}_{ns}(k)$ and the formal group $E_1(K) \cong \hat{E}(\mathcal{M})$. We are left to study the remaining piece, the quotient $E(K)/E_0(K)$.

The most important fact about this quotient is that it is finite. As the next theorem indicates, we can actually say quite a bit more. Unfortunately, a direct proof working explicitly with Weierstrass equations is quite lengthy, and even the simplifying assumption $\mathrm{char}(k) \geq 5$ leads to a long case-by-case analysis. So we do not give the proof in this volume. If the residue field k is finite, then the mere finiteness of $E(K)/E_0(K)$ can be proven by an easy compactness argument; see Exercise 7.6.

Theorem 6.1. (Kodaira, Néron) *Let E/K be an elliptic curve. If E has split multiplicative reduction over K, then $E(K)/E_0(K)$ is a cyclic group of order $v(\Delta) = -v(j)$. In all other cases, the group $E(K)/E_0(K)$ is finite and has order at most 4.*

Corollary 6.2. *The subgroup $E_0(K)$ has finite index in $E(K)$.*

PROOF. The finiteness of $E(K)/E_0(K)$ follows from the existence of the Néron model, which is a group scheme over $\mathrm{Spec}(R)$ whose generic fiber is E/K; see [266, IV §§5, 6]. The specific description of $E(K)/E_0(K)$ comes from the complete classification of the possible special fibers of a Néron model; see [266, IV §8]. Alternatively, it is possible to give an elementary, but lengthy, proof via explicit computations with Weierstrass equations. See (C §15) for further discussion. □

Our most important application of (VII.6.2) is the proof of the criterion of Néron–Ogg–Shafarevich, which we give in the next section. Another interesting application is the following result.

Proposition 6.3. *Let K be a finite extension of $\mathbb{Q}_p$, so in particular $\mathrm{char}(K) = 0$ and k is a finite field. Then $E(K)$ contains a subgroup of finite index that is isomorphic to R^+, the additive group of R.*

PROOF. From (VII.6.2) we know that $E(K)/E_0(K)$ is finite, and (VII.2.1) tells us that $E_0(K)/E_1(K)$ is isomorphic to the finite group $\tilde{E}_{\mathrm{ns}}(k)$. (This is where we use the fact that k is finite.) It thus suffices to prove that $E_1(K)$ has a subgroup of finite index that is isomorphic to R^+. We know from (VIII.2.2) that $E_1(K)$ is isomorphic to the formal group $\hat{E}(\mathcal{M})$, and (IV.3.2a) tells us that $\hat{E}(\mathcal{M})$ has a filtration

$$\hat{E}(\mathcal{M}) \subset \hat{E}(\mathcal{M}^2) \subset \hat{E}(\mathcal{M}^2) \subset \cdots .$$

Further, each quotient $\hat{E}(\mathcal{M}^i)/\hat{E}(\mathcal{M}^{i+1})$ is isomorphic to $\mathcal{M}^i/\mathcal{M}^{i+1}$, which is finite, since it is a one-dimensional k-vector space, so it suffices to prove that there is some $r \geq 1$ such that $\hat{E}(\mathcal{M}^r)$ is isomorphic to R^+. This last assertion is a consequence of (IV.6.4b), which says that if r is sufficiently large, then the formal logarithm map

$$\log_{\hat{E}} : \hat{E}(\mathcal{M}^r) \xrightarrow{\ \sim\ } \mathbb{G}_a(\mathcal{M}^r) = \pi^r R \cong R^+$$

is an isomorphism. □

VII.7 The Criterion of Néron–Ogg–Shafarevich

If an elliptic curve E/K has good reduction and $m \geq 1$ is an integer that is prime to $\mathrm{char}(k)$, then we have seen (VII.4.1) that the torsion subgroup $E[m]$ is unramified. Various partial converses to this statement were proven by Néron, Ogg, and Shafarevich, and these were vastly generalized by Serre and Tate. We follow the exposition in [239].

Theorem 7.1. (Criterion of Néron–Ogg–Shafarevich). *Let E/K be an elliptic curve. Then the following are equivalent:*
(a) E *has good reduction at* K.
(b) $E[m]$ *is unramified at* v *for all integers* $m \geq 1$ *that are relatively prime to* $\mathrm{char}(k)$.
(c) *The Tate module* $T_\ell(E)$ *is unramified at* v *for some (all) primes* ℓ *satisfying* $\ell \neq \mathrm{char}(k)$.
(d) $E[m]$ *is unramified at* v *for infinitely many integers* $m \geq 1$ *that are relatively prime to* $\mathrm{char}(k)$.

PROOF. The implication (a) $\Rightarrow$ (b) has already been proven in (VII.4.1), and the implications (b) $\Rightarrow$ (c) $\Rightarrow$ (d) are obvious. (Note that $T_\ell(E)$ is unramified if and only if $E[\ell^n]$ is unramified for every $n \geq 1$.) It remains to prove that (d) implies (a).

Assume that (d) is true. Let K^{nr} be the maximal unramified extension of K, and choose an integer m satisfying the following conditions:

(i) m is relatively prime to $\mathrm{char}(k)$.

(ii) $m > \#E(K^{\mathrm{nr}})/E_0(K^{\mathrm{nr}})$.

(iii) $E[m]$ is unramified at v.

It is clear that such an m exists, since we are assuming that (d) is true and the quotient group $E(K^{\mathrm{nr}})/E_0(K^{\mathrm{nr}})$ is finite from (VII.6.2).

We consider the two exact sequences

$$0 \longrightarrow E_0(K^{\mathrm{nr}}) \longrightarrow E(K^{\mathrm{nr}}) \longrightarrow E(K^{\mathrm{nr}})/E_0(K^{\mathrm{nr}}) \longrightarrow 0,$$

$$0 \longrightarrow E_1(K^{\mathrm{nr}}) \longrightarrow E_0(K^{\mathrm{nr}}) \longrightarrow \tilde{E}_{\mathrm{ns}}(\bar{k}) \longrightarrow 0.$$

(Note that $\bar{k}$ is the residue field of the ring of integers of K^{nr}.) Since $E[m] \subset E(K^{\mathrm{nr}})$, we see that $E(K^{\mathrm{nr}})$ has a subgroup that is isomorphic to $(\mathbb{Z}/m\mathbb{Z})^2$. But from (ii), the group $E(K^{\mathrm{nr}})/E_0(K^{\mathrm{nr}})$ has order strictly less than m. It follows from the first exact sequence that there is a prime ℓ dividing m such that $E_0(K^{\mathrm{nr}})$ contains a subgroup isomorphic to $(\mathbb{Z}/\ell\mathbb{Z})^2$. Now look at the second exact sequence. From (VII.3.1a), the group $E_1(K^{\mathrm{nr}})$ contains no nontrivial ℓ-torsion, so we conclude that $\tilde{E}_{\mathrm{ns}}(\bar{k})$ contains a subgroup isomorphic to $(\mathbb{Z}/\ell\mathbb{Z})^2$.

Suppose that E has bad reduction over K^{nr}. If the reduction is multiplicative, then (VII.5.1b) tells us that

$$\tilde{E}_{\mathrm{ns}}(\bar{k}) = \bar{k}^*,$$

in which case the ℓ-torsion $\boldsymbol{\mu}_\ell$ is isomorphic to $\mathbb{Z}/\ell\mathbb{Z}$. Hence E cannot have multiplicative reduction. Similarly, if E has additive reduction over K^{nr}, then (VII.5.1c) says that

$$\tilde{E}_{\mathrm{ns}}(\bar{k}) = \bar{k}^+,$$

so $\tilde{E}_{\mathrm{ns}}(\bar{k})$ has no ℓ-torsion. Thus E also cannot have additive reduction. Having eliminated multiplicative and additive reduction as possibilities, all that remains is for E to have good reduction over K^{nr}. Finally, since K^{nr}/K is unramified, we use (VII.5.4a) to conclude that E has good reduction over K. $\square$

Corollary 7.2. *Let E_1/K and E_2/K be elliptic curves that are isogenous over K. Then E_1 has good reduction over K if and only if E_2 has good reduction over K.*

PROOF. Let $\phi : E_1 \to E_2$ be a nonzero isogeny defined over K, and let $m \geq 2$ be an integer that is relatively prime to both $\mathrm{char}(k)$ and $\deg \phi$. Then the induced map

$$\phi : E_1[m] \longrightarrow E_2[m]$$

is an isomorphism of $G_{\bar{K}/K}$-modules, so in particular, either both $E_1[m]$ and $E_2[m]$ are unramified at v, or both are ramified at v. Now use the (a) $\Leftrightarrow$ (d) equivalence in (VII.7.1). $\square$

Another immediate corollary of (VII.7.1) is a criterion, in terms of the action of inertia, for determining whether an elliptic curve has potential good reduction.

Corollary 7.3. *Let E/K be an elliptic curve. Then E has potential good reduction if and only if the inertia group I_v acts on the Tate module $T_\ell(E)$ through a finite quotient for some (all) prime(s) $\ell \neq \mathrm{char}(k)$.*

PROOF. Suppose that E has potential good reduction, and let K'/K be a finite extension such that E has good reduction over K'. Extending K', we may assume that K'/K is a Galois extension. Let v' be the valuation on K' and let $I_{v'}$ be the inertia group of $G_{\bar{K}'/K'}$. We know from (VII.7.1) that $I_{v'}$ acts trivially on $T_\ell(E)$ for any prime $\ell \neq \mathrm{char}(k)$. Hence the action of I_v on $T_\ell(E)$ factors through the finite quotient $I_v/I_{v'}$. This proves one implication.

Assume now that for some prime $\ell \neq \mathrm{char}(k)$, the inertia group I_v acts on $T_\ell(E)$ through a finite quotient, say I_v/J. Then the fixed field of J, which we denote by $\bar{K}^J$, is a finite extension of $K^{\mathrm{nr}} = \bar{K}^{I_v}$. Hence we can find a finite extension K'/K such that $\bar{K}^J$ is the compositum

$$\bar{K}^J = K'K^{\mathrm{nr}}.$$

Then the inertia group of K' is equal to J, and by assumption J acts trivially on $T_\ell(E)$. Now (VII.7.1) implies that E has good reduction over K'. $\square$

Exercises

7.1. Assume that $\mathrm{char}(k) \neq 2, 3$.

(a) Let E/K be an elliptic curve given by a Weierstrass equation with coefficients $a_i \in R$. Prove that the equation is minimal if and only if either $v(\Delta) < 12$ or $v(c_4) < 4$.

(b) Let E/K be given by a minimal Weierstrass equation of the form

$$E : y^2 = x^3 + Ax + B.$$

Prove that E has

   (i) good reduction $\Longleftrightarrow 4A^3 + 27B^2 \in R^*$,

  (ii) multiplicative reduction $\Longleftrightarrow 4A^3 + 27B^2 \in \mathcal{M}$ and $AB \in R^*$,

 (iii) additive reduction $\Longleftrightarrow A \in \mathcal{M}$ and $B \in \mathcal{M}$.

7.2. Let E/K be an elliptic curve with j-invariant $j(E) \in R$. Prove that the minimal discriminant Δ of E satisfies

$$v(\Delta) < 12 + 12v(2) + 6v(3).$$

7.3. Describe all Weierstrass equations

$$E : y^2 + a_1 xy + a_3 y = x^3 + a_2 x^2 + a_4 x + a_6$$

with $a_i \in \mathbb{Z}$ and $\Delta \neq 0$ such that $E(\mathbb{Q})$ contains a torsion point P with $x(P) \notin \mathbb{Z}$. (*Hint.* See (VIII.3.5).)

7.4. Let E/K be an elliptic curve given by a minimal Weierstrass equation, and for each $n \geq 1$, define a subset of $E(K)$ by

$$E_n(K) = \big\{ P \in E(K) : v\big(x(P)\big) \leq -2n \big\} \cup \{O\}.$$

(a) Prove that $E_n(K)$ is a subgroup of $E(K)$.

(b) Prove that

$$E_n(K)/E_{n+1}(K) \cong k^+.$$

7.5. Show that the following elliptic curves have good reduction over a field of the indicated form by writing down a minimal equation for E over that field.

(a) $E : y^2 = x^3 + x$, $\mathbb{Q}_2(\eta, i)$, $\eta^8 = 2$, $i^2 = -1$.

(b) $E : y^2 + y = x^3$, $\mathbb{Q}_3(\pi, \eta)$, $\pi^2 = \sqrt{-3}$, $\eta^3 = 2$.

(c) $E : y^2 = x^3 + x^2 - 3x - 2$, $\mathbb{Q}_5(\pi)$, $\pi^4 = 5$.

7.6. Assume that K is locally compact for the topology induced by the discrete valuation v. (This is equivalent to the assumption that the residue field k is finite; see [42, §7].) This exercise sketches a proof of (VII.6.2) for such fields. However, we note that for applications such as (VII.7.1) we need to know the stronger statement that $E(K)/E_0(K)$ is finite when the residue field k is algebraically closed.

(a) Use v to define a topology on $\mathbb{P}^N(K)$ and show that $\mathbb{P}^N(K)$ is compact for this topology.

(b) Let E/K be an elliptic curve, let $E(K) \subset \mathbb{P}^2(K)$ be the inclusion coming from a minimal Weierstrass equation, and give $E(K)$ the topology induced from $\mathbb{P}^2(K)$. Prove that $E(K)$ is compact, and that for any $P \in E(K)$, the translation-by-P map $\tau_P : E(K) \to E(K)$ is continuous.

(c) Prove that $E_0(K)$ is an open subset of $E(K)$. (It is also a closed subset!)

(d) Prove that $E(K)/E_0(K)$ is finite.

7.7. The following examples illustrate some special cases of (VII.6.1). We assume throughout that $\operatorname{char}(k) \neq 2, 3$. Let E/K be an elliptic curve given by a Weierstrass equation

$$E : y^2 = x^3 + Ax + B.$$

(a) If $v(A) \geq 1$ and $v(B) = 1$, prove that $E(K) = E_0(K)$.

(b) If $v(A) = 1$ and $v(B) \geq 2$, prove that $E(K)/E_0(K) \cong \mathbb{Z}/2\mathbb{Z}$. (*Hint.* Suppose that $P, Q \notin E_0(K)$. Use the addition formula to show that $P + Q \in E_0(K)$.)

(c) If $v(A) \geq 2$ and $v(B) = 2$, prove that $E(K)/E_0(K)$ is either 0 or $\mathbb{Z}/3\mathbb{Z}$.

7.8. Let E/K be an elliptic curve, and let m be an integer that is relatively prime to $\operatorname{char}(k)$. Prove that

$$E_0(K^{\mathrm{nr}})/mE_0(K^{\mathrm{nr}}) = 0.$$

7.9. Let E/K be an elliptic curve with potential good reduction, let m be an integer that is relatively prime to $\operatorname{char}(k)$, and let $K\big(E[m]\big)$ be the field obtained by adjoining to K the coordinates of the points in $E[m]$.

(a) Prove that the inertia group of $K\big(E[m]\big)/K$ is independent of m. (*Hint.* For each prime $\ell \neq \operatorname{char}(k)$, let $\ell' = \ell$ if $\ell \geq 3$ and let $\ell' = 4$ if $\ell = 2$. Show that $\rho_\ell(I_v)$ has trivial intersection with the kernel of the map

$$\operatorname{Aut}\big(T_\ell(E)\big) \longrightarrow \operatorname{Aut}\big(T_\ell(E)/\ell' T_\ell(E)\big) \cong \operatorname{GL}_2(\mathbb{Z}/\ell'\mathbb{Z}).$$

Characterize the inertia group of $K\big(E[m]\big)/K$ in terms of the kernels of the various ρ_ℓ.)

(b) Prove that $K\big(E[m]\big)/K$ is unramified if and only if E has good reduction at v.

(c) If $\operatorname{char}(k) \geq 5$, prove that $K\big(E[m]\big)/K$ is tamely ramified.

7.10. Let K be a finite extension of $\mathbb{Q}_p$, let R be the ring of integers of K, and let E/K be an elliptic curve with complex multiplication. Prove that $j(E) \in R$. (*Hint.* Use the description of the maximal abelian extension K^{ab} of K provided by local class field theory to prove that the action of $G_{K^{\mathrm{ab}}/K}$ on $T_\ell(E)$ factors through a finite quotient. Then apply (VII.5.5), (VII.7.3), and Exercise 3.24.)

7.11. Use Exercise 3.23 to prove (VII.5.4c) and (VII.5.5) in characteristic 2.

7.12. Let $[K : \mathbb{Q}_p] = 2$, let E/K be an elliptic curve given by a Weierstrass equation having coefficients in R, and let $P \in E(K)$ be a point of exact order $m \geq 2$ such that $x(P) \notin R$, i.e., such that $v\big(x(P)\big) < 0$.

(a) Prove that $p = 2$ or 3 and that $m = 2, 3,$ or 4. Give examples to show that each value of m is possible.

(b) Suppose that the reduced curve $\tilde{E}/k$ is supersingular. Prove that $p = m = 2$.

7.13. Let $E/\mathbb{F}_p$ be an elliptic curve with the property that $\#E(\mathbb{F}_p) = p$. (Such curves are called *anomalous.*) Let $P, Q \in E(\mathbb{F}_p)$ be points such that Q is in the subgroup generated by P. This exercise describes an algorithm that solves the elliptic curve discrete logarithm problem (ECDLP) for anomalous curves, i.e., it finds an integer m satisfying $Q = [m]P$. (See (V.1.6) and Exercise 5.18.)

(a) Let $E'/\mathbb{Q}_p$ be an elliptic curve whose reduction modulo p is $E/\mathbb{F}_p$. Prove that there are points $P', Q' \in E'(\mathbb{Q}_p)$ whose reductions modulo p are, respectively, P and Q.

(b) Prove that $[p]P'$ and $[p]Q'$ are in the formal group $E_1'(\mathbb{Q}_p)$.

(c) Let
$$\log_{E'} : E_1'(\mathbb{Q}_p) \longrightarrow \mathbb{Q}_p$$
be the formal logarithm map (IV.6.4a), and let
$$r = \frac{\log_{E'}\left([p]Q'\right)}{\log_{E'}\left([p]P'\right)}.$$

Prove that $r \in \mathbb{Z}_p$.

(d) Let $m \in \mathbb{Z}$ be an integer satisfying $m \equiv r \pmod{p}$. Prove that $Q = [m]P$.

7.14. Let $P \in E_0(K)$ and let L be the tangent line to E at P. Prove that the reduced line $\tilde{L}$ is the tangent line to $\tilde{E}$ at $\tilde{P}$; cf. (VII.2.1.1).

7.15. Let $P_1, P_2 \in E_0(K)$ and $P_3 \in E(K)$ satisfy $P_1 + P_2 + P_3 = O$, and let L be the line intersecting E at P_1, P_2, P_3, with appropriate multiplicities. For each of the following situations, show that $\tilde{L}$ intersects $\tilde{E}$ at $\tilde{P}_1, \tilde{P}_2, \tilde{P}_3$ with appropriate multiplicities, and hence that $\tilde{P}_3 \in E_{\mathrm{ns}}(k)$, $P_3 \in E_0(K)$, and $\tilde{P}_1 + \tilde{P}_2 + \tilde{P}_3 = \tilde{O}$. Use your results to complete the proof of (VII.2.1).

(a) P_1, P_2, P_3 are distinct and $\tilde{P}_1 = \tilde{P}_2 = \tilde{P}_3$.

(b) $P_1 = P_2 \neq P_3$ and $\tilde{P}_1 = \tilde{P}_2 \neq \tilde{P}_3$.

(c) $P_1 = P_2 \neq P_3$ and $\tilde{P}_1 = \tilde{P}_2 = \tilde{P}_3$.

(d) $P_1 = P_2 = P_3$.

Chapter VIII

Elliptic Curves over Global Fields

Let K be a number field and let E/K be an elliptic curve. Our primary goal in this chapter is to prove the following result.

Mordell–Weil Theorem. *The group $E(K)$ is finitely generated.*

The proof of this theorem consists of two quite distinct parts, the so-called "weak Mordell–Weil theorem," proven in (VIII §1), and the "infinite descent" using height functions proven in (VIII §§3,5,6). We also give, in (VIII §4), a separate proof of the descent step in the simplest case, where the general theory of height functions may be replaced by explicit polynomial calculations.

The Mordell–Weil theorem tells us that the *Mordell–Weil group* $E(K)$ has the form

$$E(K) \cong E(K)_{\text{tors}} \times \mathbb{Z}^r,$$

where the torsion subgroup $E(K)_{\text{tors}}$ is finite and the *rank* r of $E(K)$ is a nonnegative integer. For a given elliptic curve, it is relatively easy to determine the torsion subgroup; see (VIII §7). The rank is much more difficult to compute, and in general there is no known procedure that is guaranteed to yield an answer. We study the question of computing the rank of $E(K)$ in more detail in Chapter X.

The following notation will be used for the next three chapters:

K a number field.

M_K a complete set of inequivalent absolute values on K.

M_K^∞ the archimedean absolute values in M_K.

M_K^0 the nonarchimedean absolute values in M_K.

$v(x)$ $= -\log|x|_v$, for an absolute value $v \in M_K$.

ord_v normalized valuation for $v \in M_K^0$, i.e., satisfying $\text{ord}_v(K^*) = \mathbb{Z}$.

J.H. Silverman, *The Arithmetic of Elliptic Curves, Second Edition*, Graduate Texts   207
in Mathematics 106, DOI 10.1007/978-0-387-09494-6_VIII,
© Springer Science+Business Media, LLC 2009

R the ring of integers of K, equal to $\{x \in K : v(x) \geq 0 \text{ for all } v \in M_K^0\}$.

R^* the unit group of R, equal to $\{x \in K : v(x) = 0 \text{ for all } v \in M_K^0\}$.

K_v the completion of K at v for $v \in M_K$.

R_v the ring of integers of K_v for $v \in M_K^0$.

$\mathcal{M}_v$ the maximal ideal of R_v for $v \in M_K^0$.

k_v the residue field of R_v for $v \in M_K^0$.

Finally, in those situations in which it is important to have the absolute values in M_K coherently normalized, such as in the theory of height functions, we always adopt the "standard normalization" as described in (VIII §5).

VIII.1   The Weak Mordell–Weil Theorem

Our goal in this section is to prove the following result.

Theorem 1.1. (Weak Mordell–Weil Theorem) *Let K be a number field, let E/K be an elliptic curve, and let $m \geq 2$ be an integer. Then*

$$E(K)/mE(K)$$

is a finite group.

For the rest of this section, E/K and m are as in the statement of (VIII.1.1). We begin with the following reduction lemma.

Lemma 1.1.1. *Let L/K be a finite Galois extension. If $E(L)/mE(L)$ is finite, then $E(K)/mE(K)$ is also finite.*

PROOF. The inclusion $E(K) \hookrightarrow E(L)$ induces a natural map

$$E(K)/mE(K) \longrightarrow E(L)/mE(L).$$

Let Φ be the kernel of this map, so

$$\Phi = \frac{E(K) \cap mE(L)}{mE(K)}.$$

Then for each $P \pmod{mE(K)}$ in Φ, we can choose a point $Q_P \in E(L)$ satisfying $[m]Q_P = P$. (The point Q_P need not be unique, of course.) Having done this, we define a map of sets (which is not, in general, a group homomorphism)

$$\lambda_P : G_{L/K} \longrightarrow E[m], \qquad \lambda_P(\sigma) = Q_P^\sigma - Q_P.$$

Note that $Q_P^\sigma - Q_P$ is in $E[m]$, since

$$[m](Q_P^\sigma - Q_P) = ([m]Q_P)^\sigma - [m]Q_P = P^\sigma - P = O.$$

(The map λ_P is an example of a 1-cocycle; see (VIII §2).)

Suppose that $P, P' \in E(K) \cap mE[L]$ satisfy $\lambda_P = \lambda_{P'}$. Then

$$(Q_P - Q_{P'})^\sigma = Q_P - Q_{P'} \qquad \text{for all } \sigma \in G_{L/K},$$

so $Q_P - Q_{P'} \in E(K)$. It follows that

$$P - P' = [m]Q_P - [m]Q_{P'} \in mE(K),$$

and hence that $P \equiv P' \pmod{mE(K)}$. This proves that the association

$$\Phi \longrightarrow \mathrm{Map}\big(G_{L/K}, E[m]\big), \qquad P \longmapsto \lambda_P,$$

is one-to-one. But $G_{L/K}$ and $E[m]$ are finite sets, so there is only a finite number of maps between them. Therefore the set Φ is finite.

Finally, the exact sequence

$$0 \longrightarrow \Phi \longrightarrow E(K)/mE(K) \longrightarrow E(L)/mE(L)$$

nests $E(K)/mE(K)$ between two finite groups, so it, too, is finite. $\qquad\square$

Using (VIII.1.1.1), we see that it suffices to prove the weak Mordell–Weil theorem (VIII.1.1) under the additional assumption that

$$E[m] \subset E(K).$$

For this remainder of this section we assume, without further comment, that this inclusion is true.

The next step is to translate the putative finiteness of $E(K)/mE(K)$ into a statement about a certain field extension of K. In order to do this, we use the following tool.

Definition. The *Kummer pairing*

$$\kappa : E(K) \times G_{\bar{K}/K} \longrightarrow E[m]$$

is defined as follows. Let $P \in E(K)$ and choose any point $Q \in E(\bar{K})$ satisfying $[m]Q = P$. Then

$$\kappa(P, \sigma) = Q^\sigma - Q.$$

The next result describes basic properties of the Kummer pairing.

Proposition 1.2. (a) *The Kummer pairing is well-defined.*
(b) *The Kummer pairing is bilinear.*
(c) *The kernel of the Kummer pairing on the left is $mE(K)$.*
(d) *The kernel of the Kummer pairing on the right is $G_{\bar{K}/L}$, where*

$$L = K\big([m]^{-1}E(K)\big)$$

is the compositum of all fields $K(Q)$ as Q ranges over the points in $E(\bar{K})$ satisfying $[m]Q \in E(K)$.

Hence the Kummer pairing induces a perfect bilinear pairing

$$E(K)/mE(K) \times G_{L/K} \longrightarrow E[m],$$

where L is the field given in (d).

Remark 1.2.1. The field L described in (VIII.1.2) is the elliptic analogue of the classical Kummer extension K'/K obtained by adjoining all m^{th} roots to K. More precisely, assuming that $\mu_m \subset K$, there is a perfect bilinear pairing

$$K^*/(K^*)^m \times G_{K'/K} \longrightarrow \mu_m, \qquad (a, \sigma) \longrightarrow \sqrt[m]{a}^{\sigma} / \sqrt[m]{a},$$

exactly analogous to the pairing $E(K)/mE(K) \times G_{L/K} \to E[m]$ in (VIII.1.2).

PROOF OF (VIII.1.2). Most of this proposition follows immediately from basic properties of group cohomology; see (VIII §2). For the convenience of the reader, we give a direct proof here.

(a) We must show that $\kappa(P, \sigma)$ is in $E[m]$ and that its value does not depend on the choice of Q. For the first statement, we observe that

$$[m]\kappa(P, \sigma) = [m]Q^{\sigma} - [m]Q = P^{\sigma} - P = O,$$

since $P \in E(K)$ and σ fixes K. For the second statement, we note that any other choice has the form $Q + T$ for some $T \in E[m]$. Then

$$(Q + T)^{\sigma} - (Q + T) = Q^{\sigma} + T^{\sigma} - Q - T = Q^{\sigma} - Q,$$

because we have assumed that $E[m] \subset E(K)$, so σ fixes T.

(b) The linearity in P is obvious. For linearity in σ, we let $\sigma, \tau \in G_{\bar{K}/K}$ and compute

$$\kappa(P, \sigma\tau) = Q^{\sigma\tau} - Q = (Q^{\sigma} - Q)^{\tau} - (Q^{\tau} - Q) = \kappa(P, \sigma)^{\tau} + \kappa(P, \tau).$$

But $\kappa(P, \sigma) \in E[m] \subset E(K)$, so $\kappa(P, \sigma)$ is fixed by τ.

(c) Suppose that $P \in mE(K)$, say $P = [m]Q$ with $Q \in E(K)$. Then Q is fixed by every $\sigma \in G_{\bar{K}/K}$, so

$$\kappa(P, \sigma) = Q^{\sigma} - Q = O.$$

Conversely, suppose that $\kappa(P, \sigma) = 0$ for all $\sigma \in G_{\bar{K}/K}$. Then choosing some point $Q \in E(\bar{K})$ with $[m]Q = P$, we have

$$Q^{\sigma} = Q \quad \text{for all } \sigma \in G_{\bar{K}/K}.$$

Therefore $Q \in E(K)$, so $P = [m]Q \in mE(K)$.

(d) If $\sigma \in G_{\bar{K}/L}$, then

$$\kappa(P, \sigma) = Q^{\sigma} - Q = O,$$

since $Q \in E(L)$ from the definition of L. Conversely, suppose that $\sigma \in G_{\bar{K}/K}$ satisfies $\kappa(P, \sigma) = O$ for all $P \in E(K)$. Then for every point $Q \in E(\bar{K})$ satisfying $[m]Q \in E(K)$ we have

$$O = \kappa([m]Q, \sigma) = Q^\sigma - Q.$$

But L is the compositum of $K(Q)$ over all such Q, so σ fixes L. Hence $\sigma \in G_{\bar{K}/L}$.

Finally, the last statement of (VIII.1.2) is clear from what precedes it, once we note that L/K is Galois because elements of $G_{\bar{K}/K}$ map $[m]^{-1}E(K)$ to itself. Alternatively, it follows from (d) that $G_{\bar{K}/L}$ is the kernel of the homomorphism

$$G_{\bar{K}/K} \longrightarrow \mathrm{Hom}(E(K), E[m]), \qquad \sigma \longmapsto \kappa(\,\cdot\,, \sigma),$$

so $G_{\bar{K}/L}$ is a normal subgroup of $G_{\bar{K}/K}$. □

It follows from (VIII.1.2) that the finiteness of $E(K)/mE(K)$ is equivalent to the finiteness of the extension L/K. The next step in the proof of the weak Mordell–Weil theorem is to analyze this extension. Our main tool will be (VII.3.1), which we restate after making the appropriate definitions.

Definition. Let K be a number field and let E/K be an elliptic curve. Let $v \in M_K^0$ be a discrete valuation. Then E is said to have *good* (respectively *bad*) *reduction at v* if E has good (respectively bad) reduction when considered over the completion K_v, cf. (VII §5). Taking a minimal Weierstrass equation for E over K_v, we denote the reduced curve over the residue field by $\tilde{E}_v/k_v$. N.B. It is not always possible to choose a single Weierstrass equation for E over K that is simultaneously minimal for all K_v. However, this can be done if $K = \mathbb{Q}$. See (VIII §8) for further details.

Remark 1.3. Take any Weierstrass equation for E/K,

$$E : y^2 + a_1 xy + a_3 y = x^3 + a_2 x^2 + a_4 x + a_6,$$

say with discriminant Δ. Then for all but finitely $v \in M_K^0$ we have

$$v(a_i) \geq 0 \quad \text{for } i = 1, \dots, 6 \quad \text{and} \quad v(\Delta) = 0.$$

For any v satisfying these conditions, the given equation is already a minimal Weierstrass equation and the reduced curve $\tilde{E}_v/k_v$ is nonsingular. This shows that E has good reduction at v for all but finitely many $v \in M_K^0$.

Proposition 1.4. (restatement of (VII.3.1b)) *Let $v \in M_K^0$ be a discrete valuation such that $v(m) = 0$ and such that E has good reduction at v. Then the reduction map*

$$E(K)[m] \longrightarrow \tilde{E}_v(k_v)$$

is injective.

We are now ready to analyze the extension L/K appearing in (VIII.1.2).

Proposition 1.5. *Let*

$$L = K([m]^{-1}E(K))$$

be the field defined in (VIII.1.2d).

(a) *The extension L/K is abelian and has exponent m, i.e., the Galois group $G_{L/K}$ is abelian and every element of $G_{L/K}$ has order dividing m.*

(b) *Let*

$$S = \{v \in M_K^0 : E \text{ has bad reduction at } v\} \cup \{v \in M_K^0 : v(m) \neq 0\} \cup M_K^\infty.$$

The L/K is unramified outside S, i.e., if $v \in M_K$ and $v \notin S$, then L/K is unramified at v.

PROOF. (a) This follows immediately from (VIII.1.2), which implies that there is an injection

$$G_{L/K} \longrightarrow \operatorname{Hom}(E(K), E[m]), \qquad \sigma \longmapsto \kappa(\,\cdot\,, \sigma).$$

(b) Let $v \in M_K$ with $v \notin S$, let $Q \in E(\bar{K})$ satisfy $[m]Q \in E(K)$, and let $K' = K(Q)$. It suffices to show that K'/K is unramified at v, since L is the compositum of all such K'. Let $v' \in M_{K'}$ be a place of K' lying above v and let $k'_{v'}/k_v$ be the corresponding extension of residue fields. The assumption that $v \notin S$ ensures that E has good reduction at v, so it also has good reduction at v', since we can take the same Weierstrass equation. Thus we have the usual reduction map

$$E(K') \longrightarrow \tilde{E}(k'_{v'}),$$

which we denote as usual by a tilde.

Let $I_{v'/v} \subset G_{\bar{K}/K}$ be the inertia group for v'/v, and take any element $\sigma \in I_{v'/v}$. By definition, an element of inertia such as σ acts trivially on $\tilde{E}(k'_{v'})$, so

$$\widetilde{Q^\sigma - Q} = \tilde{Q}^\sigma - \tilde{Q} = \tilde{O}.$$

On the other hand, the fact that $[m]Q \in E(K)$ tells us that

$$[m](Q^\sigma - Q) = ([m]Q)^\sigma - [m]Q = O.$$

Thus $Q^\sigma - Q$ is a point of order m that is in the kernel of the reduction-modulo-v' map. It follows from (VIII.1.4) that

$$Q^\sigma - Q = O.$$

This proves that Q is fixed by every element of the inertia group $I_{v'/v}$, and hence that $K' = K(Q)$ is unramified over K at v'. Since this holds for every v' lying over v and for every $v \notin S$, this completes the proof that K'/K is unramified outside of S. $\qquad\square$

All that remains to complete the proof of the weak Mordell–Weil theorem is to show that any field extension L/K satisfying the conditions of (VIII.1.5) is necessarily a finite extension. The proof of this fact relies on the two fundamental finiteness theorems of algebraic number theory, namely the finiteness of the ideal class group and the finite generation of the group of S-units.

Proposition 1.6. *Let K be a number field, let $S \subset M_K$ be a finite set of places that contains M_K^∞, and let $m \geq 2$ be an integer. Let L/K be the maximal abelian extension of K having exponent m that is unramified outside of S. Then L/K is a finite extension.*

PROOF. Suppose that we know that the proposition is true for some finite extension K' of K, where S' is the set of places of K' lying over S. Then LK'/K', being abelian of exponent m unramified outside S', would be finite, and hence L/K would also be finite. It thus suffices to prove the proposition under the assumption that K contains the mth roots of unity $\boldsymbol{\mu}_m$.

Similarly, we may increase the size of the set S, since this only has the effect of making L larger. Using the fact that the class number of K is finite, we adjoin a finite number of elements to S so that the *ring of S-integers*

$$R_S = \big\{ a \in K : v(a) \geq 0 \text{ for all } v \in M_K \text{ with } v \notin S \big\}$$

is a principal ideal domain. (Explicitly, choose integral ideals $\mathfrak{a}_1, \ldots, \mathfrak{a}_h$ representing the ideal classes of K and adjoin to S the valuations corresponding to the primes dividing $\mathfrak{a}_1 \cdots \mathfrak{a}_h$.) We also enlarge S so as to ensure that $v(m) = 0$ for all $v \notin S$.

We now apply the main theorem of Kummer theory, which says that if a field of characteristic 0 contains $\boldsymbol{\mu}_m$, then its maximal abelian extension of exponent m is obtained by adjoining the mth roots of all of its elements. For a proof of this result, see any basic textbook on field theory, for example [17, §2], [68, §17.3], or [7, Theorem 25], or do Exercise 8.4. Thus L is the largest subfield of

$$K\big(\sqrt[m]{a} : a \in K \big)$$

that is unramified outside of S.

Let $v \in M_K$ with $v \notin S$. Consider the equation

$$X^m - a = 0$$

over the local field K_v. Since $v(m) = 0$ and since the discriminant of the polynomial $X^m - a$ equals $\pm m^m a^{m-1}$, we see that $K_v\big(\sqrt[m]{a} \big)/K_v$ is unramified if and only if

$$\operatorname{ord}_v(a) \equiv 0 \pmod{m}.$$

(Recall that ord_v is the normalized valuation associated to v.) We note that when we adjoin m^{th} roots, it is necessary to take only one representative for each class in $K^*/(K^*)^m$, so if we let

$$T_S = \big\{ a \in K^*/(K^*)^m : \operatorname{ord}_v(a) \equiv 0 \pmod{m} \text{ for all } v \in M_K \text{ with } v \notin S \big\},$$

then

$$L = K\big(\sqrt[m]{a} : a \in T_S \big).$$

To complete the proof of (VIII.1.6), it suffices to show that the set T_S is finite.

Consider the natural map

$$R_S^* \longrightarrow T_S.$$

We claim that this map is surjective. To see this, suppose that $a \in K^*$ represents an element of T_S. Then the ideal aR_S is the mth power of an ideal in R_S, since the prime ideals of R_S correspond to the valuations $v \notin S$. Using the fact that R_S is a principal ideal domain, we can find a $b \in K^*$ such that $aR_S = b^m R_S$. Hence there is a $u \in R_S^*$ satisfying

$$a = ub^m.$$

Then a and u give the same element of T_S, which proves that R_S^* surjects onto T_S. Further, the kernel of the map $R_S^* \to T_S$ clearly contains $(R_S^*)^m$, which proves that there is a surjection

$$R_S^*/(R_S^*)^m \twoheadrightarrow T_S.$$

(This map is actually an isomorphism.) Finally, we apply Dirichlet's S-unit theorem [142, V §1], which says that R_S^* is a finitely generated group. It follows that T_S is finite, which completes the proof of the proposition. ☐

The preceding three propositions may now be combined to prove the main result of this section.

PROOF OF THE WEAK MORDELL–WEIL THEOREM (VIII.1.1). Let

$$L = K\big([m]^{-1}E(K)\big)$$

be the field defined in (VIII.1.2d). Since $E[m]$ is finite, the perfect pairing given in (VIII.2.1) shows that $E(K)/mE(K)$ is finite if and only if $G_{L/K}$ is finite. Now (VIII.1.5) says that L has certain properties, and (VIII.1.6) says that any extension of K having these properties is a finite extension. This gives the desired result. (Note that (VIII.1.3) ensures that the set S of (VIII.1.5b) is a finite set.) ☐

Remark 1.7. The heart of the proof of the weak Mordell–Weil theorem lies in the assertion that the field $L = K\big([m]^{-1}E(K)\big)$ is a finite extension of K. We proved this by first showing (VIII.1.5) that it is abelian of exponent m and that it is unramified outside of a certain finite set $S \subset M_K$. The desired result then followed from basic Kummer theory of fields as given in the proof of (VIII.1.6). It is worth noting that rather than using (VIII.1.6), we could have used the more general theorem of Minkowski that asserts that there are only finitely many extensions of K of bounded degree that are unramified outside of S. To apply this in the present instance, note that for any $Q \in [m]^{-1}E(K)$, the field $K(Q)$ has degree at most m^2 over K, since the Galois conjugates of Q all have the form $Q + T$ for some $T \in E[m]$ and we are assuming that $E[m] \subset E(K)$. It follows from Minkowski's theorem that as Q ranges over $[m]^{-1}E(K)$, there are only finitely many possibilities for the fields $K(Q)$. Hence their compositum $K\big([m]^{-1}E(K)\big)$ is a finite extension of K.

Remark on Effectivity 1.8. Let E/K be an elliptic curve with $E[m] \subset E(K)$, let $S \subset M_K$ be the usual set of bad places for E/K as described in (VIII.1.5b), and let L/K be the maximal abelian extension of K having exponent m such that L/K

is unramified outside of S. Then (VIII.1.2) and (VIII.1.5) tell us that the Kummer pairing induces an injection

$$E(K)/mE(K) \longhookrightarrow \mathrm{Hom}\big(G_{L/K}, E[m]\big).$$

It is possible to make the proof of (VIII.1.6) completely explicit, and hence to exactly determine the group $G_{L/K}$; see Exercise 8.1. Thus we can describe all of the elements of the group $\mathrm{Hom}\big(G_{L/K}, E[m]\big)$, so the crucial question is that of determining which of these elements come from points of $E(K)/mE(K)$. It is this last question for which there is, at present, no known effective solution. In Chapter X we examine this problem in more detail. There we will exhibit a smaller group into which $E(K)/mE(K)$ injects and discuss what can be said about the cokernel. We want to stress that this is the only stage at which the Mordell–Weil theorem is ineffective; if we know generators for $E(K)/mE(K)$, then we can effectively find generators for $E(K)$; see (VIII.3.1) and Exercise 8.18.

We also remark that there is a conditional algorithm due to Manin [156], [114, § F.4.1] that effectively computes generators for $E(K)$ if one accepts the validity of a number of standard (but very deep) conjectures, including in particular the conjecture of Birch and Swinnerton-Dyer (C.16.5).

VIII.2 The Kummer Pairing via Cohomology

In this section we reinterpret the Kummer pairing from (VIII §1) in terms of group cohomology. The methods used here will not be used again until Chapter X and may be omitted by the reader wishing to proceed directly to the proof of the Mordell–Weil theorem. For a summary of the basic facts on group cohomology that are used in this section, see Appendix B and/or the references listed there.

We start with the short exact sequence of $G_{\bar{K}/K}$-modules

$$0 \longrightarrow E[m] \longrightarrow E(\bar{K}) \xrightarrow{\ [m]\ } E(\bar{K}) \longrightarrow 0,$$

where $m \geq 2$ is a fixed integer. Taking $G_{\bar{K}/K}$-cohomology yields a long exact sequence that starts

$$0 \longrightarrow E(K)[m] \longrightarrow E(K) \xrightarrow{\ [m]\ } E(K) \overset{\delta}{}$$

$$\xrightarrow{} H^1\big(G_{\bar{K}/K}, E[m]\big) \longrightarrow H^1\big(G_{\bar{K}/K}, E(\bar{K})\big) \xrightarrow{\ [m]\ } H^1\big(G_{\bar{K}/K}, E(\bar{K})\big).$$

From the middle of this exact sequence we extract the following short exact sequence, which is called the *Kummer sequence for E/K*:

$$0 \longrightarrow \frac{E(K)}{mE(K)} \xrightarrow{\ \delta\ } H^1\big(G_{\bar{K}/K}, E[m]\big) \longrightarrow H^1\big(G_{\bar{K}/K}, E(\bar{K})\big)[m] \longrightarrow 0.$$

(As usual, for any abelian group A, we write $A[m]$ to denote the m-torsion subgroup of A.)

From general principles, the connecting homomorphism δ is computed as follows. Let $P \in E(K)$ and choose some $Q \in E(\bar{K})$ satisfying $[m]Q = P$. Then a 1-cocycle representing $\delta(P)$ is given by

$$c : G_{\bar{K}/K} \longrightarrow E[m], \qquad c_\sigma = Q^\sigma - Q.$$

But this is exactly the Kummer pairing defined in (VIII §1),

$$c_\sigma = \kappa(P, \sigma).$$

(This assumes that we use the same Q on both sides, of course.)

Now suppose that $E[m]$ is contained in $E(K)$. Then

$$H^1(G_{\bar{K}/K}, E[m]) = \operatorname{Hom}(G_{\bar{K}/K}, E[m]),$$

so under this assumption we obtain an injective homomorphism

$$E(K)/mE(K) \longleftrightarrow \operatorname{Hom}(G_{\bar{K}/K}, E[m]), \qquad P \longmapsto \kappa(P, \cdot).$$

This provides an alternative proof of (VIII.1.2abc).

Similarly, we can use the inflation–restriction sequence (B.2.4) to give a quick proof of the reduction lemma described in (VIII.1.1.1). Thus if L/K is a finite Galois extension, say satisfying $E[m] \subset E(L)$, then we have a commutative diagram

$$
\begin{array}{ccccccc}
0 \longrightarrow & \Phi & \longrightarrow & E(K)/mE(K) & \longrightarrow & E(L)/mE(L) \\
\downarrow & \downarrow & & \downarrow & & \downarrow \\
0 \longrightarrow H^1(G_{L/K}, E[m]) & \xrightarrow{\text{inf}} & H^1(G_{\bar{K}/K}, E[m]) & \xrightarrow{\text{res}} & H^1(G_{\bar{L}/L}, E[m]),
\end{array}
$$

where the vertical arrows are injections. Since $G_{L/K}$ and $E[m]$ are finite groups, the cohomology group $H^1(G_{L/K}, E[m])$ is finite, so Φ is also finite. We observe that the map $\lambda_P : G_{L/K} \to E[m]$ defined in the proof of (VIII.1.1.1) is a cocycle whose cohomology class is precisely the image of $P \in \Phi$ in $H^1(G_{L/K}, E[m])$.

Returning now to the general case, we reinterpret (VIII.1.5b) in terms of cohomology.

Definition. Let M be a $G_{\bar{K}/K}$-module, let $v \in M_K^0$ be a discrete valuation, and let $I_v \subset G_{\bar{K}/K}$ be an inertia group for v. A cohomology class $\xi \in H^r(G_{\bar{K}/K}, M)$ is said to be *unramified at* v if it is trivial when restricted to $H^r(I_v, M)$. (The inertia group I_v depends on choosing an extension of v to $\bar{K}$, but one can show that the definition of unramified cohomology class is independent of this choice; cf. (X.4.1.1) and Exercise B.6.)

Proposition 2.1. *Let*

$$S = \{v \in M_K^0 : E \text{ has bad reduction at } v\} \cup \{v \in M_K^0 : v(m) \neq 0\} \cup M_K^\infty.$$

Then the image of $E(K)$ in $H^1(G_{\bar{K}/K}, E[m])$ under the connecting homomorphism δ consists of cohomology classes that are unramified at every $v \in M_K$ with $v \notin S$.

PROOF. Let $P \in E(K)$ and, as above, let

$$c_\sigma = Q^\sigma - Q$$

be the cocycle representing $\delta(P)$ for some point Q satisfying $[m]Q = P$. Then (VIII.1.5b) says that the field $K(Q)$ is unramified at v. (N.B. The proof of (VIII.1.5b) did not use the assumption that $E[m]$ is contained in $E(K)$.) Hence I_v acts trivially on Q, so $c_\sigma = 0$ for all $\sigma \in I_v$. $\qquad\square$

The Kummer Sequence for Fields

The exact sequences that we have derived for elliptic curves are analogous to the classical exact sequences that arise in Kummer theory for fields. To make the analogy clear, we briefly recall the relevant material. The multiplication-by-m sequence for an elliptic curve E corresponds to the following exact sequence of $G_{\bar{K}/K}$-modules:

$$1 \longrightarrow \mu_m \longrightarrow \bar{K}^* \xrightarrow{z \to z^m} \bar{K}^* \longrightarrow 1.$$

Taking $G_{\bar{K}/K}$-cohomology yields a long exact sequence from which we extract the short exact sequence

$$1 \longrightarrow K^*/(K^*)^m \xrightarrow{\delta} H^1(G_{\bar{K}/K}, \mu_m) \longrightarrow H^1(G_{\bar{K}/K}, \bar{K}^*)[m] \longrightarrow 0.$$

Hilbert's famous "Theorem 90" (B.2.5) asserts that

$$H^1(G_{\bar{K}/K}, \bar{K}^*) = 0,$$

so the connecting homomorphism is an isomorphism. This is in marked contrast to the situation for elliptic curves, where the nontriviality of $H^1(G_{\bar{K}/K}, E(\bar{K}))$ provides much added complication. (See Chapter X.) Collecting this material and using an explicit computation of the connecting homomorphism gives the following result.

Proposition 2.2. *There is an isomorphism*

$$\delta : K^*/(K^*)^m \xrightarrow{\;\sim\;} H^1(G_{\bar{K}/K}, \mu_m)$$

given by the formula

$$\delta(a) = \text{cohomology class of the map } \sigma \mapsto \alpha^\sigma/\alpha,$$

where $\alpha \in \bar{K}^*$ is any element satisfying $\alpha^m = a$.

VIII.3 The Descent Procedure

Our primary goal in this chapter is to prove that $E(K)$, the group of rational points on an elliptic curve, is finitely generated. So far, we know from (VIII.1.1) that the quotient group $E(K)/mE(K)$ is finite. It is easy to see that this is not enough. For example, $\mathbb{R}/m\mathbb{R} = 0$ for every integer $m \geq 1$, yet $\mathbb{R}$ is certainly not a finitely generated group. Similarly, if $E/\mathbb{Q}_p$ is an elliptic curve, then (VII.6.3) says that $E(\mathbb{Q}_p)$ has a subgroup of finite index that is isomorphic to $\mathbb{Z}_p$. Hence $E(\mathbb{Q}_p)/mE(\mathbb{Q}_p)$ is finite, while $E(\mathbb{Q}_p)$ is not finitely generated.

An examination of these two examples shows that the problem occurs because of the large number of elements in the group that are divisible by m. The idea used to complete the proof of the Mordell–Weil theorem is to show that on an elliptic curve over a number field, the multiplication-by-m map tends to increase the "size" of a point, where there are only finitely many points whose "size" is bounded. This will bound how high a power of m may divide a point, and thus eliminate problems such as in the above examples. Of course, all of this is very vague until we explain what is meant by the "size" of a point.

In this section we axiomatize the situation and describe the type of size (or height) function needed to prove that an abelian group is finitely generated. Then, in the next section, we define such a function on an elliptic curve in the simplest case and use explicit formulas to prove that it has the desired properties. This will suffice to prove a special case of the Mordell–Weil theorem. We then turn to the general case and develop the theory of height functions in sufficient generality both to prove the Mordell–Weil theorem and to be useful for later applications.

Theorem 3.1. (Descent Theorem) *Let A be an abelian group. Suppose that there exists a* (height) *function*

$$h : A \longrightarrow \mathbb{R}$$

with the following three properties:

(i) *Let $Q \in A$. There is a constant C_1, depending on A and Q, such that*

$$h(P + Q) \leq 2h(P) + C_1 \qquad \textit{for all } P \in A.$$

(ii) *There are an integer $m \geq 2$ and a constant C_2, depending on A, such that*

$$h(mP) \geq m^2 h(P) - C_2 \qquad \textit{for all } P \in A.$$

(iii) *For every constant C_3, the set*

$$\{P \in A : h(P) \leq C_3\}$$

is finite.

Suppose further that for the integer m in (ii), *the quotient group A/mA is finite. Then A is finitely generated.*

PROOF. Choose elements $Q_1, \ldots, Q_r \in A$ to represent the finitely many cosets in A/mA, and let $P \in A$ be an arbitrary element. The idea is to show that the difference between P and an appropriate linear combination of $Q_1, \ldots, Q_r$ is a multiple of a point whose height is smaller than a constant that is *independent of P.* Then $Q_1, \ldots, Q_r$ and the finitely many points with height less than this constant are generators for A.

We begin by writing

$$P = mP_1 + Q_{i_1} \qquad \text{for some } 1 \le i_1 \le r.$$

Next we do the same thing with P_1, then with P_2, etc., which gives us a list of points

$$P = mP_1 + Q_{i_1},$$
$$P_1 = mP_2 + Q_{i_2},$$
$$\vdots$$
$$P_{n-1} = mP_n + Q_{i_n}.$$

For any index j, we have

$$h(P_j) \le \frac{1}{m^2}\big(h(mP_j) + C_2\big) \qquad \text{from (ii)},$$
$$= \frac{1}{m^2}\big(h(P_{j-1} - Q_{i_j}) + C_2\big)$$
$$\le \frac{1}{m^2}\big(2h(P_{j-1}) + C_1' + C_2\big) \qquad \text{from (i)},$$

where C_1' is the maximum of the constants from (i) for $Q \in \{-Q_1, \ldots, -Q_r\}$. Note that C_1' and C_2 do not depend on P.

We use this inequality repeatedly, starting from P_n and working back to P. This yields

$$h(P_n) \le \left(\frac{2}{m^2}\right)^n h(P) + \left(\frac{1}{m^2} + \frac{2}{m^2} + \frac{4}{m^2} + \cdots + \frac{2^{n-1}}{m^2}\right)(C_1' + C_2)$$
$$< \left(\frac{2}{m^2}\right)^n h(P) + \frac{C_1' + C_2}{m^2 - 2}$$
$$\le \frac{1}{2^n}h(P) + \frac{1}{2}(C_1' + C_2) \qquad \text{since } m \ge 2.$$

It follows that if n is sufficiently large, then

$$h(P_n) \le 1 + \frac{1}{2}(C_1' + C_2).$$

Since P is a linear combination of P_n and $Q_1, \ldots, Q_r$,

$$P = m^n P_n + \sum_{j=1}^{n} m^{j-1} Q_{i_j},$$

it follows that every P in A is a linear combination of points in the set

$$\{Q_1, \ldots, Q_r\} \cup \left\{Q \in A : h(Q) \leq 1 + \frac{1}{2}(C_1' + C_2)\right\}.$$

Property (iii) of the height function h tells us that this is a finite set, which completes the proof that A is finitely generated. $\qquad\qquad\square$

Remark 3.2. What is needed to make the descent theorem effective, i.e., to allow us to find generators for the group A? First, we must be able to calculate the constants $C_1 = C_1(Q_i)$ for each of the elements $Q_1, \ldots, Q_r \in A$ representing the cosets of A/mA. Second, we must be able to calculate the constant C_2. Third, for any constant C_3, we must be able to determine the elements in the finite set $\{P \in A : h(P) \leq C_3\}$. The reader may check (Exercise 8.18) that for the height functions used on elliptic curves (VIII §§4, 5, 6), all of these constants are effectively computable, *provided* that we can find elements of $E(K)$ that generate the finite group $E(K)/mE(K)$. Unfortunately, at present there is no known procedure that is guaranteed to give generators for $E(K)/mE(K)$. We return to this question in Chapter X.

VIII.4   The Mordell–Weil Theorem over $\mathbb{Q}$

In this section we prove the following special case of the Mordell–Weil theorem.

Theorem 4.1. *Let $E/\mathbb{Q}$ be an elliptic curve. Then the group $E(\mathbb{Q})$ is finitely generated.*

We will, of course, soon be ready to prove the general case; see (VIII.6.7). However, it seems worthwhile to first prove (VIII.4.1), since in this case the necessary height computations using explicit formulas are not too cumbersome.

Fix a Weierstrass equation for $E/\mathbb{Q}$ of the form

$$E : y^2 = x^3 + Ax + B \qquad \text{with } A, B \in \mathbb{Z}.$$

We know from (VIII.1.1) that $E(\mathbb{Q})/2E(\mathbb{Q})$ is finite, so in order to apply the descent result (VIII.3.1), we need to define a height function on $E(\mathbb{Q})$ and show that it has the requisite properties.

Definition. Let $t \in \mathbb{Q}$, and write $t = p/q$ as a fraction in lowest terms. The *height of t*, denoted by $H(t)$, is defined by

$$H(t) = \max\{|p|, |q|\}.$$

Definition. The *(logarithmic) height on $E(\mathbb{Q})$*, relative to the given Weierstrass equation, is the function

$$h_x : E(\mathbb{Q}) \longrightarrow \mathbb{R}, \qquad h_x(P) = \begin{cases} \log H\big(x(P)\big) & \text{if } P \neq O, \\ 0 & \text{if } P = O. \end{cases}$$

We note that $h_x(P)$ is always nonnegative.

The next lemma gives us the information that we need in order to apply (VIII.3.1) with the height function h_x.

Lemma 4.1. *Let $E/\mathbb{Q}$ be an elliptic curve given by a Weierstrass equation*

$$E : y^2 = x^3 + Ax + B \qquad \text{with } A, B \in \mathbb{Z}.$$

(a) *Let $P_0 \in E(\mathbb{Q})$. There is a constant C_1 that depends on P_0, A, and B such that*

$$h_x(P + P_0) \le 2h_x(P) + C_1 \qquad \text{for all } P \in E(\mathbb{Q}).$$

(b) *There is a constant C_2 that depends on A and B such that*

$$h_x([2]P) \ge 4h_x(P) - C_2 \qquad \text{for all } P \in E(\mathbb{Q}).$$

(c) *For every constant C_3, the set*

$$\{P \in E(\mathbb{Q}) : h_x(P) \le C_3\}$$

is finite.

PROOF. We may assume that $C_1 > \max\{h_x(P_0), h_x([2]P_0)\}$, which ensures that (a) is true if $P_0 = O$ or if $P \in \{O, \pm P_0\}$. In all other cases we write

$$P = (x, y) = \left(\frac{a}{d^2}, \frac{b}{d^3}\right) \qquad \text{and} \qquad P = (x_0, y_0) = \left(\frac{a_0}{d_0^2}, \frac{b_0}{d_0^3}\right),$$

where all fractions are in lowest terms. The addition formula (III.2.3d) says that

$$x(P + P_0) = \left(\frac{y - y_0}{x - x_0}\right)^2 - x - x_0.$$

Expanding this expression and using the fact that P and P_0 satisfy the given Weierstrass equation yields

$$
\begin{aligned}
x(P + P_0) &= \frac{(xx_0 + A)(x + x_0) + 2B - 2yy_0}{(x - x_0)^2} \\
&= \frac{(aa_0 + Ad^2 d_0^2)(ad_0^2 + a_0 d^2) + 2Bd^4 d_0^4 - 2bdb_0 d_0}{(ad_0^2 - a_0 d^2)^2}.
\end{aligned}
$$

In computing the height of a rational number, cancellation between numerator and denominator can only decrease the height, so we find by an easy estimation that

$$H\big(x(P + P_0)\big) \le C_1' \max\{|a|^2, |d|^4, |bd|\},$$

where C_1' has a simple expression in terms of A, B, a_0, b_0, d_0. Since $H\big(x(P)\big) = \max\{|a|, |d|^2\}$, this is almost what we want, the only possible difficulty being the presence of $|bd|$ in the maximum. To deal with this problem, we use the fact that the point P lies on the curve E, so its coordinates satisfy

$$b^2 = a^3 + Aad^4 + Bd^6.$$

Thus

$$|b| \leq C_1'' \max\{|a|^{3/2}, |d|^3\},$$

which combined with the above estimate for $H\big(x(P + P_0)\big)$ yields

$$H\big(x(P + P_0)\big) \leq C_1 \max\{|a|^2, |d|^4\} = C_1 H\big(x(P)\big)^2.$$

Taking logarithms gives the desired result.

(b) Choosing C_2 to satisfy

$$C_2 \geq 4 \max\{h_x(T) : T \in E(\mathbb{Q})[2]\},$$

we may assume that $[2]P \neq O$. Then, writing $P = (x, y)$, the duplication formula (III.2.3d) reads

$$x\big([2]P\big) = \frac{x^4 - 2Ax^2 - 8Bx + A^2}{4x^3 + 4Ax + 4B}.$$

It is convenient to define homogeneous polynomials

$$F(X, Z) = X^4 - 2AX^2Z^2 - 8BXZ^3 + A^2Z^4,$$
$$G(X, Z) = 4X^3Z + 4AXZ^3 + 4BZ^4.$$

If we write $x = x(P) = a/b$ as a fraction in lowest terms, then $x\big([2]P\big)$ may be written as a quotient of integers,

$$x\big([2]P\big) = \frac{F(a, b)}{G(a, b)}.$$

However, in contrast to the proof of (a), we are now looking for a lower bound for $H\big(x([2]P)\big)$, so it is necessary to bound how much cancellation may occur between numerator and denominator.

To do this, we use the fact that $F(X, 1)$ and $G(X, 1)$ are relatively prime polynomials, so they generate the unit ideal in $\mathbb{Q}[X]$. This implies that identities of the following sort exist.

Sublemma 4.3. *Let* $\Delta = 4A^3 + 27B^2$, *and define polynomials*

$$F(X, Z) = X^4 - 2AX^2Z^2 - 8BXZ^3 + A^2Z^4,$$
$$G(X, Z) = 4X^3Z + 4AXZ^3 + 4BZ^4,$$
$$f_1(X, Z) = 12X^2Z + 16AZ^3,$$
$$g_1(X, Z) = 3X^3 - 5AXZ^2 - 27BZ^3,$$
$$f_2(X, Z) = 4(4A^3 + 27B^2)X^3 - 4A^2BX^2Z$$
$$\qquad\qquad + 4A(3A^3 + 22B^2)XZ^2 + 12B(A^3 + 8B^2)Z^3,$$
$$g_2(X, Z) = A^2bX^2 + A(5A^3 + 32B^2)X^2Z$$
$$\qquad\qquad + 2B(13A63 + 96B^2)XZ^2 - 3A^2(A^3 + 8B^2)Z^3.$$

Then the following identities hold in $\mathbb{Z}[A, B, X, Z]$:

$$f_1(X, Z)F(X, Z) - g_1(X, Z)G(X, Z) = 4\Delta Z^7,$$
$$f_2(X, Z)F(X, Z) - g_2(X, Z)G(X, Z) = 4\Delta X^7.$$

PROOF. One can check that if $\Delta \neq 0$, then $F(X, Z)$ and $G(X, Z)$ are relatively prime homogeneous polynomials, so identities of this sort must exist. Checking the validity of the two identities is, at worst, a tedious calculation, which we leave for the reader. To actually find the polynomials f_1, g_1, f_2, g_2, one can use the Euclidean algorithm or the theory of resultants. $\square$

We return to the proof of (VIII.4.2b). Let

$$\delta = \gcd\big(F(a, b), G(a, b)\big)$$

denote the cancellation in our fraction for $x\big([2]P\big)$. From the equations

$$f_1(a, b)F(a, b) - g_1(a, b)G(a, b) = 4\Delta b^7,$$
$$f_2(a, b)F(a, b) - g_2(a, b)G(a, b) = 4\Delta a^7,$$

we see that δ divides 4Δ. This gives the bound

$$|\delta| \leq |4\Delta|,$$

and hence

$$H\big(x([2]P)\big) \geq \frac{\max\{|F(a, b)|, |G(a, b)|\}}{|4\Delta|}.$$

On the other hand, the same identities give the estimates

$$|4\Delta b^7| \leq 2\max\{|f_1(a, b)|, |g_1(a, b)|\}\max\{|F(a, b)|, |G(a, b)|\},$$
$$|4\Delta a^7| \leq 2\max\{|f_2(a, b)|, |g_2(a, b)|\}\max\{|F(a, b)|, |G(a, b)|\}.$$

Looking at the expressions for f_1, f_2, g_1, g_2 in (VIII.4.3), we have

$$\max\{|f_1(a, b)|, |g_1(a, b)|, |f_2(a, b)|, |g_2(a, b)|\} \leq C\max\{|a|^3, |b|^3\},$$

where C is a constant depending on A and B. Combining the last three inequalities yields

$$\max\{|4\Delta a^7|, |4\Delta b^7|\} \leq 2C\max\{|a|^3, |b|^3\}\max\{|F(a, b)|, |G(a, b)|\}.$$

Canceling $\max\{|a|^3, |b|^3\}$ from both sides, we obtain the estimate

$$\frac{\max\{|F(a, b)|, |G(a, b)|\}}{|4\Delta|} \geq (2C)^{-1}\max\{|a|^4, |b|^4\},$$

and then using the fact that $\max\{|a|, |b|\} = H\big(x(P)\big)$ gives the desired result,

$$H\big(x([2]P)\big) \geq (2C)^{-1} H\big(x(P)\big)^4.$$

(c) For any constant C, the set

$$\{t \in \mathbb{Q} : H(t) \leq C\}$$

is clearly finite. Indeed, it has at most $(2C+1)^2$ elements, since the numerator and denominator of t are integers restricted to lie between $-C$ and C. Further, given any value for x, there are at most two values of y for which (x, y) is a point of E. Therefore

$$\{P \in E(\mathbb{Q}) : h_x(P) \leq C_3\}$$

is also a finite set. $\qquad\qquad\qquad\qquad\qquad\qquad\qquad\qquad\qquad\qquad\qquad\qquad\qquad\Box$

The proof of (VIII.4.1) is now simply a matter of fitting together what we have already done.

PROOF OF (VIII.4.1). We know from (VIII.1.1) that $E(\mathbb{Q})/2E(\mathbb{Q})$ is finite. It follows from (VIII.4.2) that the height function

$$h_x : E(\mathbb{Q}) \longrightarrow \mathbb{R}$$

satisfies the conditions needed to apply the descent procedure (VIII.3.1) with $m = 2$. The conclusion of (VIII.3.1) is that $E(\mathbb{Q})$ is finitely generated. $\qquad\qquad\Box$

VIII.5 Heights on Projective Space

In order to use the descent theorem (VIII.3.1) to prove the Mordell–Weil theorem in general, we need to define a height function on the K-rational points of an elliptic curve. It is possible to proceed in an ad hoc manner using explicit equations, as was done in the last section, but we instead develop a general theory of height functions. From this general theory will follow all of the necessary properties, plus considerably more. Elliptic curves are given as subsets of projective space, so in this section we study height function defined on all of projective space, and then in the next section we examine its properties when restricted to the points of an elliptic curve.

Example 5.1. Let $P \in \mathbb{P}^N(\mathbb{Q})$ be a point with rational coordinates. Since $\mathbb{Z}$ is a principal ideal domain, we can find homogeneous coordinates

$$P = [x_0, \ldots, x_N]$$

satisfying

$$x_0, \ldots, x_N \in \mathbb{Z} \qquad \text{and} \qquad \gcd(x_0, \ldots, x_N) = 1.$$

Then a natural measure of the *height of P* is

$$H(P) = \max\{|x_0|, \ldots, |x_N|\}.$$

With this definition, it is clear that for any constant C, the set

$$\{P \in \mathbb{P}^N(\mathbb{Q}) : H(P) \leq C\}$$

is a finite set. Indeed, it has at most $(2C+1)^N$ elements. This is the sort of finiteness property that is needed for the descent procedure described in (VIII.3.1).

If we try to generalize (VIII.5.1) to arbitrary number fields, we run into the difficulty that the ring of integers need not be a principal ideal domain. We thus take a somewhat different approach, for which purpose we now specify more precisely how the absolute values in M_K are normalized.

Definition. The *set of standard absolute values on* $\mathbb{Q}$, which we denote by $M_\mathbb{Q}$, consists of the following:

(i) $M_\mathbb{Q}$ contains one archimedean absolute value, defined by

$$|x|_\infty = \text{usual absolute value} = \max\{x, -x\}.$$

(ii) For each prime $p \in \mathbb{Z}$, the set $M_\mathbb{Q}$ contains one nonarchimedean (p-adic) absolute value defined by

$$\left| p^n \frac{a}{b} \right|_p = p^{-n} \qquad \text{for } a, b \in \mathbb{Z} \text{ satisfying } p \nmid ab.$$

The *set of standard absolute values* on a number field K, denoted by M_K, is the set of all absolute values on K whose restriction to $\mathbb{Q}$ is one of the absolute values in $M_\mathbb{Q}$.

Definition. Let $v \in M_K$. The *local degree at* v, denoted by n_v, is

$$n_v = [K_v : \mathbb{Q}_v],$$

where K_v and $\mathbb{Q}_v$ denote the completions of K and $\mathbb{Q}$ with respect to the absolute value v.

With the preceding definitions, we state two basic facts from algebraic number theory that will be needed later.

Extension Formula 5.2. *Let* $L/K/\mathbb{Q}$ *be a tower of number fields, and let* $v \in M_K$. *Then*

$$\sum_{\substack{w \in M_L \\ w|v}} n_w = [L : K]n_v.$$

(*Here* $w \mid v$ *means that* w *restricted to* K *is equal to* v.)

Product Formula 5.3. *Let* $x \in K^*$. *Then*

$$\prod_{v \in M_K} |x|_v^{n_v} = 1.$$

For proofs of (VIII.5.2) and (VIII.5.3), see any standard text on algebraic number theory, for example [142, II §1 and V §1].

We are now ready to define the height of a point in projective space.

Definition. Let $P \in \mathbb{P}^N(K)$ be a point with homogeneous coordinates

$$P = [x_0, \ldots, x_N], \qquad x_0, \ldots, x_N \in K.$$

The *height of P* (*relative to K*) is

$$H_K(P) = \prod_{v \in M_K} \max\{|x_0|_v, \ldots, |x_N|_v\}^{n_v}.$$

Proposition 5.4. *Let $P \in \mathbb{P}^N(K)$.*

(a) *The height $H_K(P)$ does not depend on the choice of homogeneous coordinates for P.*

(b) *The height satisfies*

$$H_K(P) \geq 1.$$

(c) *Let L/K be a finite extension. Then*

$$H_L(P) = H_K(P)^{1/[K:\mathbb{Q}]}.$$

PROOF. (a) Any other choice of homogeneous coordinates for P has the form $[\lambda x_0, \ldots, \lambda x_N]$ for some $\lambda \in K^*$. Using the product formula (VIII.5.3), we have

$$\prod_{v \in M_K} \max\{|\lambda x_0|_v, \ldots, |\lambda x_N|_v\}^{n_v} = \prod_{v \in M_K} |\lambda|^{n_v} \max\{|x_0|_v, \ldots, |x_N|_v\}^{n_v}$$

$$= \prod_{v \in M_K} \max\{|x_0|_v, \ldots, |x_N|_v\}^{n_v}.$$

(b) Given any point P in projective space, we can always find homogeneous coordinates for P such that one of the coordinates is 1. Then every factor in the product defining $H_K(P)$ is at least 1.

(c) We compute

$$H_L(P) = \prod_{w \in M_L} \max\{|x_i|_w\}^{n_w}$$

$$= \prod_{v \in M_K} \prod_{\substack{w \in M_L \\ w|v}} \max\{|x_i|_v\}^{n_w} \qquad \text{since } x_i \in K,$$

$$= \prod_{v \in M_K} \max\{|x_i|_v\}^{[L:K]n_v} \qquad \text{from (VIII.5.2),}$$

$$= H_K(P)^{[L:K]}. \qquad \qquad \square$$

Remark 5.5. If $K = \mathbb{Q}$, then $H_{\mathbb{Q}}$ agrees with the more intuitive height function given in (VIII.5.1). To see this, let $P \in \mathbb{P}^N(\mathbb{Q})$ and choose homogeneous coordinates $[x_0, \ldots, x_N]$ for P with $x_i \in \mathbb{Z}$ and $\gcd(x_0, \ldots, x_N) = 1$. Then, for any nonarchimedean absolute value $v \in M_{\mathbb{Q}}$, we have $|x_i|_v \leq 1$ for all i and $|x_i|_v = 1$ for at least one i. Hence in the product defining $H_{\mathbb{Q}}(P)$, only the factor for the archimedean absolute value contributes, so

$$H_{\mathbb{Q}}(P) = \max\{|x_0|_\infty, \ldots, |x_N|_\infty\}.$$

In particular, it follows that for any constant C, the set

$$\{P \in \mathbb{P}^N(\mathbb{Q}) : H_{\mathbb{Q}}(P) \leq C\}$$

is finite. One of our goals is to extend this statement to H_K. We will actually prove something stronger; see (VIII.5.11).

Sometimes it is easier to work with a height function that is not relative to a particular number field. We use (VIII.5.4c) to create such a function.

Definition. Let $P \in \mathbb{P}^N(\bar{\mathbb{Q}})$. The *(absolute) height of* P, denoted by $H(P)$, is defined as follows. Choose a number field K such that $P \in \mathbb{P}^N(K)$. Then

$$H(P) = H_K(P)^{1/[K:\mathbb{Q}]},$$

where we take the positive root. We see from (VIII.5.4c) that $H(P)$ is well-defined, independent of the choice of K, and (VIII.5.4b) implies that $H(P) \geq 1$.

We next investigate how the height changes under mappings between projective spaces. We recall the following definition; cf. (I.3.3).

Definition. A *morphism of degree d* between projective spaces is a map

$$F : \mathbb{P}^N \longrightarrow \mathbb{P}^M, \qquad F(P) = [f_0(P), \ldots, f_M(P)],$$

where $f_0, \ldots, f_M \in \bar{\mathbb{Q}}[X_0, \ldots, X_N]$ are homogeneous polynomials of degree d having no common zero in $\bar{\mathbb{Q}}^N$ other than $X_0 = \cdots = X_N = 0$. If F can be written using polynomials f_i having coefficients in K, then F is said to be *defined over K*.

Theorem 5.6. *Let*

$$F : \mathbb{P}^N \longrightarrow \mathbb{P}^M$$

be a morphism of degree d. Then there are positive constants C_1 and C_2, depending on F, such that

$$C_1 H(P)^d \leq H\big(F(P)\big) \leq C_2 H(P)^d \qquad \text{for all } P \in \mathbb{P}^N(\bar{\mathbb{Q}}).$$

PROOF. Write $F = [f_0, \ldots, f_M]$ using homogeneous polynomials f_i having no common zeros, and let $P = [x_0, \ldots, x_N] \in \mathbb{P}^N(\bar{\mathbb{Q}})$ be a point with algebraic coordinates. Choose some number field K that contains $x_0, \ldots, x_N$ and also contains all of the coefficients of all of the f_i. For each absolute value $v \in M_K$, we let

$$|P|_v = \max_{0 \le i \le N} |x_i|_v \qquad \text{and} \qquad \left|F(P)\right|_v = \max_{0 \le j \le M} \left|f_j(P)\right|_v,$$

and we also define

$$|F|_v = \max\{|a|_v : a \text{ is a coefficient of some } f_i\}.$$

Then, from the definition of height, we have

$$H_K(P) = \prod_{v \in M_K} |P|_v^{n_v} \qquad \text{and} \qquad H_K\big(F(P)\big) = \prod_{v \in M_K} \left|F(P)\right|_v^{n_v},$$

so it makes sense to define

$$H_K(F) = \prod_{v \in M_K} |F|_v^{n_v}.$$

In other words, $H_K(F) = H\big([a_0, a_1, \ldots]\big)$, where the a_j are the coefficients of the f_i. Finally, we let $C_1, C_2, \ldots$ denote constants that depend only on M, N, and d, and we set

$$\epsilon(v) = \begin{cases} 1 & \text{if } v \in M_K^\infty, \\ 0 & \text{if } v \in M_K^0. \end{cases}$$

To illustrate the utility of the function ϵ, we observe that the triangle inequality may be concisely written as

$$|t_1 + \cdots + t_n|_v \le n^{\epsilon(v)} \max\{|t_1|_v, \ldots, |t_n|_v\}$$

for all $v \in M_K$, both archimedean and nonarchimedean.

Having set notation, we turn to the proof of (VIII.5.6). The upper bound is relatively easy. Let $v \in M_K$. The triangle inequality yields

$$\left|f_i(P)\right|_v \le C_1^{\epsilon(v)} |F|_v |P|_v^d,$$

since f_i is homogeneous of degree d. Here C_1 could equal the number of terms in f_i, which is at most $\binom{N+d}{N}$, i.e., the number of monomials of degree d in $N+1$ variables. Since this estimate holds for every i, we find that

$$\left|F(P)\right|_v \le C_1^{\epsilon(v)} |F|_v |P|_v^d.$$

Now raise to the n_v power, multiply over all $v \in M_K$, and take the $[K : \mathbb{Q}]^{\text{th}}$ root. This yields the desired upper bound

$$H\big(F(P)\big) \le C_1 H(F) H(P)^d,$$

where we have used the formula (VIII.5.2),

$$\sum_{v \in M_K} \epsilon(v) n_v = \sum_{v \in M_K^\infty} n_v = [K : \mathbb{Q}].$$

It is worth mentioning that in proving this upper bound, we did not use the fact that the f_i have no common nontrivial zeros. However, we will certainly need to use this property to prove the lower bound, since without it there are easy counterexamples; see Exercise 8.10.

Thus we now assume that the set

$$\{Q \in \mathbb{A}^{N+1}(\bar{\mathbb{Q}}) : f_0(Q) = \cdots = f_M(Q) = 0\}$$

consists of the single point $(0, \ldots, 0)$. It follows from the Nullstellensatz ([111, I.1.3A], [73, Theorem 1.6]) that the ideal generated by $f_0, \ldots, f_M$ in $\bar{\mathbb{Q}}[X_0, \ldots, X_N]$ contains some power of each of $X_0, \ldots, X_N$, since each X_i also vanishes at the point $(0, \ldots, 0)$. Thus there are polynomials $g_{ij} \in \bar{\mathbb{Q}}[X_0, \ldots, X_N]$ and an integer $e \geq 1$ such that

$$X_i^e = \sum_{j=0}^{M} g_{ij} f_j \qquad \text{for each } 0 \leq i \leq N.$$

Replacing K by a finite extension if necessary, we may assume that each $g_{ij} \in K[X_0, \ldots, X_N]$, and discarding all terms on the right-hand side except those that are homogeneous of degree e, we may assume that each g_{ij} is homogeneous of degree $e - d$. We further set the following reasonable notation:

$$|G|_v = \max\{|b|_v : b \text{ is a coefficient of some } g_{ij}\},$$
$$H_K(G) = \prod_{v \in M_K} |G|_v^{n_v}.$$

We observe that e and $H_K(G)$ may be bounded in terms of M, N, d, and $H_K(F)$, although finding a good bound is not an easy task. See (VIII.5.7) for a discussion. For our purposes it is enough to know that e and $H_K(G)$ do not depend on the point P.

Recalling that $P = [x_0, \ldots, x_N]$, we see that the formula for X_i^e implies that

$$|x_i|_v^e = \left| \sum_{j=0}^{M} g_{ij}(P) f_j(P) \right|_v \leq C_2^{\epsilon(v)} \max_{0 \leq j \leq M} |g_{ij}(P) f_j(P)|_v$$

$$\leq C_2^{\epsilon(v)} \max_{0 \leq j \leq M} |g_{ij}(P)| \, |F(P)|_v.$$

We now take the maximum over i to obtain

$$|P|_v^e \leq C_2^{\epsilon(v)} \max_{\substack{0 \leq j \leq M \\ 0 \leq i \leq N}} |g_{ij}(P)|_v \, |F(P)|_v.$$

Each g_{ij} is homogeneous of degree $e - d$, so the usual application of the triangle inequality yields

$$|g_{ij}(P)|_v \leq C_3^{\epsilon(v)} |G|_v |P|_v^{e-d}.$$

Here C_3 may depend on e, but as noted earlier, we can bound e in terms of M, N, and d. Substituting this estimate into the earlier one and multiplying by $|P|^{d-e}$ gives

$$|P|_v^d \leq C_v^{\epsilon(v)} |G|_v |F(P)|_v,$$

and now the usual raising to the n_v power, multiplying over $v \in M_K$, and taking the $[K : \mathbb{Q}]^{\text{th}}$ root yields the desired lower bound.   □

Remark 5.7. As indicated during the proof of (VIII.5.6), the dependence of C_1 on F in the inequality

$$C_1 H(P)^d \leq H(F(P))$$

is not at all straightforward. It is possible to express C_1 in terms of the coefficients of certain polynomials whose existence is guaranteed by the Nullstellensatz, and the Nullstellensatz can be made completely explicit by the use of elimination theory, but this method leads to a very poor estimate. For an explicit version of the Nullstellensatz in which an effort has been made to give good estimates for the coefficients, see [162].

We also record the special case of (VIII.5.6) for an automorphism of $\mathbb{P}^N$.

Corollary 5.8. Let $A \in \mathrm{GL}_{N+1}(\bar{\mathbb{Q}})$, so multiplication by the matrix A induces an automorphism $A : \mathbb{P}^N \to \mathbb{P}^N$. There are positive constants C_1 and C_2, depending on the entries of the matrix A, such that

$$C_1 H(P) \leq H(AP) \leq C_2 H(P) \qquad \text{for all } P \in \mathbb{P}^N(\bar{\mathbb{Q}}).$$

PROOF. This is (VIII.5.6) for morphisms of degree one.   □

We next investigate the relationship between the coefficients of a polynomial and the height of its roots.

Notation. For $x \in \bar{\mathbb{Q}}$, let

$$H(x) = H([x, 1]),$$

and similarly for $x \in K$, let

$$H_K(x) = H_K([x, 1]).$$

Theorem 5.9. Let

$$f(T) = a_0 T^d + a_1 T^{d-1} + \cdots + a_d = a_0 (T - \alpha_1) \cdots (T - \alpha_d) \in \bar{\mathbb{Q}}[T]$$

be a polynomial of degree d. Then

$$2^{-d} \prod_{j=1}^{d} H(\alpha_j) \leq H([a_0, \ldots, a_d]) \leq 2^{d-1} \prod_{j=1}^{d} H(\alpha_j).$$

PROOF. First note that the inequality to be proven remains unchanged if $f(T)$ is multiplied by a nonzero constant. It thus suffices to prove the result for monic polynomials, so we may assume that $a_0 = 1$.

Let $K = \mathbb{Q}(\alpha_1, \ldots, \alpha_d)$, and for $v \in M_K$, set

$$\epsilon(v) = \begin{cases} 2 & \text{if } v \in M_K^\infty, \\ 1 & \text{if } v \in M_K^0. \end{cases}$$

Note that this notation differs from the notation used in the proof of (VIII.5.6). In the present instance, the triangle inequality reads

$$|x + y|_v \le \epsilon(v) \max\{|x|_v, |y|_v\} \qquad \text{for } v \in M_K \text{ and } x, y \in K.$$

Of course, if $v \in M_K^0$ and $|x|_v \ne |y|_v$, then the triangle inequality becomes an equality.

We are going to prove that

$$\epsilon(v)^{-d} \prod_{j=1}^d \max\{|\alpha_j|_v, 1\} \le \max_{0 \le i \le d}\{|a_i|_v\} \le \epsilon(v)^{d-1} \prod_{j=1}^d \max\{|\alpha_j|_v, 1\}.$$

Once we have done this, raising to the n_v power, multiplying over all $v \in M_K$, and taking the $[K : \mathbb{Q}]^{\text{th}}$ root gives the desired result.

The proof is by induction on $d = \deg(f)$. For $d = 1$ we have $f(T) = T - \alpha_1$, so the inequalities are clear. Assume now that we know the result for all polynomials (with roots in K) of degree $d - 1$. Choose an index k such that

$$|\alpha_k|_v \ge |\alpha_j|_v \qquad \text{for all } 0 \le j \le d,$$

and define a polynomial

$$\begin{aligned} g(T) &= (T - \alpha_1) \cdots (T - \alpha_{k-1})(T - \alpha_{k+1}) \cdots (T - \alpha_d) \\ &= b_0 T^{d-1} + b_1 T^{d-2} + \cdots + b_{d-1}. \end{aligned}$$

Thus $f(T) = (T - \alpha_k)g(T)$, so comparing coefficients yields

$$a_i = b_i - \alpha_k b_{i-1}.$$

(This holds for all $0 \le i \le d$ if we set $b_{-1} = b_d = 0$.)

We begin with the upper bound:

$$\begin{aligned} \max_{0 \le i \le d}\{|a_i|_v\} &= \max_{0 \le i \le d}\{|b_i - \alpha_k b_{i-1}|_v\} \\ &\le \epsilon(v) \max_{0 \le i \le d}\{|b_i|_v, |\alpha_k b_{i-1}|_v\} && \text{triangle inequality,} \\ &\le \epsilon(v) \max_{0 \le i \le d}\{|b_i|_v\} \max\{|\alpha_k|_v, 1\} \\ &\le \epsilon(v)^{d-1} \prod_{j=1}^d \max\{|\alpha_j|_v, 1\} && \begin{aligned}&\text{induction hypothesis}\\&\text{applied to } g.\end{aligned} \end{aligned}$$

Next, to prove the lower bound, we consider two cases. First, if $|\alpha_k|_v \le \epsilon(v)$, then by the choice of the index k we have

$$\prod_{j=1}^{d} \max\{|\alpha_j|_v, 1\} \le \max\{|\alpha_k|_v, 1\}^d \le \epsilon(v)^d,$$

so the result is clear. (Remember that $a_0 = 1$.) Next, suppose that $|\alpha_k|_v > \epsilon(v)$. Then

$$\max_{0 \le i \le d}\{|a_i|_v\} = \max_{0 \le i \le d}\{|b_i - \alpha_k b_{i-1}|_v\} \ge \epsilon(v)^{-1} \max_{0 \le i \le d-1}\{|b_i|_v\}\{|\alpha_k|_v, 1\}.$$

Here the last line is an equality for $v \in M_K^0$, while for $v \in M_K^\infty$ we are using the calculation

$$\max_{0 \le i \le d}\{|b_i - \alpha_k b_{i-1}|_v\} \ge \left(|\alpha_k|_v - 1\right) \max_{0 \le i \le d-1}\{|b_i|_v\}$$
$$> \epsilon(v)^{-1}|\alpha_k|_v \max_{0 \le i \le d-1}\{|b_i|_v\} \quad \text{since } |\alpha_k|_v > \epsilon(v) = 2.$$

Applying the induction hypothesis to g gives the desired lower bound, which completes the proof of (VIII.5.9). $\square$

Our first application of (VIII.5.9) is to show that there are only finitely many points of bounded height in projective space. To do this, we first need to show that the action of the Galois group does not affect the height of a point.

Theorem 5.10. Let $P \in \mathbb{P}^N(\bar{\mathbb{Q}})$ and let $\sigma \in G_{\bar{\mathbb{Q}}/\mathbb{Q}}$. Then

$$H(P^\sigma) = H(P).$$

PROOF. Let $K/\mathbb{Q}$ be a field such that $P \in \mathbb{P}^N(K)$. The field K may not be Galois over $\mathbb{Q}$, but in any case σ gives an isomorphism $\sigma : K \xrightarrow{\sim} K^\sigma$, and σ likewise identifies the sets of absolute values of K and K^σ,

$$\sigma : M_K \xrightarrow{\sim} M_{K^\sigma}, \qquad v \longmapsto v^\sigma.$$

Here, if $x \in K$ and $v \in M_K$, then the associated absolute value v^σ satisfies $|x^\sigma|_{v^\sigma} = |x|_v$. It is clear that σ also induces an isomorphism $K_v \xrightarrow{\sim} K_{v^\sigma}^\sigma$, so the local degrees satisfy $n_v = n_{v^\sigma}$. We now compute

$$H_{K^\sigma}(P^\sigma) = \prod_{w \in M_{K^\sigma}} \max\{|x_i^\sigma|_w\}^{n_w}$$
$$= \prod_{v \in M_K} \max\{|x_i^\sigma|_{v^\sigma}\}^{n_{v^\sigma}}$$
$$= \prod_{v \in M_K} \max\{|x_i|_v\}^{n_v}$$
$$= H_K(P).$$

Since $[K : \mathbb{Q}] = [K^\sigma : \mathbb{Q}]$, this is the desired result. $\square$

Theorem 5.11. *Let C and d be constants. Then the set*

$$\{P \in \mathbb{P}^N(\bar{\mathbb{Q}}) : H(P) \leq C \text{ and } [\mathbb{Q}(P) : \mathbb{Q}] \leq d\}$$

is a finite set of points, where we recall from (I §2) that $\mathbb{Q}(P)$ is the minimal field of definition of P. In particular, for any number field K,

$$\{P \in \mathbb{P}^N(K) : H_K(P) \leq C\}$$

is a finite set.

PROOF. Let $P \in \mathbb{P}^N(\bar{\mathbb{Q}})$. We choose homogeneous coordinates for P, say

$$P = [x_0, \ldots, x_N],$$

with some $x_j = 1$. Then $\mathbb{Q}(P) = \mathbb{Q}(x_0, \ldots, x_N)$, and we have the easy estimate

$$H_{\mathbb{Q}(P)}(P) = \prod_{v \in M_{\mathbb{Q}(P)}} \max_{0 \leq i \leq N} \{|x_i|_v\}^{n_v}$$

$$\geq \max_{0 \leq i \leq N} \left(\prod_{v \in M_{\mathbb{Q}(P)}} \max\{|x_i|_v, 1\}^{n_v} \right)$$

$$= \max_{0 \leq i \leq N} H_{\mathbb{Q}(P)}(x_i).$$

Thus if $H(P) \leq C$ and $[\mathbb{Q}(P) : \mathbb{Q}] \leq d$, then

$$\max_{0 \leq i \leq N} H_{\mathbb{Q}(P)}(x_i) \leq C \qquad \text{and} \qquad \max_{0 \leq i \leq N} [\mathbb{Q}(x_i) : \mathbb{Q}] \leq d.$$

It thus suffices to prove that the set

$$\{x \in \bar{\mathbb{Q}} : H(x) \leq C \text{ and } [\mathbb{Q}(x) : \mathbb{Q}] \leq d\}$$

is finite. In other words, we have reduced to the case that $N = 1$.

Suppose that $x \in \bar{\mathbb{Q}}$ is in this set, and let $e = [\mathbb{Q}(x) : \mathbb{Q}]$, so $e \leq d$. Further, let $x_1, \ldots, x_e \in \bar{\mathbb{Q}}$ be the conjugates of x, where we take $x_1 = x$. The minimal polynomial of x over $\mathbb{Q}$ is

$$f_x(T) = (T - x_1) \cdots (T - x_e) = T^e + a_1 T^{e-1} + \cdots + a_e \in \mathbb{Q}[T].$$

We estimate

$$H([1, a_1, \ldots, a_e]) \leq 2^{e-1} \prod_{j=1}^e H(x_j) \qquad \text{from (VIII.5.9)},$$

$$= 2^{e-1} H(x)^e \qquad \text{from (VIII.5.10)},$$

$$\leq (2C)^d \qquad \text{since } H(x) \leq C \text{ and } e \leq d.$$

Since the a_i are in $\mathbb{Q}$, it follows that for a given C and d, there are only finitely many possibilities for the polynomial $f_x(T)$. (We are using the easy-to-prove case of (VIII.5.11) with $K = \mathbb{Q}$; see (VIII.5.1) and (VIII.5.3).) Since each polynomial $f_x(T)$ has a most d roots in K, and thus contributes at most d elements to our set, this completes the proof that the set is finite. $\qquad \square$

Remark 5.12. Tracing through the proof of (VIII.5.11), it is easy to give an upper bound, in terms of C and D, for the number of points in the set

$$\{P \in \mathbb{P}^N(\bar{\mathbb{Q}}) : H(P) \le C \text{ and } [\mathbb{Q}(P) : \mathbb{Q}] \le d\}.$$

(See Exercise 8.7a.) A formula due to Schanuel gives a precise asymptotic formula for

$$\#\{P \in \mathbb{P}^N(K) : H_K(P) \le C\}$$

as a function of C as $C \to \infty$. See [139, Chapter 3, Section 5] or [220] for details.

VIII.6   Heights on Elliptic Curves

In this section we use the general theory of heights as developed in the previous section to define height functions on elliptic curves. The main theorems that we prove, (VIII.6.2) and (VIII.6.4), highlight the interplay between the height function and the addition law on the elliptic curve. As an immediate corollary, we deduce the remaining results needed to prove the Mordell–Weil theorem for arbitrary number fields (VIII.6.7).

It is convenient to use "big-O" notation.

Notation. Let f and g be real-valued functions on a set $\mathcal{S}$. We write

$$f = g + O(1)$$

if there are constants C_1 and C_2 such that

$$C_1 \le f(P) - g(P) \le C_2 \qquad \text{for all } P \in \mathcal{S}.$$

If only the lower inequality is satisfied, then we write $f \ge g + O(1)$, and similarly if only the upper inequality is true, then we write $f \le g + O(1)$.

Let E/K be an elliptic curve. Recall from (II.2.2) that any nonconstant function $f \in \bar{K}(E)$ determines a surjective morphism, which we also denote by f,

$$f : E \longrightarrow \mathbb{P}^1, \qquad P \longmapsto \begin{cases} [1, 0] & \text{if } P \text{ is a pole of } f, \\ [f(P), 1] & \text{otherwise.} \end{cases}$$

It would be reasonable to use f to define a height function on $E(\bar{K})$ by setting $H_f(P) = H(f(P))$. However, the height function H tends to behave multiplicatively, as for example in (VIII.5.6), while for our purposes it is more convenient to have a height function that behaves additively. This prompts the following definitions.

Definition. The (*absolute logarithmic*) *height* on projective space is the function

$$h : \mathbb{P}^N(\bar{\mathbb{Q}}) \longrightarrow \mathbb{R}, \qquad h(P) = \log H(P).$$

Note that (VIII.5.4b) tells us that $h(P) \ge 0$ for all P.

Definition. Let E/K be an elliptic curve, and let $f \in \bar{K}(E)$ be a function. The *height on E (relative to f)* is the function

$$h_f : E(\bar{K}) \longrightarrow \mathbb{R}, \qquad h_f(P) = h\big(f(P)\big).$$

We start by transcribing the finiteness result from (VIII §5) into the current setting.

Proposition 6.1. *Let E/K be an elliptic curve, and let $f \in K(E)$ be a nonconstant function. Then for any constant C, the set*

$$\{P \in E(K) : h_f(P) \le C\}$$

is a finite set of points.

PROOF. The function $f \in K(E)$ is defined over K, so it maps points $P \in E(K)$ to points $f(P) \in \mathbb{P}^1(K)$. Hence f gives a finite-to-one map from the set in question to the set

$$\{Q \in \mathbb{P}^1(K) : H(Q) \le e^C\}.$$

Finally, we know from (VIII.5.11) that this last set is finite. $\qquad\qquad\square$

The next theorem gives a fundamental relationship between height functions and the addition law on an elliptic curve.

Theorem 6.2. *Let E/K be an elliptic curve, and let $f \in K(E)$ be an even function, i.e., a function satisfying $f \circ [-1] = f$. Then for all $P, Q \in E(\bar{K})$ we have*

$$h_f(P+Q) + h_f(P-Q) = 2h_f(P) + 2h_f(Q) + O(1).$$

The constants inherent in the $O(1)$ depend on the elliptic curve E and the function f, but are independent of the points P and Q.

PROOF. Choose a Weierstrass equation for E/K of the form

$$E : y^2 = x^3 + Ax + B.$$

We start by proving the theorem for the particular function $f = x$. The general case is then an easy corollary.

Since $h_x(O) = 0$ and $h_x(-P) = h_x(P)$, the desired result is clear if $P = O$ or if $Q = O$. We now assume that $P \ne O$ and $Q \ne O$, and we write

$$x(P) = [x_1, 1], \qquad\qquad x(Q) = [x_2, 1],$$
$$x(P+Q) = [x_3, 1], \qquad\quad x(P-Q) = [x_4, 1].$$

Here x_3 or x_4 may equal ∞ if $P = \pm Q$. The addition formula (III.2.3d) and a little bit of algebra yield the relations

$$x_3 + x_4 = \frac{2(x_1 + x_2)(A + x_1 x_2) + 4B}{(x_1 + x_2)^2 - 4x_1 x_2},$$

$$x_3 x_4 = \frac{(x_1 x_2 - A)^2 - 4B(x_1 + x_2)}{(x_1 + x_2)^2 - 4x_1 x_2}.$$

Define a map $g : \mathbb{P}^2 \to \mathbb{P}^2$ by

$$g([t, u, v]) = [u^2 - 4tv, 2u(At + v) + 4Bt^2, (v - At)^2 - 4Btu].$$

Then the formulas for x_3 and x_4 show that there is a commutative diagram

$$
\sigma \left(
\begin{array}{ccc}
E \times E & \xrightarrow{\ G\ } & E \times E \\
\downarrow & & \downarrow \\
\mathbb{P}^1 \times \mathbb{P}^1 & & \mathbb{P}^1 \times \mathbb{P}^1 \\
\downarrow & & \downarrow \\
\mathbb{P}^2 & \xrightarrow{\ g\ } & \mathbb{P}^2
\end{array}
\right) \sigma
$$

where

$$G(P, Q) = (P + Q, P - Q),$$

and where the vertical map σ is the composition of the two maps

$$E \times E \longrightarrow \mathbb{P}^1 \times \mathbb{P}^1, \qquad (P, Q) \longmapsto \big(x(P), x(Q)\big),$$

and

$$\mathbb{P}^1 \times \mathbb{P}^1 \longrightarrow \mathbb{P}^2, \qquad \big([\alpha_1, \beta_1], [\alpha_2, \beta_2]\big) \longmapsto [\beta_1 \beta_2, \alpha_1 \beta_2 + \alpha_2 \beta_1, \alpha_1 \alpha_2].$$

The idea here is that we are viewing t, u, and v as representing 1, $x_1 + x_2$, and $x_1 x_2$, so $g([t, u, v])$ becomes $[1, x_3 + x_4, x_3 x_4]$.

The next step is to show that g is a morphism, which will allow us to apply (VIII.5.6). By definition (cf. (I.3.3)), we must show that the three homogeneous polynomials defining g have no common zeros other than $t = u = v = 0$. Suppose that $g([t, u, v]) = 0$. If $t = 0$, then from

$$u^2 - 4tv = 0 \qquad \text{and} \qquad (v - At)^2 - 4Btu = 0$$

we see that $u = v = 0$. Thus we may assume that $t \neq 0$, so we may define a new quantity $x = u/2t$. [*Intuition:* If we identify

$$t = 1, \qquad u = x_1 + x_2, \qquad v = x_1 x_2,$$

then the equation $u^2 - 4tv = 0$ becomes $(x_1 - x_2)^2 = 0$, so $x_1 = x_2 = u/2t$. In other words, we are now dealing with the case that $P = \pm Q$.]

Using the new quantity x, the equation $u^2 - 4tv = 0$ can be written as $x^2 = v/t$. Now dividing the equalities

$$2u(At + v) + 4Bt^2 = 0 \qquad \text{and} \qquad (v - At)^2 - 4Btu = 0$$

by t^2 and rewriting them in terms of x yields the two equations

$$\psi(x) = 4x(A + x^2) + 4B = 4x^3 + 4Ax + 4B = 0,$$
$$\phi(x) = (x^2 - A)^2 - 8Bx = x^4 - 2Ax^2 - 8Bx + A^2 = 0.$$

These polynomials should be familiar, since their ratio is the rational function that appears in the duplication formula (III.2.3d). In order to show that $\psi(X)$ and $\phi(X)$ have no common root, it suffices to verify the following formal identity that we already used in the proof of (VIII.4.3),

$$(12X^2 + 16A)\phi(X) - (3X^3 - 5AX - 27B)\psi(X) = 4(4A^3 + 27B^2) \neq 0.$$

Note how the nonsingularity of the Weierstrass equation plays a crucial role here. This completes the proof that g is a morphism.

We return to our commutative diagram and compute

$$
\begin{aligned}
h\big(\sigma(P + Q, P - Q)\big) &= h\big(\sigma \circ G(P, Q)\big) \\
&= h\big(g \circ \sigma(P, Q)\big) \\
&= 2h\big(\sigma(P, Q)\big) + O(1) \qquad \text{from (VIII.5.6),}
\end{aligned}
$$

since g is a morphism of degree 2. To complete the proof of (VIII.6.2) for $f = x$, we will show that

$$h\big(\sigma(R_1, R_2)\big) = h_x(R_1) + h_x(R_2) + O(1) \qquad \text{for all } R_1, R_2, \in E(\bar{K}).$$

Then, applying this relation to each side of the equation

$$h\big(\sigma(P + Q, P - Q)\big) = 2h\big(\sigma(P, Q)\big) + O(1)$$

gives the desired result.

It is clear that if either $R_1 = O$ or $R_2 = O$, then $h\big(\sigma(R_1, R_2)\big)$ is equal to $h_x(R_1) + h_x(R_2)$. Otherwise we write

$$x(R_1) = [\alpha_1, 1] \qquad \text{and} \qquad x(R_2) = [\alpha_2, 1],$$

and then

$$h\big(\sigma(R_1, R_2)\big) = h\big([1, \alpha_1 + \alpha_2, \alpha_1\alpha_2]\big) \quad \text{and} \quad h_x(R_1) + h_x(R_2) = h(\alpha_1) + h(\alpha_2).$$

We apply (VIII.5.9) to the polynomial $(T + \alpha_1)(T + \alpha_2)$ to obtain the desired estimate

$$h(\alpha_1) + h(\alpha_2) - \log 4 \leq h\big([1, \alpha_1 + \alpha_2, \alpha_1\alpha_2]\big) \leq h(\alpha_1) + h(\alpha_2) + \log 2.$$

Finally, in order to deal with an arbitrary even function $f \in K(E)$, we prove in the next lemma (VIII.6.2) that

$$h_f = \frac{1}{2}(\deg f)h_x + O(1).$$

Then (VIII.6.2) follows immediately on multiplying the proven relation for h_x by $\frac{1}{2} \deg f$. $\qquad \square$

Lemma 6.3. *Let $f, g \in K(E)$ be even functions. Then*

$$(\deg g)h_f = (\deg f)h_g + O(1).$$

PROOF. Let $x, y \in K(E)$ be Weierstrass coordinates for E/K. We know from (III.2.3.1) that the subfield of $K(E)$ consisting of even functions is exactly $K(x)$, so we can find a rational function $r(X) \in K(X)$ such that there is a commutative diagram

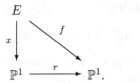

Hence, using (VIII.5.6) and the fact (II.2.1) that r is a morphism, we deduce that

$$h_f = h_x \circ r = (\deg r)h_x + O(1).$$

The diagram tells us that

$$\deg f = (\deg x)(\deg r) = 2 \deg r,$$

so we find that

$$2h_f = (\deg f)h_x + O(1).$$

The same reasoning applied to g yields

$$2h_g = (\deg g)h_x + O(1),$$

and combining these last two equalities gives the desired result. $\qquad\square$

Corollary 6.4. *Let E/K be an elliptic curve, and let $f \in K(E)$ be an even function.*
(a) *Let $Q \in E(\bar{K})$. Then*

$$h_f(P + Q) \le 2h_f(P) + O(1) \qquad \text{for all } P \in E(\bar{K}),$$

where the $O(1)$ depends on E, f, and Q.
(b) *Let $m \in \mathbb{Z}$. Then*

$$h_f([m]P) = m^2 h_f(P) + O(1) \qquad \text{for all } P \in E(\bar{K}),$$

where the $O(1)$ depends on E, f, and m.

PROOF. (a) This follows immediately from (VIII.6.2), since $h_f(P - Q) \ge 0$.
(b) Since f is even, it suffices to consider $m \ge 0$. Further, the result is trivial for $m = 0$ and $m = 1$. We use induction to complete the proof. Suppose that the desired result is known for $m - 1$ and for m. Replacing P and Q in (VIII.6.2) by $[m]P$ and P, respectively, we find that

$$h_f([m+1]P) = -h_f([m-1]P) + 2h_f([m]P) + 2h_f(P) + O(1)$$
$$= (-(m-1)^2 + 2m^2 + 2)h_f(P) + O(1) \qquad \text{by the induction}$$
$$\text{hypothesis,}$$
$$= (m+1)^2 h_f(P) + O(1).$$

This completes the induction proof. $\qquad\qquad\qquad\qquad\qquad\qquad\qquad\qquad$ □

Remark 6.5. It is clear that (VIII.6.2), (VIII.6.3), and (VIII.6.4) are also true for odd functions f, since then f^2 is even, and it is easy to check that $h_{f^2} = 2h_f$. More generally, although we do not give the proof, our results are true for arbitrary $f \in K(E)$ to "within ϵ." Precisely, say for (VIII.6.4b), for every $\epsilon > 0$ it is true that

$$(1 - \epsilon)m^2 h_f + O(1) \le h_f \circ [m] \le (1 + \epsilon)m^2 h_f + O(1),$$

where now the $O(1)$ depends on E, f, m, and ϵ. See Exercise 9.14c or, for a general result, see [139, Chapter 4, Corollary 3.5].

Remark 6.6. We can interprest (VIII.6.2) as saying that the height function h_f is more or less a quadratic form. We will see later (VIII §9) that there is an actual quadratic form, called the *canonical height*, that differs from h_f by a bounded amount.

We now have all of the tools needed to complete the proof of the Mordell–Weil theorem.

Theorem 6.7. (Mordell–Weil theorem) *Let K be a number field, and let E/K be an elliptic curve. Then the group $E(K)$ is finitely generated.*

PROOF. Choose any even nonconstant function $f \in K(E)$, for example, f could be the x-coordinate on a Weierstrass equation. The Mordell–Weil theorem follows immediately from the weak Mordell–Weil theorem (VIII.1.1) with $m = 2$ and the descent theorem (VIII.3.1) as soon as we show that the height function

$$h_f : E(K) \longrightarrow \mathbb{R}$$

has the following three properties:

(i) Let $Q \in E(K)$. There is a constant C_1, depending on E, f, and Q, such that

$$h_f(P + Q) \le 2h_f(P) + C_1 \qquad \text{for all } P \in E(K).$$

(ii) There is a constant C_2, depending on E and f, such that

$$h_f([2]P) \ge 4h_f(P) - C_2 \qquad \text{for all } P \in E(K).$$

(iii) For every constant C_3, the set

$$\{P \in E(K) : h_f(P) \le C_3\}$$

is a finite set of points.

Here (i) is a restatement of (VIII.6.4a), while (ii) is immediate from the $m = 2$ case of (VIII.6.4b), and (iii) is (VIII.6.1). This completes the proof of the Mordell–Weil theorem. $\qquad\qquad$ □

VIII.7   Torsion Points

The Mordell–Weil theorem implies that the group of rational torsion points on an elliptic curve is finite. Of course, this also follows from the corresponding result for local fields. Since we may view an elliptic curve defined over a number field K as being defined over the completion K_v for each $v \in M_K$, the local integrality conditions for torsion points (VII.3.4) can be pieced together to give the following global statement.

Theorem 7.1. *Let E/K be an elliptic curve with Weierstrass equation*

$$y^2 + a_1 xy + a_3 y = x^3 + a_2 x^2 + a_4 x + a_6,$$

and assume that $a_1, \ldots, a_6$ are all in the ring of integers R of K. Let $P \in E(K)$ be a torsion point of exact order $m \geq 2$.
(a) *If m is not a prime power, then*

$$x(P), y(P) \in R.$$

(b) *If $m = p^n$ is a prime power, then for each $v \in M_K^0$ we let*

$$r_v = \left[\frac{\operatorname{ord}_v(p)}{p^n - p^{n-1}} \right],$$

where $[\,\cdot\,]$ denotes the greatest integer. Then

$$\operatorname{ord}_v\big(x(P)\big) \geq -2r_v \qquad \text{and} \qquad \operatorname{ord}_v\big(y(P)\big) \geq -3r_v.$$

In particular, if $\operatorname{ord}_v(p) = 0$, then $x(P)$ and $y(P)$ are v-integral.

The next corollary was proven independently by Lutz and Nagell, who had discovered divisibility conditions somewhat weaker than those given in (VIII.7.1).

Corollary 7.2. ([152], [190]) *Let $E/\mathbb{Q}$ be an elliptic curve with Weierstrass equation*

$$y^2 = x^3 + Ax + B, \qquad A, B \in \mathbb{Z}.$$

Suppose that $P \in E(\mathbb{Q})$ is a nonzero torsion point.
(a) $x(P), y(P) \in \mathbb{Z}$.
(b) *Either $[2]P = O$ or else $y(P)^2$ divides $4A^3 + 27B^2$.*

PROOF. (a) Let P have exact order m. If $m = 2$, then $y(P) = 0$, so $x(P) \in \mathbb{Z}$, since it is the root of a monic polynomial with integer coefficients. If $m > 2$, the desired result follows immediately from (VIII.7.1), since the quantity r_v in (VIII.7.1b) is necessarily 0.
(b) We assume that $[2]P \neq O$, so $y(P) \neq 0$. Then applying (a) to both P and $[2]P$, we deduce that $x(P), y(P), x([2]P) \in \mathbb{Z}$. Let

$$\phi(X) = X^4 - 2AX^2 - 8BX + A^2 \qquad \text{and} \qquad \psi(X) = X^3 + AX + B.$$

Then the duplication formula (III.2.3d) reads

$$x([2]P) = \frac{\phi(x(P))}{4\psi(x(P))}.$$

On the other hand, we have the usual polynomial identity (VIII.4.3)

$$f(X)\phi(X) - g(X)\psi(X) = 4A^3 + 27B^2,$$

where $f(X) = 3X^2 + 4A$ and $g(X) = 3X^3 - 5AX - 27B$. Setting $X = x(P)$ and using the duplication formula and the fact that $y(P)^2 = \psi(x(P))$ yields

$$y(P)^2\Big(4f(x(P))x([2]P) - g(x(P))\Big) = 4A^3 + 27B^2.$$

Since all of the quantities in this equation are integers, the desired result follows. □

Remark 7.3.1. A glance at the proof of (VIII.7.2b) shows that we have proved that any point $P \in E(\mathbb{Q})$ such that $x(P)$ and $x([2]P)$ are both integers has the property that $y(P)^2$ divides $4A^3 + 27B^2$. The same argument works for number fields. Further, even if $x(P)$ or $x([2]P)$ is not integral, any bound on their denominators, for example as in (VIII.7.1b), gives a corresponding bound for $y(P)$; see Exercise 8.11.

Remark 7.3.2. Recall from (VII.3.2) that in practice, one of the fastest methods to bound the torsion in $E(K)$ is to choose various finite places v for which E has good reduction and use the injection (VII.3.1)

$$E(K_v)[m] \longrightarrow \tilde{E}(k_v),$$

which is valid for integers m that are prime to $\operatorname{char}(k_v)$.

Example 7.4. The Weierstrass equation

$$E : y^2 = x^3 - 43x + 166$$

has

$$4A^3 + 27B^2 = 425984 = 2^{15} \cdot 13.$$

Hence any torsion point in $E(\mathbb{Q})$ has its y-coordinate in the set

$$\{0, \pm 1, \pm 2, \pm 4, \pm 8, \pm 16, \pm 32, \pm 64, \pm 128\}.$$

A little bit of work with a calculator reveals the points

$$\{(3, \pm 8), (-5, \pm 16), (11, \pm 32)\}.$$

On the other hand, since E has good reduction modulo 3, we know that $E_{\text{tors}}(\mathbb{Q})$ injects into $\tilde{E}(\mathbb{F}_3)$ (cf. VII.3.5), and it is easy to check that $\#\tilde{E}(\mathbb{F}_3) = 7$. This still does not prove anything, since the divisibility condition in (VIII.7.2b) is only necessary, not sufficient. However, using the doubling formula for $P = (3, 8)$ yields

$$x(P) = 3, \qquad x([2]P) = -5, \qquad x([4]P) = 11, \qquad x([8]P) = 3.$$

Hence $[8]P = \pm P$, so P is a torsion point of exact order 7 or 9. (Note that it doesn't have order 3, since $x(P) \neq x([2]P)$.) From above, the only possibility is order 7, so we conclude that $E_{\text{tors}}(\mathbb{Q})$ is a cyclic group of order 7 consisting of the six listed points, together with O.

Our discussion thus far has focused on characterizing the torsion subgroup of a given elliptic curve. Another type of question that one might ask is the following: given a prime p, does there exist an elliptic curve $E/\mathbb{Q}$ such that $E(\mathbb{Q})$ contains a point of order p? The answer for most primes is no. For example, $E(\mathbb{Q})$ can never contain a point of order 11, a fact that is by no means obvious. Such a statement, which deals uniformly with the set of all elliptic curves, naturally tends to be more difficult to prove than does a result such as (VIII.7.2) in which the bound changes as the elliptic curve is varied. The definitive characterization of torsion subgroups over $\mathbb{Q}$ is given by the following theorem due to Mazur; the proof is unfortunately far beyond the scope of this book.

Theorem 7.5. (Mazur [165], [166]) *Let $E/\mathbb{Q}$ be an elliptic curve. Then the torsion subgroup $E_{\text{tors}}(\mathbb{Q})$ of $E(\mathbb{Q})$ is isomorphic to one of the following fifteen groups:*

$$\mathbb{Z}/N\mathbb{Z} \qquad \text{with } 1 \leq N \leq 10 \text{ or } N = 12,$$
$$\mathbb{Z}/2\mathbb{Z} \times \mathbb{Z}/2N\mathbb{Z} \quad \text{with } 1 \leq N \leq 4.$$

Further, each of these groups occurs as $E_{\text{tors}}(\mathbb{Q})$ for some elliptic curve $E/\mathbb{Q}$. (See Exercise 8.12 for an example of each possible group.)

Mazur's theorem was generalized to number fields of degree up to 14 by Kamienny and others [2, 121, 122], and then the general case was settled by Merel.

Theorem 7.5.1. (Merel [170]) *For every integer $d \geq 1$ there is a constant $N(d)$ such that for all number fields $K/\mathbb{Q}$ of degree at most d and all elliptic curves E/K,*

$$\left| E_{\text{tors}}(K) \right| \leq N(d).$$

Remark 7.6. Prior to the proof of Merel's theorem (VIII.7.5.1), Manin [155] used a completely different method to show that for any fixed prime p, the p-primary component of $E_{\text{tors}}(K)$ may be bounded in terms of K and p.

Remark 7.8. For those torsion subgroups that are allowed by Mazur's theorem (VIII.7.5), it is a classical result that the elliptic curves having the specified torsion subgroup lie in a one-parameter family. For example, the curves E/K with a point $P \in E(K)$ of order 7 all have Weierstrass equations of the form

$$y^2 + (1 + d - d^2)xy + (d^2 - d^3)y = x^3 + (d^2 - d^3)x^2, \qquad P = (0,0),$$

with

$$d \in K \quad \text{and} \quad \Delta = d^7(d-1)^7(d^3 - 8d^2 + 5d + 1) \neq 0.$$

See Exercise 8.13a,b for a derivation and [132] for a complete list of such formulas. In general, the elliptic curves E/K with a point $P \in E(K)$ of order $m \geq 4$ are parametrized by the K-rational points of another curve, called a *modular curve*; see Exercise 8.13c and (C §13).

VIII.8 The Minimal Discriminant

Let E/K be an elliptic curve. For each nonarchimedean absolute value $v \in M_K^0$ we choose a Weierstrass equation for E,

$$y_v^2 + a_{1,v} x_v y_v + a_{3,v} y_v = x_v^3 + a_{2,v} x_v^2 + a_{4,v} x_v + a_{6,v},$$

that is a minimal equation for E at v. In other words, all of the $a_{i,v}$ satisfy

$$\mathrm{ord}_v(a_{i,v}) \geq 0,$$

and subject to this condition, the discriminant Δ_v of the equation has valuation $\mathrm{ord}_v(\Delta_v)$ that is as small as possible.

Definition. The *minimal discriminant of E/K*, denoted by $\mathcal{D}_{E/K}$, is the (integral) ideal of K given by

$$\mathcal{D}_{E/K} = \prod_{v \in M_K^0} \mathfrak{p}_v^{\mathrm{ord}_v(\Delta_v)}.$$

Here $\mathfrak{p}_v$ is the prime ideal of R associated to v. Thus $\mathcal{D}_{E/K}$ catalogs the valuation of the minimal discriminant of E at every place $v \in M_K^0$. It measures, in some sense, the arithmetic complexity of the elliptic curve E.

We now ask whether it is possible to find a single Weierstrass equation that is simultaneously minimal for every $v \in M_K^0$. Let

$$y^2 + a_1 xy + a_3 y = x^3 + a_2 x^2 + a_4 x + a_6$$

be any Weierstrass equation for E/K, say with discriminant Δ. For each $v \in M_K^0$ we can find a change of coordinates

$$x = u_v^2 x_v + r_v, \qquad y = u_v^3 y_v + s_v u_v^2 x_v + t_v,$$

that transforms the initial equation into an equation that is minimal at v. As usual, the discriminants of the two equations are related by

$$\Delta = u_v^{12} \Delta_v.$$

Hence if we define an ideal

$$\mathfrak{a}_\Delta = \prod_{v \in M_K^0} \mathfrak{p}_v^{-\mathrm{ord}_v(u_v)},$$

then the minimal discriminant is related to Δ via the formula

$$\mathcal{D}_{E/K} = (\Delta) \mathfrak{a}_\Delta^{12}.$$

Lemma 8.1. *With notation as above, the ideal class in K of the ideal $\mathfrak{a}_\Delta$ is independent of Δ.*

PROOF. Suppose that we take a different Weierstrass equation for E over K, say with discriminant Δ'. Then $\Delta = u^{12}\Delta'$ for some $u \in K^*$, so directly from the definitions we see that

$$(\Delta')\mathfrak{a}_{\Delta'}^{12} = \mathcal{D}_{E/K} = (\Delta)\mathfrak{a}_{\Delta}^{12} = (\Delta')\big((u)\mathfrak{a}_{\Delta}\big)^{12}.$$

Hence $\mathfrak{a}_{\Delta'} = (u)\mathfrak{a}_{\Delta}$, so $\mathfrak{a}_{\Delta'}$ and $\mathfrak{a}_{\Delta}$ are in the same ideal class. $\qquad\square$

Definition. The *Weierstrass class of E/K*, denoted by $\bar{\mathfrak{a}}_{E/K}$, is the ideal class in K corresponding to any ideal $\mathfrak{a}_{\Delta}$ as above.

Definition. A *global minimal Weierstrass equation for E/K* is a Weierstrass equation

$$y^2 + a_1 xy + a_3 y = x^3 + a_2 x^2 + a_4 x + a_6$$

for E/K such that $a_1, a_2, a_3, a_4, a_6 \in R$ and such that the discriminant Δ of the equation satisfies $\mathcal{D}_{E/K} = (\Delta)$.

Proposition 8.2. *There exists a global minimal Weierstrass equation for E/K if and only if $\bar{\mathfrak{a}}_{E/K} = (1)$.*

PROOF. Suppose that E/K has a global minimal Weierstrass equation, say with discriminant Δ. Then $\mathcal{D}_{E/K} = (\Delta)$, so with notation as above, for any $v \in M_K^0$ we have

$$12 \operatorname{ord}_v(\mathfrak{a}_{\Delta}) = \operatorname{ord}_v(\mathcal{D}_{E/K}) - \operatorname{ord}_v(\Delta) = 0.$$

Hence $\mathfrak{a}_{\Delta} = (1)$, so $\bar{\mathfrak{a}}_{E/K} = (\text{class of } \mathfrak{a}_{\Delta}) = (1)$.

Conversely, suppose that $\bar{\mathfrak{a}}_{E/K} = (1)$. Choose any Weierstrass equation for E/K having $a_1, \ldots, a_6 \in R$, and let Δ be the discriminant of this chosen equation. For each $v \in M_K^0$, let

$$x = u_v^2 x_v + r_v, \qquad y = u_v^3 y_v + s_v u_v^2 x_v + t_v,$$

be a change of variables that produces a minimal equation at v, say with coefficients $a_{1,v}, \ldots, a_{6,v}$ and discriminant Δ_v. Letting

$$S = \big\{ v \in M_K^0 : \operatorname{ord}_v(\Delta) \neq 0 \big\},$$

the chosen equation is already minimal for all $v \notin S$, so we may take $u_v = 1$ and $r_v = s_v = t_v = 0$ for $v \notin S$. Note that S is a finite set. Further, from (VII.1.3d), we see that u_v, r_v, s_v, t_v are v-integral for all $v \in M_K^0$.

The assumption that $\bar{\mathfrak{a}}_{E/K} = (1)$ means that the ideal

$$\prod_{v \in M_K^0} \mathfrak{p}_v^{\operatorname{ord}_v(u_v)}$$

is principal, say generated by $u \in K^*$. This means that

$$\operatorname{ord}_v(u) = \operatorname{ord}_v(u_v) \qquad \text{for all } v \in M_K^0.$$

We use the Chinese remainder theorem [142, Chapter I, Section 4] to find elements $r, s, t \in R$ such that for all $v \in S$ we have

$$\operatorname{ord}_v(r - r_v), \operatorname{ord}_v(s - s_v), \operatorname{ord}_v(t - t_v) > \max_{i=1,2,3,4,6} \operatorname{ord}_v(u_v^i a_{i,v}).$$

Now consider the new Weierstrass equation for E/K given by the change of coordinates

$$x = u^2 x' + r, \qquad y = u^3 y' + s u^2 x' + t,$$

having coefficients $a_1', \ldots, a_6'$ and discriminant Δ'. Then $\Delta = u^{12} \Delta'$, so

$$\operatorname{ord}_v(\Delta') = \operatorname{ord}_v(u^{-12}\Delta) = \operatorname{ord}_v\big((u_v/u)^{12}\Delta_v\big) = \operatorname{ord}_v(\Delta_v).$$

Thus the discriminant of the new equation is minimal at all $v \in M_K^0$, so in order to verify that it is a global minimal equation, we must show that all of its coefficients are integral. This is easily checked using the coefficient transformation formulas (III.1.2). If $v \notin S$, then $\operatorname{ord}_v(u) = 0$, so each a_i' is v-integral since it is a polynomial in $r, s, t, a_1, \ldots, a_6$. For $v \in S$ we illustrate the argument for a_2', the other coefficients being done similarly. Thus

$$\begin{aligned}
\operatorname{ord}_v(u^2 a_2') &= \operatorname{ord}_v(a_2 - s a_1 + 3r - s^2) \\
&= \operatorname{ord}_v\big(u_v^2 a_{2,v} - (s - s_v)(a_1 + s + s_v) + 3(r - r_v)\big) \\
&= \operatorname{ord}_v(u_v^2 a_{2,v}),
\end{aligned}$$

where the last line follows from the previous one by our choice of r and s and the nonarchimedean nature of v. Since

$$\operatorname{ord}_v(u) = \operatorname{ord}_v(u_v) \qquad \text{and} \qquad \operatorname{ord}_v(a_{2,v}) \geq 0,$$

this gives the desired result. $\qquad\qquad\qquad\qquad\qquad\qquad\qquad\qquad\qquad\qquad$ $\square$

Corollary 8.3. *If K has class number one, then every elliptic curve E/K has a global minimal Weierstrass equation. In particular, this is true for $K = \mathbb{Q}$.*

The converse to (VIII.8.3) is also true; see Exercise 8.14.

Example 8.4. The Weierstrass equation

$$E : y^2 = x^3 + 16$$

has discriminant $\Delta = -2^{12}3^3$ and it is not minimal at 2. The substitution

$$x = 4x', \qquad y = 8y' + 4,$$

gives the global minimal equation

$$(y')^2 + y' = (x')^3.$$

Example 8.5. Let $K = \mathbb{Q}(\sqrt{-10}\,)$, so K has class number 2, the class group being generated by the prime ideal $\mathfrak{p} = (5, \sqrt{-10}\,)$. Let E/K be the elliptic curve given by the equation

$$E : y^2 = x^3 + 125.$$

This equation has discriminant $\Delta = -2^4 3^3 5^6$, so (VII.1.1) tells us that it is already minimal at every prime of K except possibly at the prime $\mathfrak{p}$ lying over (5). For $\mathfrak{p}$, the change of coordinates

$$x = (\sqrt{-10}\,)^2 x', \qquad y = (\sqrt{-10}\,)^3 y'$$

gives an equation

$$(y')^2 = (x')^3 - \frac{1}{8}$$

that has good reduction at $\mathfrak{p}$. Hence

$$\mathcal{D}_{E/K} = (2^4 3^3) \qquad \text{and} \qquad \bar{\mathfrak{a}}_{E/K} = (\text{ideal class of } \mathfrak{p}).$$

Since $\bar{\mathfrak{a}}_{E/K}$ is not principal, (VIII.8.2) tells us that E/K does not have a global minimal Weierstrass equation.

Remark 8.6. If K has class number one and E/K is an elliptic curve, then we can construct a global minimal Weierstrass equation for E/K by finding local minimal equations, e.g., by using Tate's algorithm [266, IV §9], [283], and then following the proof of (VIII.8.2). There is also an algorithm, due to Laska [146], that is fast and easy to implement on a computer.

Even if R has class number greater than one, it is often useful to know that an elliptic curve E/K has a global Weierstrass equation that is, in some sense, "almost minimal." The following proposition gives one possibility; see Exercise 8.14c for another.

Proposition 8.7. Let $S \subset M_K$ be a finite set of absolute values containing M_K^∞ and all finite places dividing 2 and 3. Assume further that the ring of S-integers R_S is a principal ideal domain. Then every elliptic curve E/K has a Weierstrass equation of the form

$$E : y^2 = x^3 + Ax + B$$

with $A, B \in R_S$ and discriminant $\Delta = -16(4A^3 + 27B^2)$ satisfying

$$\mathcal{D}_{E/K} R_S = \Delta R_S.$$

(Such a Weierstrass equation might be called S-minimal.)

PROOF. Choose any Weierstrass equation for E/K of the form

$$E : y^2 = x^3 + Ax + B,$$

and let $\Delta = -16(4A^3 + 27B^2)$. For each $v \in M_K$ with $v \notin S$, choose $u_v \in K^*$ such that the substitution

$$x = u_v^2 x', \qquad y = u_v^3 y',$$

gives a minimal equation at v. Then

$$v(\mathcal{D}_{E/K}) = v(\Delta) - 12v(u_v) \qquad \text{for all } v \in M_K \text{ with } v \notin S.$$

Since R_S is a principal ideal domain, we can find an element $u \in K^*$ satisfying

$$v(u) = v(u_v) \qquad \text{for all } v \in M_K \text{ with } v \notin S.$$

Then the equation

$$E : y^2 = x^3 + u^{-4} A x + u^{-6} B$$

has the desired property. $\qquad\qquad\qquad\qquad\qquad\qquad\qquad\qquad\qquad$ $\square$

VIII.9 The Canonical Height

Let E/K be an elliptic curve, and let $f \in K(E)$ be an even function. We saw in (VIII.6.1) and (VIII.6.4) that the height function h_f is more or less a quadratic form, at least "up to $O(1)$." André Néron asked whether one could find an actual quadratic form that differs from h_f by a bounded amount. He constructed such a function by writing it as a sum of "quasi-quadratic" local functions [194]. At the same time, John Tate gave a simpler global definition. In this section we describe Tate's construction. (For a discussion of local height functions, see (C §18) or [266, Chapter VI].)

Proposition 9.1. (Tate) *Let E/K be an elliptic curve, let $f \in K(E)$ be a nonconstant even function, and let $P \in E(\bar{K})$. Then the limit*

$$\frac{1}{\deg(f)} \lim_{N \to \infty} 4^{-N} h_f([2^N]P)$$

exists and is independent of f.

PROOF. We prove that the limit exists by showing that the sequence is Cauchy. Applying (VIII.6.4b) with $m = 2$, there is a constant C such that for all $Q \in E(\bar{K})$,

$$\left| h_f([2]Q) - 4h_f(Q) \right| \leq C.$$

For integers $N \geq M \geq 0$ we use a telescoping sum argument to estimate

$$\left| 4^{-N} h_f([2^N]P) - 4^{-M} h_f([2^M]P) \right|$$

$$= \left| \sum_{n=M}^{N-1} 4^{-n-1} h_f([2^{n+1}]P) - 4^{-n} h_f([2^n]P) \right|$$

$$\leq \sum_{n=M}^{N-1} 4^{-n-1} \left| h_f([2^{n+1}]P) - 4 h_f([2^n]P) \right|$$

$$\leq \sum_{n=M}^{N-1} 4^{-n-1} C \qquad \text{taking } Q = [2^n]P \text{ above,}$$

$$\leq 4^{-M} C.$$

This shows that the sequence $4^{-N} h_f([2^N]P)$ is Cauchy, hence it converges.

Next let $g \in K(E)$ be another nonconstant even function. Then from (VIII.6.3) we have

$$(\deg g) h_f = (\deg f) h_g + O(1),$$

so

$$\frac{4^{-N} h_f([2^N]P)}{\deg(f)} - \frac{4^{-N} h_g([2^N]P)}{\deg(g)} = O(4^{-N}) \xrightarrow[N\to\infty]{} 0.$$

Hence the limit does not depend on the choice of the function f.   □

Definition. The *canonical* (or *Néron–Tate*) *height on* E/K, denoted by $\hat{h}$ or $\hat{h}_E$, is the function

$$\hat{h} : E(\bar{K}) \longrightarrow \mathbb{R}$$

defined by

$$\hat{h}(P) = \frac{1}{\deg(f)} \lim_{N\to\infty} 4^{-N} h_f([2^N]P),$$

where $f \in K(E)$ is any nonconstant even function.

Remark 9.2. From (VIII.9.1), the canonical height is well-defined and independent of the choice of f. We remark that some authors use a canonical height that is equal to $2\hat{h}$. This is more natural in some contexts, for example it eliminates a power of 2 in the statement of the conjecture of Birch and Swinnerton-Dyer (C.16.5).

Theorem 9.3. (Néron, Tate) *Let* E/K *be an elliptic curve, and let* $\hat{h}$ *be the canonical height on* E.
(a) *For all* $P, Q \in E(\bar{K})$ *we have*

$$\hat{h}(P+Q) + \hat{h}(P-Q) = 2\hat{h}(P) + 2\hat{h}(Q) \qquad \text{(parallelogram law)}.$$

(b) *For all* $P \in E(\bar{K})$ *and all* $m \in \mathbb{Z}$,

$$\hat{h}([m]P) = m^2 \hat{h}(P).$$

(c) *The canonical height $\hat{h}$ is a quadratic form on E, i.e., $\hat{h}$ is an even function, and the pairing*

$$\langle \cdot, \cdot \rangle : E(\bar{K}) \times E(\bar{K}) \longrightarrow \mathbb{R},$$
$$\langle P, Q \rangle = \hat{h}(P + Q) - \hat{h}(P) - \hat{h}(Q),$$

is bilinear.

(d) *Let $P \in E(\bar{K})$. Then $\hat{h}(P) \geq 0$, and*

$$\hat{h}(P) = 0 \quad \text{if and only if} \quad P \text{ is a torsion point.}$$

(See also Exercise 8.6.)

(e) *Let $f \in K(E)$ be an even function. Then*

$$(\deg f)\hat{h} = h_f + O(1),$$

where the $O(1)$ depends on E and f.

Further, if $\hat{h}' : E(\bar{K}) \to \mathbb{R}$ is any other function satisfying (e) for some nonconstant even function f and satisfying (b) for some integer $m \geq 2$, then $\hat{h}' = \hat{h}$.

PROOF. We start with (e) and then return to (a)–(d).

(e) In the course of proving (VIII.9.1) we found a constant C, depending on f, such that for all integers $N \geq M \geq 0$ and all points $P \in E(\bar{K})$,

$$\left| 4^{-N} h_f([2^N]P) - 4^{-M} h_f([2^M]P) \right| \leq 4^{-M} C.$$

Taking $M = 0$ and letting $N \to \infty$ gives the desired estimate

$$\left| (\deg f)\hat{h}(P) - h_f(P) \right| \leq C.$$

(a) From (VIII.6.2) we have

$$h_f(P + Q) + h_f(P - Q) = 2h_f(P) + 2h_f(Q) + O(1).$$

We replace P and Q by $[2^N]P$ and $[2^N Q]$, respectively, divide by $(\deg f)4^N$, and let $N \to \infty$. The $O(1)$ term disappears and we obtain

$$\hat{h}(P + Q) + \hat{h}(P - Q) = 2\hat{h}(P) + 2\hat{h}(Q).$$

(b) From (VIII.6.4b) we have

$$h_f([m]P) = m^2 h_f(P) + O(1).$$

As usual, we replace P by $[2^N]P$, divide by 4^N, and let $N \to \infty$. (Alternative proof: Use (a) and induction on m.)

(c) It is a standard fact from linear algebra that a function satisfying the parallelogram law is quadratic. For completeness, we include a proof.

Putting $P = O$ in the parallelogram law (a) shows that $\hat{h}(-Q) = \hat{h}(Q)$, so $\hat{h}$ is even. By symmetry, it suffices to prove that

$$\langle P + R, Q \rangle = \langle P, Q \rangle + \langle R, Q \rangle,$$

which in terms of $\hat{h}$ is

$$\hat{h}(P+Q+R) - \hat{h}(P+R) - \hat{h}(P+Q) - \hat{h}(R+Q) + \hat{h}(P) + \hat{h}(Q) + \hat{h}(R) = 0.$$

Four applications of the parallelogram law and the evenness of $\hat{h}$ yield

$$\hat{h}(P+R+Q) + \hat{h}(P+R-Q) - 2\hat{h}(P+R) - 2\hat{h}(Q) = 0,$$
$$\hat{h}(P-R+Q) + \hat{h}(P+R-Q) - 2\hat{h}(P) - 2\hat{h}(R-Q) = 0,$$
$$\hat{h}(P-R+Q) + \hat{h}(P+R+Q) - 2\hat{h}(P+Q) - 2\hat{h}(R) = 0,$$
$$2\hat{h}(R+Q) + 2\hat{h}(R-Q) - 4\hat{h}(R) - 4\hat{h}(Q) = 0.$$

The alternating sum of these four equations is the desired result.

(d) The first conclusion is clear, since $h_f(P) \geq 0$ for all functions f and all points P, so $\hat{h}(P)$ is a limit of nonnegative values. For the second, we observe that one implication is immediate, since if P is a torsion point, then $[2^N]P$ takes on only finitely many values as N varies, so $4^{-N} h_f([2^N]P) \to 0$ as $N \to \infty$.

Conversely, let $P \in E(K')$ for some finite extension K'/K, and suppose that $\hat{h}(P) = 0$. Then

$$\hat{h}([m]P) = m^2 \hat{h}(P) = 0 \qquad \text{for every integer } m,$$

so from (e) there is a constant C such that for all $m \in \mathbb{Z}$,

$$h_f([m]P) = \left| (\deg f)\hat{h}([m]P) - h_f([m]P) \right| \leq C.$$

Thus the set $\{P, [2]P, [3]P, \dots\}$ is contained in

$$\{Q \in E(K') : h_f(Q) \leq C\}.$$

Now (VIII.6.1) tells us that this set of bounded height is a finite set, so P must have finite order.

This completes the proof of (a)–(e). Finally, to prove uniqueness, suppose that there are an integer $m \geq 2$ and a nonconstant even function f such that $\hat{h}'$ satisfies

$$\hat{h}' \circ [m] = m^2 \hat{h}' \qquad \text{and} \qquad (\deg f)\hat{h}' = h_f + O(1).$$

Repeated application of the first equality yields

$$\hat{h}' \circ [m^N] = m^{2N} \hat{h}' \quad \text{for } N = 1, 2, 3, \dots.$$

Further, since $\hat{h}$ satisfies (e), we have

$$\hat{h}' - \hat{h} = O(1).$$

Hence for any point $P \in E(\bar{K})$ we have

$$\begin{aligned} \hat{h}'(P) &= m^{-2N} \hat{h}'([m^N]P) \\ &= m^{-2N} \left(\hat{h}([m^N]P) + O(1) \right) \\ &= \hat{h}(P) + O(m^{-2N}) \qquad \text{since } \hat{h} \text{ satisfies (b).} \end{aligned}$$

Letting $N \to \infty$ yields $\hat{h}'(P) = \hat{h}(P)$. $\qquad\qquad\qquad\qquad\qquad$ $\square$

Remark 9.4. The Mordell–Weil theorem implies that $E(K) \otimes \mathbb{R}$ is a finite-dimensional real vector space, and (VIII.9.3cd) implies that $\hat{h}$ is a positive definite quadratic form on the quotient space $E(K)/E_{\text{tors}}(K)$, where $E_{\text{tors}}(K)$ denotes the torsion subgroup of $E(K)$. The quotient $E(K)/E_{\text{tors}}(K)$ sits as a lattice in the vector space $E(K) \otimes \mathbb{R}$, so it would appear to be clear that the extension of $\hat{h}$ to $E(K) \otimes \mathbb{R}$ is also positive definite. This is true, but as was pointed out by Cassels, one must use more than just (VIII.9.3cd).

Lemma 9.5. *Let V be a finite-dimensional real vector space and let $L \subset V$ be a lattice, i.e., L is a discrete subgroup of V containing a basis for V. Let $q : V \to \mathbb{R}$ be a quadratic form, and suppose that q has the following properties:*
(i) *For $P \in L$, we have $q(P) = 0$ if and only if $P = 0$.*
(ii) *For every constant C, the set*

$$\{P \in L : q(P) \le C\}$$

is finite.
Then q is positive definite on V.

PROOF. Choose a basis for V such that for a vector $\mathbf{x} = (x_1, \ldots, x_r) \in V$, the quadratic form q has the form

$$q(\mathbf{x}) = \sum_{i=1}^{s} x_i^2 - \sum_{i=1}^{t} x_{s+i}^2,$$

where $s + t \le r = \dim(V)$. For the existence of such a basis, see for example [143, Chapter XIV, §§3,7] or [296, §12.7]. Using this basis to identify $V \cong \mathbb{R}^n$ as $\mathbb{R}$-vector spaces, we let μ be the measure on V corresponding to the usual measure on $\mathbb{R}^n$. We apply the following basic result due to Minkowski:

> Let $B \subset V$ be a convex set that is symmetric about the origin. If $\mu(B)$ is sufficiently large, then B contains a nonzero lattice point.

For a proof of Minkowski's result, see for example [108, Theorem 447] or [142, Chapter 5, Section 3]. Now consider the set

$$B(\epsilon, \delta) = \left\{ \mathbf{x} = (x_1, \ldots, x_r) \in V : \sum_{i=1}^{s} x_i^2 \le \epsilon \quad \text{and} \quad \sum_{i=1}^{t} x_{s+i}^2 \le \delta \right\}.$$

The set $B(\epsilon, \delta)$ is convex and symmetric about the origin. Further, let

$$\lambda = \inf\{q(P) : P \in L, \, P \neq 0\}.$$

From (i) and (ii) we have $\lambda > 0$.

Now suppose that q is not positive definite on V, so $s < r$. Then Minkowski's theorem tells us that if δ is sufficiently large, then $B\left(\frac{1}{2}\lambda, \delta\right)$ contains a nonzero lattice point P. (The volume of $B\left(\frac{1}{2}\lambda, \delta\right)$ is infinite if $s + t < r$, and it grows like $\delta^{t/2}$ as $\delta \to \infty$ if $s + t = r$.) But the point P satisfies

$$q(P) = \sum_{i=1}^{s} x_i^2 - \sum_{i=1}^{t} x_{i+s}^2 \le \frac{1}{2}\lambda,$$

contradicting the definition of λ. Therefore q is positive definite on V. $\qquad\square$

Proposition 9.6. *The canonical height extends to a positive definite quadratic form on the real vector space $E(K) \otimes \mathbb{R}$.*

PROOF. We consider the lattice $E(K)/E_{\text{tors}}(K)$ inside the vector space $E(K) \otimes \mathbb{R}$ and apply (VIII.9.5) to get the desired result. Condition (i) of (VIII.9.5) is exactly (VIII.9.3cd). Condition (ii) of (VIII.9.5) follows from (VIII.9.3e), which says that bounding $\hat{h}$ is the same as bounding h_f, and then applying (VIII.6.1). $\qquad\square$

We now have the following quantities associated to E/K:

$E(K) \otimes \mathbb{R}$ a finite-dimensional vector space.

$\hat{h}$ a positive definite quadratic form on $E(K) \otimes \mathbb{R}$.

$E(K)/E_{\text{tors}}(K)$ a lattice in $E(K) \otimes \mathbb{R}$.

In such a situation, an extremely important invariant is the volume of a fundamental domain for the lattice, computed with respect to the metric induced by the quadratic form. For example, the discriminant of a number field K is the volume of its ring of integers with respect to the quadratic form $x \mapsto \text{Trace}_{K/\mathbb{Q}}(x^2)$. Similarly, the regulator of K is the volume of its unit group via the logarithm mapping and the usual metric on Euclidean space.

Definition. The *canonical height* (or *Néron–Tate*) *pairing* on E/K is the bilinear form

$$\langle \cdot, \cdot \rangle : E(\bar{K}) \times E(\bar{K}) \longrightarrow \mathbb{R},$$

defined by

$$\langle P, Q \rangle = \hat{h}(P + Q) - \hat{h}(P) - \hat{h}(Q).$$

Definition. The *elliptic regulator of E/K*, denoted by $R_{E/K}$, is the volume of a fundamental domain for $E(K)/E_{\text{tors}}(K)$ computed using the quadratic form $\hat{h}$. In other words, choose points $P_1, \ldots, P_r \in E(K)$ that generate $E(K)/E_{\text{tors}}(K)$, and then

$$R_{E/K} = \det\big(\langle P_i, P_j \rangle\big)_{\substack{1 \le i \le r \\ 1 \le j \le r}}.$$

(If $r = 0$, we set $R_{E/K} = 1$ by convention.)

An immediate corollary of (VIII.9.6) is the following result.

Corollary 9.7. *The elliptic regulator satisfies $R_{E/K} > 0$.*

Remark 9.8. We have defined the elliptic regulator using the absolute height, but there are situations in which it is more convenient to define the height relative to a given number field K. The regulator relative to K differs from $R_{E/K}$ by a factor of $[K : \mathbb{Q}]^r$.

Since $\hat{h}(P) > 0$ for all nontorsion points $P \in E(K)$, it is natural to ask how small $\hat{h}(P)$ can be if it is not zero. One might guess that $\hat{h}(P)$ must be large if the elliptic curve is "complicated" in some sense. The following precise conjecture is a strengthened version of a conjecture of Lang [135, page 92].

Conjecture 9.9. *Let E/K be an elliptic curve with j-invariant j_E and minimal discriminant $\mathcal{D}_{E/K}$. There is a constant $C > 0$, depending only on $[K : \mathbb{Q}]$, such that for all nontorsion points $P \in E(K)$ we have*

$$\hat{h}(P) > C \max\{h(j_E), \log \mathcal{N}_{K/\mathbb{Q}}\, \mathcal{D}_{E/K}, 1\}.$$

Note that the strength of the conjecture lies in the fact that the constant c is independent of both the elliptic curve E and the point P. Such estimates have applications to counting integral points on elliptic curves; see (IX.3.6). We briefly summarize what is currently known about (VIII.9.9).

Theorem 9.10. *Let E/K, j_E, and $\mathcal{D}_{E/K}$ be as in (VIII.9.9). Then the height inequality*

$$\hat{h}(P) > C \max\{h(j_E), \log \mathcal{N}_{K/\mathbb{Q}}\, \mathcal{D}_{E/K}, 1\}$$

is valid for the following choices of C:

(a) (Silverman [254], [260]) *Let $\nu(E)$ be the number of places $v \in M_K^0$ such that $\text{ord}_v(j_E) < 0$, i.e., the number of primes dividing the denominator of j_E. Then $C > 0$ may be chosen to depend only on $[K : \mathbb{Q}]$ and $\nu(E)$.*

(b) (Hindry–Silverman [113]) *Assume that the ABC conjecture[1] is true for the field K. Then $C > 0$ may be chosen to depend only on $[K : \mathbb{Q}]$ and on the exponent and constant appearing in the ABC conjecture.*

The proof of (VIII.9.10) is beyond the scope of this book, but see Exercise 8.17 for a special case.

---

[1] The ABC conjecture is described in (VIII.11.4), (VIII.11.6). It suffices to assume that the ABC conjecture is true for some fixed exponent, or equivalently, that Szpiro's conjecture (VIII.11.1) is true for some fixed exponent.

VIII.10   The Rank of an Elliptic Curve

The Mordell–Weil theorem (VIII.6.7) says that the *Mordell–Weil group* $E(K)$ of an elliptic curve can be written in the form

$$E(K) \cong E_{\text{tors}}(K) \times \mathbb{Z}^r.$$

As we have seen in (VIII §7), the torsion subgroup $E_{\text{tors}}(K)$ is relatively easy to compute, both in theory and in practice. The *rank* r is much more mysterious, and an effective procedure for determining it in all cases is still being sought. There are very few general facts known concerning the rank of elliptic curves, but there are a large number of fascinating conjectures. In Chapter X we describe some of the methods that have been developed for actually computing the group $E(K)$.

The rank of a "randomly chosen" elliptic curve over $\mathbb{Q}$ tends to be quite small, and it is difficult to produce curves $E/\mathbb{Q}$ having even moderately high rank. Nonetheless, there is the following folklore conjecture:

Conjecture 10.1. *There exist elliptic curves $E/\mathbb{Q}$ of arbitrarily large rank.*

A key piece of evidence for this conjecture comes from work of Shafarevich and Tate [244], who show that the analogous result is true for function fields, i.e., with $\mathbb{Q}$ replaced by the field of rational functions $\mathbb{F}_p(T)$. The Shafarevich–Tate construction leads to curves with constant j-invariant $j_E \in \mathbb{F}_p$, but subsequent constructions by Shioda [251] for $\bar{\mathbb{F}}_p(T)$ and Ulmer [295] for $\mathbb{F}_p(T)$ give examples with nonconstant j-invariant.

Néron constructed an infinite family of elliptic curves over $\mathbb{Q}$ having rank at least 10 [192], and later authors have constructed families of rank up to 19; see for example [76, 85, 188]. Within these families, clever search techniques due to Mestre [171] and others have yielded individual curves of higher rank. For example, Elkies [76] has produced the elliptic curve

$$
\begin{aligned}
y^2 + xy + y = x^3 &- x^2 \\
&- 20067762415575526585033208209338542750930230312178956502x \\
&+ 34481611795030556467032985690390720374855944435931918 \\
&\qquad\qquad 03612660082962919394487322434 29
\end{aligned}
$$

with rank $E(\mathbb{Q}) \geq 28$.

Attached to an elliptic curve E/K is a certain Dirichlet series $L_{E/K}(s)$ called the *L-series of E/K*; see Exercise 8.19. or (C §16). For the moment, it is enough to know that the definition of $L_{E/K}(s)$ involves only the number of points on the reductions $\tilde{E}(k_v)$ for the finite places $v \in M_K^0$. There is a famous conjecture of Birch and Swinnerton-Dyer that says that the order of vanishing of $L_{E/K}(s)$ at $s = 1$ is exactly equal to the rank of $E(K)$. The conjecture further asserts that the leading coefficient in the Taylor series expansion of $L_{E/K}(s)$ around $s = 1$ should be expressible in terms of various global arithmetic quantities associated to $E(K)$, including the elliptic regulator $R_{E/K}$. Thus in some sense, the conjecture of Birch and Swinnerton-Dyer is a local–global principle for elliptic curves, since it hypothetically shows how information about the v-adic behavior of E for all places $v \in M_K$ determines global information such as the rank of $E(K)$ and the elliptic regulator $R_{E/K}$. For further discussion of L-series and the conjecture of Birch and Swinnerton-Dyer, including some progress toward proving it, see (C §16).

In addition to wanting an effective method for computing the rank of an elliptic curve, it would be good to have a theoretical bound for the size of a generating set. Based partly on an analogy with the problem of computing generators for the unit group in a number field and partly on a number of deep conjectures in analytic number theory, Serge Lang suggested the following estimate.

Conjecture 10.2. (Lang [138], [141, Conjecture III.6.4]) *Let $\epsilon > 0$ and let $E/\mathbb{Q}$ be an elliptic curve of rank r. Then there is a basis $P_1, \ldots, P_r$ for the free part of $E(\mathbb{Q})$ satisfying*

$$\max_{1 \le i \le r} \hat{h}(P_i) \le C_\epsilon^{r^2} |\mathcal{D}_{E/\mathbb{Q}}|^{\frac{1}{12} + \epsilon}.$$

Here $\hat{h}$ is the canonical height on E (VIII §9), $\mathcal{D}_{E/\mathbb{Q}}$ is the minimal discriminant of $E/\mathbb{Q}$ (VIII §8), and C_ϵ is a constant depending only on ϵ.

Lang's conjecture is actually more precise than (VIII.10.2); see [138] or [141, Conjecture III.6.4].

Since $\hat{h}$ is a *logarithmic* height, the conjecture says that the x-coordinates of the generators may grow exponentially with the discriminant of the curve. This is similar to the way in which the height $H(u)$ of a generator for the unit group in a real quadratic field often grows exponentially with the discriminant of the field. Of course, it is easy to chose a sequence of fields such that $H(u)$ grows polynomially, but on average, one expects the growth to be exponential. The following example of Bremner and Cassels illustrates this exponential behavior. They show that the curve

$$y^2 = x^3 + 877x$$

has rank 1 and that the x-coordinate of the smallest generator P is

$$x(P) = \left(\frac{612776083187947368101}{78841535860683900210} \right)^2.$$

We compute

$$\frac{\log \hat{h}(P)}{\log |\mathcal{D}_{E/\mathbb{Q}}|} \approx 0.158,$$

so this example is roughly in the range suggested by Lang's conjecture.

VIII.11   Szpiro's Conjecture and ABC

For ease of exposition, we restrict attention in this section to elliptic curves defined over $\mathbb{Q}$. Let $E/\mathbb{Q}$ be such a curve, and let

$$y^2 + a_1 xy + a_3 y = x^3 + a_2 x^2 + a_4 x + a_6$$

be a global minimal Weierstrass equation (VIII.8.3) for $E/\mathbb{Q}$. The discriminant Δ_E of this equation is then the minimal discriminant of $E/\mathbb{Q}$, or, more properly, the minimal discriminant of $E/\mathbb{Q}$ is the ideal generated by Δ_E.

The primes dividing Δ_E are the primes for which E has bad reduction. There is another quantity associated to E that also encodes the primes of bad reduction. It is called the *conductor of E* and is denoted by N_E. The following definition of N_E is not quite correct, but suffices for our purposes. We write N_E as a product

$$N_E = \prod_{p \text{ prime}} p^{f_p(E)},$$

where

$$f_p(E) = \begin{cases} 0 & \text{if } E \text{ has good reduction at } p, \\ 1 & \text{if } E \text{ has multiplicative reduction at } p, \\ 2 & \text{if } E \text{ has additive reduction at } p. \end{cases}$$

(For $p = 2$ or 3, if E has additive reduction, then $f_p(E)$ may be greater than 2, but in any case it always satisfies $f_3(E) \leq 3$ and $f_2(E) \leq 5$. See [266, IV §10] for further information about the conductor of an elliptic curve.)

Roughly speaking, the conductor N_E is the product of the primes at which E has bad reduction raised to small powers, while the discriminant Δ_E is a product of the same primes, but they may sometimes appear to large powers. A deep conjecture made by Szpiro in 1983 says that although an occasional prime may appear in Δ_E to a high power, most primes do not.

Conjecture 11.1. (Szpiro's conjecture) *For every $\epsilon > 0$ there exists a κ_ϵ such that for all elliptic curves $E/\mathbb{Q}$,*

$$|\Delta_E| \leq \kappa_\epsilon N_E^{6+\epsilon}.$$

Although the statement of (VIII.11.1) seems relatively innocuous, the next result gives some indication of its strength.

Proposition 11.2. *Szpiro's conjecture (easily) implies Fermat's last theorem for all sufficiently large exponents, i.e., if n is sufficiently large, then the Fermat equation $a^n + b^n = c^n$ has no solutions with $a, b, c \in \mathbb{Z}$ and $abc \neq 0$.*

PROOF. Suppose that $a^n + b^n = c^n$ with $a, b, c \in \mathbb{Z}$ and $abc \neq 0$. We consider the elliptic curve (sometimes called a *Frey curve*)

$$E : y^2 = x(x + a^n)(x - b^n).$$

This Weierstrass equation for E has discriminant

$$\Delta_{a,b,c} = 16a^{2n}b^{2n}(a^n + b^n)^2 = 16(abc)^{2n}.$$

The minimal discriminant of $E/\mathbb{Q}$, which for notational clarity we denote by $\Delta_E^{\min}$, may be somewhat smaller than $\Delta_{a,b,c}$, but it cannot be too much smaller. More precisely, we prove below (VIII.11.3a) that the minimal discriminant of $E/\mathbb{Q}$ satisfies

$$|\Delta_E^{\min}| \geq \frac{|abc|^{2n}}{2^8}.$$

Szpiro's conjecture (VIII.11.1) relates the minimal discriminant $\Delta_E^{\min}$ to the conductor N_E, where we observe that the conductor has the trivial upper bound

$$N_E = \prod_{p|2abc} p^{f_p(E)} \leq \prod_{p|2abc} p^2 \leq |2abc|^2.$$

Szpiro's conjecture with $\epsilon = 1$ gives

$$\frac{|abc|^{2n}}{2^8} \leq |\Delta_E^{\min}| \leq \kappa N_E^7 \leq \kappa|2abc|^{14}$$

for an absolute constant κ. Thus

$$|abc|^{2n-14} \leq 2^{22}\kappa,$$

and since we certainly have $|abc| \geq 2$, this inequality yields an absolute upper bound for n. Hence if n is sufficiently large, then the equation $a^n + b^n = c^n$ has no solutions in nonzero integers. $\qquad\square$

Lemma 11.3. *Let $A, B, C \in \mathbb{Z}$ be nonzero integers satisfying*

$$A + B = C \qquad and \qquad \gcd(A, B, C) = 1,$$

and let $E/\mathbb{Q}$ be the elliptic curve

$$E : y^2 = x(x + A)(x - B).$$

(a) *The minimal discriminant Δ_E of E is given by either*

$$|\Delta_E| = 2^4|ABC|^2 \qquad or \qquad |\Delta_E| = 2^{-8}|ABC|^2.$$

In particular,
$$|\Delta_E| \geq 2^{-8}|ABC|^2.$$

(b) *The curve E has multiplicative reduction modulo p for all odd primes dividing ABC.*

PROOF. (a) The given Weierstrass equation for E has discriminant

$$\Delta = 16A^2B^2(A + B)^2 = 16A^2B^2C^2$$

and associated quantities

$$c_4 = 16(A^2 + AB + B^2) \qquad and \qquad c_6 = -32(2A^3 + 3A^2B + 3AB^2 + 2B^3).$$

Let $x = u^2x' + r$ and $y = u^3y' + u^2sx' + t$ be a change of variables that creates a global minimal Weierstrass equation for E; see (VIII.8.3). Applying (VII.1.3d) one prime at a time, we deduce that $u, r, s, t \in \mathbb{Z}$. The change of variable formulas in (III §1) then imply that

$$u^4 \mid c_4 \quad \text{and} \quad u^6 \mid c_6.$$

A simple resultant or Euclidean algorithm calculation gives the identities

$$(22A^2 - 8AB - 8B^2)c_4 + (A + 2B)c_6 = 288A^2,$$
$$-(8A^2 + 8AB - 22B^2)c_4 - (2A + B)c_6 = 288B^2.$$

Hence, using the assumption that $\gcd(A, B) = 1$, we find that

$$u^4 \mid \gcd(288A^4, 288B^4) = 288 = 2^5 \cdot 3^2,$$

from which it follows that $u = 1$ or 2. Therefore the absolute value of the minimal discriminant Δ_E of $E/\mathbb{Q}$,

$$|\Delta_E| = |u^{-12}\Delta| = |u^{-12}(4ABC)^2|,$$

is equal to either $16|ABC|^2$ or $2^{-8}|ABC|^2$.

(b) We recall from (a) that the c_4 value and the discriminant Δ of the Weierstrass equation $y^2 = x(x + A)(x - B)$ are

$$c_4 = 16(A^2 + AB + B^2) \quad \text{and} \quad \Delta = 16A^2B^2C^2.$$

For any prime p, we have from (VII.5.1) that

E has good reduction if $p \nmid \Delta$,

E has multiplicative reduction if $p \mid \Delta$ and $p \nmid c_4$,

E has additive reduction if $p \mid \Delta$ and $p \mid c_4$.

Let p be an odd prime dividing Δ. If $p \mid A$ or $p \mid B$, then the assumption that $\gcd(A, B) = 1$ implies that $p \nmid c_4$, so E has multiplicative reduction at p. Similarly, if $p \mid C$, so $A + B \equiv 0 \pmod{p}$, then $c_4 \equiv 16A^2 \pmod{p}$, and hence again $p \nmid c_4$ and E has multiplicative reduction at p.   □

Szpiro's conjecture is closely related to the ABC conjecture that was proposed by Masser and Oesterlé in 1985; see [196, Part I].

The ABC Conjecture 11.4. (Masser–Oesterlé) *For every $\epsilon > 0$ there exists a constant κ_ϵ such that for all nonzero integers $A, B, C \in \mathbb{Z}$ satisfying*

$$A + B = C \quad \text{and} \quad \gcd(A, B, C) = 1,$$

we have

$$\max\{|A|, |B|, |C|\} \le \kappa_\epsilon \left(\prod_{p \mid ABC} p \right)^{1+\epsilon}.$$

(The product is over all primes dividing ABC.)

The intuition behind the ABC conjecture is that in any sum of three relatively prime integers, it is not possible for all three terms to be divisible by many high prime powers. It is not hard to show that the ABC conjecture implies Szpiro's conjecture, and the converse is also true if one allows a slightly larger exponent.

Proposition 11.5. (a) *If Szpiro's conjecture* (VIII.11.1) *is true, then the ABC conjecture* (VIII.11.4) *is true with exponent* $\frac{3}{2}$. *(See also Exercise 8.20.)*
  (b) *The ABC conjecture implies Szpiro's conjecture.*

PROOF. (a) Let A, B, C be as in the statement of Szpiro's conjecture. Relabeling if necessary, we may assume that $C > B > A > 0$, so in particular

$$2B > A + B = C.$$

We consider the elliptic curve

$$E : y^2 = x(x + A)(x - B).$$

From (VIII.11.3a) we know that the minimal discriminant of E satisfies

$$|\Delta_E| \geq 2^{-8}(ABC)^2.$$

On the other hand, we know from (VIII.11.3b) that E has multiplicative reduction at all odd primes of bad reduction, so directly from the definition of the conductor,

$$N_E = 2^e \prod_{\substack{p \geq 3 \\ p | ABC}} p \qquad \text{for some } e \leq 2.$$

Applying Szpiro's conjecture to E, we deduce that for every $\epsilon > 0$ there is a $\kappa_\epsilon > 0$ such that

$$2^{-8}(ABC)^2 \leq |\Delta_E| \leq \kappa_\epsilon N_E^{6+\epsilon} \leq \kappa_\epsilon 2^{12+2\epsilon} \prod_{p | ABC} p^{6+\epsilon}.$$

Using the fact that $A \geq 1$ and $B > \frac{1}{2}C$ yields

$$2^{-10}C^4 \leq \kappa_\epsilon 2^{12+2\epsilon} \prod_{p | ABC} p^{6+\epsilon},$$

and taking fourth roots gives the ABC conjecture with exponent $\frac{3}{2}$.
  (b) Let $E/\mathbb{Q}$ be an elliptic curve given by a minimal Weierstrass equation. Then as described in (III §2), the discriminant and associated quantities c_4 and c_6 are related by the formula

$$1728\Delta = c_4^3 - c_6^2.$$

We will prove (b) under the assumption that $\gcd(c_4^3, c_6^2) = 1$ and leave the general case as an exercise for the reader; see Exercise 8.21. This assumption allows us to apply the ABC conjecture with

$$A = c_4^3, \qquad B = -c_6^2, \qquad \text{and} \qquad C = \Delta,$$

which yields

$$\max\{|c_4^3|, |c_6^2|, |\Delta|\} \le \kappa_\epsilon \prod_{p \mid c_4 c_6 \Delta} p^{1+\epsilon}.$$

The product on the right is clearly smaller than $|c_4 c_6 N_E|^{1+\epsilon}$, so we obtain the following three inequalities:

$$|c_4|^{2-\epsilon} \le \kappa_\epsilon |c_6 N_E|^{1+\epsilon},$$
$$|c_6|^{1-\epsilon} \le \kappa_\epsilon |c_4 N_E|^{1+\epsilon},$$
$$|\Delta| \le \kappa_\epsilon |c_4 c_6 N_E|^{1+\epsilon}.$$

We are going to take an appropriate (multiplicative) linear combination of these inequalities to eliminate c_4 and c_6. To do this, we raise the first inequality to the $2 + 2\epsilon$ power, raise the second inequality to the $3 + 3\epsilon$ power, raise the third inequality to the $1 - 5\epsilon$ power, and multiply the resulting three inequalities. Canceling $|c_4|^{4+2\epsilon-2\epsilon^2} |c_6|^{3-3\epsilon^2}$ from both sides yields

$$|\Delta|^{1-5\epsilon} \le \kappa_\epsilon^6 N_E^{6+6\epsilon}.$$

This is Szpiro's conjecture, up to adjusting the ϵ.   □

Remark 11.6. It is not difficult to formulate versions of Szpiro's conjecture and the ABC conjecture over a number fields. For example, if E/K is an elliptic curve defined over a number field K, we define the (naive) conductor of E/K to be the ideal

$$\mathfrak{N}_{E/K} = \prod_{\mathfrak{p}} \mathfrak{p}^{f_{\mathfrak{p}}(E)},$$

where $f_{\mathfrak{p}}(E)$ is 0, 1, or 2 according to whether E has good, multiplicative, or additive reduction at $\mathfrak{p}$. Then Szpiro's conjecture says that for every $\epsilon > 0$ there is a constant $\kappa = \kappa(\epsilon, K)$, depending only on ϵ and K, such that

$$N_{K/\mathbb{Q}} \mathcal{D}_{E/K} \le \kappa (N_{K/\mathbb{Q}} \mathfrak{N}_{E/K})^{6+\epsilon}.$$

Next suppose that $A, B, C \in R_K$ satisfy $A + B = C$. Then the ABC conjecture says that for every $\epsilon > 0$ there is a constant $\kappa = \kappa(\epsilon, K)$, depending only on ϵ and K, such that

$$H_K([A, B, C]) \le \kappa \prod_{\mathfrak{p} \mid ABC} (N_{K/\mathbb{Q}} \mathfrak{p})^{1+\epsilon}.$$

(There is no relative primality condition on $A, B,$ and C, since any common "factors" leave the left-hand side unchanged while increasing the right-hand side.)

It is very interesting to ask how the constants κ appearing in these conjectures depend on the field K.

Remark 11.7. Let k be a field of characteristic 0. There are analogues of Szpiro's conjecture and the ABC conjecture in which $\mathbb{Q}$ is replaced by a rational function field $k(T)$, or more generally, the number field K is replaced by the function field $k(C)$ of an algebraic curve C. Somewhat surprisingly, both conjectures are quite easy to prove in the function field setting, and indeed considerably stronger results are known. For example, the three-term sum in the ABC conjecture may be replaced by a sum having more terms. See [157, 258, 278] for $A + B = C$ and [31, 158, 300] for $A_1 + \cdots + A_n = 0$.

Remark 11.8. Frey has noted that Szpiro's conjecture (VIII.11.1) implies the uniform boundedness of torsion on elliptic curves (VIII.7.5), (VIII.7.5.1). The idea is as follows. Suppose that $P \in E(K)$ is a point of exact order N, and let $\phi : E \to E'$ be the isogeny whose kernel is the subgroup generated by P. Assuming that N is sufficiently large (depending only on the field K), an elementary calculation using Tate curves (see (C §14) or [266, Chapter V]) shows that there are ideals $\mathfrak{a}$ and $\mathfrak{b}$ such that the minimal discriminants of E and E' have the form

$$\mathcal{D}_E = \mathfrak{a}\mathfrak{b}^N \qquad \text{and} \qquad \mathcal{D}_{E'} = \mathfrak{a}^N\mathfrak{b}.$$

Since the primes of bad reduction divide the discriminant, we see that the conductors $\mathfrak{N}_E$ and $\mathfrak{N}_{E'}$ divide $\mathfrak{a}^2\mathfrak{b}^2$. We apply Szpiro's conjecture to E and E' to obtain

$$N_{K/\mathbb{Q}}(\mathcal{D}_E\mathcal{D}_{E'}) \leq \kappa_\epsilon N_{K/\mathbb{Q}}(\mathfrak{N}_E\mathfrak{N}_{E'})^{6+\epsilon},$$

and then substituting the discriminants' and conductors' values gives

$$N_{K/\mathbb{Q}}(\mathfrak{a}\mathfrak{b})^{N+1} \leq \kappa_\epsilon N_{K/\mathbb{Q}}(\mathfrak{a}\mathfrak{b})^{12+2\epsilon}.$$

Discarding the finitely many elliptic curves defined over K with everywhere good reduction (IX.6.1), we may assume that $N_{K/\mathbb{Q}}(\mathfrak{a}\mathfrak{b}) \geq 2$, and then the last inequality gives a bound for N that is independent of the curve E. See [89, 90, 113] for further details.

Exercises

8.1. Let E/K be an elliptic curve, let $m \geq 2$ be an integer, let $\mathcal{H}_K$ be the ideal class group of K, and let

$$S = \{v \in M_K^0 : E \text{ has bad reduction at } v\} \cup \{v \in M_K^0 : v(m) \neq 0\} \cup M_K^\infty.$$

Assume that $E[m] \subset E(K)$. Prove the following quantitative version of the weak Mordell–Weil theorem:

$$\text{rank}_{\mathbb{Z}/m\mathbb{Z}} E(K)/mE(K) \leq 2\#S + 2\,\text{rank}_{\mathbb{Z}/m\mathbb{Z}}\,\mathcal{H}_K[m].$$

8.2. For each integer $d \geq 1$, let E_d be the elliptic curve

$$E : y^2 = x^3 - d^2x.$$

Prove that

$$E_d(\mathbb{Q}) \cong (\text{finite group}) \times \mathbb{Z}^r$$

for an integer r satisfying

$$r \leq 2\nu(2d),$$

where $\nu(N)$ denotes the number of distinct primes dividing N. (*Hint.* Use Exercise 8.1.)

8.3. Let E/K be an elliptic curve and let L/K be an (infinite) algebraic extension. Suppose that the rank of $E(M)$ is bounded as M ranges over all finite extensions M/K such that M is contained in L, i.e., assume that

$$\sup_{\substack{K \subset M \subset L \\ [M:K] \text{ finite}}} \text{rank } E(M)$$

is finite.
(a) Prove that $E(L) \otimes \mathbb{Q}$ is a finite-dimensional $\mathbb{Q}$-vector space.
(b) Assume further that L/K is Galois and that $E_{\text{tors}}(L)$ is finite. Prove that $E(L)$ is finitely generated.

8.4. Assume that $\mu_m \subset K$. Prove that the maximal abelian extension of K of exponent m is the field

$$K(a^{1/m} : a \in K).$$

(*Hint.* Use (VIII.2.2), which in this case says that every homomorphism $\chi : G_{\bar{K}/K} \to \mu_m$ has the form $\chi(\sigma) = \alpha^\sigma/\alpha$ for some $\alpha \in \bar{K}^*$ satisfying $\alpha^m \in K$.)

8.5. Let $\xi \in H^1(G_{\bar{K}/K}, M)$ be unramified at v. Prove that the cohomology class of ξ contains a 1-cocycle $c : G_{\bar{K}/K} \to M$ satisfying $c_\sigma = 0$ for all $\sigma \in I_v$. (*Hint.* Use the inflation–restriction sequence (B.2.4) for $I_v \subset G_{\bar{K}/K}$.)

8.6. Prove *Kronecker's theorem*: Let $x \in \bar{\mathbb{Q}}^*$. Then $H(x) = 1$ if and only if x is a root of unity. (This is the multiplicative group version of (VIII.9.3d).)

8.7. (a) Give an explicit upper bound, in terms of N, C, and d, for the number of points in the set

$$\{P \in \mathbb{P}^N(\bar{\mathbb{Q}}) : H(P) \leq C \text{ and } [\mathbb{Q}(P) : \mathbb{Q}] \leq d\}.$$

(b) Let

$$\nu_K(N, C) = \#\{P \in \mathbb{P}^N(K) : H_K(P) \leq C\}.$$

Prove that

$$\lim_{C \to \infty} \frac{\nu_{\mathbb{Q}}(N, C)}{C^{N+1}} = \frac{2^N}{\zeta(N+1)},$$

where $\zeta(s)$ is the Riemann zeta function. (For further information about $\nu_K(N, C)$, see (VIII.5.12).)

8.8. Prove the following basic properties of height functions.
(a) $H(x_1 x_2 \cdots x_N) \leq H(x_1) H(x_1) \cdots H(x_N)$.
(b) $H(x_1 + x_2 + \cdots + x_N) \leq N H(x_1) H(x_2) \cdots H(x_N)$.

(c) For $P = [x_0, \ldots, x_N] \in \mathbb{P}^N(\bar{\mathbb{Q}})$ and $Q = [y_0, \ldots, y_M] \in \mathbb{P}^M(\bar{\mathbb{Q}})$, define

$$P \star Q = [x_0 y_0, x_0 y_1, \ldots, x_i y_j, \ldots, x_N y_M] \in \mathbb{P}^{MN+M+N}(\bar{\mathbb{Q}}).$$

Prove that

$$H(P \star Q) = H(P)H(Q).$$

(The map $(P, Q) \mapsto P \star Q$ is the *Segre embedding* of $\mathbb{P}^N \times \mathbb{P}^M$ into $\mathbb{P}^{MN+M+N}$. See [111, exercise I.2.14].)

(d) Let $M = \binom{N+d}{N} - 1$ and let $f_0(X), \ldots, f_M(X)$ be the M distinct monomials of degree d in the $N + 1$ variables $X_0, \ldots, X_N$. For any point $P = [x_0, \ldots, x_N] \in \mathbb{P}^N(\bar{\mathbb{Q}})$, let

$$P^{(d)} = \left[f_0(P), \ldots, f_M(P) \right] \in \mathbb{P}^M(\bar{\mathbb{Q}}).$$

Prove that

$$H(P^{(d)}) = H(P)^d = H\left([x_0^d, \ldots, x_N^d] \right).$$

(The map $P \mapsto P^{(d)}$ is the *d-uple embedding* of $\mathbb{P}^n$ into $\mathbb{P}^M$. See [111, exercise I.2.12].)

8.9. Let $x_0, \ldots, x_N \in K$ and let $\mathfrak{b}$ be the fractional ideal of K generated by $x_0, \ldots, x_N$. Prove that

$$H_K\left([x_0, \ldots, x_N] \right) = (\mathrm{N}_{K/\mathbb{Q}}\, \mathfrak{b})^{-1} \prod_{v \in M_K^\infty} \max_{0 \le i \le N} \left\{ |x_i|_v \right\}^{n_v}.$$

8.10. Let F be the rational map

$$F : \mathbb{P}^2 \longrightarrow \mathbb{P}^2, \qquad [x, y, z] \longmapsto [x^2, xy, z^2],$$

from (I.3.6). Note that F is a morphism at every point except at $[0, 1, 0]$, where it is not defined. Prove that there are infinitely many points $P \in \mathbb{P}^2(\mathbb{Q})$ such that

$$H\left(F(P) \right) = H(P).$$

In particular, (VIII.5.6) is false if the map F is merely required to be a rational map.

8.11. Prove the following generalization of (VIII.7.2) to arbitrary number fields. Let E/K be an elliptic curve given by an equation

$$y^2 = x^3 + Ax + B$$

with $A, B \in R$, and let $\Delta = 4A^3 + 27B^2$. Let $P \in E(K)$ be a point of exact order $m \ge 3$, and let $v \in M_K^0$.

(a) If $m = p^n$ is a prime power, prove that

$$-6r_v \le \mathrm{ord}_v\left(y(P)^2 \right) \le 6r_v + \mathrm{ord}_v(\Delta),$$

where

$$r_v = \left[\frac{\mathrm{ord}_v(p)}{p^n - p^{n-1}} \right].$$

(b) If $m = 2p^n$ is twice a prime power, prove that

$$0 \le \mathrm{ord}_v\left(y(P)^2 \right) \le 2r_v + \mathrm{ord}_v(\Delta),$$

where r_v is as in (a).

(c) If m is not of the form p^n or $2p^n$, prove that

$$0 \le \mathrm{ord}_v\big(y(P)^2\big) \le \mathrm{ord}_v(\Delta).$$

8.12. Calculate $E(\mathbb{Q})_{\mathrm{tors}}$ for each of the following elliptic curves.

(a) $y^2 = x^3 - 2$ (i) $y^2 + xy + y = x^3 - x^2 - 14x + 29$

(b) $y^2 = x^3 + 8$ (j) $y^2 + xy = x^3 - 45x + 81$

(c) $y^2 = x^3 + 4$ (k) $y^2 + 43xy - 210y = x^3 - 210x^2$

(d) $y^2 = x^3 + 4x$ (l) $y^2 = x^3 - 4x$

(e) $y^2 - y = x^3 - x^2$ (m) $y^2 = x^3 + 2x^2 - 3x$

(f) $y^2 = x^3 + 1$ (n) $y^2 + 5xy - 6y = x^3 - 3x^2$

(g) $y^2 = x^3 - 43x + 166$ (o) $y^2 + 17xy - 120y = x^3 - 60x^2$

(h) $y^2 + 7xy = x^3 + 16x$

8.13. (a) Let E/K be an elliptic curve and let $P \in E(K)$ be a point of order at least 4. Prove that there is a change of coordinates such that E has a Weierstrass equation of the form

$$E : y^2 + uxy + vy = x^3 + vx^2$$

with $u, v \in K$ and $P = (0,0)$.

(b) Prove that there is a one-parameter family of elliptic curves E/K having a K-rational point of order 6. (*Hint.* Set $[3]P = [-3]P$ in (a) and find a relation between u and v.) Same question for points of order 7, order 9, and order 12.

(c) Prove that the elliptic curves E/K having a K-rational point of order 11 are parametrized by the K-rational points of a certain curve of genus one.

8.14. (a) Generalize (VIII.8.2) as follows. Let E/K be an elliptic curve and let $\mathfrak{a}$ be any *integral* ideal in the ideal class $\bar{\mathfrak{a}}_{E/K}$. Prove that there is a Weierstrass equation of E/K having coefficients $a_i \in R$ and discriminant Δ satisfying

$$(\Delta) = \mathcal{D}_{E/K}\mathfrak{a}^{12}.$$

(b) Suppose that E/K has everywhere good reduction and that the class number of K is relatively prime to 6. Prove that E/K has a global minimal Weierstrass equation.

(c) Prove that every elliptic curve E/K has a Weierstrass equation with coefficients $a_i \in R$ and discriminant Δ satisfying

$$|\,\mathrm{N}_{K/\mathbb{Q}}\,\Delta| \le |\,\mathrm{Disc}\,K/\mathbb{Q}|^6|\,\mathrm{N}_{K/\mathbb{Q}}\,\mathcal{D}_{E/K}|.$$

Qualitatively, this says that there is a Weierstrass equation for E whose nonminimality is bounded solely in terms of K. Such an equation might be called *quasiminimal*.

(d) Let $\bar{\mathfrak{b}}$ be an ideal class of K. Prove that there is an elliptic curve E/K such that $\bar{\mathfrak{a}}_{E/K} = \bar{\mathfrak{b}}$. In particular, if K does not have class number one, then there exist elliptic curves over K that do not have global minimal Weierstrass equations. This gives a converse to (VIII.8.3). (See also [15] for an estimate of how many E/K have $\bar{\mathfrak{a}}_{E/K}$ equal to $\bar{\mathfrak{b}}$.)

8.15. Prove that there are no elliptic curves $E/\mathbb{Q}$ having everywhere good reduction. (*Hint.* Suppose that there is a Weierstrass equation with integer coefficients and discriminant $\Delta = \pm 1$. Use congruences modulo 8 to show that a_1 is odd, and hence $c_4 \equiv 1 \pmod 8$. Substitute $c_4 = u \pm 12$ into the formula $c_4^3 - c_6^2 = \pm 1728$. Show that u is either a square or three times a square. Rule out both cases by reducing modulo 8.)

8.16. Show that the conclusion of (VIII.9.5) is false if the quadratic form q is not required to satisfy the finiteness condition (ii).

8.17. Fix nonzero integers A and B with $4A^3 + 27B^2 \neq 0$. For each integer $d \neq 0$, let $E_d/\mathbb{Q}$ be the elliptic curve

$$E_d : y^2 = x^3 + d^2 Ax + d^3 B.$$

Assuming that d is squarefree, prove the following properties of E_d:
  (a) j_E is independent of d.
  (b) $\log |\mathcal{D}_{E/\mathbb{Q}}| = 6 \log |d| + O(1)$.
  (c) Every $P \in E_d(\mathbb{Q})$ satisfies either $[2]P = 0$ or $\hat{h}(P) > \frac{1}{8} \log |d| + O(1)$.
  (d) For all but finitely many squarefree integers d, the torsion subgroup of $E_d(\mathbb{Q})$ is one of $\{0\}$, $\mathbb{Z}/2\mathbb{Z}$, and $(\mathbb{Z}/2\mathbb{Z})^2$.
Note that the $O(1)$ bounds in (b) and (c) may depend on A and B, but they should be independent of d. In particular, (c) provides a proof of (VIII.9.9) for the family of curves E_d. (*Hint for* (c). If $P = (r, s) \in E_d(\mathbb{Q})$, then $P' = (r/d, s/d^{3/2}) \in E_1(\bar{\mathbb{Q}})$. Prove the following facts: (i) $\hat{h}(P) = \hat{h}(P')$; (ii) either $s = 0$ or $h_y(P')$ is greater than $\frac{3}{8} \log |d|$; and (iii) $|\hat{h} - \frac{1}{3} h_y|$ is bounded.)

8.18. Let E/K be an elliptic curve given by a Weierstrass equation

$$y^2 = x^3 + Ax + B.$$

  (a) Prove that there are *absolute constants* c_1 and c_2 such that for all points $P \in E(\bar{K})$ we have

$$\left| h_x([2]P) - 4h_x(P) \right| \leq c_1 h([A, B, 1]) + c_2.$$

  Find explicit values for c_1 and c_2. (*Hint.* Combine the proofs of (VIII.4.2) and (VIII.5.6), keeping track of the dependence on the constants. In particular, note that the use of the Nullstellensatz in (VIII.5.6) can be replaced by the explicit identities given in (VIII.4.3).)
  (b) Find *absolute constants* c_3 and c_4 such that for all points $P \in E(\bar{K})$ we have

$$\left| \frac{1}{2} h_x(P) - \hat{h}(P) \right| \leq c_3 h([A, B, 1]) + c_4.$$

  (*Hint.* Use (a) and the proof of (VIII.9.1).)
  (c) Prove that for all integers $m \geq 1$ and all points $P, Q \in E(\bar{K})$ we have

$$\left| h_x([m]P) - m^2 h_x(P) \right| \leq 2(m^2 + 1)\left(c_3 h([A, B, 1]) + c_4\right)$$

  and

$$h_x(P + Q) \leq 2h_x(P) + 2h_x(Q) + 5\left(c_3 h([A, B, 1]) + c_4\right).$$

  (*Hint.* Use (b) and (VIII.9.3).)
  (d) Let $Q_1, \ldots, Q_r \in E(K)$ be a set of generators for $E(K)/2E(K)$. Find *absolute constants* c_5, c_6, and c_7 such that the set of points $P \in E(K)$ satisfying

$$h_x(P) \leq c_5 \max_{1 \leq i \leq r} h_x(Q_i) + c_6 h([A, B, 1]) + c_7$$

  contains a complete set of generators for $E(K)$. (*Hint.* Follow the proof of (VIII.3.1), using (c) to evaluate the constants that appear.)

8.19. *The L-Series Attached to an Elliptic Curve.* Let $E/\mathbb{Q}$ be an elliptic curve and choose a global minimal Weierstrass equation for $E/\mathbb{Q}$,

$$E : y^2 + a_1 xy + a_3 y = x^3 + a_2 x^2 + a_4 x + a_6.$$

(See (VIII.8.3).) For each prime p, let $\tilde{E}$ denote the reduction of the Weierstrass equation modulo p, and let

$$t_p = p + 1 - \#\tilde{E}(\mathbb{F}_p).$$

The *L-series associated to $E/\mathbb{Q}$* is defined by the Euler product

$$L_E(s) = \prod_{p | \Delta(E)} (1 - t_p p^{-s})^{-1} \prod_{p \nmid \Delta(E)} (1 - t_p p^{-2} + p^{1-2s})^{-1}.$$

(a) If $L_E(s)$ is expanded as a Dirichlet series $\sum c_n n^{-s}$, show that for all primes p, its pth coefficient satisfies $c_p = t_p$.

(b) If E has bad reduction at p, so $p \mid \Delta(E)$, prove that t_p equals 1, -1, or 0 according to whether the reduced curve $\tilde{E} \pmod p$ has a node with tangents whose slopes are rational over $\mathbb{F}_p$ (split multiplicative reduction), a node with tangents whose slopes are quadratic over $\mathbb{F}_p$ (nonsplit multiplicative reduction), or a cusp (additive reduction). (Cf. Exercise 3.5).

(c) Prove that the Euler product for $L_E(s)$ converges for all $s \in \mathbb{C}$ with $\text{Re}(s) > \frac{3}{2}$. (*Hint.* Use (V.1.1).)

There are many important theorems and conjectures concerning the L-series of elliptic curves; see (C §16).

8.20. We proved in (VIII.11.5a) that Szpiro's conjecture implies a weaker form of the ABC conjecture with exponent $\frac{3}{2}$. This exercise explains how to reduce the exponent to $\frac{6}{5}$.

Relabeling A, B, C if necessary, we may assume that $C > B > A > 0$. Let E be the curve $y^2 = x(x + A)(x - B)$ used in the proof of (VIII.11.5a).

(a) Prove that there is an isogeny of degree 2 from E to the elliptic curve

$$E' : y^2 = x^3 - 2(A - B)x^2 + C^2 x.$$

Show that the discriminant of the equation for E' is $\Delta' = -2^8 ABC^4$.

(b) Prove a version of (VIII.11.3) for E'. In particular, prove that E' has multiplicative reduction modulo p for all odd primes dividing ABC and that its minimal discriminant satisfies

$$|\Delta_{E'}| \geq 2^{-28} |ABC^4|.$$

(c) Apply Szpiro's conjecture to E' and deduce that

$$C \leq \kappa_\epsilon \prod_{p | ABC} p^{\frac{6}{5} + \epsilon},$$

where the constant κ_ϵ depends only on ϵ.

8.21. We proved (VIII.11.5b) that the ABC conjecture (VIII.11.4) implies Szpiro's conjecture (VIII.11.1) under the assumption that $\gcd(c_4, c_6) = 1$. Prove that this implication is still true when $\gcd(c_4, c_6) > 1$. (*Hint.* Let $G = \gcd(c_4^3, c_6^2)$ and apply the ABC conjecture with $A = c_4^3/G$, $B = -c_6^2/G$, and $C = \Delta/G$. Use the minimality of the equation to bound the powers of the primes p dividing G. Also show that if $p \geq 5$ divides G, then E has additive reduction at p, so $p^2 \mid N_E$.)

8.22. Let m, n, ℓ be positive integers and consider the equation

$$x^m + y^n = z^\ell. \tag{*}$$

Assuming the ABC conjecture (VIII.11.4), prove the following two statements (see also Exercise 9.17):

(a) If $m^{-1} + n^{-1} + \ell^{-1} < 1$, then (*) has only finitely many solutions $x, y, z \in \mathbb{Z}$ with $\gcd(x, y, z) = 1$.

(b) There is a constant κ', depending only on the constant appearing in the ABC conjecture, such that if (*) has a solution in relatively prime integers satisfying $|x|, |y|, |z| \geq 2$, then

$$\max\{m, n, \ell\} \leq \kappa'.$$

8.23. Let $A, B, C \in \mathbb{Z}$ be as in the statement of the ABC conjecture (VIII.11.4), and let

$$E : y^2 = x(x + A)(x - B)$$

be the elliptic curve used in the proof of (VIII.11.5a). Assume further that

$$A \equiv 0 \pmod{16} \qquad \text{and} \qquad B \equiv 3 \pmod 4.$$

(a) Prove that the substitutions $x \mapsto 4x$ and $y \mapsto 8y + 4x$ give a global minimal Weierstrass equation for E,

$$y^2 + xy = x^3 + \frac{A - B - 1}{4}x^2 - \frac{AB}{16}x.$$

(b) Verify that the Weierstrass equation in (a) satisfies

$$c_4 = A^2 + AB + B^2, \qquad c_6 = \frac{(B - A)(A + C)(B + C)}{2}, \qquad \text{and} \qquad \Delta = \left(\frac{ABC}{16}\right)^2.$$

(c) Prove that E has multiplicative reduction for every prime p dividing Δ.

Chapter IX

Integral Points on Elliptic Curves

Many elliptic curves have infinitely many rational points, although the Mordell–Weil theorem assures us that the group of rational points is finitely generated. Another natural Diophantine question is that of determining how many of the rational points on a given (affine) Weierstrass equation have integral coordinates. In this chapter we prove a theorem of Siegel that says that there are only finitely many such integral points. Siegel gave two proofs of his theorem, which we present in (IX §3) and (IX §4). Both proofs make use of techniques from the theory of Diophantine approximation, and thus do not provide an effective procedure for actually finding all of the integral points. However, Siegel's second proof reduces the problem to that of solving the so-called unit equation, which in turn can be effectively resolved using methods from transcendence theory. We discuss effective solutions, without giving proofs, in (IX §5).

Unless otherwise specified, the notation and conventions for this chapter are the same as those for Chapter VIII. In addition, we set the following notation:

H, H_K height functions, see (VIII §5).

n_v $= [K_v : \mathbb{Q}_v]$, the local degree for $v \in M_K$, see (VIII §5).

S $\subset M_K$, generally a finite set of absolute values containing M_K^∞.

R_S the ring of S-integers of K,

$$R_S = \{x \in K : v(x) \geq 0 \text{ for all } v \in M_K \text{ with } v \notin S\}.$$

R_S^* the unit group of R_S.

J.H. Silverman, *The Arithmetic of Elliptic Curves, Second Edition*, Graduate Texts in Mathematics 106, DOI 10.1007/978-0-387-09494-6_IX,
© Springer Science+Business Media, LLC 2009

IX.1 Diophantine Approximation

The fundamental problem in the subject of Diophantine approximation is the question of how closely an irrational number can be approximated by a rational number.

Example 1.1. For any rational number p/q, we know that the quantity $\left| p/q - \sqrt{2} \right|$ is strictly positive, and since $\mathbb{Q}$ is dense in $\mathbb{R}$, an appropriate choice of p/q makes it as small as desired. The problem is to make it small without taking p and q to be too large. The next two elementary results illustrate this idea.

Proposition 1.2. (Dirichlet) *Let $\alpha \in \mathbb{R}$ with $\alpha \notin \mathbb{Q}$. Then there are infinitely many rational numbers $p/q \in \mathbb{Q}$ such that*

$$\left| \frac{p}{q} - \alpha \right| \le \frac{1}{q^2}.$$

PROOF. Let Q be a (large) integer and look at the set of real numbers

$$\{ q\alpha - [q\alpha] : q = 0, 1, \ldots, Q \},$$

where $[\,\cdot\,]$ denotes greatest integer. Since α is irrational, this set contains $Q+1$ distinct numbers in the interval between 0 and 1. Dividing the interval $[0, 1]$ into Q equal-sized pieces and applying the pigeonhole principle, we find that there are integers $0 \le q_1 < q_2 \le Q$ satisfying

$$\left| (q_1\alpha - [q_1\alpha]) - (q_2\alpha - [q_2\alpha]) \right| \le \frac{1}{Q}.$$

Hence

$$\left| \frac{[q_2\alpha] - [q_1\alpha]}{q_2 - q_1} - \alpha \right| \le \frac{1}{(q_2 - q_1)Q} \le \frac{1}{(q_2 - q_1)^2}.$$

This provides one rational approximation to α having the desired property.

Finally, having obtained a list of such approximations, let p/q be the one for which $|p/q - \alpha|$ is smallest. Then taking $Q > |p/q - \alpha|^{-1}$ ensures that we get a new approximation that is not already in our list. Hence there exist infinitely many rational numbers satisfying the conditions of the proposition. □

Remark 1.2.1. A result of Hurwitz says that the $1/q^2$ on the right-hand side of (IX.1.2) may be replaced by $1/(\sqrt{5}\, q^2)$, and that this result is best possible. See, e.g., [108, Theorem 195].

Proposition 1.3. (Liouville [151]) *Let $\alpha \in \bar{\mathbb{Q}}$ have degree $d \ge 2$ over $\mathbb{Q}$, i.e., $[\mathbb{Q}(\alpha) : \mathbb{Q}] = d$. There is a constant $C > 0$, depending on α, such that for all rational numbers p/q we have*

$$\left| \frac{p}{q} - \alpha \right| \ge \frac{C}{q^d}.$$

PROOF. Let

$$f(T) = a_0 T^d + a_1 T^{d-1} + \cdots + a_d \in \mathbb{Z}[T]$$

be a minimal polynomial for α, and let

$$C_1 = \sup\{f'(t) : \alpha - 1 \leq t \leq \alpha + 1\}.$$

Then the mean value theorem tells us that

$$\left| f\left(\frac{p}{q}\right) \right| = \left| f\left(\frac{p}{q}\right) - f(\alpha) \right| \leq C_1 \left| \frac{p}{q} - \alpha \right|.$$

On the other hand, we know that $q^d f(p/q) \in \mathbb{Z}$, and further that $f(p/q) \neq 0$, since f has no rational roots. Hence

$$\left| q^d f\left(\frac{p}{q}\right) \right| \geq 1.$$

Setting $C = \min\{C_1^{-1}, 1\}$ and combining the last two inequalities yields

$$\left| \frac{p}{q} - \alpha \right| \geq \frac{C}{q^d} \qquad \text{for all } p/q \in \mathbb{Q}. \qquad \square$$

Remark 1.3.1. Liouville used his theorem to prove the existence of transcendental numbers; see Exercise 9.2. Note that in Liouville's theorem it is quite easy to find a value for the constant C explicitly in terms of α. This is in marked contrast to the results that we consider in the rest of this section.

    Dirichlet's theorem (IX.1.2) says that every real number can be approximated by rational numbers to within $1/q^2$, while Liouville's result (IX.1.3) says that algebraic numbers of degree d can be approximated no closer than C/q^d. For quadratic irrationalities there is little more to say, but if $d \geq 3$, then it is natural to ask for the best exponent on q. There is no particular reason to restrict the approximating values to $\mathbb{Q}$, so we allow them to vary over any fixed number field K. Finally, in measuring the closeness of the approximation, we may use any absolute value on K.

Definition. Let $\tau(d)$ be a positive real-valued function on the natural numbers. A number field K is said to have *approximation exponent* τ if it has the following property:

    Let $\alpha \in \bar{K}$, let $d = [K(\alpha) : K]$, and let $v \in M_K$ be an absolute value on K that has been extended to $K(\alpha)$ in some fashion. Then for any constant C there exist only finitely many $x \in K$ satisfying the inequality

$$|x - \alpha|_v < C H_K(x)^{-\tau(d)}.$$

    Liouville's elementary estimate (IX.1.3) says that $\mathbb{Q}$ has approximation exponent $\tau(d) = d + \epsilon$ for any $\epsilon > 0$. This result has been successively improved by a number of mathematicians:

Liouville	1851	$\tau(d) = d + \epsilon$
Thue	1909	$\tau(d) = \frac{1}{2}d + 1 + \epsilon$
Siegel	1921	$\tau(d) = 2\sqrt{d} + \epsilon$
Gelfond, Dyson	1947	$\tau(d) = \sqrt{2d} + \epsilon$
Roth	1955	$\tau(d) = 2 + \epsilon$

In view of (IX.1.2), Roth's result is essentially best possible, although it has been conjectured that the ϵ can be replaced by some function $\epsilon(d)$ such that $\epsilon(d) \to 0$ as $d \to \infty$. We should also mention that Mahler showed how to handle several absolute values at once, and W. Schmidt [221, Chapter VI] dealt with the more difficult problem of simultaneously approximating several irrationals.

The main ideas that go into the proof of Roth's theorem are quite beautiful, and at least in theory, relatively elementary. Unfortunately, to develop these ideas fully would take us rather far afield. Hence rather than including a complete proof, we are content to state here the result that we will need. In (IX §8) we briefly sketch the proof of Roth's theorem without giving any of the myriad details.

Theorem 1.4. (Roth's Theorem) *For every $\epsilon > 0$, every number field K of degree d has approximation exponent*

$$\tau(d) = 2 + \epsilon.$$

PROOF. See (IX §8) for a brief sketch of the proof. A nice exposition for $K = \mathbb{Q}$ and the usual archimedean absolute value is given in [221, Chapter V]. For the general case, see [114, Part D] or [139, Chapter 7]. $\square$

Example 1.5. How do theorems on Diophantine approximation lead to results about Diophantine equations? Consider the simple example of trying to solve the equation

$$x^3 - 2y^3 = a$$

in integers $x, y \in \mathbb{Z}$, where $a \in \mathbb{Z}$ is fixed. Suppose that (x, y) is a solution with $y \neq 0$. Let ζ be a primitive cube root of unity, and factor the equation as

$$\left(\frac{x}{y} - \sqrt[3]{2} \right) \left(\frac{x}{y} - \zeta\sqrt[3]{2} \right) \left(\frac{x}{y} - \zeta^2\sqrt[3]{2} \right) = \frac{a}{y^3}.$$

The second and third factors in the product are bounded away from 0, so we obtain an estimate of the form

$$\left| \frac{x}{y} - \sqrt[3]{2} \right| \leq \frac{C}{y^3},$$

where the constant C is independent of x and y. Now (XI.1.4), or even Thue's original theorem with $\tau(d) = \frac{1}{2}d + 1 + \epsilon$, shows that there are only finitely many possibilities for x and y. Hence the equation

$$x^3 - 2y^3 = a$$

has only finitely many solutions in integers. This type of argument will reappear in the proof of (IX.4.1); see also Exercise 9.6.

Remark 1.6. The statement of (IX.1.4) says that *there exist* only finitely many elements of K having a certain property. This phrasing is felicitous because the proof of (IX.1.4) is not effective. In other words, the proof does not give an effective procedure that is guaranteed to produce all of the elements in the finite set. (See (IX.8.1) for a discussion of why this is so.) We note that as a consequence, all of the finiteness results that we prove in (IX §§2, 3) are ineffective, since they rely on (IX.1.4). Similarly, the proof in (IX.1.5) yields no explicit bound for $|x|$ and $|y|$ in terms of a. However, there are other methods, based on estimates for linear forms in logarithms, that are effective. We discuss such methods, without proof, in (IX §5).

IX.2 Distance Functions

A Diophantine inequality such as

$$|x - \alpha|_v < C H_K(x)^{-\tau(d)}$$

consists of two pieces. First, there is the height function $H_K(x)$, which measures the *arithmetic* size of x. We have already studied height functions and their transformation properties in some detail (VIII, §§5, 6). Second, there is the quantity $|x - \alpha|_v$, which is a *topological* or *metric* measure of the distance from x to α, i.e., it measures distance in the v-adic topology. In this section we define a notion of v-adic distance on curves, deduce some of its basic properties, and reinterpret the main Diophantine approximation result from (IX §1) in terms of this distance function.

Definition. Let C/K be a curve, let $v \in M_K$, and fix a point $Q \in C(K_v)$. Choose a function $t_Q \in K_v(C)$ that has a zero of order $e \geq 1$ at Q and no other zeros.[1] Then for $P \in C(K_v)$, we define the (*v-adic*) *distance from* P *to* Q by

$$d_v(P, Q) = \min \left\{ |t_Q(P)|_v^{1/e}, 1 \right\}.$$

(If t_Q has a pole at P, we formally set $|t_Q(P)| = \infty$, so $d_v(P, Q) = 1$.)

Remark 2.1. In practice, we fix the point Q and use the distance function $d_v(P, Q)$ to measure the distance from P to Q as P varies. It is clear that the distance function d_v has the right qualitative property, i.e., $d_v(P, Q)$ is small if P is v-adically close to Q. On the other hand, the value of $d_v(P, Q)$ certainly depends on the choice of the function t_Q, so possibly a better notation would be $d_v(P, t_Q)$. However, since we will use d_v only to measure the rate at which a varying point approaches a fixed point, the next result shows that the choice of t_Q is irrelevant for the statements of our theorems.

Proposition 2.2. *Let* $Q \in C(K_v)$ *and let* $F \in K_v(C)$ *be a function that vanishes at* Q. *Then the limit*

---

[1]To see that t_Q exists, we use the the the Riemann–Roch theorem. Thus (II.5.5c) tells us that if C has genus g and if $e \geq g + 1$, then $\ell(e(Q)) \geq 2$, so there is a nonconstant function $f \in \mathcal{L}(e(Q))$. This function f has a pole at Q and no other poles, and we can take $t_Q = 1/f$.

$$\lim_{\substack{P \in C(K_v) \\ P \xrightarrow{v} Q}} \frac{\log|F(P)|_v}{\log d_v(P, Q)} = \operatorname{ord}_Q(F)$$

exists and is independent of the choice of the function t_Q used to define $d_v(P,Q)$.

Here $\operatorname{ord}_Q(F)$ *is the order of vanishing of F at Q as in* (II §2), *while the notation $P \xrightarrow{v} Q$ means that $P \in C(K_v)$ approaches Q in the v-adic topology, i.e., $d_v(P,Q) \to 0$.*

PROOF. Let t_Q be the function vanishing only at Q that we are using to define $d_v(\,\cdot\,,Q)$. Let $e = \operatorname{ord}_Q(t_Q)$ and $f = \operatorname{ord}_Q(F)$. Then the function $\phi = F^e/t_Q^f$ has neither a zero nor a pole at Q, so $|\phi(P)|_v$ is bounded away from 0 and ∞ as $P \xrightarrow{v} Q$. Hence

$$\lim_{\substack{P \in C(K_v) \\ P \xrightarrow{v} Q}} \frac{\log|F(P)|_v}{\log d_v(P, t_Q)} = \lim_{\substack{P \in C(K_v) \\ P \xrightarrow{v} Q}} \frac{\log|F(P)|_v}{\log|t_Q(P)|_v^{1/e}}$$

$$= f + \lim_{\substack{P \in C(K_v) \\ P \xrightarrow{v} Q}} \frac{1}{e} \cdot \frac{\log|\phi(P)|_v}{\log|t_Q(P)|_v}$$

$$= f. \qquad \square$$

Remark 2.2.1. The use of the function t_Q in the definition of distance is somewhat artificial and does not generalize well to higher-dimensional varieties. An alternative definition that does generalize uses a finite list of functions $t_1, \ldots, t_r \in K(E)$ with the property that each t_i vanishes at Q and such that $t_1, \ldots, t_r$ have no other common zeros. Then, if we let e_i denote the order of vanishing of t_i at Q, a distance function d_v may be defined by

$$d_v(P, Q) = \min\left\{ \max\left\{ |t_1(P)|_v^{1/e_1}, \ldots, |t_r(P)|_v^{1/e_r} \right\}, 1 \right\}.$$

This function is an example of a local height function; see [139, Chapter 10], [114, §B.8], or [261] for further details.

Next we examine the effect of finite maps on the distance between points. The crucial observation is that this effect depends on the ramification of the map, not on its degree. To see the difference, compare (IX.2.3) with (VIII.5.6).

Proposition 2.3. *Let C_1/K and C_2/K be curves, and let $\phi : C_1 \to C_2$ be a finite map defined over K. Let $Q \in C_1(K_v)$, and let $e_\phi(Q)$ be the ramification index of ϕ at Q (II §2). Then*

$$\lim_{\substack{P \in C_1(K_v) \\ P \xrightarrow{v} Q}} \frac{\log d_v\big(\phi(P), \phi(Q)\big)}{\log d_v(P, Q)} = e_\phi(Q).$$

PROOF. Let $t_Q \in K_v(C_1)$ be a function that vanishes to order $e_1 \geq 1$ at Q and has no other zeros, and similarly let $t_{\phi(Q)} \in K_v(C_2)$ be a function that vanishes to order $e_2 \geq 1$ at $\phi(Q)$ and has no other zeros. It follows from the definition of ramification index that

$$\mathrm{ord}_Q \, t_{\phi(Q)} \circ \phi = e_\phi(P) \, \mathrm{ord}_{\phi(Q)} \, t_{\phi(Q)} = e_\phi(P)e_2,$$

so the functions $(t_{\phi(Q)} \circ \phi)^{e_1}$ and $t_Q^{e_\phi(P)e_2}$ vanish to the same order at Q. Hence the function

$$f = \frac{(t_{\phi(Q)} \circ \phi)^{e_1}}{t_Q^{e_\phi(Q)e_2}} \in K_v(C_1)$$

has neither a zero nor a pole at Q. It follows that $\left|f(P)\right|_v$ is bounded away from 0 and ∞ as $P \underset{v}{\to} Q$. Therefore

$$\frac{\log d_v\big(\phi(P), \phi(Q)\big)}{\log d_v(P, Q)} = \frac{\log\left|t_{\phi(Q)}\big(\phi(P)\big)\right|_v^{1/e_2}}{\log\left|t_Q(P)\right|_v^{1/e_1}}$$

$$= \frac{e_\phi(Q) \log\left|t_Q(P)\right|_v^{1/e_1} + \log\left|f(P)\right|_v}{\log\left|t_Q(P)\right|_v^{1/e_1}}$$

$$\underset{v}{\longrightarrow} e_\phi(Q) \qquad \text{as } P \underset{v}{\to} Q. \qquad \square$$

Finally, we reinterpret Roth's theorem (IX.1.4) in terms of distance functions.

Corollary 2.4. (of (IX.1.4)) *Fix an absolute value $v \in M_K$. Let C/K be a curve, let $f \in K(C)$ be a nonconstant function, and let $Q \in C(\bar{K})$. Then*

$$\liminf_{\substack{P \in C(K) \\ P \underset{v}{\to} Q}} \frac{\log d_v(P, Q)}{\log H_K\big(f(P)\big)} \geq -2.$$

(If Q is not a v-adic accumulation point of $C(K)$, then we define the lim inf to be 0.)

PROOF. Replacing f by $1/f$ if necessary, we may assume that $f(Q) \neq \infty$. (Note that $H_K\big((1/f)(P)\big) = H_K\big(f(P)\big)$.) The function $f - f(Q)$ vanishes at Q, say to order e, so (IX.2.2) tells us that

$$\liminf_{\substack{P \in C(K) \\ P \underset{v}{\to} Q}} \frac{\log\left|f(P) - f(Q)\right|_v}{d_v(P, Q)} = e.$$

Hence

$$\liminf_{\substack{P \in C(K) \\ P \underset{v}{\to} Q}} \frac{\log d_v(P, Q)}{\log H_K\big(f(P)\big)} = \liminf_{\substack{P \in C(K) \\ P \underset{v}{\to} Q}} \frac{\log\left|f(P) - f(Q)\right|_v}{e \log H_K\big(f(P)\big)}$$

$$= \frac{1}{e} \liminf_{\substack{P \in C(K) \\ P \underset{v}{\to} Q}} \left(\frac{\log\big(H_K\big(f(P)\big)^\tau \left|f(P) - f(Q)\right|_v\big)}{\log H_K\big(f(P)\big)} - \tau \right).$$

We now set $\tau = 2 + \epsilon$. Then (IX.1.4) implies that

$$H_K\big(f(P)\big)^\tau \big|f(P) - f(Q)\big|_v \geq 1$$

for all but finitely many $P \in C(K)$. Therefore

$$\liminf_{\substack{P \in C(K) \\ P \xrightarrow{v} Q}} \frac{\log d_v(P, Q)}{\log H_K\big(f(P)\big)} \geq -\frac{\tau}{e} \geq -\frac{2 + \epsilon}{e}.$$

Since $\epsilon > 0$ is arbitrary and $e \geq 1$, this is the desired result.   □

IX.3   Siegel's Theorem

In this section we prove a result of Siegel that represents a significant improvement on the Diophantine approximation result (IX.2.4).

Theorem 3.1. (Siegel) *Let E/K be an elliptic curve with $\#E(K) = \infty$. Fix a point $Q \in E(\bar{K})$, a nonconstant even function $f \in E(K)$, and an absolute value $v \in M_{K(Q)}$. Then*

$$\lim_{\substack{P \in E(K) \\ h_f(P) \to \infty}} \frac{\log d_v(P, Q)}{h_f(P)} = 0.$$

Remark 3.1.1. Although we prove (IX.3.1) only for even functions, it is in fact true in general; see Exercise 9.14d.

Before proving (IX.3.1), we give some indication of its power.

Corollary 3.2.1. *Let E/K be an elliptic curve with Weierstrass coordinate functions x and y, let $S \subset M_K$ be a finite set of places containing M_K^∞, and let R_S be the ring of S-integers of K. Then*

$$\{P \in E(K) : x(P) \in R_S\}$$

is a finite set.

PROOF. We apply (IX.3.1) with the function $f = x$. Suppose that there is a sequence of distinct points $P_1, P_2, \ldots \in E(K)$ with every $x(P_i) \in R_S$. The definition of height then tells us that

$$h_x(P_i) = \frac{1}{[K : \mathbb{Q}]} \sum_{v \in S} \log \max\{1, |x(P_i)|_v^{n_v}\},$$

since the terms with $v \notin S$ have $|x(P_i)|_v \leq 1$. Hence we can find a particular $v \in S$ and a subsequence of the P_i (which we relabel as $P_1, P_2, \ldots$) such that

$$h_x(P_i) \leq \#S \cdot \log|x(P_i)|_v \qquad \text{for all } i = 1, 2, \ldots.$$

(Note that $n_v \leq [K : \mathbb{Q}]$.) In particular, we see that $\left|x(P_i)\right|_v \to \infty$, and since O is the only pole of x, it follows that $d_v(P_i, O) \to \infty$.

The function x has a pole of order 2 at O and no other poles, so we may take as our distance function

$$d_v(P_i, O) = \min\{\left|x(P_i)\right|_v^{-1/2}, 1\}.$$

Then, for all sufficiently large i, we have

$$\frac{-\log d_v(P_i, O)}{h_x(P_i)} \geq \frac{1}{2\#S}.$$

This contradicts (IX.3.1), which says that the left-hand side approaches 0 as $i \to \infty$. □

It is clear that the proof of (IX.3.2.1) works for any even function, not just x, since (IX.3.1) is given for all even functions. However, it is possible to reduce the case of arbitrary (not necessarily even) functions to the special case given in (IX.3.2.1). This reduction step, which we now give, is important in its own right, since it is used both in Siegel's second proof of finiteness (IX.4.3.1) and with the effective methods provided by linear forms in logarithms (IX.5.7).

Corollary 3.2.2. *Let C/K be a curve of genus one, let $f \in K(C)$ be a nonconstant function, and let S and R_S be as in (IX.3.2.1). Then*

$$\{P \in C(K) : f(P) \in R_S\}$$

is a finite set. Further, (IX.3.2.2) follows formally from (IX.3.2.1).

PROOF. We are clearly proving something stronger if we extend the field K and enlarge the set S. We may thus assume that $C(K)$ contains a pole Q of f, and taking Q to be the identity element, we view (C, Q) as an elliptic curve defined over K. Let x and y be coordinates on a Weierstrass equation for (C, Q), which we may take in the form

$$y^2 = x^3 + Ax + B.$$

We have $f \in K(C) = K(x, y)$ and $\left[K(x, y) : K(x)\right] = 2$, so we can write

$$f(x, y) = \frac{\phi(x) + \psi(x)y}{\eta(x)}$$

with polynomials $\phi(x), \psi(x), \eta(x) \in K[x]$. Further, since

$$\mathrm{ord}_Q(x) = -2, \qquad \mathrm{ord}_Q(y) = -3, \qquad \text{and} \qquad \mathrm{ord}_Q(f) < 0,$$

it follows that

$$2 \deg \eta < \max\{2 \deg \phi, 2 \deg \psi + 3\}.$$

(This is the condition for f to have a pole at Q.) Next we compute

$$\left(f\eta(x) - \phi(x)\right)^2 = \left(\psi(x)y\right)^2 = \psi(x)^2(x^3 + Ax + B).$$

Writing this out as a polynomial in x with coefficients in $K[f]$, we see that the highest power of x comes from one of the three terms $f^2\eta(x)^2$, $\phi(x)^2$, $\psi(x)^2 x^3$. From above, the first of these has lower degree in x than the latter two, while the leading terms of $\phi(x)^2$ and $\psi(x)^2 x^3$ cannot cancel, since they have different degrees. (One has even degree, the other odd degree.) It follows that x satisfies a *monic* polynomial with coefficients in $K[f]$, i.e., x is integral over $K[f]$. Multiplying this monic polynomial by an appropriate element of K to "clear denominators," we have shown that x satisfies a relation

$$a_0 x^N + a_1(f)x^{N-1} + \cdots + a_{N-1}(f)x + a_N(f) = 0,$$

where $a_0 \in R_S$ is nonzero and $a_i(f) \in R_S[f]$ for $1 \le i \le N$. Enlarging the set S, we may assume that $a_0 \in R_S^*$, and then dividing the polynomial by a_0, we may assume that $a_0 = 1$.

Now suppose that $P \in C(K)$ satisfies $f(P) \in R_S$. Then P is not a pole of x, and the relation

$$x(P)^N + a_1\bigl(f(P)\bigr)x(P)^{N-1} + \cdots + a_{N-1}\bigl(f(P)\bigr)x(P) + a_N\bigl(f(P)\bigr) = 0$$

shows that $x(P)$ is integral over R_S. Since also $x(P) \in K$ and R_S is integrally closed, it follows that $x(P) \in R_S$. This proves that

$$\{P \in C(K) : f(P) \in R_S\} \subset \{P \in C(K) : x(P) \in R_S\},$$

and thus the finiteness assertion in (IX.3.2.1) implies the desired finiteness result described in (IX.3.2.2).   □

Example 3.3. Consider the Diophantine equation

$$y^2 = x^3 + Ax + B,$$

where $A, B \in \mathbb{Z}$ and $4A^3 + 27B^2 \neq 0$. The corollary (IX.3.2.1) says that this equation has only finitely many solutions $x, y \in \mathbb{Z}$. What does (IX.3.1) say in this situation, say if we take $Q = O$, $f = x$, and v the archimedean absolute value on $\mathbb{Q}$?

Label the nonzero rational points $P_1, P_2, \ldots \in E(\mathbb{Q})$ in order of nondecreasing height, and write

$$x(P_i) = \frac{a_i}{b_i} \in \mathbb{Q}$$

as a fraction in lowest terms. Then

$$\log d_v(P_i, O) = \frac{1}{2} \log \min\left\{\left|\frac{b_i}{a_i}\right|, 1\right\},$$

$$h_x(P_i) = \log \max\{|a_i|, |b_i|\}.$$

(Note that the $\frac{1}{2}$ appears because x^{-1} has a zero of order 2 at O.) We see from (IX.3.1) that

$$\lim_{i \to \infty} \frac{\min\{\log|b_i/a_i|, 0\}}{\max\{\log|a_i|, \log|b_i|\}} = 0.$$

Next let Q_1 and Q_2 be the zeros of the function x, where we allow $Q_1 = Q_2$. Then it is not hard to check that

$$\log\min\{|x(P)|_v, 1\} = d_v(P, Q_1) + d_v(P, Q_2) + O(1) \qquad \text{for all } P \in E(K_v),$$

where the $O(1)$ depends on the choice of the distance functions $d_v(\cdot, Q_i)$, but is independent of P; see Exercise 9.16. Writing $v \in M_{\mathbb{Q}}^\infty$ for the usual archimedean absolute value on $\mathbb{Q}$, we use (IX.3.1) twice to obtain

$$\lim_{i \to \infty} \frac{\min\{\log|a_i/b_i|, 0\}}{\max\{\log|a_i|, \log|b_i|\}} = \lim_{i \to \infty} \frac{\log\min\{|x(P_i)|, 1\}}{h_x(P_i)}$$

$$= \lim_{i \to \infty} \frac{d_v(P_i, Q_1) + d_v(P_i, Q_2) + O(1)}{h_x(P_i)}$$

$$= 0.$$

Finally, combining the limit involving b_i/a_i with the limit involving a_i/b_i, it is easy to deduce that

$$\lim_{i \to \infty} \frac{\log|a_i|}{\log|b_i|} = 1.$$

In other words, when looking at the x-coordinates of the rational points on an elliptic curve, we will see that the numerators and the denominators tend to have about the same number of digits. This is a much stronger assertion than (IX.3.2.1), which merely says that there are only finitely many points whose denominator is 1.

Remark 3.4. Siegel's theorem (IX.3.2.1) is not effective, which means that the proof does not give an explicitly computable upper bound for the height of all integral points. However, Siegel's proof can be made quantitative in the following sense; see for example [81]:

> Given a nonsingular Weierstrass equation with coefficients in a number field K and given a finite set of absolute values S, there is a constant N, which can be explicitly calculated in terms of the field K, the set S, and the coefficients of the equation, such that the equation has no more than N integral solutions.

A subtler Diophantine problem, motivated by work of Dem'janenko and posed as a general conjecture by Serge Lang, is to give an intrinsic relationship between the number of integral points and the rank of the Mordell–Weil group.

Conjecture 3.5. (Lang [135, page 140]) *Let E/K be an elliptic curve, and choose a quasiminimal Weierstrass equation for E/K,*

$$E : y^2 = x^3 + Ax + B.$$

(See Exercise 8.14c.) Let $S \subset M_K$ be a finite set of places containing M_K^∞, and let R_S be the ring of S-integers of K. There exists a constant C, depending only on K, such that

$$\#\{P \in E(K) : x(P) \in R_S\} \leq C^{\#S + \operatorname{rank} E(K)}.$$

This conjecture is known to be true if one restricts attention to elliptic curves having integral j-invariant. More generally, the following is known.

Theorem 3.6. *Let E/K, S, and R_S be as in (IX.3.5).*

(a) (Silverman [104, 262]) *There is a constant C, depending only on $[K : \mathbb{Q}]$ and on the number of places $v \in M_K^0$ with $\mathrm{ord}_v(j_E) < 0$, such that*

$$\#\{P \in E(K) : x(P) \in R_S\} \leq C^{\#S + \mathrm{rank}\, E(K)}.$$

(b) (Hindry–Silverman [113]) *Assume that the ABC conjecture (with any expo-nent) (VIII.8.4), (VIII.8.6) is true for the field K. Then there is a constant C, depending only on $[K : \mathbb{Q}]$ and on the constants appearing in the ABC conjec-ture, such that*

$$\#\{P \in E(K) : x(P) \in R_S\} \leq C^{\#S + \mathrm{rank}\, E(K)}.$$

We turn now to the proof of (IX.3.1). In broad outline, the argument goes as follows. Our theorem on Diophantine approximation (IX.2.4) gives us a bound, in terms of the height of P, on how fast P can approach Q. Suppose now that we write $P = [m]P' + R$ and $Q = [m]Q' + R$. Then (IX.2.3) tells us that the dis-tance from P' to Q' is about the same as the distance from P to Q, since the map $P \mapsto [m]P + R$ is unramified. On the other hand, the height of P' is much smaller than the height of P. Now applying (IX.2.4) to P' and Q' gives an improved estimate, and taking m sufficiently large gives the desired result.

PROOF OF (IX.3.1). Choose a sequence of distinct points $P_i \in E(K)$ satisfying

$$\lim_{i \to \infty} \frac{\log d_{\ell}(P_i, Q)}{h_f(P_i)} = L = \liminf_{\substack{P \in E(K) \\ h_f(P) \to \infty}} \frac{\log d_v(P, Q)}{h_f(P)}.$$

Since $d_v(P, Q) \leq 1$ and $h_f(P) \geq 0$ for all points $P \in E(K)$, we have $L \leq 0$. It thus suffices to prove that $L \geq 0$.

Let m be a large integer. From the weak Mordell–Weil theorem (VIII.1.1), the quotient group $E(K)/mE(K)$ is finite. Hence some coset contains infinitely many of the P_i. Replacing $\{P_i\}$ by a subsequence, we may assume that

$$P_i = [m]P_i' + R,$$

where $P_i, R \in E(K)$ and where R does not depend on i. We use standard properties of height functions to compute

$$\begin{aligned}
m^2 h_f(P_i') &= h_f([m]P_i') + O(1) &&\text{using (VIII.6.4b),} \\
&= h_f(P_i - R) + O(1) \\
&\leq 2h_f(P_i) + O(1) &&\text{using (VIII.6.4a).}
\end{aligned}$$

Note that the $O(1)$ is independent of i.

We next do an analogous computation with distance functions. If P_i is bounded away from Q in the v-adic topology, then $\log d_v(P_i, Q)$ is bounded, so clearly $L = 0$. Otherwise we can replace P_i with a subsequence such that $P_i \xrightarrow{v} Q$. It follows that $[m]P_i' \xrightarrow{v} Q - R$, so the sequence P_i' accumulates to at least one of the m^2 possible m^{th} roots of $Q - R$. Again taking a subsequence, we can find a point $Q \in E(\bar{K})$ satisfying

$$P_i' \xrightarrow{v} Q' \qquad \text{and} \qquad Q = [m]Q' + R.$$

We next observe that the map $E \to E$ defined by $P \mapsto [m]P + R$ is everywhere unramified (III.4.10c), so (IX.2.3) tells us that

$$\lim_{i \to \infty} \frac{\log d_v(P_i, Q)}{\log d_v(P_i', Q')} = 1.$$

Combining this with the height inequality yields

$$L = \lim_{i \to \infty} \frac{\log d_v(P_i, Q)}{h_f(P_i)} \geq \lim_{i \to \infty} \frac{\log d_v(P_i', Q')}{\frac{1}{2}m^2 h_f(P_i') + O(1)}.$$

(Note that the $\log d_v$ expressions are negative, which reverses the inequality.)

We now apply the theorem on Diophantine approximation (IX.2.4) to the sequence $\{P_i'\} \subset E(K)$ as it converges v-adically to $Q' \in E(\bar{K})$. This yields

$$\liminf_{i \to \infty} \frac{\log d_v(P_i', Q')}{[K : \mathbb{Q}]h_f(P_i')} \geq -2.$$

(The factor of $[K : \mathbb{Q}]$, which in any case is not important, arises because h_f is the absolute height, while (IX.2.4) is stated using the relative height H_K.) Combining the last two inequalities yields

$$L \geq -\frac{4[K : \mathbb{Q}]}{m^2}.$$

The field K is fixed, while the value of m is arbitrary, which completes the proof that $L \geq 0$. $\square$

IX.4   The S-Unit Equation

The finiteness of S-integral points on elliptic curves (IX.3.2.1) is a special case of Siegel's general result that an (affine) curve C/K of genus at least one has only finitely many S-integral points; see [114, Theorem D.9.1] or [139, Chapter 8, Theorem 2.4]. Of course, for curves C of genus two or greater, Siegel's result is superseded by Faltings' theorem [82, 84], which asserts that the full set of rational points $C(K)$ is finite.

Siegel gave a second proof of his theorem that applies to a restricted set of curves, but that does include all elliptic curves. This second method is important because, when combined with results from linear forms in logarithms (XI §5), it leads to an effective procedure for finding all S-integral points. In this section we describe Siegel's alternative proof.

The idea is to reduce the problem of solving for S-integral points on a curve to the problem of solving several equations of the form

$$ax + by = 1$$

in S-units. We start with a quick sketch of how solving this S-unit equation can be reduced to a Diophantine approximation theorem such as (IX.1.4). This ineffective theorem can then be replaced by an effective estimate as described in (IX §5).

Theorem 4.1. *Let $S \subset M_K$ be a finite set of places, and let $a, b \in K^*$. Then the equation*

$$ax + by = 1$$

has only finitely many solutions in S-units $x, y \in R_S^*$.

INEFFECTIVE PROOF (SKETCH). Let m be a large integer. Dirichlet's S-unit theorem [142, V §1] implies that the quotient group $R_S^*/(R_S^*)^m$ is finite, so we can choose a finite set of coset representatives $c_1, \ldots, c_r \in R_S^*$. Then any solution (x, y) to the original equation can be written as

$$x = c_i X^m, \qquad y = c_j Y^m,$$

for some $X, Y \in R_S^*$ and some choice of c_i and c_j, and thus (X, Y) is a solution to the equation

$$ac_i X^m + bc_j Y^m = 1.$$

Since there are only finitely many choices for c_i and c_j, it suffices to prove that for any $\alpha, \beta \in K^*$, the equation

$$\alpha X^m + \beta Y^m = 1$$

has only finitely many solutions $X, Y \in R_S$.

Suppose that there are infinitely many such solutions. Then, since

$$H_K(Y) = \prod_{v \in S} \max\{1, |Y|_v^{n_v}\},$$

we can choose some $v \in S$ so that there are infinitely many solutions satisfying

$$|Y|_v \geq H_K(Y)^{1/([K:\mathbb{Q}]\#S)}.$$

(Note that $n_v \leq [K : \mathbb{Q}]$.) Let $\gamma \in \bar{K}$ be a solution to

$$\gamma^m = -\beta/\alpha.$$

We will specify later which m^{th} root to take. The idea is that if m is large enough, then X/Y provides too close an approximation to γ.

We factor the left-hand side of the equation $\alpha X^m + \beta Y^m = 1$ to obtain

$$\prod_{\zeta \in \mu_m} \left(\frac{X}{Y} - \zeta \gamma \right) = \frac{1}{\alpha Y^m}.$$

Since there are supposed to be infinitely many solutions, we may assume that $H_K(Y)$ is large, so also $|Y|_v$ is large. Then from the equality

$$\prod_{\zeta \in \mu_m} \left| \frac{X}{Y} - \zeta \gamma \right|_v = \frac{1}{|\alpha Y^m|_v},$$

we see that X/Y must be close to one of the $\zeta \gamma$ values. Replacing γ by the appropriate $\zeta \gamma$, we may assume that $|X/Y - \gamma|_v$ is quite small. But then $|X/Y - \zeta \gamma|_v$ cannot be too small for $\zeta \neq 1$, since

$$\left| \frac{X}{Y} - \zeta \gamma \right|_v \geq |\gamma(1 - \zeta)|_v - \left| \frac{X}{Y} - \gamma \right|_v.$$

Hence we can find a constant C_1, independent of X/Y, such that

$$\left| \frac{X}{Y} - \gamma \right| \leq \frac{C_1}{|Y|_v^m}.$$

(See Exercise 9.5.) Finally, from the expression

$$\alpha \left(\frac{X}{Y} \right)^m = \left(\frac{1}{Y} \right)^m - \beta,$$

one easily deduces that

$$H_K \left(\frac{X}{Y} \right) \leq C_2 H_K(Y),$$

where C_2 depends on only α, β, and m. Combining all of the above estimates yields

$$\left| \frac{X}{Y} - \gamma \right|_v \leq C H_K \left(\frac{X}{Y} \right)^{-m/([K:\mathbb{Q}]\#S)}.$$

But if we take $m > 2[K : \mathbb{Q}]\#S$, then Roth's theorem (IX.1.4) says that there are only finitely many possibilities for X/Y. Further, since

$$Y^m = \left(\alpha \left(\frac{X}{Y} \right)^m + \beta \right)^{-1} \qquad \text{and} \qquad X = \left(\frac{X}{Y} \right) Y,$$

each ratio X/Y corresponds to at most m possible pairs (X, Y). This contradicts our original assumption that there are infinitely many solutions, which completes the proof of (IX.4.1). $\square$

Remark 4.2.1. There is a great similarity in the methods of proof for Siegel's theorem (IX.3.1) and the S-unit equation (IX.4.1). In both cases, we start with a point in a finitely generated group, namely $P \in E(K)$ for the former and $(x, y) \in R_S^* \times R_S^*$ for the latter. Next we pull back using the multiplication-by-m map in the group to produce a new solution whose height is much smaller than the original solution but that closely approximates another point defined over a finite extension of K. Finally, we invoke a theorem on Diophantine approximation, such as (IX.1.4), to complete the proof.

Remark 4.2.2. The proof that we have given for (IX.4.1) is ineffective because it makes use of Roth's theorem (IX.1.4). However, just as for Siegel's theorem, it is possible to make (IX.4.1) *quantitative*, i.e., to give an upper bound on the number of solutions. One might expect, a priori, that such a bound would depend on the field K and on the set of primes S, but Evertse proved the following uniform result for the S-unit equation that is an analogue of Lang's conjecture (IX.3.5) for elliptic curves. The proof, which we omit, is quite intricate.

Theorem 4.2.3. (Evertse [80]) *Let* $S \subset M_K$ *be a finite set of places containing* M_K^∞, *and let* $a, b \in K^*$. *Then the equation*

$$ax + by = 1$$

has at most $3 \times 7^{[K:\mathbb{Q}]+2\#S}$ *solutions in S-units* $x, y \in R_S^*$.

To see the analogy with (IX.3.5), note that R_S^* is a finitely generated group of rank $\#S - 1$. Thus the bound in (IX.3.5) has the form $C^{\text{rank } R_S^* + \text{rank } E(K) + 1}$, while the bound in (IX.4.2.3) may be written as $C^{\text{rank } R_S^* + 1}$.

We next describe Siegel's reduction of S-integral points on hyperelliptic curves to solutions of the S-unit equation. Although we do not do so, the reader should note that every step in this reduction process can be made effective.

Theorem 4.3. (Siegel) *Let* $f(x) \in K[x]$ *be a polynomial of degree* $d \geq 3$ *with distinct roots in* $\bar{K}$. *Then the equation*

$$y^2 = f(x)$$

has only finitely many solutions in S-integers $x, y \in R_S$.

PROOF. We are clearly proving something stronger if we take a finite extension of K and enlarge the set S. Thus we may assume that f splits over K, say

$$f(x) = a(x - \alpha_1)(x - \alpha_2) \cdots (x - \alpha_d) \qquad \text{with } \alpha_1, \ldots, \alpha_d \in K.$$

Enlarging S, we may assume that the following statements are true:

    (i) $a \in R_S^*$.

    (ii) $\alpha_i - \alpha_j \in R_S^*$ for all $i \neq j$.

    (iii) R_S is a principal ideal domain.

Now suppose that $x, y \in R_S$ satisfy $y^2 = f(x)$. Let $\mathfrak{p}$ be a prime ideal of R_S. Then $\mathfrak{p}$ divides at most one $x - \alpha_i$, since if it divides both $x - \alpha_i$ and $x - \alpha_j$, then it divides $\alpha_i - \alpha_j$, contradicting (ii). Further, we see from (i) that $\mathfrak{p}$ does not divide a. It follows from the equation

$$y^2 = a(x - \alpha_1)(x - \alpha_2) \cdots (x - \alpha_d)$$

that $\mathrm{ord}_{\mathfrak{p}}(x - \alpha_i)$ is even, and since this is true for all primes, the ideal $(x - \alpha_i)R_S$ is the square of an ideal in R_S. From (iii) we know that R_S is a principal ideal domain, so there are elements $z_i \in R_S$ and units $b_i \in R_S^*$ such that

$$x - \alpha_i = b_i z_i^2 \qquad \text{for} \quad i = 1, 2, \ldots, d.$$

Now let L/K be the extension of K obtained by adjoining to K the square root of every element of R_S^*. Note that L/K is a finite extension, since Dirichlet's S-unit theorem tells us that $R_S^*/(R_S^*)^2$ is finite. Let $T \subset M_L$ be the set of places of L lying over elements of S, and let R_T be the ring of T-integers in L. By construction, each b_i is a square in R_T, say $b_i = \beta_i^2$, so

$$x - \alpha_i = (\beta_i z_i)^2.$$

Taking the difference of any two of these equations yields

$$\alpha_j - \alpha_i = (\beta_i z_i - \beta_j z_j)(\beta_i z_i + \beta_j z_j).$$

Note that $\alpha_j - \alpha_i \in R_T^*$, while each of the two factors on the right is in R_T. It follows that each of these factors is a unit,

$$\beta_i z_i \pm \beta_j z_j \in R_T^* \qquad \text{for } i \neq j.$$

To complete the proof we use *Siegel's identity*:

$$\frac{\beta_1 z_1 \pm \beta_2 z_2}{\beta_1 z_1 - \beta_3 z_3} \mp \frac{\beta_2 z_2 \pm \beta_3 z_3}{\beta_1 z_1 - \beta_3 z_3} = 1.$$

This gives two elements of R_T^* that sum to 1, so (IX.4.1) says that there are only finitely many choices for

$$\frac{\beta_1 z_1 + \beta_2 z_2}{\beta_1 z_1 - \beta_3 z_3} \quad \text{and} \quad \frac{\beta_1 z_1 - \beta_2 z_2}{\beta_1 z_1 - \beta_3 z_3}.$$

Multiplying these two numbers, we find that there are only finitely many possibilities for

$$\frac{\alpha_2 - \alpha_1}{(\beta_1 z_1 - \beta_3 z_3)^2},$$

hence only finitely many for

$$\beta_1 z_1 - \beta_3 z_3,$$

and thus only finitely many for

$$\beta_1 z_1 = \frac{1}{2}\left((\beta_1 z_1 - \beta_3 z_3) + \frac{\alpha_3 - \alpha_1}{\beta_1 z_1 - \beta_3 z_3}\right).$$

Finally, since

$$x = \alpha_1 + (\beta_1 z_1)^2,$$

there are only finitely many possible values for x, and each x value gives at most two y values. $\qquad\square$

Corollary 4.3.1. *Let C/K be a curve of genus one and let $f \in K(C)$ be a nonconstant function. Then there are only finitely many points $P \in C(K)$ such that $f(P) \in R_S$.*

PROOF. The reduction procedure described in (IX.3.2.2) says that it suffices to consider the case that f is the x-coordinate of a Weierstrass equation. The case $f = x$ is covered by (IX.4.3). $\qquad\square$

IX.5   Effective Methods

In 1949, Gelfond and Schneider independently solved Hilbert's problem concerning the transcendence of $2^{\sqrt{2}}$. They actually proved the following strong transcendence criterion.

Theorem 5.1. (Gelfond, Schneider) *Let $\alpha, \beta \in \bar{\mathbb{Q}}$ with $\alpha \neq 0, 1$ and $\beta \notin \mathbb{Q}$. Then α^β is transcendental.*

Gelfond rephrased his result in terms of logarithms: If $\alpha_1, \alpha_2 \in \bar{\mathbb{Q}}^*$ and if $\log \alpha_1$ and $\log \alpha_2$ are linearly independent over $\mathbb{Q}$, then they are linearly independent over $\bar{\mathbb{Q}}$. He further showed that it is possible to give an explicit lower bound for

$$|\beta_1 \log \alpha_1 + \beta_2 \log \alpha_2|$$

whenever this quantity is nonzero, and he noted that many Diophantine problems could be solved effectively if one knew an analogous result for sums of arbitrarily many logarithms. Alan Baker proved such a theorem in 1966. The proof is quite involved, so we are content to quote the following version.

Theorem 5.2. (Baker) *Let $\alpha_1, \ldots, \alpha_n \in K^*$ and let $\beta_1, \ldots, \beta_n \in K$. For any constant κ, define*

$$\tau(\kappa) = \tau(\kappa; \alpha_1, \ldots, \alpha_n, \beta_1, \ldots, \beta_n) = h([1, \beta_1, \ldots, \beta_n]) h([1, \alpha_1, \ldots, \alpha_n])^\kappa.$$

N.B. These are logarithmic height functions. Fix an embedding $K \subset \mathbb{C}$ and let $|\cdot|$ be the corresponding absolute value. Assume that

$$\beta_1 \log \alpha_1 + \cdots + \beta_n \log \alpha_n \neq 0.$$

Then there are effectively computable constants $C > 0$ and $\kappa > 0$, depending only on n and $[K : \mathbb{Q}]$, such that

$$|\beta_1 \log \alpha_1 + \cdots + \beta_n \log \alpha_n| > C^{-\tau(\kappa)}.$$

PROOF. See [11] or [135, VIII, Theorem 1.1]. □

Remark 5.2.1. We have restricted ourselves in (XI.5.2) to the case of the archimedean absolute value. There are analogous results in the nonarchimedean case, although minor technical difficulties arise due to the fact that the p-adic logarithm is defined only in a neighborhood of 1. See (IX.5.6) for a further discussion.

It is not immediately clear how Baker's theorem (IX.5.2) can be applied to give a bound for the solutions of the S-unit equation. We start with an elementary lemma; see also Exercise 9.8.

Lemma 5.3. *Let V be a finite-dimensional vector space over $\mathbb{R}$. Given any basis $\mathbf{e} = \{e_1, \dots, e_n\}$ for V, let $\|\cdot\|_{\mathbf{e}}$ be the sup norm with respect to $\mathbf{e}$, i.e.,*

$$\|x\|_{\mathbf{e}} = \left\|\sum x_i e_i\right\|_{\mathbf{e}} = \max\{|x_i|\}.$$

Let $\mathbf{f} = \{f_1, \dots, f_n\}$ be another basis for V. There are positive constants c_1 and c_2, depending on $\mathbf{e}$ and $\mathbf{f}$, such that for all $x \in V$,

$$c_1\|x\|_{\mathbf{e}} \le \|x\|_{\mathbf{f}} \le c_2\|x\|_{\mathbf{e}}.$$

PROOF. Let $A = (a_{ij})$ be the change of basis matrix from $\mathbf{e}$ to $\mathbf{f}$, so $e_i = \sum_j a_{ij} f_j$, and let $\|A\| = \max\{|a_{ij}|\}$. Then for any $x = \sum_i x_i e_i \in V$ we have $x = \sum_{i,j} x_i a_{ij} f_j$, so

$$\|x\|_{\mathbf{f}} = \max_j \left\{\left|\sum_i x_i a_{ij}\right|\right\} \le n\max_{i,j}\{|a_{ij}|\}\max_i\{|x_i|\} = n\|A\|\,\|x\|_{\mathbf{e}}.$$

This gives one inequality, and the other follows by symmetry. □

We apply (IX.5.3) to the following situation. Let $S \subset M_K$ be a finite set of places containing M_K^∞, let $s = \#S$, and choose a basis $\alpha_1, \dots, \alpha_{s-1}$ for the free part of R_S^*. Then every $\alpha \in R_S^*$ can be written uniquely as

$$\alpha = \zeta\alpha_1^{m_1} \cdots \alpha_{s-1}^{m_{s-1}}$$

with integers $m_1, \dots, m_{s-1}$ and a root of unity ζ. Define the *size of α* (*relative to* $\{\alpha_1, \dots, \alpha_{s-1}\}$) by

$$m(\alpha) = \max\{|m_i|\}.$$

Lemma 5.4. *With notation as above, there are positive constants c_1 and c_2, depending only on K and S, such that every $\alpha \in R_S^*$ satisfies*

$$c_1 h(\alpha) \le m(\alpha) \le c_2 h(\alpha).$$

PROOF. Let $S = \{v_1, \dots, v_s\}$ and, to ease notation, let $n_i = n_{v_i}$ be the local degree corresponding to v_i. We consider the *S-regulator homomorphism*

$$\rho_S : R_S^* \longrightarrow \mathbb{R}^s, \qquad \alpha \longmapsto \big(n_1 v_1(\alpha), \dots, n_s v_s(\alpha)\big).$$

Note that the image of ρ_S lies in the hyperplane $H = \{x_1 + \cdots + x_s = 0\}$, and Dirichlet's S-unit theorem says that the image of ρ_S spans H. Let $\| \cdot \|_1$ be the sup norm on $\mathbb{R}^s$ relative to the standard basis, and let $\| \cdot \|_2$ be the sup norm relative to the basis

$$\{\rho_S(\alpha_1), \ldots, \rho_S(\alpha_{s-1}), (1, 1, \ldots, 1)\}.$$

Here $\rho_S(\alpha_1), \ldots, \rho_S(\alpha_{s-1})$ span H, and we have added one extra vector in order to span all of $\mathbb{R}^s$. From (IX.5.3) we find positive constants c_1 and c_2 such that

$$c_1 \|x\|_1 \leq \|x\|_2 \leq c_2 \|x\|_1 \qquad \text{for all } x \in \mathbb{R}^s.$$

Now let $\alpha \in R_S^*$ and write $\rho_S(\alpha) = \sum m_i \rho_S(\alpha_i)$. Then directly from the definitions we have

$$\|\rho_S(\alpha)\|_2 = \max\{|m_i|\} = m(\alpha),$$
$$\|\rho_S(\alpha)\|_1 = \max\{n_i |v_i(\alpha)|\},$$
$$h_K(\alpha) = \sum \max\{0, -n_i v_i(\alpha)\}.$$

(Note that the sum for $h_K(\alpha)$ needs to include only the absolute values in S, since by assumption $v(\alpha) = 0$ for all $v \notin S$.) It remains to compare $\|\rho_S(\alpha)\|_1$ and $h_K(\alpha)$.

In general, for any $x = (x_1, \ldots, x_s) \in H$, we can compare $\|x\|_1$ to the height $h(x) = \sum \max\{0, -x_i\}$. First, since $\max\{0, -x_i\} \leq |x_i|$, we have the obvious estimate

$$h(x) \leq s\|x\|_1.$$

On the other hand, if we sum the identity

$$x_i = \max\{0, x_i\} - \max\{0, -x_i\}$$

for $1 \leq i \leq s$ and use the fact that $x \in H$, i.e., $\sum x_i = 0$, we obtain

$$0 = h(-x) - h(x),$$

and hence $h(-x) = h(x)$. This allows us to compute

$$\begin{aligned} 2h(x) &= h(x) + h(-x) \\ &= \sum \left(\max\{0, -x_i\} + \max\{0, x_i\} \right) \\ &= \sum |x_i| \\ &\geq \max\{|x_i|\} \\ &= \|x\|_1. \end{aligned}$$

Thus $\frac{1}{2}\|x\|_1 \leq h(x) \leq s\|x\|_1$, and combining this with the earlier estimates gives the desired result,

$$(c_1/s) h_K(\alpha) \leq m(\alpha) \leq 2c_2 h_K(\alpha). \qquad \square$$

We now have the tools needed to show how solving the S-unit equation can be reduced to the problem of giving bounds for linear forms in logarithms.

Theorem 5.5. *Fix $a, b \in K^*$. There exists an effectively computable constant $C = C(K, S, a, b)$ such that any solution $(\alpha, \beta) \in R_S^* \times R_S^*$ to the S-unit equation*

$$a\alpha + b\beta = 1$$

satisfies $H(\alpha) < C$.

PROOF. Let (α, β) be a solution and choose the absolute value v in S for which $|\alpha|_v$ is largest. Then, since $|\alpha|_w = 1$ for all $w \notin S$, we have

$$|\alpha|_v^{[K:\mathbb{Q}]s} \geq \prod_{w \in S} \max\{1, |\alpha|_w^{n_w}\} = H_K(\alpha),$$

and hence

$$|\alpha|_v \geq H(\alpha)^{1/s}.$$

(Here, as usual, $s = \#S$.)

To simplify our discussion, we will assume that v is archimedean, which is certainly true if, for example, $S = M_K^\infty$. (For arbitrary S, see the discussion in (IX.5.6).) The mean value theorem applied to the function $\log(x)$ yields

$$\left| \frac{\log x - \log y}{x - y} \right| \leq \frac{1}{\min\{|x|, |y|\}}.$$

We apply this inequality with $x = a\alpha$ and $y = -b\beta$, so $x - y = 1$, and we obtain

$$|\log a\alpha - \log b\beta| \leq \min\{|a\alpha|, |a\alpha - 1|\}^{-1}$$
$$\leq 2\big(|a|H(\alpha)^{1/s}\big)^{-1}.$$

(For the last line, we have assumed that $|\alpha| > 2/|a|$, since otherwise we have the excellent bound $H(\alpha) \leq |\alpha|^s \leq (2/|a|)^s$.)

Let $\alpha_1, \ldots, \alpha_{s-1}$ be a basis for R_S^*, and write

$$\alpha = \zeta \alpha_1^{m_1} \cdots \alpha_{s-1}^{m_{s-1}} \qquad \text{and} \qquad \beta = \zeta' \alpha_1^{m_1'} \cdots \alpha_{s-1}^{m_{s-1}'}.$$

Substituting this into the previous inequality yields

$$\left| \sum_{i=1}^{s-1} (m_i - m_i') \log \alpha_i + \log\left(\frac{a\zeta}{b\zeta'}\right) \right| \leq \frac{c_1}{H(\alpha)^{1/s}},$$

where here and in what follows, the constants $c_1, c_2, \ldots$ are effectively computable and depend only on K, S, a, and b.

From the equality $a\alpha + b\beta = 1$, it is easy to obtain an estimate

$$\big|h(\alpha) - h(\beta)\big| \leq c_2,$$

and applying (IX.5.4) yields

$$c_3 m(\alpha) \leq m(\beta) \leq c_4 m(\alpha).$$

(Clearly we may assume that $m(\alpha) \geq 1$ and $m(\beta) \geq 1$.) In particular,

$$|m_i - m_i'| \leq m(\alpha) + m(\beta) \leq c_5 h(\alpha).$$

Letting $q_i = m_i - m_i'$ and $\gamma = a\zeta / b\zeta'$ to ease notation, we have the inequality

$$|q_1 \log \alpha_1 + \cdots + q_{s-1} \log \alpha_{s-1} + \log \gamma| \leq c_1 H(\alpha)^{-1/s}.$$

We now apply Baker's theorem (IX.5.2). This gives a lower bound of the form

$$|q_1 \log \alpha_1 + \cdots + q_{s-1} \log \alpha_{s-1} + \log \gamma| \geq c_6^{-\tau},$$

where

$$\tau = h\big([1, q_1, \ldots, q_{s-1}]\big) h\big([1, \alpha_1, \ldots, \alpha_{s-1}, \gamma]\big)^\kappa$$

and κ is a constant depending only on K and s. But from above,

$$h\big([1, q_1, \ldots, q_{s-1}]\big) = \log \max\{1, |q_1|, \ldots, |q_{s-1}|\} \leq \log\big(c_5 h(\alpha)\big).$$

Combining the upper and lower bounds for the linear form in logarithms and using this estimate yields

$$c_7^{\log(c_5 h(\alpha))} \leq c_1 H(\alpha)^{1/s}.$$

(Note that the basis $\alpha_1, \ldots, \alpha_{s-1}$ depends only on the field K and the set S, so we have absorbed the $h\big([1, \alpha_1, \ldots, \alpha_{s-1}, \gamma]\big)^\kappa$ into c_7.) Now a little bit of algebra gives

$$H(\alpha) \leq c_8 h(\alpha)^{c_9},$$

and since $h(\alpha) = \log H(\alpha)$, this implies the desired bound for $H(\alpha)$.   □

Remark 5.6. In order to apply the argument given in (IX.5.5) to a nonarchimedean absolute value, it is necessary to make some minor technical alterations. The main difficulty is that the logarithm function in the p-adic setting converges only in a neighborhood of 1. What one does is to take a subgroup of finite index in R_S^* that is generated by S-units that are p-adically close to 1, together with a uniformizer at p. Then, assuming that $|\alpha|_p$ is sufficiently large, one shows that $a\alpha/b\beta$ is p-adically close to 1. Now applying the above argument to some power of $a\alpha/b\beta$ gives a well-defined linear form in p-adic logarithms, and from then on the argument goes just the same. For the final step, of course, one must use a p-adic analogue of Baker's theorem. For further details on the reduction step, see for example [135, VI,§1].

Remark 5.7. In order to obtain an effective bound for the points on an elliptic curve satisfying $f(P) \in R_S$, where f is an arbitrary nonconstant function, it is necessary to make the reduction step given in (IX.3.2.2) effective. This essentially involves giving an effective version of the Riemann–Roch theorem, which has been done by Coates [48]. As the reader might guess from the number of reduction steps involved, the effective bounds that come out of the proofs are quite large. To indicate the magnitudes involved, we quote two results; see also (IX.7.2), and (IX.7.4).

Theorem 5.8. (a) (Baker [11, page 45]) *Let* $A, B, C, D \in \mathbb{Z}$ *satisfy*

$$\max\{|A|, |B|, |C|, |D|\} \leq H,$$

and assume that

$$E : Y^2 = AX^3 + BX^2 + CX + D$$

is an elliptic curve. Then any point $P = (x, y) \in E(\mathbb{Q})$ *with* $x, y \in \mathbb{Z}$ *satisfies*

$$\max\{|x|, |y|\} < \exp\left((10^6 H)^{10^6}\right).$$

(b) (Baker–Coates [12]) *Let* $F(X, Y) \in \mathbb{Z}[X, Y]$ *be an absolutely irreducible polynomial such that the curve* $F(X, Y) = 0$ *has genus one. Let* n *be the degree of* F, *and assume that the coefficients of* F *all have absolute value at most* H. *Then any solution to* $F(x, y) = 0$ *with* $x, y \in \mathbb{Z}$ *satisfies*

$$\max\{|x|, |y|\} < \exp\exp\exp\left((2H)^{10^{n^{10}}}\right).$$

Remark 5.8.1. There is an extensive literature on effective bounds for S-integral solutions to equations of the form $y^m = f(x)$; see for example [32, 96, 131, 268, 279, 301]. To quote one instance, we mention that [301] improves (IX.5.8a) to

$$\max\{|x|, |y|\} \leq \exp\left(cH^{270}(\log H)^{54}\right)$$

for an absolute constant c.

Linear Forms in Elliptic Logarithms

Rather than reducing the problem of integral points on an elliptic curve to the question of solutions to the S-unit equation, and thence as above to bounds for linear forms in logarithms, one can instead work directly with the analytic parametrization of the elliptic curve. We briefly indicate how this is done in the simplest case.

Let $E/\mathbb{Q}$ be an elliptic curve given by a Weierstrass equation

$$E : y^2 = 4x^3 - g_2 x - g_3 \qquad \text{with} \quad g_2, g_3 \in \mathbb{Z}.$$

We are interested in bounding the height of points $P \in E(\mathbb{Q})$ that satisfy $x(P) \in \mathbb{Z}$. Let

$$\phi : \mathbb{C}/\Lambda \longrightarrow E(\mathbb{C})$$

be the analytic parametrization of $E(\mathbb{C})$ given by the Weierstrass $\wp$-function and its derivative (VI.5.1.1). We fix a basis $\{\omega_1, \omega_2\}$ for the lattice Λ. Let

$$\psi : E(\mathbb{C}) \longrightarrow \mathbb{C}$$

be the map that is inverse to ϕ and takes values in the fundamental parallelogram spanned by ω_1 and ω_2, shifted to be centered at 0. The map ϕ is the *elliptic exponential map*, and choosing a fundamental domain for the *elliptic logarithm map* ψ is analogous to choosing a principal value for the ordinary logarithm function $\log : \mathbb{C}^* \to \mathbb{C}$. (The analogy becomes even clearer if we identify $\mathbb{C}^*$ with $\mathbb{C}/\mathbb{Z}$.)

Fix a basis $P_1, \ldots, P_r$ for the free part of $E(\mathbb{Q})$. Given any point $P \in E(\mathbb{Q})$, we can write

$$P = q_1 P_1 + \cdots + q_r P_r + T$$

with integers $q_1, \ldots, q_r$ and a torsion point $T \in E_{\text{tors}}(\mathbb{Q})$. It follows that

$$\psi(P) = q_1 \psi(P_1) + \cdots + q_r \psi(P_r) + \psi(T) \pmod{\Lambda},$$

so there are integers m_1 and m_2 such that

$$\psi(P) = q_1 \psi(P_1) + \cdots + q_r \psi(P_r) + \psi(T) + m_1 \omega_1 + m_2 \omega_2.$$

Suppose now that P is a large integral point, i.e., $x(P) \in \mathbb{Z}$ and $|x(P)|$ is large. Then P is close to O in the complex topology on $E(\mathbb{C})$, so $\psi(P)$ is close to 0. More precisely, since $\wp(z) = x(\phi(z))$ behaves like z^{-2} for z close to 0, we see that

$$\left|\psi(P)\right|^2 \leq c_1 \left|x(P)\right|^{-1} = c_1 H\big(x(P)\big)^{-1}.$$

We are using the fact that if $x \in \mathbb{Z}$ with $x \neq 0$, then $H(x) = |x|$. The constant c_1 depends on g_2 and g_3, but not on P.

On the other hand, since the canonical height is quadratic and positive definite from (VIII.9.3) and (VIII.9.6), we can estimate

$$\log H\big(x(P)\big) = h_x(P) = 2\hat{h}(P) + O(1)$$
$$= 2\hat{h}\left(\sum q_i P_i + T\right) + O(1)$$
$$\geq c_2 \max\{|q_i|\}^2,$$

where c_2 depends on E and the choice of the basis $P_1, \ldots, P_r$. (See Exercise 9.8.) Substituting this above, we obtain an upper bound for our linear form in elliptic logarithms,

$$\left|q_1 \psi(P_1) + \cdots + q_r \psi(P_r) + \psi(T) + m_1 \omega_1 + m_2 \omega_2\right| \leq c_3^{-\max\{|q_i|\}^2}.$$

Further, since ω_1 and ω_2 are $\mathbb{R}$-linearly independent, it is easy to see that

$$\max\{|m_1|, |m_2|\} \leq c_4 \max\{|q_i|\},$$

where c_4 depends on E, $\{P_i\}$, ω_1, and ω_2. Thus, if we let

$$q = \max\{|q_1|, \ldots, |q_r|, |m_1|, |m_2|\},$$

then we obtain the estimate

$$\left| q_1\psi(P_1) + \cdots + q_r\psi(P_r) + \psi(T) + m_1\omega_1 + m_2\omega_2 \right| \le c_5^{-q^2}.$$

Now the desired finiteness result follows if we can find a lower bound for the left-hand side having the form $C^{-\tau(q)}$ with $\tau(q)/q^2 \to 0$ as $q \to \infty$. The first effective estimate of this sort was proven by Masser [159] in the case that E has complex multiplication. The general case was proven by Wüstholz [313, 314], who had to overcome significant technical difficulties associated with the necessary zero and multiplicity estimates.

It remains to discuss the question of effectivity. The reduction to linear forms in ordinary logarithms via the S-unit equation is fully effective. It is possible to give an explicit upper bound for the height of any S-integral point of $E(K)$ in easily computed quantities associated to K, S, and E. One of these quantities, for example, is a bound for the heights of generators of the unit group R_S^*. In the analogous reduction to linear forms in elliptic logarithms, we similarly use a set of generators of the Mordell–Weil group $E(K)$, and the bound for the integral points depends on the heights of these generators. Unfortunately, as we have noted in (VIII.3.2) (see also Chapter X), the proof of the Mordell–Weil theorem is not effective. Thus although the approach to integral points on elliptic curves via elliptic logarithms is more natural than the roundabout route through the S-unit equation, it is likely to remain ineffective until an effective proof of the Mordell–Weil theorem is found. On the other hand, we should mention that if one is able to find a basis for the Mordell–Weil group, for example using the techniques in Chapter X, then the method of elliptic logarithms often provides the best known algorithm for finding the integral points on a given elliptic curve. See for example [58, 59, 96, 268, 279, 315].

IX.6 Shafarevich's Theorem

Recall that an elliptic curve E/K has good reduction at a finite place $v \in M_K$ if it has a Weierstrass equation whose coefficients are v-integral and whose discriminant is a v-adic unit (VII §5).

Theorem 6.1. (Shafarevich [242]) *Let $S \subset M_K$ be a finite set of places containing M_K^∞. Then up to isomorphism over K, there are only finitely many elliptic curves E/K having good reduction at all primes not in S.*

PROOF. Clearly we are proving something stronger if we enlarge S, so we may assume that S contains all primes of K lying over 2 and 3. Enlarging S further, we may also assume that the ring of S-integers R_S has class number one.

Under these assumptions, we see from (VIII.8.7) that any elliptic curve E/K has a Weierstrass equation of the form

$$E : y^2 = x^3 + Ax + B, \qquad A, B \in R_S,$$

with discriminant $\Delta = -16(4A^3 + 27B^2)$ satisfying

$$\Delta R_S = \mathcal{D}_{E/K} R_S.$$

Here $\mathcal{D}_{E/K}$ is the minimal discriminant of E/K; see (VIII §8). If we further assume that E has good reduction outside S, then $\mathrm{ord}_v(\mathcal{D}_{E/K}) = 0$ for all places $v \notin S$, so Δ is in R_S^*.

Assume now that we are given a list of elliptic curves $E_1/K, E_2/K, \ldots$, each of which has good reduction outside of S. We associate to each E_i a Weierstrass equation as above, say with coefficients $A_i, B_i \in R_S$ and discriminant $\Delta_i \in R_S^*$. Breaking the sequence of E_i into finitely many subsequences according to the residue class of Δ_i in the finite group $R_S^*/(R_S^*)^{12}$, we may replace the original sequence with an infinite subsequence satisfying $\Delta_i = CD_i^{12}$ for a fixed C and with $D_i \in R_S^*$.

The relation $\Delta = -16(4A^3 + 27B^2)$ implies that for each i, the point

$$\left(-\frac{12A_i}{D_i^4}, \frac{72B_i}{D_i^6} \right)$$

is an S-integral point on the elliptic curve

$$Y^2 = x^3 + 27C.$$

Siegel's theorem (IX.3.2.1) says that there are only finitely many such points, and thus only finitely many possibilities for A_i/D_i^4 and B_i/D_i^6. However, if

$$\frac{A_i}{D_i^4} = \frac{A_j}{D_j^4} \quad \text{and} \quad \frac{B_i}{D_i^6} = \frac{B_j}{D_j^6},$$

then the change of variables

$$x = (D_i/D_j)^2 x', \qquad y = (D_i/D_j)^3 y',$$

gives a K-isomorphism from E_i to E_j. Hence the sequence $E_1, E_2, \ldots$ contains only finitely many K-isomorphism classes of elliptic curves. $\qquad\square$

Example 6.1.1. There are no elliptic curves $E/\mathbb{Q}$ having everywhere good reduction; see Exercise 8.15. There are 24 curves $E/\mathbb{Q}$ having good reduction outside of $\{2\}$ and 784 curves $E/\mathbb{Q}$ having good reduction outside of $\{2,3\}$; for the complete list, see [19, Table 4]. Similar lists have been compiled for various quadratic fields; see for example [147] or [204].

Shafarevich's theorem (IX.6.1) has a number of important applications. We content ourselves with the following two corollaries.

Corollary 6.2. *Fix an elliptic curve E/K. Then there are only finitely many elliptic curves E'/K that are K-isogenous to E.*

PROOF. If E and E' are isogenous over K, then (VII.7.2) says that E and E' have the same set of primes of bad reduction. Now apply (IX.6.1). $\qquad\square$

Corollary 6.3. (Serre) *Let E/K be an elliptic curve with no complex multiplication. Then for all but finitely many primes ℓ, the group of ℓ-torsion points $E[\ell]$ has no nontrivial $G_{\bar{K}/K}$-invariant subgroups. (In other words, the representation of $G_{\bar{K}/K}$ on $E[\ell]$ is irreducible.)*

PROOF. Suppose that $\Phi_\ell \subset E[\ell]$ is a nontrivial $G_{\bar{K}/K}$-invariant subgroup of $E[\ell]$. We know that $E[\ell] \cong (\mathbb{Z}/\ell\mathbb{Z})^2$, so Φ_ℓ is necessarily cyclic of order ℓ. We apply (III.4.12) to produce an elliptic curve E_ℓ/K and an isogeny $\phi_\ell : E \to E_\ell$ with $\ker(\phi_\ell) = \Phi$. The Galois invariance of Φ ensures that the curve E_ℓ and the isogeny ϕ_ℓ are defined over K.

Each E_ℓ is K-isogenous to E, so (IX.6.2) says that the E_ℓ fall into finitely many K-isomorphism classes. Suppose that $E_\ell \cong E_{\ell'}$ for two primes ℓ and ℓ'. Then the composition

$$E \xrightarrow{\phi_\ell} E_\ell \cong E_{\ell'} \xrightarrow{\hat{\phi}_{\ell'}} E$$

defines an endomorphism of E of degree

$$(\deg \phi_\ell)(\deg \hat{\phi}_{\ell'}) = \ell\ell'.$$

By assumption, $\mathrm{End}(E) = \mathbb{Z}$, so every endomorphism of E has degree n^2 for some $n \in \mathbb{Z}$. This shows that $\ell = \ell'$, and thus that $E_\ell \not\cong E_{\ell'}$ for $\ell \neq \ell'$. Therefore there are only finitely many primes ℓ for which such a subgroup Φ_ℓ and curve E_ℓ can exist. $\square$

Example 6.4. For $K = \mathbb{Q}$, results of Mazur [166] and Kenku [125] give a statement that is far more precise than (IX.6.2). They show that for a given elliptic curve $E/\mathbb{Q}$, there are at most eight $\mathbb{Q}$-isomorphism classes of elliptic curves $E'/\mathbb{Q}$ that are $\mathbb{Q}$-isogenous to E. Further, if $\phi : E \to E'$ is a $\mathbb{Q}$-isogeny whose kernel is a cyclic group, then either

$$1 \leq \deg \phi \leq 19 \qquad \text{or} \qquad \deg \phi \in \{21, 25, 27, 37, 43, 67, 163\}.$$

It is no coincidence that the possibilities for $\deg \phi$ are values of d for which $\mathbb{Q}(\sqrt{-d})$ has class number one. The class number one condition means that the elliptic curve corresponding to the lattice

$$\mathbb{Z} + \mathbb{Z}\left(\tfrac{1}{2} + \tfrac{1}{2}\sqrt{-d}\right)$$

via (VI.5.1.1) is isomorphic to an elliptic curve defined over $\mathbb{Q}$. (See (C.11.3.1) for details.) Now we need merely observe that multiplication by $\sqrt{-d}$ gives an isogeny from E to itself that is defined over $\mathbb{Q}$ and whose kernel Φ is invariant under the action of $G_{\bar{\mathbb{Q}}/\mathbb{Q}}$. Then $E \to E/\Phi$ is a cyclic isogeny of degree d between elliptic curves defined over $\mathbb{Q}$.

Remark 6.5. An examination of the proof of (IX.6.1) reveals an interesting possibility. If we had some other proof of (IX.6.1) that did not use either Siegel's theorem or Diophantine approximation techniques, then we could deduce that the equation

$$Y^2 = X^3 + D$$

has only finitely many solutions $X, Y \in R_S$. For given such a solution, the equation

$$y^2 = x^3 - Xx - Y$$

defines an elliptic curve with good reduction outside of the set

$$S \cup \{\text{primes dividing 2 and 3}\}.$$

Hence, assuming (IX.6.1), there can be only finitely many such curves, and we could argue back to the finiteness of the number of pairs (X, Y). Building on this idea, Parshin [203] showed how a generalization of (IX.6.1) to curves of higher genus (which had already been conjectured by Shafarevich [242]) could be used to prove Mordell's conjecture that curves of genus at least 2 have only finitely many *rational* points. The subsequent proof of Shafarevich's conjecture by Faltings [82, 84] completed this chain of reasoning. Faltings' proof, together with Parshin's idea, also gives a proof of Siegel's theorem (IX.3.2) that does not involve the use of Diophantine approximation. Subsequent to Faltings' proof of the Mordell conjecture, Vojta [299] gave a somewhat more natural proof based on Diophantine approximation methods. For an exposition of this latter proof, see for example [114, Part E].

IX.7   The Curve $Y^2 = X^3 + D$

Many of the general results known and conjectured about the arithmetic of elliptic curves were originally noticed and tested on various special sorts of equations, such as the one given in the title of this section. For example, long before the work of Mordell and Siegel led to general finiteness results such as (IX.3.2.1), many special cases had been proven by a variety of methods. (See, e.g., [185, Chapter 26].) The next result gives two examples in which the complete set of integral solutions can be obtained by relatively elementary means.

Proposition 7.1. (a) (V.A. Lebesgue) *The equation*

$$y^2 = x^3 + 7$$

has no solutions in integers $x, y \in \mathbb{Z}$.

(b) (Fermat) *The only integral solutions to the equation*

$$y^2 = x^3 - 2$$

are $(x, y) = (3, \pm 5)$.

PROOF. (a) Suppose that $x, y \in \mathbb{Z}$ satisfy $y^2 = x^3 + 7$. We first observe that x must be odd, since no integer of the form $8k + 7$ is a square. Next we rewrite the equation as

$$y^2 + 1 = x^3 + 8 = (x + 2)(x^2 - 2x + 4).$$

Since x is odd,

$$x^2 - 2x + 4 = (x - 1)^2 + 3 \equiv 3 \pmod{4},$$

so there exists at least one prime $p \equiv 3 \pmod 4$ that divides $x^2 - 2x + 4$. But then $y^2 + 1 \equiv 0 \pmod p$, which is not possible.

(b) Suppose that we have a solution $x, y \in \mathbb{Z}$ to $y^2 = x^3 - 2$. We factor the equation as

$$(y + \sqrt{-2})(y - \sqrt{-2}) = x^3.$$

The ring $R = \mathbb{Z}[\sqrt{-2}]$ is a principal ideal domain, and the greatest common divisor of $y + \sqrt{-2}$ and $y - \sqrt{-2}$ in R divides $2\sqrt{-2}$, so we see that $y + \sqrt{-2}$ has one of the following forms:

$$y + \sqrt{-2} = \zeta^3 \quad \text{or} \quad \sqrt{-2}\zeta^3 \quad \text{or} \quad 2\zeta^3 \quad \text{for some } \zeta \in R.$$

Applying complex conjugation gives

$$y - \sqrt{-2} = \bar{\zeta}^3 \quad \text{or} \quad -\sqrt{-2}\bar{\zeta}^3 \quad \text{or} \quad 2\bar{\zeta}^3,$$

and taking the product yields

$$x^3 = y^2 + 2 = (\zeta\bar{\zeta})^3 \quad \text{or} \quad 2(\zeta\bar{\zeta})^3 \quad \text{or} \quad 4(\zeta\bar{\zeta}^3).$$

Since $x \in \mathbb{Z}$ and $\zeta\bar{\zeta} \in \mathbb{Z}$, only the first case is possible, so

$$y + \sqrt{-2} = \zeta^3 \qquad \text{and} \qquad y - \sqrt{-2} = \bar{\zeta}^3.$$

Subtracting these two equations gives

$$2\sqrt{-2} = \zeta^3 - \bar{\zeta}^3 = (\zeta - \bar{\zeta})(\zeta^2 + \zeta\bar{\zeta} + \bar{\zeta}^2).$$

We write $\zeta = a + b\sqrt{-2}$ with $a, b \in \mathbb{Z}$ and substitute to obtain

$$2\sqrt{-2} = 2\sqrt{-2}\, b(3a^2 - 2b^2).$$

Since a and b are in $\mathbb{Z}$, we must have

$$b = \pm 1 \qquad \text{and} \qquad 3a^2 - 2b^2 = \pm 1,$$

where the signs are the same. It follows that $(a, b) = (\pm 1, 1)$, and working back through the various substitutions yields the values $(x, y) = (3, \pm 5)$. $\qquad \square$

Remark 7.1.1. It is worth remarking that the result in (IX.7.1b) is far more interesting than that in (IX.7.1a). The reason is that the Mordell–Weil group over $\mathbb{Q}$ of the elliptic curve $y^2 = x^3 + 7$ is trivial, so (IX.7.1a) reflects the fact that the equation has no rational solutions. On the other hand, the Mordell–Weil group of $y^2 = x^3 - 2$ is infinite cyclic (see Exercise 10.19), so (IX.7.1b) says that among the infinitely many rational points, only two have integer coordinates.

Baker applied his effective estimate for linear forms in logarithms to give an explicit upper bound, in terms of D, for the integral solutions to $y^2 = x^3 + D$. This bound was refined by Stark, who proved the following result.

Theorem 7.2. (Stark [273]) *For every $\epsilon > 0$ there is an effectively computable constant C_ϵ, depending only on ϵ, such that if $D \in \mathbb{Z}$ with $D \neq 0$ and if $x, y \in \mathbb{Z}$ are solutions to the equation*

$$y^2 = x^3 + D,$$

then

$$\log \max\{|x|, |y|\} \leq C_\epsilon |D|^{1+\epsilon}.$$

Example 7.3. Stark's estimate (IX.7.2) gives a bound for x and y that is slightly worse than exponential in D. It is natural to ask whether this bound is of the correct order of magnitude. Various people have conducted computer searches for large solutions, see for example [75, 106, 134]. Among the interesting examples found, we mention:

$$378,661^2 = 5234^3 + 17,$$

$$911,054,064^2 = 939,787^3 - 307,$$

$$149,651,610,621^2 = 28,187,351^3 + 1090,$$

$$447,884,928,428,402,042,307,918^2 = 5,853,886,516,781,223^3 - 1641843.$$

Although these examples show that x and y may be quite large in comparison to D, a close examination of the data led M. Hall to make the following conjecture, which was partly generalized by Lang.

Conjecture 7.4. (a) (Hall [106]) *For every $\epsilon > 0$ there is a constant C_ϵ, depending only on ϵ, such that for all $D \in \mathbb{Z}$ with $D \neq 0$ and for all $x, y \in \mathbb{Z}$ satisfying*

$$y^2 = x^3 + D,$$

we have

$$|x| \leq C_\epsilon D^{2+\epsilon}.$$

(b) (Hall–Lang [138]) *There are absolute constants C and κ such that for every elliptic curve $E/\mathbb{Q}$ given by a Weierstrass equation*

$$y^2 = x^3 + Ax + B \qquad \text{with } A, B \in \mathbb{Z}$$

and for every integral point $P \in E(\mathbb{Q})$, i.e., satisfying $x(P) \in \mathbb{Z}$, we have

$$|x(P)| \leq C \max\{|A|, |B|\}^\kappa.$$

The evidence for these conjectures is fragmentary. They are true for function fields, for which Davenport [57] proved (IX.7.4a) and Schmidt proved (IX.7.4b). Vojta [298, 4 §4] has shown that (IX.7.4a) over number fields is a consequence of his very general Nevanlinna-type conjectures for algebraic varieties. It is also easy to deduce (IX.7.4a) from the ABC conjecture; see Exercise 9.17. However, both Vojta's conjectures and the ABC conjecture are well beyond the reach of current techniques. (See also Exercise 9.10 for a proof that the exponent in (IX.7.4a) cannot be improved.) Aside from these few facts, very little is known. It is worth pointing out that the effective techniques from (IX §5) seem intrinsically incapable of leading to estimates as strong as those described in (IX.7.4). We briefly explain the problem for the equation $y^2 = x^3 + D$.

When performing the reduction to the S-unit equation, we use a number field K whose discriminant looks like a power of D. The Brauer–Siegel theorem says that $\log(h_K \mathcal{R}_K) \sim \frac{1}{2} \log d_K$ as $[K : \mathbb{Q}]/\log d_K \to 0$, where h_K is the class number, $\mathcal{R}_K$ the regulator, and d_K the absolute discriminant of K. (See, e.g., [142, Chapter XVI].) In general there is no reason to expect the class number of K to be large, so the best that we can hope for is to find a bound for the regulator that is a power of $|D|$. Since the regulator is the determinant of the *logarithms* of a basis for the unit group R^*, the resulting bounds for the heights $H(\alpha_i)$ of generators $\alpha_i \in R^*$ will be exponential in $|D|$. This eventually leads to an exponential bound for x and y as in (IX.7.2).

There is a similar problem if we try to prove (IX.7.4) using linear forms in elliptic logarithms or by following Siegel's method of proof as in (IX.3.1), even assuming that we could prove strong effective versions of Roth's theorem and the Mordell–Weil theorem. The difficulty is that it is likely that the best possible upper bound for generators of the Mordell–Weil group of $y^2 = x^3 + D$ has the form $\hat{h}(P) \leq C|D|^\kappa$, cf. (VIII.10.2). Here $\hat{h}$ is a logarithmic height, so this again leads to a bound for the x-coordinate of integral points that is exponential in D.

The problem in both cases can be explained most clearly by the analogy given in (IX.4.2.1). When solving the S-unit equation or when finding integral points on elliptic curves, one is initially given a finitely generated group ($R_S^* \times R_S^*$, respectively $E(K)$) and a certain exceptional subset (solutions to $ax + by = 1$, respectively points with $x(P) \in R_S$). The first step is to choose a basis for the finitely generated group and express the exceptional points in terms of the basis. The difficulty that arises in trying to prove (IX.7.4) or the analogous estimate for the S-unit equation is that in general, the best (conjectural) upper bound for the heights of the basis elements is exponentially larger than the desired upper bound for the exceptional points! The moral of this story, assuming the validity of various conjectures, is that a randomly chosen elliptic curve $E/\mathbb{Q}$ is unlikely to have any integral points at all.

IX.8   Roth's Theorem—An Overview

In this section we give a brief sketch of the principal steps that go into the proof of Roth's theorem (IX.1.4). None of the steps are tremendously deep, but the details required to make them rigorous are quite lengthy. For the full proof, see for example [114, Part D], [139, Chapter 7], or [221].

We assume that we are given an $\alpha \in \bar{K}$, an absolute value $v \in M_K$, and positive real numbers ϵ and C. We then want to prove that there are only finitely many $x \in K$ satisfying

$$|x - \alpha|_v \leq CH_K(x)^{-2-\epsilon}.$$

Step I: An Auxiliary Polynomial

For any given integers $m, d_1, \ldots, d_m$, one uses elementary estimates and the pigeon-hole principle to construct a polynomial

$$P(X_1, \ldots, X_m) \in R[X_1, \ldots, X_m]$$

of degree d_i in X_i such that P vanishes to fairly high order (in terms of m and the d_i) at the point $(\alpha, \ldots, \alpha)$. Further, one shows that P may be chosen with coefficients having fairly small heights, the bound for the heights being given explicitly in terms of α, m, and the d_i.

Step II: An Upper Bound for P

Suppose now that we are given elements $x_1, \ldots, x_m \in K$ satisfying

$$|x_i - \alpha|_v \leq CH_K(x_i)^{-2-\epsilon} \qquad \text{for } 1 \leq i \leq m.$$

Using the Taylor series expansion for $P(X_1, \ldots, X_m)$ around $(\alpha, \ldots, \alpha)$ and the fact that P vanishes to high order at $(\alpha, \ldots, \alpha)$, one shows that $\left| P(x_1, \ldots, x_m) \right|_v$ is fairly small.

Step III: A Nonvanishing Result (Roth's Lemma)

Suppose that the degrees $d_1, \ldots, d_m$ are fairly rapidly decreasing, where the rate of decrease depends on m, and suppose that $x_1, \ldots, x_m \in K$ have the property that their heights are fairly rapidly increasing, the rate of increase depending on m and $d_1, \ldots, d_m$. Suppose further that $P(X_1, \ldots, X_m) \in R[X_1, \ldots, X_m]$ has degree d_i in X_i and coefficients whose heights are bounded in terms of d_1 and $h(x_1)$. Then one shows that P does not vanish to too high an order at $(x_1, \ldots, x_m)$.

This is the hardest step in Roth's theorem. In Thue's original theorem, he used a polynomial of the form $P(X, Y) = f(X) + g(X)Y$ and obtained an approximation exponent $\tau(d) = \frac{1}{2}d + \epsilon$. The improvements of Siegel, Gelfond, and Dyson used a general polynomial in two variables. It was clear at that time that the way to obtain $\tau(d) = 2 + \epsilon$ was to use polynomials in more variables; the only stumbling block was the lack of a nonvanishing result such at the one that we have just described.

The proof of Roth's lemma is by induction on m, the number of variables in the polynomial P. If P factors as

$$P(X_1, \ldots, X_m) = F(X_1)G(X_2, \ldots, X_m),$$

then the induction proceeds fairly smoothly. Of course, such a factorization is unlikely to happen. What one does is to construct differential operators $\mathcal{D}_{ij}$ such that the generalized Wronskian determinant $\det(\mathcal{D}_{ij}P)$ is a nonzero polynomial that does factor in the above fashion. It is then a delicate matter to estimate the degrees and heights of the coefficients of the resulting polynomial and to show that they have not grown too large to allow the inductive hypothesis to be applied.

Step IV: The Final Estimate

Suppose that the inequality

$$|x - \alpha|_v \le C H_K(x)^{-2-\epsilon}$$

has infinitely many solutions $x \in K$. We derive a contradiction as follows.

First choose a value for m, depending on ϵ, C, and $[K(\alpha) : K]$. Second, choose $x_1, \ldots, x_m \in K$ in succession satisfying

$$|x_i - \alpha|_v \le C H_K(x_i)^{-2-\epsilon},$$

such that $H_K(x_1)$ is large, depending on m, and such that $H_K(x_{i+1}) > H_K(x_i)^\kappa$ for some constant κ depending on m. Third, choose a large integer d_1, depending on m and the $H_K(x_i)$, and then choose $d_2, \ldots, d_m$ in terms of d_1 and the $H_K(x_i)$. We are now ready to apply the initial three steps.

Using Step I, choose a polynomial $P(X_1, \ldots, X_m)$ of degree d_i in X_i such that P vanishes to high order at $(\alpha, \ldots, \alpha)$. The order of vanishing depends on m and $d_1, \ldots, d_m$. From Step III, we know that P does not vanish to too high an order at $(x_1, \ldots, x_m)$, so we can choose a low-order partial derivative that does not vanish,

$$z = \frac{\partial^{i_1+\cdots+i_m}}{\partial X_1^{i_1} \cdots \partial X_m^{i_m}} P(x_1, \ldots, x_m) \ne 0.$$

From Step II, we know that $|z|_v$ is fairly small. On the other hand, since $z \ne 0$, we can use the product formula to show that $|z|_v$ cannot be too small. Specifically, we have $|z|_v \ge H_K(z)^{-1}$; see Exercise 9.9. Next, using elementary triangle inequality estimates, we find a lower bound for $H_K(z)^{-1}$. Combining this lower bound with the earlier upper bound, some algebra gives a contradiction. It follows that the inequality

$$|x - \alpha|_v \le C H_K(x)^{-2-\epsilon}$$

has only finitely many solutions.

Remark 8.1. In examining the proof sketch of Roth's theorem, especially the sequence of choices in Step IV, it is clear why we do not obtain an effective procedure for finding all $x \in K$ satisfying $|x - \alpha|_v \le C H_K(x)^{-2-\epsilon}$. What the proof shows is that we cannot find a long sequence of x_i whose heights grow sufficiently rapidly, where the terms "long sequence" and "sufficiently rapidly" can be made completely explicit in terms of K, α, ϵ, and C. The difficulty is that the required growth of the height of each x_i is given in terms of its predecessor. What this boils down to is that if we can find a large number of good approximations to α whose heights are sufficiently large, then we can obtain a bound for all other good approximations to α in terms of the approximations that we already know. Unfortunately, the bounds that come out of Roth's theorem are so large that it is highly unlikely that there exists even a single good approximation to α having the requisite height.

Using an elaboration of the above argument, one can prove quantitative versions of Roth's theorem such as in the following result.

Theorem 8.2. ([173, 103]) *Let $K/\mathbb{Q}$ be a number field, let $\alpha \in \bar{K} \smallsetminus K$, and let $S \subset M_K$ be a finite set of absolute values, each of which is extended in some way to $\mathbb{Q}(\alpha)$. Let $\epsilon > 0$. There are constants C_1 and C_2, depending only on ϵ and $\big[K(\alpha) : K\big]$, such that the inequality*

$$\prod_{v \in S} \min\{|x - \alpha|_v^{n_v}, 1\} \le CH_K(x)^{-2-\epsilon}$$

has at most $4^{\#S}C_1$ solutions $x \in K$ satisfying $H_K(x) > \big(2H_K(\alpha)\big)^{C_2}$.

Of course, the constant C_2 in (IX.8.2) turns out to be sufficiently large that it is highly unlikely that there are any $x \in K$ satisfying the two conditions of the theorem. But the proof of Roth's theorem does not preclude the existence of large solutions, and it provides no tools with which to find them if they do exist!

Exercises

9.1. Let $\big(\phi(n)\big)_{n=1,2,\ldots}$ be a sequence of positive numbers. We say that a number $\alpha \in \mathbb{R}$ is ϕ-*approximable (over $\mathbb{Q}$)* if there are infinitely many $p/q \in \mathbb{Q}$ satisfying

$$\left|\alpha - \frac{p}{q}\right| < \frac{1}{q\phi(q)}.$$

For example, Roth's theorem says that no element of $\bar{\mathbb{Q}}$ is $n^{1+\epsilon}$-approximable.
  (a) Prove that for any $\epsilon > 0$, the set

$$\{\alpha \in \mathbb{R} : \alpha \text{ is } n^{1+\epsilon}\text{-approximable}\}$$

   is a set of measure 0.
  (b) More generally, prove that if the series $\sum_{n \ge 1} 1/\phi(n)$ converges, then the set

$$\{\alpha \in \mathbb{R} : \alpha \text{ is } \phi\text{-approximable}\}$$

   is a set of measure 0.

9.2. (a) Use Liouville's theorem (IX.1.3) to prove that the number $\sum_{n \ge 1} 2^{-n!}$ is transcendental.
  (b) More generally, let $\big(e(n)\big)_{n=1,2,\ldots}$ be a sequence of real numbers with the property that for every $d > 0$ there is a constant $C_d > 0$ such that

$$e(n) \ge C_d n^d \qquad \text{for all } n = 1, 2, \ldots.$$

   (In complexity theory terminology, one says that the growth rate of the function $e(n)$ is faster than polynomial.) Let $b \ge 2$ be an integer. Prove that the number $\sum_{n \ge 1} b^{-e(n)}$ is transcendental.

(c) Use (b) to prove that there are uncountably many transcendental numbers.

9.3. For each integer $m \neq 0$, let

$$N(m) = \#\{(x, y) \in \mathbb{Z} : y^2 = x^3 + m\}.$$

Note that (IX.3.2) tells us that $N(m)$ is finite.

(a) Prove that $N(m)$ can be arbitrarily large. (*Hint.* Choose an m_0 such that $y^2 = x^3 + m_0$ has infinitely many rational solutions. Then clear the denominators of a lot of them.)

(b) More precisely, prove that there is an absolute constant $c > 0$ such that

$$N(m) > c\big(\log|m|\big)^{1/3}$$

for infinitely many $m \in \mathbb{Z}$. (*Hint.* Use height functions to estimate the size of the denominators cleared in (a).)

(c) ** Prove or disprove that $N(m)$ is unbounded as m ranges over sixth-power-free integers, i.e., integers divisible by no nontrivial sixth power.

(d) Suppose that there is a value of m_0 such that the Mordell–Weil group $E_0(\mathbb{Q})$ of the elliptic curve $E_0 : y^2 = x^3 + m_0$ has rank r. Using an elaboration of the argument in (b), prove that there is an absolute constant $c > 0$ such that

$$N(m) > c\big(\log|m|\big)^{r/(r+2)}$$

for infinitely many $m \in \mathbb{Z}$.

(e) ** Let $\epsilon > 0$. Prove or disprove that

$$\lim_{|m| \to \infty} \frac{N(m)}{\big(\log|m|\big)^{1+\epsilon}} = 0.$$

9.4. Let $E/\mathbb{Q}$ be an elliptic curve and let $P \in E(\mathbb{Q})$ be a point of infinite order.

(a) For each prime $p \in \mathbb{Z}$ at which E has good reduction, let n_p be the order of the reduced point $\tilde{P}$ in the finite group $\tilde{E}(\mathbb{F}_p)$. Prove that the set

$$\{n_p : p \text{ prime}\}$$

contains all but finitely many positive integers. (*Hint.* You will need the strong form of Siegel's theorem; see (IX.3.3).)

(b) An alternative formulation for (a) is to write $x(nP) = a_n/d_n^2$ as a fraction in lowest terms. The sequence $(d_n)_{n \geq 1}$ is an *elliptic divisibility sequence.*[2] A prime p is called a *primitive divisor of d_n* if $p \mid d_n$ and $p \nmid d_m$ for all $m < n$. Prove that all but finitely many terms in the sequence d_n have a primitive divisor. (This is an analogue for elliptic curves of a classical result for the multiplicative group that is due to Bang and Zsigmondy [317].)

9.5. (a) Let $f(T) = a_0 T^n + \cdots + a_n \in \mathbb{Z}[T]$ be a polynomial with $a_0 a_n \neq 0$ and with distinct roots $\xi_1, \ldots, \xi_n \in \mathbb{C}$. Let $A = \max\{|a_0|, \ldots, |a_n|\}$. Prove that for every rational number $t \in \mathbb{Q}$,

$$\big|f(t)\big| \geq (2n^2 A)^{-n} \min\big\{|t - \xi_1|, \ldots, |t - \xi_n|\big\}.$$

---

[2] This definition differs from that given in exercises 3.34–3.36. In general, it may be necessary to take a subsequence $(d_{nk})_{n \geq 1}$ in order to obtain a sequence satisfying the recurrence described in Exercise 3.34.

(b) Let $f(T) = a_0 T^n + \cdots + a_n \in K[T]$ be a polynomial with $a_0 a_n \neq 0$ and with distinct roots $\xi_1, \ldots, \xi_n \in \bar{K}$. Let $S \subset M_K$ be a finite set of places of K, each extended in some fashion to $\bar{K}$. Prove that there is a constant $C_f > 0$, depending only on f, such that for every $t \in K$,

$$\prod_{v \in S} \min\left\{1, |f(t)|_v^{n_v}\right\} \geq C_f \prod_{v \in S} \min_{1 \leq i \leq n} \left\{1, |t - \xi_i|_v^{n_v}\right\}.$$

(c) Find an explicit expression for the constant C_f appearing in (b), where your value for C_f should depend only on n and $H_K([a_0, \ldots, a_n])$.

9.6. (a) Let $F(X, Y) \in \mathbb{Z}[X, Y]$ be a homogeneous polynomial of degree $d \geq 3$ with nonzero discriminant. Prove that for every nonzero integer b, *Thue's equation*

$$F(X, Y) = b$$

has only finitely many solutions $(x, y) \in \mathbb{Z}^2$. (*Hint.* Let $f(T) = F(T, 1)$, and write $b = F(x, y) = y^d f(x/y)$. Now use Exercise 9.5a and (IX.1.4).)

(b) More generally, let $F(X, Y) \in K[X, Y]$ be a homogeneous polynomial of degree $d \geq 3$ with nonzero discriminant, and let $S \subset M_K$ be a finite set of places containing M_K^∞. Prove that for every $b \in K^*$, the equation

$$F(X, Y) = b$$

has only finitely many solutions $(x, y) \in R_S \times R_S$.

(c) Let $f(X) \in K[X]$ be a polynomial with at least two distinct roots in $\bar{K}$, let $S \subset M_K$ be as in (b), and let $n \geq 3$ be an integer. Prove that the equation

$$Y^n = f(X)$$

has only finitely many solutions $(x, y) \in R_S \times R_S$. (*Hint.* Mimic the proof of (IX.4.3) until you end up with a number of equations of the form $aW^n + bZ^n = c$, and then use (b).)

9.7. Let E/K be an elliptic curve without complex multiplication. Prove that for every prime ℓ, the representation of $G_{\bar{K}/K}$ on the $\mathbb{Q}_\ell$-vector space $T_\ell(E) \otimes \mathbb{Q}_\ell$ is irreducible.

9.8. (a) Let $\|\cdot\|$ be the usual Euclidean norm on $\mathbb{R}^n$, and let $\{v_1, \ldots, v_n\}$ be a basis for $\mathbb{R}^n$. Prove that there is a constant $c > 0$, depending only on n and $\{v_1, \ldots, v_n\}$, such that

$$\left\| \sum_{i=1}^n a_i v_i \right\| \geq c \max\{|a_i|\} \qquad \text{for all } a_1, \ldots, a_n \in \mathbb{R}.$$

(b) Let $\Lambda \subset \mathbb{R}^n$ be a lattice. Prove that there exist a basis $\{v_1, \ldots, v_n\}$ for Λ and a constant $c_n > 0$ *depending only on n* such that

$$\left\| \sum_{i=1}^n a_i v_i \right\| \geq c_n \sum_{i=1}^n \|a_i v_i\| \qquad \text{for all } a_1, \ldots, a_n \in \mathbb{R}.$$

(*Hint.* Ideally, one would like to choose an orthogonal basis for Λ. This is not generally possible, but mimic the Gram–Schmidt process to find a basis that is reasonably orthogonal.)

(c) Let $\| \cdot \|_1$ and $\| \cdot \|_2$ be norms on $\mathbb{R}^n$, i.e., they satisfy $\|v\| \geq 0$, $\|v\| = 0$ if and only if $v = 0$, $\|av\| \leq |a|\,\|v\|$, and $\|v + w\| \leq \|v\| + \|w\|$. Prove that there are constants $c_1, c_2 > 0$ such that

$$c_1\|v\|_1 \leq \|v\|_2 \leq c_2\|v\|_1 \qquad \text{for all } v \in \mathbb{R}^n.$$

(d) Let Q be a positive definite quadratic form on $\mathbb{R}^n$. Prove that there is a constant $c > 0$, depending on n and Q, such that for any integral lattice point $(a_1, \ldots, a_n) \in \mathbb{Z}^n \subset \mathbb{R}^n$,

$$Q(a_1, \ldots, a_n) \geq c\max\{|a_1|, \ldots, |a_n|\}^2.$$

(e) Let E/K be an elliptic curve and let $P_1, \ldots, P_r$ be a basis for the free part of $E(K)$. Prove that there is a constant $c > 0$, depending on E and $P_1, \ldots, P_r$, such that for all integers $m_1, \ldots, m_r$,

$$\hat{h}(m_1 P_1 + \cdots + m_r P_r) \geq c\max\{|m_1|, \ldots, |m_r|\}^2.$$

9.9. Let $z \in K$ with $z \neq 0$.

(a) Prove that for any $v \in M_K$,
$$|z|_v \geq H_K(z)^{-1}.$$

(b) More generally, prove that for any (not necessarily finite) set of absolute values $S \subset M_K$,
$$\prod_{v \in S} \min\{1, |z|_v^{n_v}\} \geq H_K(z)^{-1}.$$

(This lemma, as trivial as it appears, lies at the heart of all known proofs in Diophantine approximation and transcendence theory. In its simplest guise, namely for $K = \mathbb{Q}$, it asserts nothing more than the fact that there are no positive integers smaller than one!)

9.10. Prove that there is an (absolute) constant $C > 0$ such that the inequality

$$0 < |y^2 - x^3| < C\sqrt{|x|}$$

has infinitely many solutions $(x, y) \in \mathbb{Z}$. (*Hint.* Verify the identity

$$(t^2 - 5)^2\big((t + 9)^2 + 4\big) - (t^2 + 6t - 11)^3 = -1728(t - 2).$$

Take solutions to $y^2 - 2v^2 = -1$ and set $t = 2u - 9$. Show that this leads to a value $C = 432\sqrt{2} + \epsilon$ for any $\epsilon > 0$.)

9.11. (a) Let $d \equiv 2 \pmod 4$ and let $D = d^3 - 1$. Prove that the equation

$$y^2 = x^3 + D$$

has no solution in integers $x, y \in \mathbb{Z}$.

(b) For each of the primes p in the set $\{11, 19, 43, 67, 163\}$, find all solutions $x, y \in \mathbb{Z}$ to the equation

$$y^2 = x^3 - p.$$

(*Hint.* Work in the ring $R = \mathbb{Z}\left[\frac{1}{2}(1 + \sqrt{-p})\right]$. Note that R is a principal ideal domain and that 2 does not split in R.)

9.12. Let $E/\mathbb{Q}$ be an elliptic curve given by a Weierstrass equation

$$E : y^2 + a_1 xy + a_3 y = x^3 + a_2 x^2 + a_4 x + a_6$$

with $a_1, \ldots, a_6 \in \mathbb{Z}$, and let $P \in E(\mathbb{Q})$ be a point of infinite order.

(a) Suppose that $x([m]P) \in \mathbb{Z}$ for some integer $m \geq 1$. Prove that $x(P) \in \mathbb{Z}$. (This result is often useful when one is searching for integral points on elliptic curves of rank 1. See Exercise 9.13 for an example.)

(b) More generally, for any $m \geq 1$, write $x(mP) = a_m/d_m^2 \in \mathbb{Q}$ as a fraction in lowest terms. Prove that

$$m \mid n \Longrightarrow d_m \mid d_n.$$

Thus the sequence $(d_m)_{m \geq 1}$ is a *divisibility sequence*; see Exercise 3.36.

9.13. Let $E/\mathbb{Q}$ be the elliptic curve

$$E : y^2 + y = x^3 - x.$$

For this exercise you may assume that $E(\mathbb{Q})$ has rank 1. (For a proof that rank $E(\mathbb{Q}) = 1$, see Exercise 10.9.)

(a) Prove that $E_{\mathrm{tors}}(\mathbb{Q}) = \{O\}$, and hence that $E(\mathbb{Q}) \cong \mathbb{Z}$.

(b) Prove that $(0,0)$ is a generator for $E(\mathbb{Q})$. (*Hint.* Make a sketch of $E(\mathbb{R})$ and show that $(0,0)$ is not on the identity component. Use Exercise 9.12 to conclude that a generator for $E(\mathbb{Q})$ must be a point with *integer* coordinates on the nonidentity component, and find all such points.)

(c) Find all of the integer points in $E(\mathbb{Q})$. (*Hint.* Let $P = (0,0)$. Suppose that $[m]P$ is integral. Write $m = 2^\alpha n$ with n odd and use Exercise 9.12 to show that $[n]P$ is integral. Use an argument as in (b) to find all possible values of n, and then do some computations to find the possible a values.)

(d) Solve the following classical number theory problem: Find all positive integers that are simultaneously the product of two consecutive integers and the product of three consecutive integers.

9.14. Let C/K be a curve and let $f, g \in K(C)$ be nonconstant functions.

(a) * Prove that

$$\lim_{\substack{P \in C(\bar{K}) \\ h_f(P) \to \infty}} \frac{h_f(P)}{h_g(P)} = \frac{\deg f}{\deg g}.$$

(b) Prove that for every $\epsilon > 0$ there exists a constant $c = c(f, g, \epsilon)$ such that

$$\left| \frac{1}{\deg f} h_f(P) - \frac{1}{\deg g} h_g(P) \right| < \epsilon h_f(P) + c \qquad \text{for all } P \in C(\bar{K}).$$

(c) Let C be an elliptic curve. Prove that there is a constant $c = c(f, m, \epsilon)$ such that

$$\left| h_f([m]P) - m^2 h_f(P) \right| < \epsilon h_f(P) + c \qquad \text{for all } P \in C(\bar{K}).$$

(d) Prove that (IX.3.1) is true for all nonconstant functions $f \in K(E)$. Use this to prove the finiteness result (IX.3.2.2) directly, without first reducing to (IX.3.2.1).

9.15. For a given $Q \in C(K_v)$, let d_v be the distance function defined in (IX §2), and let d_v^{alt} denote the distance function given by the alternative definition in (IX.2.2.1). Prove that the ratio $d_v^{\mathrm{alt}}(P, Q)/d_v(P, Q)$ is bounded for $P \in C(K_v)$.

9.16. Let C/K be a curve, let $f \in K(C)$ be a nonconstant function, and write the divisor of zeros of f as

$$\mathrm{div}_0(f) = \sum_{\substack{Q \in C(\bar{K}) \\ \mathrm{ord}_Q(f) > 0}} \mathrm{ord}_Q(f)(Q) = n_1(Q_1) + n_2(Q_2) + \cdots + n_r(Q_r).$$

Replacing K by an extension field, we assume that $Q_1, \ldots, Q_r \in C(K)$. Let $v \in M_K$. Prove that

$$\log \min\{|f(P)|_v, 1\} = n_1 d_v(P, Q_1) + \cdots + n_r d_v(P, Q_r) + O(1) \qquad \text{for all } P \in C(K_v),$$

where the $O(1)$ depends on f and the choice of distance functions, but is independent of P.

9.17. Let $\epsilon > 0$, and let m and n be positive integers satisfying $nm > n + m$. Assuming that the ABC conjecture (VIII.11.4) is true, prove the following assertions (see also Exercise 8.22):
  (a) There is a constant $C = C(\epsilon, m, n)$ such that if

$$y^m = x^n + D \qquad \text{with } x, y, D \in \mathbb{Z} \text{ and } D \neq 0,$$

  then
$$|x|^{nm-n-m} \le C|D|^{m+\epsilon} \qquad \text{and} \qquad |y|^{nm-n-m} \le C|D|^{n+\epsilon}.$$

  (This is a generalized version of Hall's conjecture (IX.7.4).)
  (b) Suppose now that $D \neq 0$ is fixed. If $\max\{m, n\}$ is sufficiently large, then the equation $y^m = x^n + D$ has no solutions $x, y \in \mathbb{Z}$ with $x, y \notin \{0, \pm 1\}$. (*Hint.* You'll need to keep track of how the constant in (a) depends on m and n.)

9.18. Let E be the elliptic curve $y^2 = x^3 + 2089$.
  (a) Prove that the points

$$P_1 = (-12, 19), \quad P_2 = (-10, 33), \quad P_3 = (-4, 45), \quad P_4 = (3, 46),$$

  are independent points in $E(\mathbb{Q})$.
  (b) * Prove that $E(\mathbb{Q}) \cong \mathbb{Z}^4$ and that P_1, P_2, P_3, P_4 are a basis for $E(\mathbb{Q})$.
  (c) Find 10 more points (x, y) in $E(\mathbb{Q})$ with $x, y \in \mathbb{Z}$ and $y > 0$. Express these integral points in terms of the basis listed in (a).

Chapter X

Computing the Mordell–Weil Group

A better title for this chapter might be "Computing the *Weak* Mordell–Weil Group," since we will be concerned solely with the problem of computing generators for the group $E(K)/mE(K)$. However, given generators for $E(K)/mE(K)$, a finite amount of computation yields generators for $E(K)$; see (VIII.3.2) and Exercise 8.18. Unfortunately, there is no comparable algorithm currently known that is guaranteed to give generators for $E(K)/mE(K)$ in a finite amount of time!

We start in (X §1) by taking the proof of the weak Mordell–Weil theorem given in (VIII §1) and making it quite explicit. In this way the computation of the quotient $E(K)/mE(K)$ (in a special case) is reduced to the problem of determining whether each of a certain finite set of auxiliary curves, called *homogeneous spaces*, has a single rational point. The question whether a given homogeneous space has a rational point may often be answered either affirmatively by finding a point or negatively by showing that it has no points in some completion K_v of K.

The subsequent two sections develop the general theory of homogeneous spaces (for elliptic curves). Then, in (X §4), we apply this theory to the problem of computing $E(K)/mE(K)$ or, more generally, $E'(K)/\phi\big(E(K)\big)$ for an isogeny

$$\phi : E \to E'.$$

Again this computation is reduced to the problem of the existence of a single rational point on certain homogeneous spaces. The only impediment to solving this latter problem occurs if some homogeneous space has a K_v-rational point for every completion K_v of K, yet fails to have a K-rational point. Unfortunately, this precise situation, the failure of the so-called Hasse principle, does occur. The extent of its failure is quantified by the elements of a certain group, called the *Shafarevich–Tate group*. The question of an effective algorithm for the computation of $E(K)/mE(K)$ is thus finally reduced to the problem of giving a bound for divisibility in the Shafarevich–Tate group, or, even better, proving the conjecture that the Shafarevich–Tate group is finite.

J.H. Silverman, *The Arithmetic of Elliptic Curves, Second Edition*, Graduate Texts in Mathematics 106, DOI 10.1007/978-0-387-09494-6_X,
© Springer Science+Business Media, LLC 2009

In the last section we illustrate the general theory by studying in some detail the family of elliptic curves given by the equations

$$E_D : Y^2 = X^3 + DX, \qquad D \in \mathbb{Q}.$$

In particular, we compute the torsion subgroups and give an upper bound for the rank of $E_D(\mathbb{Q})$, we give a large class of examples for which $E_D(\mathbb{Q})$ has rank 0, and we show that in certain cases $E_D(\mathbb{Q})$ has an associated homogeneous space that violates the Hasse principle, i.e., the homogeneous space has points defined over $\mathbb{R}$ and over $\mathbb{Q}_p$ for every prime p, but it has no $\mathbb{Q}$-rational points.

Unless explicitly stated to the contrary, the notation for this chapter is the same as for Chapter VIII. In particular, K is a number field and M_K is a complete set of inequivalent absolute values on K. However, as indicated in the text, this specification is dropped in (X §§2,3,5), where K is allowed to be an arbitrary (perfect) field.

X.1 An Example

For this section we let E/K be an elliptic curve and $m \geq 2$ an integer, and we assume that

$$E[m] \subset E(K).$$

Recall from (VIII §1) that under this assumption there is a pairing

$$\kappa : E(K) \times G_{\bar{K}/K} \longrightarrow E[m]$$

defined by

$$\kappa(P, \sigma) = Q^\sigma - Q,$$

where $Q \in E(\bar{K})$ is chosen to satisfy $[m]Q = P$. Further, (VIII.1.2) says that the kernel on the left is $mE(K)$, so we may view κ as a homomorphism

$$\delta_E : E(K)/mE(K) \longrightarrow \mathrm{Hom}\big(G_{\bar{K}/K}, E[m]\big),$$
$$\delta_E(P)(\sigma) = \kappa(P, \sigma).$$

(This is the connecting homomorphism for a group cohomology long exact sequence; see (VIII §2).)

We also observe from (III.8.1.1) that our assumption $E[m] \subset E(K)$ implies that $\mu_m \subset K^*$. This follows from basic properties of the Weil pairing (III.8.1.1),

$$e_m : E[m] \times E[m] \longrightarrow \mu_m.$$

The Weil pairing will play a prominent role in this section.

Finally, since $\mu_m \subset K^*$, Hilbert's Theorem 90 (B.2.5c) says that every homomorphism $G_{\bar{K}/K} \to \mu_m$ has the form

$$\sigma \longmapsto \frac{\beta^\sigma}{\beta} \qquad \text{for some } \beta \in \bar{K}^* \text{ satisfying } \beta^m \in K^*.$$

In other words, there is an isomorphism (cf. VIII §2)

$$\delta_K : K^*/(K^*)^m \longrightarrow \mathrm{Hom}(G_{\bar{K}/K}, \mu_m)$$

defined by

$$\delta_K(b)(\sigma) = \beta^\sigma/\beta,$$

where $\beta \in \bar{K}^*$ is chosen to satisfy $\beta^m = b$. Note the close resemblance in the definitions of δ_E and δ_K. This is no coincidence. The map δ_E is the connecting homomorphism for the Kummer sequence associated to the group variety E/K, and δ_K is the connecting homomorphism for the Kummer sequence associated to the group variety $\mathbb{G}_m/K$.

Using these maps, we can make the proof of the weak Mordell–Weil theorem much more explicit, and by doing so, derive formulas that allow us to compute the Mordell–Weil group in certain cases. We start with a theoretical description of the method.

Theorem 1.1. (a) *With notation as above, there is a bilinear pairing*

$$b : E(K)/mE(K) \times E[m] \longrightarrow K^*/(K^*)^m$$

satisfying

$$e_m\big(\delta_E(P), T\big) = \delta_K\big(b(P, T)\big).$$

(b) *The pairing in (a) is nondegenerate on the left.*

(c) *Let $S \subset M_K$ be the union of the set of infinite places, the set of finite primes at which E has bad reduction, and the set of finite primes dividing m. Then the image of the pairing in (a) lies in the following subgroup of $K^*/(K^*)^m$:*

$$K(S, m) = \big\{ b \in K^*/(K^*)^m : \mathrm{ord}_v(b) \equiv 0 \ (\mathrm{mod}\ m) \text{ for all } v \notin S \big\}.$$

(d) *The pairing in (a) may be computed as follows. For each $T \in E[m]$, choose functions $f_T, g_T \in \bar{K}(E)$ satisfying the conditions*

$$\mathrm{div}(f_T) = m(T) - m(O) \qquad and \qquad f_T \circ [m] = g_T^m$$

(cf. the definition of the Weil pairing (III §8)). Then for any point $P \neq T$,

$$b(P, T) \equiv f_T(P) \quad (\mathrm{mod}\ (K^*)^m).$$

(If $P = T$, we can compute $b(T, T)$ using linearity. For example, if $[2]T \neq O$, then $b(T, T) = f_T(-T)^{-1}$. More generally, let $Q \in E(K)$ be any point with $Q \neq T$; then $b(T, T) = f_T(T + Q)f_T(Q)^{-1}$.)

Remark 1.2. Why do we say that (X.1.1) provides formulas that help us to compute the Mordell–Weil group? First, the group $K(S, m)$ in (c) is finite (see the proof of (VIII.1.6)), and in fact it is reasonably easy to explicitly compute $K(S, m)$. Second, the functions f_T in (d) are also fairly easy to compute from the equation of the curve. (This is true even for quite large values of m; see (XI.8.1).) Then the fact that the pairing in (a) is nondegenerate on the left means that in order to compute $E(K)/mE(K)$, it is necessary to do "only" the following:

Fix generators T_1 and T_2 for $E[m]$. For each of the finitely many pairs

$$(b_1, b_2) \in K(S, m) \times K(S, m),$$

check whether the simultaneous equations

$$b_1 z_1^m = f_{T_1}(P) \qquad \text{and} \qquad b_2 z_2^m = f_{T_2}(P)$$

have a solution $(P, z_1, z_2) \in E(K) \times K^* \times K^*$. We can be even more explicit if we express the function f_T in terms of Weierstrass coordinates x and y. Then we are looking for a solution $(x, y, z_1, z_2) \in K \times K \times K^* \times K^*$ satisfying

$$y^2 + a_1 xy + a_3 y = x^3 + a_2 x^2 + a_4 x + a_6,$$

$$b_1 z_1^m = f_{T_1}(x, y), \qquad b_2 z_2^m = f_{T_2}(x, y).$$

These equations define a new curve, called a *homogeneous space for* E/K. (We discuss homogeneous spaces in more detail in (X §3).) What we have done is reduce the problem of calculating $E(K)/mE(K)$ to the problem of the existence or non-existence of a single rational point on each of an explicitly given finite set of curves. Frequently, many of these curves can be immediately eliminated from consideration because they have no points over some completion K_v of K, which is an easy thing to check. On the other hand, a short search by hand or with a computer often uncovers rational points on some of the others. If, in this way, we can deal with all of the homogeneous spaces in question, then the determination of $E(K)/mE(K)$ is complete. The problem that arises is that occasionally there is a homogeneous space having points defined over every completion K_v, yet having no K-rational points. It is this situation, the failure of the Hasse principle, that makes the Mordell–Weil theorem ineffective.

Remark 1.3. Notice that the condition $\operatorname{div}(f_T) = m(T) - m(O)$ in (X.1.1d) is enough to specify f_T only up to multiplication by an arbitrary element of K^*. However, the equality $f_T \circ [m] = g_T^m$ with $g_T \in K(E)$ means that in fact f_T is well-determined up to multiplication by an element of $(K^*)^m$. Thus the value of $f_T(P)$ in (X.1.1d) is a well-defined element of $K^*/(K^*)^m$.

We now give the proof of (X.1.1), after which we study the case $m = 2$ in more detail and use it to compute $E(K)/2E(K)$ for an example.

PROOF OF (X.1.1). (a) Hilbert's Theorem 90 (B.2.5c) shows that the pairing is well-defined. Bilinearity follows from bilinearity of the Kummer pairing (VIII.1.2b) and bilinearity of the Weil e_m-pairing (III.8.1a).

(b) In order to prove nondegeneracy on the left, we suppose that $b(P, T) = 1$ for all $T \in E[m]$. This means that for all $T \in E[m]$ and all $\sigma \in G_{\bar{K}/K}$,

$$e_m\big(\kappa(P, \sigma), T\big) = 1.$$

The nondegeneracy of the Weil pairing (III.8.1c) implies that $\kappa(P, \sigma) = 0$ for all σ, and then (VIII.1.2c) tells us that $P \in mE(K)$.

(c) Let $\beta = b(P, T)^{1/m}$. Tracing through the definitions, we see that the field $K(\beta)$ is contained in the field $L([m]^{-1}E(K))$ described in (VIII.1.2d). Further, (VIII.1.5b) says that the extension L/K is unramified outside S. But it is easy to see that if $v \in M_K$ is a finite place with $v(m) = 0$, then the extension $K(\beta)/K$ is unramified at v if and only if

$$\operatorname{ord}_v(\beta^m) \equiv 0 \pmod{m}.$$

(Here $\operatorname{ord}_v : K^* \twoheadrightarrow \mathbb{Z}$ is the normalized valuation associated to v.) This says precisely that $b(P, T) \in K(S, m)$.

(d) Choose $Q \in E(\bar{K})$ and $\beta \in \bar{K}^*$ satisfying

$$P = [m]Q \quad\text{and}\quad b(P, T) = \beta^m.$$

Then for all $\sigma \in G_{\bar{K}/K}$ we have by definition

$$\begin{aligned}
e_m\big(\delta(P)(\sigma), T\big) &= \delta_K\big(b(P, T)\big)(\sigma), \\
e_m(Q^\sigma - Q, T) &= \beta^\sigma/\beta, \\
g_T(X + Q^\sigma - Q)/g_T(X) &= \beta^\sigma/\beta, \\
g_T(Q)^\sigma/g_T(Q) &= \beta^\sigma/\beta \qquad \text{putting } X = Q.
\end{aligned}$$

Since δ_K is an isomorphism, it follows that $g_T(Q)^m \equiv \beta^m \pmod{(K^*)^m}$. (Note that $g_T(Q)^m = f_T(P)$ is in K^*.) Therefore

$$f_T(P) = f_T \circ [m](Q) = g_T(Q)^m \equiv \beta^m = b(P, T) \pmod{(K^*)^m}. \qquad \square$$

We now consider the special case $m = 2$, which is by far the easiest with which to work. Under our assumption $E[m] \subset E(K)$, we may take a Weierstrass equation for E of the form

$$y^2 = (x - e_1)(x - e_2)(x - e_3) \qquad \text{with } e_1, e_2, e_3 \in K.$$

The three nontrivial 2-torsion points are

$$T_1 = (e_1, 0), \quad T_2 = (e_2, 0), \quad T_3 = (e_3, 0).$$

Letting $T = (e, 0)$ represent any one of these points, we claim that the associated function f_T specified in (X.1.1d) is $f_T = x - e$. It is clear that this function has the correct divisor,

$$\operatorname{div}(x - e) = 2(T) - 2(O).$$

It is then a calculation to check that

$$x \circ [2] = \left(\frac{x^2 - 2ex - 2e^2 + 2(e_1 + e_2 + e_3)e - (e_1 e_2 + e_1 e_3 + e_2 e_3)}{2y} \right)^2,$$

so $x - e$ has both of the properties needed to be f_T.

Now suppose that we have chosen a pair $(b_1, b_2) \in K(S, m) \times K(S, m)$ and that we want to determine whether there is a point $P \in E(K)/2E(K)$ satisfying

$$b(P, T_1) = b_1 \quad \text{and} \quad b(P, T_2) = b_2.$$

Such a point exists if and only if there is a solution

$$(x, y, z_1, z_2) \in K \times K \times K^* \times K^*$$

to the system of equations

$$y^2 = (x - e_1)(x - e_2)(x - e_3), \qquad b_1 z_1^2 = x - e_1, \qquad b_2 z_2^2 = x - e_2.$$

We substitute the latter two equations into the former and define a new variable z_3 by $y = b_1 b_2 z_1 z_2 z_3$, which is permissible since b_1, b_2, z_1, and z_2 take only nonzero values. This yields the three equations

$$b_1 b_2 z_3^2 = x - e_3, \qquad b_1 z_1^2 = x - e_1, \qquad b_2 z_2^2 = x - e_2.$$

Finally, eliminating x gives the pair of equations

$$b_1 z_1^2 - b_2 z_2^2 = e_2 - e_1, \qquad b_1 z_1^2 - b_1 b_2 z_3^2 = e_3 - e_1.$$

This gives a finite collection of equations, one for each pair (b_1, b_2), and we may use whatever techniques are at our disposal (e.g., v-adic, computer search) to determine whether they have a solution. Notice that if we do find a solution (z_1, z_2, z_3), then we immediately recover the corresponding point in $E(K)/2E(K)$ using the formulas

$$x = b_1 z_1^2 + e_1, \qquad y = b_1 b_2 z_1 z_2 z_3.$$

Finally we must deal with the fact that the definition $b(P, T) = f_T(P)$ cannot be used if it happens that $P = T$. In other words, there are two pairs (b_1, b_2) that do not arise from the above procedure, namely the pairs $\big(b(T_1, T_1), b(T_1, T_2)\big)$ and $\big(b(T_2, T_1), b(T_2, T_2)\big)$. These values may be computed using linearity as

$$\begin{aligned}
b(T_1, T_1) &= b(T_1, T_1 + T_2) b(T_1, T_2)^{-1} \\
&= b(T_1, T_3) b(T_1, T_2)^{-1} \\
&= \frac{e_1 - e_3}{e_1 - e_2},
\end{aligned}$$

and similarly

$$b(T_2, T_2) = \frac{e_2 - e_3}{e_2 - e_1}.$$

We summarize this entire procedure in the following proposition.

Proposition 1.4. (Complete 2-Descent) *Let E/K be an elliptic curve given by a Weierstrass equation*

$$y^2 = (x - e_1)(x - e_2)(x - e_3) \qquad \text{with } e_1, e_2, e_3 \in K.$$

Let $S \subset M_K$ be a finite set of places of K including all archimedean places, all places dividing 2, and all places at which E has bad reduction. Further let

$$K(S, 2) = \{b \in K^*/(K^*)^2 : \mathrm{ord}_v(b) \equiv 0 \ (\mathrm{mod}\ 2) \text{ for all } v \notin S\}.$$

Then there is an injective homomorphism

$$E(K)/2E(K) \longrightarrow K(S, 2) \times K(S, 2)$$

defined by

$$P = (x, y) \longmapsto \begin{cases} (x - e_1, x - e_2) & \text{if } x \neq e_1, e_2, \\ \left(\dfrac{e_1 - e_3}{e_1 - e_2}, e_1 - e_2 \right) & \text{if } x = e_1, \\ \left(e_2 - e_1, \dfrac{e_2 - e_3}{e_2 - e_1} \right) & \text{if } x = e_2, \\ (1, 1) & \text{if } x = \infty, \text{ i.e., if } P = O. \end{cases}$$

Let $(b_1, b_2) \in K(S, 2) \times K(S, 2)$ be a pair that is not the image of one of the three points O, $(e_1, 0)$, $(e_2, 0)$. Then (b_1, b_2) is the image of a point

$$P = (x, y) \in E(K)/2E(K)$$

if and only if the equations

$$b_1 z_1^2 - b_2 z_2^2 = e_2 - e_1,$$
$$b_1 z_1^2 - b_1 b_2 z_3^2 = e_3 - e_1,$$

have a solution $(z_1, z_2, z_3) \in K^* \times K^* \times K$. If such a solution exists, then we can take

$$P = (x, y) = (b_1 z_1^2 + e_1, b_1 b_2 z_1 z_2 z_3).$$

PROOF. As explained above, this is a special case of (X.1.1). □

Example 1.5. We use (X.1.4) to compute $E(\mathbb{Q})/2E(\mathbb{Q})$ for the elliptic curve

$$E : y^2 = x^3 - 12x^2 + 20x = x(x - 2)(x - 10).$$

This equation has discriminant

$$\Delta = 409600 = 2^{14}5^2,$$

so it has good reduction except at 2 and 5. Reducing the equation modulo 3, we easily check that $\#\tilde{E}(\mathbb{F}_3) = 4$. Since $E[2] \subset E_{\mathrm{tors}}(\mathbb{Q})$ and $E_{\mathrm{tors}}(\mathbb{Q})$ injects into $\tilde{E}(\mathbb{F}_3)$ from (VII.3.5), we see that

$$E_{\text{tors}}(\mathbb{Q}) = E[2].$$

Let $S = \{2, 5, \infty\} \subset M_{\mathbb{Q}}$. Then a complete set of representatives for

$$\mathbb{Q}(S, 2) = \{b \in \mathbb{Q}^*/(\mathbb{Q}^*)^2 : \text{ord}_p(b) \equiv 0 \ (\text{mod } 2) \text{ for all } p \notin S\}$$

is given by the set

$$\{\pm 1, \pm 2, \pm 5, \pm 10\}.$$

We identify this set with $\mathbb{Q}(S, 2)$. Now consider the map given in (X.1.4),

$$E(\mathbb{Q})/2E(\mathbb{Q}) \longrightarrow \mathbb{Q}(S, 2) \times \mathbb{Q}(S, 2),$$

say with

$$e_1 = 0, \qquad e_2 = 2, \qquad \text{and} \qquad e_3 = 10.$$

There are 64 pairs $(b_1, b_2) \in \mathbb{Q}(S, 2) \times \mathbb{Q}(S, 2)$, and for each pair, we must check to see whether it comes from an element of $E(\mathbb{Q})/2E(\mathbb{Q})$. For example, using (X.1.4), we can compute the image of $E[2]$ in $\mathbb{Q}(S, 2) \times \mathbb{Q}(S, 2)$:

$$O \mapsto (1, 1), \qquad (0, 0) \mapsto (5, -2), \qquad (2, 0) \mapsto (2, -1), \qquad (10, 0) \mapsto (10, 2).$$

It remains to determine, for every other pair (b_1, b_2), whether the equations

$$b_1 z_1^2 - b_2 z_2^2 = 2, \qquad b_1 z_1^2 - b_1 b_2 z_3^2 = 10, \qquad (*)$$

have a solution $z_1, z_2, z_3 \in \mathbb{Q}$. For example, if $b_1 < 0$ and $b_2 > 0$, then $(*)$ clearly has no rational solutions, since the first equation does not even have a solution in $\mathbb{R}$.

Proceeding systematically, we list our results in Table 10.1. The entry for each pair (b_1, b_2) consists of either a point of $E(\mathbb{Q})$ that maps to (b_1, b_2), or else a (local) field over which the equations listed in $(*)$ have no solution. (Note that if (z_1, z_2, z_3) is a solution to $(*)$, then the corresponding point of $E(\mathbb{Q})$ is $(b_1 z_1^2 + e_1, b_1 b_2 z_1 z_2 z_3)$.) The circled numbers in the table refer to the notes that explain each entry. Finally, we note that since the map $E(\mathbb{Q})/2E(\mathbb{Q}) \to \mathbb{Q}(S, 2) \times \mathbb{Q}(S, 2)$ is a *homomorphism*, it is not necessary to check every pair (b_1, b_2). For example, if both (b_1, b_2) and (b_1', b_2') come from $E(\mathbb{Q})$, then so does $(b_1 b_1', b_2 b_2')$. Similarly, if (b_1, b_2) does and (b_1', b_2') does not, then $(b_1 b_1', b_2 b_2')$ does not. This observation substantially reduces the number of cases of $(*)$ that must be considered.

1. If $b_1 < 0$ and $b_2 > 0$, then $b_1 z_1^2 - b_2 z_2^2 = 2$ has no solutions in $\mathbb{R}$.

2. If $b_1 < 0$ and $b_2 < 0$, then $b_1 z_1^2 - b_1 b_2 z_3^2 = 10$ has no solutions in $\mathbb{R}$.

3. The four 2-torsion points $\{O, (0, 0), (2, 0), (10, 0)\}$ map respectively to the four points $(1, 1), (5, -2), (2, -1),$ and $(10, 2)$.

4. $(b_1, b_2) = (1, -1)$: By inspection, the equations

$$z_1^2 + z_2^2 = 2 \qquad \text{and} \qquad z_1^2 + z_3^2 = 10$$

have the solution $(1, 1, 3)$. This gives the point $(1, -3) \in E(\mathbb{Q})$.

b_1 / b_2	1	2	5	10	-1	-2	-5	-10
1	0	$(18,-48)$ [5]	$\mathbb{Q}_5$ [9]		$\mathbb{R}$ [1]			
2	$\mathbb{Q}_5$ [8]	$\mathbb{Q}_5$ [9]	$(20,60)$ [5]	$(10,0)$ [3]				
5	$\mathbb{Q}_5$ [6]		$\mathbb{Q}_5$ [7]					
10								
-1	$(1,-3)$ [4]	$(2,0)$ [3]	$\mathbb{Q}_5$ [9]		$\mathbb{R}$ [2]			
-2	$\mathbb{Q}_5$ [9]		$(0,0)$ [3]	$\left(\frac{10}{9}, -\frac{80}{27}\right)$ [5]				
-5	$\mathbb{Q}_5$ [6]		$\mathbb{Q}_5$ [7]					
-10								

Table 10.1: Computing $E(\mathbb{Q})$ for $E : y^2 = x^3 - 12x^2 + 20x$.

5. Adding $(1,-3) \in E(\mathbb{Q})$ to the nontrivial two-torsion points corresponds to multiplying their (b_1, b_2) values. This gives three pairs $(5,2)$, $(2,1)$, and $(10,-2)$ in $\mathbb{Q}(S,2) \times \mathbb{Q}(S,2)$, which correspond to the three rational points $(20,60)$, $(18,-48)$, and $(10/9, -80/27)$ in $E(\mathbb{Q})$.

6. $b_1 \not\equiv 0 \pmod 5$ and $b_2 \equiv 0 \pmod 5$: The first equation in $(*)$ implies that z_1 and z_2 must be 5-adically integral. Then the second equation shows that $z_1 \equiv 0 \pmod 5$, and so from the first equation we obtain $0 \equiv 2 \pmod 5$. Therefore $(*)$ has no solutions in $\mathbb{Q}_5$.

7. The eight pairs in (6) are $\mathbb{Q}_5$-nontrivial, i.e., there are no $\mathbb{Q}_5$ solutions to $(*)$. If we multiply these eight pairs by the $\mathbb{Q}$-trivial pair $(5,2)$, we obtain eight more $\mathbb{Q}_5$-nontrivial pairs.

8. $(b_1, b_2) = (1,2)$: The two equations in $(*)$ are

$$z_1^2 - 2z_2^2 = 2 \quad \text{and} \quad z_1^2 - 2z_3^2 = 10.$$

Since 2 is a quadratic nonresidue modulo 5, the second equation implies that $z_1 \equiv z_3 \equiv 0 \pmod 5$. But then the second equation says that $0 \equiv 10 \pmod{25}$. Therefore there are no solutions in $\mathbb{Q}_5$.

9. Taking the $\mathbb{Q}_5$-nontrivial pair $(1,2)$ from (8) and multiplying it by the seven $\mathbb{Q}$-trivial pairs already in the table gives seven new $\mathbb{Q}_5$-nontrivial pairs that fill the remaining entries in the table.

Conclusion. $E(\mathbb{Q}) \cong \mathbb{Z} \times \mathbb{Z}/2\mathbb{Z} \times \mathbb{Z}/2\mathbb{Z}$.

X.2 Twisting—General Theory

For this section and the next we drop the requirement that K be a number field, so K will be an arbitrary (perfect) field. As we saw in (X §1), computation of the Mordell–Weil group of an elliptic curve E leads naturally to the problem of the existence or nonexistence of a single rational point on various other curves. These other curves are certain *twists* of E that are called *homogeneous spaces*. In this section we study the general question of twisting which, since it is no more difficult, we develop for curves of arbitrary genus. Then, in the next section, we look at the homogeneous spaces associated to elliptic curves.

Definition. Let C/K be a smooth projective curve. The *isomorphism group of C*, denoted by $\mathrm{Isom}(C)$, is the group of $\bar{K}$-isomorphisms from C to itself. We denote the subgroup of $\mathrm{Isom}(C)$ consisting of isomorphisms defined over K by $\mathrm{Isom}_K(C)$. To ease notation, we write composition of maps multiplicatively, thus $\alpha\beta$ instead of $\alpha \circ \beta$.

Remark 2.1. The group that we are denoting by $\mathrm{Isom}(C)$ is usually called the *automorphism group of C* and denoted by $\mathrm{Aut}(C)$. However, if E is an elliptic curve, then we have defined $\mathrm{Aut}(E)$ to be the group of isomorphisms from E to E that take O to O. Thus $\mathrm{Aut}(E) \neq \mathrm{Isom}(E)$, since for example, the group $\mathrm{Isom}(E)$ contains translation maps $\tau_P : E \to E$. We describe $\mathrm{Isom}(E)$ more fully in (X §5).

Definition. A *twist of C/K* is a smooth curve C'/K that is isomorphic to C over $\bar{K}$. We treat two twists as equivalent if they are isomorphic over K. The set of twists of C/K, modulo K-isomorphism, is denoted by $\mathrm{Twist}(C/K)$.

Let C'/K be a twist of C/K. Thus there is an isomorphism $\phi : C' \to C$ defined over $\bar{K}$. To measure the failure of ϕ to be defined over K, we consider the map

$$\xi : G_{\bar{K}/K} \longrightarrow \mathrm{Isom}(C), \qquad \xi_\sigma = \phi^\sigma \phi^{-1}.$$

It turns out that ξ is a 1-cocycle and that the cohomology class of ξ is uniquely determined by the K-isomorphism class of C'. Further, every cohomology class comes from some twist of C/K. In this way $\mathrm{Twist}(C/K)$ may be identified with a certain cohomology set. We now prove these assertions.

Theorem 2.2. *Let C/K be a smooth projective curve. For each twist C'/K of C/K, choose a $\bar{K}$-isomorphism $\phi : C' \to C$ and define a map $\xi_\sigma = \phi^\sigma \phi^{-1} \in \mathrm{Isom}(C)$ as above.*

(a) *The map ξ is a 1-cocycle, i.e.,*

$$\xi_{\sigma\tau} = (\xi_\sigma)^\tau \xi_\tau \qquad \text{for all } \sigma, \tau \in G_{\bar{K}/K}.$$

The associated cohomology class in $H^1\big(G_{\bar{K}/K}, \mathrm{Isom}(C)\big)$ is denoted by $\{\xi\}$.

(b) *The cohomology class $\{\xi\}$ is determined by the K-isomorphism class of C' and is independent of the choice of ϕ. We thus obtain a natural map*

$$\mathrm{Twist}(C/K) \longrightarrow H^1\big(G_{\bar{K}/K}, \mathrm{Isom}(C)\big).$$

(c) *The map in (b) is a bijection. In other words, the twists of C/K, up to K-isomorphism, are in one-to-one correspondence with the elements of the cohomology set $H^1(G_{\bar{K}/K}, \mathrm{Isom}(C))$.*

Remark 2.3. We emphasize that the group $\mathrm{Isom}(C)$ is often nonabelian, and indeed, it is always nonabelian for elliptic curves. Hence $H^1(G_{\bar{K}/K}, \mathrm{Isom}(C))$ is generally only a *pointed set*, not a group. See (B §3) for details. However, if $\mathrm{Isom}(C)$ has a $G_{\bar{K}/K}$-invariant abelian subgroup A, then $H^1(G_{\bar{K}/K}, A)$ is a group, and its image in $H^1(G_{\bar{K}/K}, \mathrm{Isom}(C))$ gives a natural group structure to some subset of $\mathrm{Twist}(C)$. We apply this observation in (X §3) when C is an elliptic curve, taking for A the group of translations, and in (X §5) we do the same with $A = \mathrm{Aut}(E)$.

PROOF. (a) We compute

$$\xi_{\sigma\tau} = \phi^{\sigma\tau}\phi^{-1} = (\phi^\sigma\phi^{-1})^\tau(\phi^\tau\phi^{-1}) = (\xi_\sigma)^\tau\xi_\tau.$$

(b) Let C''/K be another twist of C/K that is K-isomorphic to C'. Choose a $\bar{K}$-isomorphism $\psi : C'' \to C$. We must show that the cocycles $\phi^\sigma\phi^{-1}$ and $\psi^\sigma\psi^{-1}$ are cohomologous. By assumption there is a K-isomorphism $\theta : C'' \to C'$. Consider the element $\alpha = \phi\theta\psi^{-1} \in \mathrm{Isom}(C)$. We compute

$$\begin{aligned}
(\alpha^\sigma)(\psi^\sigma\psi^{-1}) &= (\phi\theta\psi^{-1})^\sigma(\psi^\sigma\psi^{-1}) = \phi^\sigma\theta^\sigma\psi^{-1} \\
&= \phi^\sigma\theta\psi^{-1} = (\phi^\sigma\phi^{-1})(\phi\theta\psi^{-1}) = (\phi^\sigma\phi^{-1})\alpha.
\end{aligned}$$

The proves that $\phi^\sigma\phi^{-1}$ and $\psi^\sigma\psi^{-1}$ are cohomologous when viewed as elements of $H^1(G_{\bar{K}/K}, \mathrm{Isom}(C))$.

(c) Suppose that C'/K and C''/K are twists of C/K that give the same cohomology class in $H^1(G_{\bar{K}/K}, \mathrm{Isom}(C))$. This means that if we choose $\bar{K}$-isomorphisms $\phi : C' \to C$ and $\psi : C'' \to C$, then there is a map $\alpha \in \mathrm{Isom}(C)$ such that

$$\alpha^\sigma(\psi^\sigma\psi^{-1}) = (\phi^\sigma\phi^{-1})\alpha \qquad \text{for all } \sigma \in G_{\bar{K}/K}.$$

In other words, the cocycles associated to ϕ and ψ are cohomologous. We now consider the map $\theta : C'' \to C'$ defined by $\theta = \phi^{-1}\alpha\psi$. It is clearly a $\bar{K}$-isomorphism, and we claim that it is, in fact, defined over K. To prove this, for any $\sigma \in G_{\bar{K}/K}$ we compute

$$\theta^\sigma = (\phi^\sigma)^{-1}(\alpha^\sigma\psi^\sigma) = (\phi^\sigma)^{-1}(\phi^\sigma\phi^{-1}\alpha\psi) = \phi^{-1}\alpha\psi = \theta.$$

Therefore C'' and C' are K-isomorphic, and thus they give the same element of $\mathrm{Twist}(C/K)$. This proves that the map

$$\mathrm{Twist}(C/K) \to H^1(G_{\bar{K}/K}, \mathrm{Isom}(C))$$

is injective.

To prove surjectivity, we start with a 1-cocycle

$$\xi : G_{\bar{K}/K} \to \mathrm{Isom}(C)$$

and use it to construct a curve C'/K and an isomorphism $\phi : C' \to C$ satisfying $\xi_\sigma = \phi^\sigma \phi^{-1}$. To do this, we consider a field, denoted by $\bar{K}(C)_\xi$, that is isomorphic, as an abstract field extension of $\bar{K}$, to $\bar{K}(C)$, say by an isomorphism that we denote by $Z : \bar{K}(C) \to \bar{K}(C)_\xi$. The difference between $\bar{K}(C)$ and $\bar{K}(C)_\xi$ lies in the action of the Galois group $G_{\bar{K}/K}$; the action on $\bar{K}(C)_\xi$ is *twisted by* ξ. What this means is that

$$Z(f)^\sigma = Z(f^\sigma \xi_\sigma) \qquad \text{for all } f \in \bar{K}(C) \text{ and all } \sigma \in G_{\bar{K}/K}.$$

In this equality we are viewing f as a map $f : C \to \mathbb{P}^1$ as in (II.2.2), and $f^\sigma \xi_\sigma$ is composition of maps. Equivalently, the map $\xi_\sigma : C \to C$ of curves induces a map $\xi_\sigma^* : \bar{K}(C) \to \bar{K}(C)$ of fields, and $f^\sigma \xi_\sigma$ is an alternative notation for $\xi_\sigma^*(f^\sigma)$.

For this action of $G_{\bar{K}/K}$ on $\bar{K}(C)_\xi$, we consider the subfield field $\mathcal{F} \subset \bar{K}(C)_\xi$ consisting of the elements of $\bar{K}(C)_\xi$ that are fixed by $G_{\bar{K}/K}$. We now show, in several steps, that the field $\mathcal{F}$ is the function field of the desired twist of C.

Step (i): $\mathcal{F} \cap \bar{K} = K$

Suppose that $Z(f) \in \mathcal{F} \cap \bar{K}$. In particular, since Z induces the identity on $\bar{K}$, we have $f \in \bar{K}$. Now the fact that $Z(f) \in \mathcal{F}$, combined with the fact that f is a constant function and thus unaffected by isomorphisms of C, implies that

$$Z(f) = Z(f)^\sigma = Z(f^\sigma \xi_\sigma) = Z(f^\sigma).$$

This holds for all $\sigma \in G_{\bar{K}/K}$, and hence $f \in K$.

Step (ii): $\bar{K}\mathcal{F} = \bar{K}(C)_\xi$

This is an immediate consequence of (II.5.8.1) applied to the $\bar{K}$-vector space $\bar{K}(C)_\xi$.

It follows from Step (ii) that $\mathcal{F}$ has transcendence degree one over K, and thus using Step (i) and (II.2.4c), we deduce that there exists a smooth curve C'/K such that $\mathcal{F} \cong K(C')$. Further, Step (ii) implies that

$$\bar{K}(C') = \bar{K}\mathcal{F} = \bar{K}(C)_\xi \cong \bar{K}(C),$$

so (II.2.4.1) says that C' and C are isomorphic over $\bar{K}$. In other words, C' is a twist of C, and the final step in proving surjectivity is to show that C' gives the cohomology class $\{\xi\}$.

Let $\phi : C' \to C$ be a $\bar{K}$-isomorphism, as described in (II.2.4b), whose associated map ϕ^* is the isomorphism of fields

$$Z : \bar{K}(C) \longrightarrow \bar{K}(C)_\xi = \bar{K}\mathcal{F} = \bar{K}(C').$$

Step (iii): $\xi_\sigma = \phi^\sigma \phi^{-1}$ for all $\sigma \in G_{\bar{K}/K}$

Having identified ϕ^* with Z, the relation $Z(f)^\sigma = Z(f^\sigma \xi_\sigma)$ used to define the map Z can be rewritten as $(f\phi)^\sigma = f^\sigma \xi_\sigma \phi$. In other words,

$$f^\sigma \phi^\sigma = (f\phi)^\sigma = f^\sigma \xi_\sigma \phi \qquad \text{for all } f \in \bar{K}(C).$$

This implies that $\phi^\sigma = \xi_\sigma \phi$, which is exactly the desired result. $\square$

Example 2.4. Let E/K be an elliptic curve, let $K(\sqrt{d})$ be a quadratic extension of K, and let

$$\chi : G_{\bar{K}/K} \to \{\pm 1\}, \qquad \chi(\sigma) = \sqrt{d}^{\,\sigma}/\sqrt{d},$$

be the quadratic character associated to $K(\sqrt{d})/K$. (Note that $\operatorname{char}(K) \neq 2$.) We use χ to define a 1-cocycle

$$\xi : G_{\bar{K}/K} \longrightarrow \operatorname{Isom}(E), \qquad \xi_\sigma = [\chi(\sigma)].$$

Let C/K be the corresponding twist of E/K. We are going to derive an equation for C/K.

We choose a Weierstrass equation for E/K of the form $y^2 = f(x)$ and we write $\bar{K}(E) = \bar{K}(x,y)$ and $\bar{K}(C) = \bar{K}(x,y)_\xi$. Since $[-1](x,y) = (x,-y)$, the action of $\sigma \in G_{\bar{K}/K}$ on $\bar{K}(x,y)_\xi$ is determined by the formulas

$$\sqrt{d}^{\,\sigma} = \chi(\sigma)\sqrt{d}, \qquad x^\sigma = x, \qquad y^\sigma = \chi(\sigma)y.$$

Notice that the functions $x' = x$ and $y' = y/\sqrt{d}$ in $\bar{K}(x,y)_\xi$ are fixed by $G_{\bar{K}/K}$, and they satisfy the equation

$$dy'^2 = f(x'),$$

which is the equation of an elliptic curve defined over K. Further, the identification $(x',y') \mapsto (x',y'\sqrt{d})$ shows that this curve is isomorphic to E over $K(\sqrt{d})$. It is now an easy matter to check that the associated cocycle is ξ, and thus to verify that we have found an equation for C/K. The curve C is a *quadratic twist of E*; more precisely, it is the *twist of E by the quadratic character* χ. We will return to this example in more detail in (X §5).

X.3 Homogeneous Spaces

We recall from (VIII §2) that associated to an elliptic curve E/K is a Kummer sequence

$$0 \longrightarrow \frac{E(K)}{mE(K)} \longrightarrow H^1\big(G_{\bar{K}/K}, E[m]\big) \longrightarrow H^1\big(G_{\bar{K}/K}, E\big)[m] \longrightarrow 0.$$

The proof of the weak Mordell–Weil theorem hinges on the essential fact that the image of the first term in the second consists of elements that are unramified outside of a certain finite set of primes. In this section we analyze the third term in the sequence by associating to each element of $H^1(G_{\bar{K}/K}, E)$ a certain twist of E called a *homogeneous space*. However, rather than starting with cohomology, we instead begin by directly defining homogeneous spaces and describing their basic properties. We follow this with the cohomological interpretation, which says that homogeneous spaces are those twists that correspond to cocycles taking values in the group of translations.

Definition. Let E/K be an elliptic curve. A (*principal*) *homogeneous space for* E/K is a smooth curve C/K together with a simply transitive algebraic group action of E on C defined over K. In other words, a homogeneous space for E/K consists of a pair (C, μ), where C/K is a smooth curve and

$$\mu : C \times E \longrightarrow C$$

is a morphism defined over K having the following three properties:

(i) $\mu(p, O) = p$ for all $p \in C$.

(ii) $\mu\big(\mu(p, P), Q\big) = \mu(p, P + Q)$ for all $p \in C$ and $P, Q \in E$.

(iii) For all $p, q \in C$ there is a unique $P \in E$ satisfying $\mu(p, P) = q$.

We will often replace $\mu(p, P)$ with the more intuitive notation $p + P$. Then property (ii) is just the associative law $(p + P) + Q = p + (P + Q)$. Of course, one must determine from context whether $+$ means addition on E or the action of E on C.

In view of the simple transitivity of the action, we may define a *subtraction map* on C by the rule

$$\nu : C \times C \longrightarrow E,$$
$$\nu(q, p) = (\text{the unique } P \in E \text{ satisfying } \mu(p, P) = q).$$

It is not clear, a priori, that the map ν is even a rational map, but we will soon see that ν is a morphism and is defined over K. (This fact also follows from elementary intersection theory on $C \times C$.) In conjunction with our addition notation for μ, we often write $\nu(q, p)$ as $q - p$.

We now verify that addition and subtraction on a homogeneous space have the right properties.

Lemma 3.1. *Let C/K be a homogeneous space for E/K. Then for all $p, q \in C$ and all $P, Q \in E$:*

(a) $\qquad\qquad \mu(p, O) = p \quad$ and $\quad \nu(p, p) = O.$

(b) $\qquad\qquad \mu\big(p, \nu(q, p)\big) = q \quad$ and $\quad \nu\big(\mu(p, P), p\big) = P.$

(c) $\qquad\qquad \nu\big(\mu(q, Q), \mu(p, P)\big) = \nu(q, p) + Q - P.$

Equivalently, using the alternative "addition" and "subtraction" notation:

(a) $\qquad\qquad p + O = p \quad$ and $\quad p - p = O.$

(b) $\qquad\qquad p + (q - p) = q \quad$ and $\quad (p + P) - p = P.$

(c) $\qquad\qquad (q + Q) - (p + P) = (q - p) + Q - P.$

In other words, using $+$ and $-$ signs provides the right intuition.

PROOF. (a) The equality $\mu(p, O) = p$ is part of the definition of homogeneous space. Next, the definition of ν says that $\nu(p, p)$ is the unique point $P \in E$ satisfying $\mu(p, P) = p$. We know that this last equation is true for $P = O$, so $\nu(p, p) = O$.
(b) The relation $\mu\big(p, \nu(q, p)\big) = q$ is the definition of ν. Then, from

$$\mu\big(p, \nu(\mu(p, P), p)\big) = \mu(p, P),$$

we conclude that $\nu(\mu(p, P), p) = P$.

(c) We start with

$$q = \mu\big(p, \nu(q, p)\big).$$

Adding Q to both sides gives

$$\begin{aligned}
\mu(q, Q) &= \mu\big(p, \nu(q, p) + Q\big) \\
&= \mu\big(p, P + \nu(q, p) + Q - P\big) \\
&= \mu\big(\mu(p, P), \nu(q, p) + Q - P\big).
\end{aligned}$$

From the definition of ν, this is equivalent to

$$\nu\big(\mu(q, Q), \mu(p, P)\big) = \nu(q, p) + Q - P. \qquad \square$$

Next we show that a homogeneous space C/K for E/K is a twist of of E/K as described in (X §2). We also describe addition and subtraction on C in terms of a given $\bar{K}$-isomorphism $E \to C$.

Proposition 3.2. *Let E/K be an elliptic curve, and let C/K be a homogeneous space for E/K. Fix a point $p_0 \in C$ and define a map*

$$\theta : E \longrightarrow C, \qquad \theta(P) = p_0 + P.$$

(a) *The map θ is an isomorphism defined over $K(p_0)$. In particular, the curve C/K is a twist of E/K.*

(b) *For all $p \in C$ and all $P \in E$,*

$$p + P = \theta\big(\theta^{-1}(p) + P\big).$$

(N.B. The first $+$ is the action of E on C, while the second $+$ is addition on E.)

(c) *For all $p, q \in C$,*

$$q - p = \theta^{-1}(q) - \theta^{-1}(p).$$

(d) *The subtraction map*

$$\nu : C \times C \longrightarrow E, \qquad \nu(q, p) = q - p,$$

is a morphism and is defined over K.

PROOF. (a) The action of E on C is defined over K. Hence for any $\sigma \in G_{\bar{K}/K}$ satisfying $p_0^\sigma = p_0$, we have

$$\theta(P)^\sigma = (p_0 + P)^\sigma = p_0^\sigma + P^\sigma = p_0 + P^\sigma = \theta(P^\sigma).$$

This shows that θ is defined over $K(p_0)$. Further, the simple transitivity of the action tells us that θ has degree one, and then (II.2.4.1) allows us to conclude that θ is an isomorphism.

(b) We compute

$$\theta\big(\theta^{-1}(p) + P\big) = p_0 + \theta^{-1}(p) + P = p + P.$$

Note that we are using the fact that $\theta^{-1}(p)$ is the unique point of E that gives p when it is added to p_0.

(c) We compute

$$\theta^{-1}(q) - \theta^{-1}(p) = \big(p_0 + \theta^{-1}(q)\big) - \big(p_0 + \theta^{-1}(p)\big) = q - p.$$

(d) The fact that ν is a morphism follows from (c), since (III.3.6) says that subtraction on E is a morphism. To check that ν is defined over K, we let $\sigma \in G_{\bar{K}/K}$ and use (c) to compute

$$
\begin{aligned}
(q - p)^\sigma &= \big(\theta^{-1}(q) - \theta^{-1}(p)\big)^\sigma \\
&= \theta^{-1}(q)^\sigma - \theta^{-1}(p)^\sigma && \text{since subtraction on } E \text{ is} \\
&&& \text{defined over } K, \\
&= \big(p_0 + \theta^{-1}(q)\big)^\sigma - \big(p_0 + \theta^{-1}(p)\big)^\sigma && \text{since the action of } E \text{ on} \\
&&& C \text{ is defined over } K, \\
&= q^\sigma - p^\sigma.
\end{aligned}
$$

This completes the proof that ν is defined over K. $\square$

Definition. Two homogeneous spaces C/K and C'/K for E/K are *equivalent* if there is an isomorphism $\theta : C \to C'$ defined over K that is compatible with the action of E on C and C'. In other words,

$$\theta(p + P) = \theta(p) + P \qquad \text{for all } p \in C \text{ and all } P \in E.$$

The equivalence class containing E/K, acting on itself by translation, is called the *trivial class*. The collection of equivalence classes of homogeneous spaces for E/K is called the *Weil–Châtelet group for E/K* and is denoted by $\mathrm{WC}(E/K)$. (We will see later why $\mathrm{WC}(E/K)$ is a group.)

The next result explains which homogeneous spaces are trivial.

Proposition 3.3. *Let C/K be a homogeneous space for E/K. Then C/K is in the trivial class if and only if $C(K)$ is not the empty set.*

PROOF. Suppose that C/K is in the trivial class. Then there is a K-isomorphism $\theta : E \to C$, and thus $\theta(O) \in C(K)$.

Conversely, suppose that $p_0 \in C(K)$. Then from (X.3.2a), the map

$$\theta : E \longrightarrow C, \qquad \theta(P) = p_0 + P,$$

is an isomorphism defined over $K(p_0) = K$. The required compatibility condition on θ is

$$p_0 + (P + Q) = (p_0 + P) + Q,$$

which is part of the definition of homogeneous space. $\square$

Remark 3.4. Notice that (X.3.3) says that the problem of checking the triviality of a homogeneous space is exactly equivalent to answering the fundamental Diophantine question whether the given curve has any rational points. Thus our next step, namely the identification of $WC(E/K)$ with a certain cohomology group, may be regarded as the development of a tool to help us study this difficult Diophantine problem.

Lemma 3.5. *Let* $\theta : C/K \to C'/K$ *be an equivalence of homogeneous spaces for* E/K. *Then*

$$\theta(q) - \theta(p) = q - p \qquad \text{for all } p, q \in C.$$

PROOF. This is just a matter of grouping points so that the additions and subtractions are well-defined. Thus

$$\theta(q) - \theta(p) = \Big(\big(\theta(q) - (p - q) \big) - \theta(p) \Big) - (q - p)$$

$$= \Big(\theta \big(q + (p - q) \big) - \theta(p) \Big) + (q - p)$$

$$= q - p. \qquad \qquad \square$$

Theorem 3.6. *Let* E/K *be an elliptic curve. There is a natural bijection*

$$WC(E/K) \longrightarrow H^1(G_{\bar{K}/K}, E)$$

defined as follows:

 Let C/K *be a homogeneous space for* E/K *and choose any point* $p_0 \in C$. *Then*

$$\{C/K\} \longmapsto \{\sigma \mapsto p_0^\sigma - p_0\}.$$

(The braces indicate that we are taking the equivalence class of C/K *and the cohomology class of the* 1-*cocycle* $\sigma \mapsto p_0^\sigma - p_0$.*)*

Remark 3.6.1. Since $H^1(G_{\bar{K}/K}, E)$ is a group, we can use (X.3.6) to define a group structure on the set $WC(E/K)$. It is also possible to describe the group law on $WC(E/K)$ geometrically, without using cohomology, which in fact is the way that it was originally defined. See Exercise 10.2 and [307].

PROOF. First we check that the map is well-defined. It is easy to see that the map $\sigma \mapsto p_0^\sigma - p_0$ is a cocycle:

$$p_0^{\sigma\tau} - p_0 = (p_0^{\sigma\tau} - p_0^\tau) - (p_0^\tau - p_0) = (p_0^\sigma - p_0)^\tau - (p_0^\tau - p_0).$$

Now suppose that C'/K is another homogeneous space that is equivalent to C/K. Let $\theta : C \to C'$ be a K-isomorphism giving the equivalence, and let $p_0' \in C'$. We use (X.3.5) to compute

$$p_0^\sigma - p_0 = \theta(p_0^\sigma) - \theta(p_0)$$

$$= (p_0'^\sigma - p_0') + \Big(\big(\theta(p_0) - p_0' \big)^\sigma - \big(\theta(p_0) - p_0' \big) \Big).$$

Hence the cocycles $p_0^\sigma - p_0$ and $p_0'^\sigma - p_0'$ differ by the coboundary generated by the point $\theta(p_0) - p_0' \in E$, so they give the same cohomology class in $H^1(G_{\bar{K}/K}, E)$.

Next we check injectivity. Suppose that the cocycles $p_0^\sigma - p_0$ and $p_0'^\sigma - p_0'$ corresponding to C/K and C'/K are cohomologous. This means that there is a point $P_0 \in E$ satisfying

$$p_0^\sigma - p_0 = p_0'^\sigma - p_0' + (P_0^\sigma - P_0) \qquad \text{for all } \sigma \in G_{\bar{K}/K}.$$

Consider the map

$$\theta : C \longrightarrow C', \qquad \theta(p) = p_0' - (p - p_0) + P_0.$$

It is clear that θ is a $\bar{K}$-isomorphism and that it is compatible with the action of E on C and C'. We claim that θ is defined over K. In order to prove this, we compute

$$
\begin{aligned}
\theta(p)^\sigma &= p_0'^\sigma + (p^\sigma - p_0^\sigma) + P_0^\sigma \\
&= p_0' + (p^\sigma - p_0) + P_0 + \left((p_0'^\sigma - p_0') + P_0^\sigma - P_0 - (p_0^\sigma - p_0) \right) \\
&= \theta(p^\sigma).
\end{aligned}
$$

This proves that C and C' are equivalent.

It remains to prove surjectivity. Let $\xi : G_{\bar{K}/K} \to E$ be a 1-cocycle representing an element in $H^1(G_{\bar{K}/K}, E)$. We embed E into $\mathrm{Isom}(E)$ by sending $P \in E$ to the translation map $\tau_P \in \mathrm{Isom}(E)$, and then we may view ξ as living in the cohomology group $H^1(G_{\bar{K}/K}, \mathrm{Isom}(E))$. From (X.2.2), there are a curve C/K and a $\bar{K}$-isomorphism $\phi : C \to E$ such that for all $\sigma \in G_{\bar{K}/K}$,

$$\phi^\sigma \circ \phi^{-1} = (\text{translation by } -\xi_\sigma).$$

(The reason for using $-\xi$, rather than ξ, will soon become apparent.)

Define a map

$$\mu : C \times E \longrightarrow C, \qquad \mu(p, P) = \phi^{-1}\big(\phi(p) + P\big).$$

We now show that μ gives C/K the structure of a homogeneous space over E/K and that its associated cohomology class is $\{\xi\}$.

First we check that μ is simply transitive. Let $p, q \in C$. Then by definition we have

$$\mu(p, P) = q \quad \text{if and only if} \quad \phi^{-1}\big(\phi(p) + P\big) = q,$$

so the only choice for P is $P = \phi(q) - \phi(p)$. Second we verify that μ is defined over K. We take $\sigma \in G_{\bar{K}/K}$ and compute

$$
\begin{aligned}
\mu(p, P)^\sigma &= (\phi^{-1})^\sigma\big(\phi^\sigma(p^\sigma) + P^\sigma\big) \\
&= \phi^{-1}\Big(\big(\phi(p^\sigma) - \xi_\sigma + P^\sigma\big) + \xi_\sigma\Big) \\
&= \mu(p^\sigma, P^\sigma).
\end{aligned}
$$

Finally, we compute the cohomology class associated to C/K. To do this, we may choose *any* point $p_0 \in C$ and take the class of the cocycle $\sigma \mapsto p_0^\sigma - p_0$. In particular, if we take $p_0 = \phi^{-1}(O)$, then

$$
\begin{aligned}
p_0^\sigma - p_0 &= (\phi^\sigma)^{-1}(O) - \phi^{-1}(O) \\
&= \phi^{-1}(O + \xi_\sigma) - \phi^{-1}(O) \\
&= \xi_\sigma.
\end{aligned}
$$

This completes the proof of (X.3.6). $\qquad\qquad\qquad\qquad\qquad\qquad\qquad$ □

Remark 3.7. Let E/K be an elliptic curve and let $K(\sqrt{d})/K$ be a quadratic extension, so in particular $\mathrm{char}(K) \neq 2$. Let $T \in E(K)$ be a nontrivial point of order 2. Then the homomorphism

$$
\xi : G_{\bar{K}/K} \longrightarrow E,
$$

$$
\sigma \longmapsto \begin{cases} O & \text{if } \sqrt{d}^\sigma = \sqrt{d}, \\ T & \text{if } \sqrt{d}^\sigma = -\sqrt{d}, \end{cases}
$$

is a 1-cocycle. We now construct the homogeneous space corresponding to the element $\{\xi\} \in H^1(G_{\bar{K}/K}, E)$.

Since $T \in E(K)$, we may choose a Weierstrass equation for E/K in the form

$$
E : y^2 = x^3 + ax^2 + bx \qquad \text{with} \quad T = (0,0).
$$

Then the translation-by-T map has the simple form

$$
\tau_T(P) = (x,y) + (0,0) = \left(\frac{b}{x}, -\frac{by}{x^2} \right).
$$

Thus if we let $\sigma \in G_{\bar{K}/K}$ be the nontrivial automorphism of $K(\sqrt{d})/K$, then the action of σ on the twisted field $\bar{K}(E)_\xi$ may be summarized by

$$
\sqrt{d}^\sigma = -\sqrt{d}, \qquad x^\sigma = \frac{b}{x}, \qquad y^\sigma = -\frac{by}{x^2}.
$$

We need to find the subfield of $K(\sqrt{d})(x,y)_\xi$ that is fixed by σ.

The functions

$$
\frac{\sqrt{d}\,x}{y} \quad \text{and} \quad \sqrt{d}\left(x - \frac{b}{x} \right)
$$

are easily seen to be invariant. Anticipating the form of our final equation, we consider instead the functions

$$
z = \frac{\sqrt{d}\,x}{y} \quad \text{and} \quad w = \sqrt{d}\left(x - \frac{b}{x} \right)\left(\frac{x}{y} \right)^2.
$$

We find a relation between z and w by computing

$$d\left(\frac{w}{z^2}\right)^2 = \left(x - \frac{b}{x}\right)^2 = \left(x + \frac{b}{x}\right)^2 - 4b$$

$$= \left(\left(\frac{y}{x}\right)^2 - a\right)^2 - 4b = \left(\frac{d}{z^2} - a\right)^2 - 4b.$$

Thus (z, w) are affine coordinates for the hyperelliptic curve

$$C : dw^2 = d^2 - 2adz^2 + (a^2 - 4b)z^4.$$

(See (II.2.5.1) and Exercise 2.14 for general properties of hyperelliptic curves.) We claim that C/K is the twist of E/K corresponding to the cocycle ξ.

First, we recall from (II.2.5.1) that C is a smooth *affine* curve provided that the polynomial $d^2 - 2adz^2 + (a^2 - 4b)z^4$ has four distinct roots in $\bar{K}$. Further, (II.2.5.2) says that if the quartic polynomial has distinct roots, then there is a smooth curve in $\mathbb{P}^3$ that has an affine piece isomorphic to C. This smooth curve consists of C together with the two points

$$\left[0, 0, \pm\sqrt{\frac{a^2 - 4b}{d}}, 1\right]$$

at infinity. (N.B. The projective closure of C in $\mathbb{P}^2$ is always singular.) It is easy to check that the quartic has distinct roots if and only if $b(a^2 - 4b) \neq 0$. On the other hand, since E is nonsingular, we know that $\Delta(E) = 16b^2(a^2 - 4b) \neq 0$. Therefore C is an affine piece of a smooth curve in $\mathbb{P}^3$. To ease notation, we also use C to denote this smooth curve $C \subset \mathbb{P}^3$.

There is a natural map defined over $K(\sqrt{d}\,)$,

$$\phi : E \longrightarrow C,$$

$$(x, y) \longmapsto (z, w) = \left(\frac{\sqrt{d}\,x}{y}, \sqrt{d}\left(x - \frac{b}{x}\right)\left(\frac{x}{y}\right)^2\right).$$

Note that since

$$\frac{x}{y} = \frac{xy}{y^2} = \frac{y}{x^2 + ax + b},$$

the map ϕ may also be written as

$$\phi(x, y) = \left(\frac{\sqrt{d}\,y}{x^2 + ax + b}, \frac{\sqrt{d}\,(x^2 - b)}{x^2 + ax + b}\right).$$

This allows us to evaluate

$$\phi(0, 0) = (0, -\sqrt{d}\,) \quad \text{and} \quad \phi(O) = (0, \sqrt{d}\,).$$

To show that ϕ is an isomorphism, we compute its inverse:

$$\frac{\sqrt{d}\,w}{z^2} = x - \frac{b}{x} = 2x - \left(x + \frac{b}{x}\right)$$

$$= 2x - \left(\left(\frac{y}{x}\right)^2 - a\right) = 2x - \left(\frac{d}{z^2} - a\right).$$

This gives x in terms of z and w, and then $y = \sqrt{d}\,x/z$. Thus

$$\phi^{-1} : C \longrightarrow E,$$

$$(z, w) \longmapsto \left(\frac{\sqrt{d}\,w - az^2 + d}{2z^2}, \frac{dw - a\sqrt{d}\,z^2 + d\sqrt{d}}{2z^3}\right).$$

Since C and E are smooth, it follows from (II.2.4.1) that ϕ is an isomorphism.

Finally, in order to compute the element of $H^1(G_{\bar{K}/K}, E)$ corresponding to the curve C/K, we may choose *any* point $p \in C$ and compute the cocycle

$$\sigma \longmapsto p^\sigma - p = \phi^{-1}(p^\sigma) - \phi^{-1}(p).$$

For instance, we may take $p = (0, \sqrt{d}) \in C$. It is clear that if σ fixes $\sqrt{d}$, then $p^\sigma - p = O$. On the other hand, if $\sqrt{d}^\sigma = -\sqrt{d}$, then

$$p^\sigma - p = \phi^{-1}(0, -\sqrt{d}) - \phi^{-1}(0, \sqrt{d}) = (0, 0).$$

Therefore $p^\sigma - p = \xi_\sigma$ for all $\sigma \in G_{\bar{K}/K}$, so $\{C/K\} \in \mathrm{WC}(E/K)$ maps to $\{\xi\} \in H^1(G_{\bar{K}/K}, E)$. Of course, it was just "luck" that we obtained an equality $p^\sigma - p = \xi_\sigma$. In general, the difference of these two cocycles would be some coboundary.

We conclude this section by showing that if C/K is a homogeneous space for E/K, then $\mathrm{Pic}^0(C)$ may be canonically identified with E. This means that E is the *Jacobian variety of C/K*. Since every curve C/K of genus one is a homogeneous space for some elliptic curve E/K (Exercise 10.3), this shows that the abstract group $\mathrm{Pic}^0(C)$ can always be represented as the group of points of an elliptic curve. The analogous result for curves of higher genus, in which $\mathrm{Pic}^0(C)$ is represented by an abelian variety of dimension equal to the genus of C, is considerably harder to prove.

Theorem 3.8. *Let C/K be a homogeneous space for an elliptic curve E/K. Choose a point $p_0 \in C$ and consider the summation map*

$$\mathrm{sum} : \mathrm{Div}^0(C) \longrightarrow E,$$

$$\sum n_i(p_i) \longmapsto \sum [n_i](p_i - p_0).$$

(a) *There is an exact sequence*

$$1 \longrightarrow \bar{K}^* \longrightarrow \bar{K}(C)^* \xrightarrow{\mathrm{div}} \mathrm{Div}^0(C) \xrightarrow{\mathrm{sum}} E \longrightarrow 0.$$

(b) *The summation map is independent of the choice of the point p_0.*

(c) *The summation map commutes with the natural actions of the Galois group $G_{\bar{K}/K}$ on $\mathrm{Div}^0(C)$ and on E. Hence it induces an isomorphism of $G_{\bar{K}/K}$-modules (also denoted by* sum*)*

$$\mathrm{sum} : \mathrm{Pic}^0(C) \xrightarrow{\ \sim\ } E.$$

In particular,

$$\mathrm{Pic}^0_K(C) \cong E(K).$$

PROOF. (a) Using (II.3.4), we see that we must check that the summation map is a surjective homomorphism and that its kernel is the set of principal divisors. It is clear that it is a homomorphism. Let $P \in E$ and $D = (p_0 + P) - (p_0) \in \mathrm{Div}^0(C)$. Then

$$\mathrm{sum}(D) = ((p_0 + P) - p_0) - (p_0 - p_0) = P,$$

so sum is surjective.

Next suppose that $D = \sum n_i(p_i) \in \mathrm{Div}^0(C)$ satisfies $\mathrm{sum}(D) = O$. Then the divisor $\sum n_i(p_i - p_0) \in \mathrm{Div}^0(E)$ sums to O, so (III.3.5) tells us that it is principal, say

$$\sum n_i(p_i - p_0) = \mathrm{div}(f) \quad \text{for some } f \in \bar{K}(E)^*.$$

We have an isomorphism

$$\phi : C \longrightarrow E, \qquad \phi(p) = p - p_0,$$

and hence applying (II.3.6b),

$$\mathrm{div}(\phi^* f) = \phi^* \,\mathrm{div}(f) = \sum n_i \phi^*\big((p_i - p_0)\big) = \sum n_i(p_i) = D.$$

Therefore D is principal.

Finally, if $D = \mathrm{div}(g)$ is principal, then

$$\sum n_i(p_i - p_0) = (\phi^{-1})^* \,\mathrm{div}(g) = \mathrm{div}\big((\phi^{-1})^* g\big),$$

and hence $\mathrm{sum}(D) = O$. This shows that the kernel of the summation map is the set of principal divisors.

(b) Let $\mathrm{sum}' : \mathrm{Div}^0(C) \to E$ be the summation map defined using the base point $p_0' \in C$. Then

$$\mathrm{sum}(D) - \mathrm{sum}'(D) = \sum [n_i]\big((p_i - p_0) - (p_i - p_0')\big)$$

$$= \sum [n_i](p_0' - p_0)$$

$$= O,$$

since $\sum n_i = \deg(D) = 0$.

(c) Let $\sigma \in G_{\bar{K}/K}$. Then

$$\text{sum}(D)^\sigma = \sum [n_i](p_i^\sigma - p_0^\sigma) = \text{sum}(D^\sigma),$$

since we know from (b) that the sum is the same if we use p_0^σ as our base point instead of p_0. Now (a) and the definition of $\text{Pic}^0(C)$ tell us that we have a group isomorphism sum : $\text{Pic}^0(C) \to E$, and the fact that the summation map commutes with the action of $G_{\bar{K}/K}$ says precisely that it is an isomorphism of $G_{\bar{K}/K}$-modules. Finally, the last statement in (X.3.8c) follows by taking $G_{\bar{K}/K}$-invariants. $\qquad\square$

X.4 The Selmer and Shafarevich–Tate Groups

We return now to the problem of calculating the Mordell–Weil group of an elliptic curve E/K defined over a number field K. As we have seen in (VIII.3.2) and Exercise 8.18, it is enough to find generators for the finite group $E(K)/mE(K)$ for any integer $m \geq 2$.

Suppose that we are given another elliptic curve E'/K and a nonzero isogeny $\phi : E \to E'$ defined over K. For example, we could take $E = E'$ and $\phi = [m]$. Then there is an exact sequence of $G_{\bar{K}/K}$-modules

$$0 \longrightarrow E[\phi] \longrightarrow E \overset{\phi}{\longrightarrow} E' \longrightarrow 0,$$

where $E[\phi]$ denotes the kernel of ϕ. Taking Galois cohomology yields the long exact sequence

$$0 \longrightarrow E(K)[\phi] \longrightarrow E(K) \overset{\phi}{\longrightarrow} E'(K) \longrightarrow$$
$$\overset{\delta}{}$$
$$\longrightarrow H^1(G_{\bar{K}/K}, E[\phi]) \longrightarrow H^1(G_{\bar{K}/K}, E) \overset{\phi}{\longrightarrow} H^1(G_{\bar{K}/K}, E'),$$

and from this we form the fundamental short exact sequence

$$0 \to E'(K)/\phi(E(K)) \overset{\delta}{\longrightarrow} H^1(G_{\bar{K}/K}, E[\phi]) \to H^1(G_{\bar{K}/K}, E)[\phi] \to 0. \quad (*)$$

Note that (X.3.6) says that the last term in $(*)$ may be identified with the ϕ-torsion in the Weil–Châtelet group $\text{WC}(E/K)$.

The next step is to replace the second and third terms of $(*)$ with certain finite groups. This is accomplished by local considerations. For each $v \in M_K$ we fix an extension of v to $\bar{K}$, which serves to fix an embedding $\bar{K} \subset \bar{K}_v$ and a decomposition group $G_v \subset G_{\bar{K}/K}$. Now G_v acts on $E(\bar{K}_v)$ and $E'(\bar{K}_v)$, and repeating the above argument yields exact sequences

$$0 \to E'(K_v)/\phi(E(K_v)) \overset{\delta}{\longrightarrow} H^1(G_v, E[\phi]) \to H^1(G_v, E)[\phi] \to 0. \quad (*_v)$$

The natural inclusions $G_v \subset G_{\bar{K}/K}$ and $E(\bar{K}) \subset E(\bar{K}_v)$ give restriction maps on cohomology, and we thus end up with the following commutative diagram, in which we have replaced each $H^1(G, E)$ with the corresponding Weil–Châtelet group:

$$0 \to \quad E'(K)/\phi(E(K)) \quad \xrightarrow{\delta} H^1(G_{\bar{K}/K}, E[\phi]) \to \quad \mathrm{WC}(E/K)[\phi] \quad \to 0$$

$$\downarrow \qquad\qquad\qquad \downarrow \qquad\qquad\qquad \downarrow$$

$$0 \to \prod_{v \in M_K} E'(K_v)/\phi(E(K_v)) \xrightarrow{\delta} \prod_{v \in M_K} H^1(G_v, E[\phi]) \to \prod_{v \in M_K} \mathrm{WC}(E/K_v)[\phi] \to 0$$

$$(**)$$

Our ultimate goal is to compute the image of $E'(K)/\phi(E(K))$ in the cohomology group $H^1(G_{\bar{K}/K}, E[\phi])$, or equivalently, to compute the kernel of the map

$$H^1(G_{\bar{K}/K}, E[\phi]) \to \mathrm{WC}(E/K)[\phi].$$

Using (X.3.3), we see that this last problem is the same as determining whether certain homogeneous spaces possess a K-rational point, which may be a very difficult question to answer. On the other hand, by the same reasoning, the determination of each local kernel

$$\ker\Big(H^1(G_v, E[\phi]) \longrightarrow \mathrm{WC}(E/K_v)[\phi]\Big)$$

is straightforward, since the question whether a curve has a point over a complete local field K_v reduces, by Hensel's lemma, to checking whether the curve has a point in some finite ring $R_v/\mathcal{M}_v^e$ for some easily computable integer e, which clearly requires only a finite amount of computation. This prompts the following definitions.

Definition. Let $\phi : E/K \to E'/K$ be an isogeny. The ϕ-*Selmer group of* E/K is the subgroup of $H^1(G_{\bar{K}/K}, E[\phi])$ defined by

$$S^{(\phi)}(E/K) = \ker\left\{ H^1(G_{\bar{K}/K}, E[\phi]) \longrightarrow \prod_{v \in M_K} \mathrm{WC}(E/K_v) \right\}.$$

The *Shafarevich–Tate group of* E/K is the subgroup of $\mathrm{WC}(E/K)$ defined by

$$\text{Ш}(E/K) = \ker\left\{ \mathrm{WC}(E/K) \longrightarrow \prod_{v \in M_K} \mathrm{WC}(E/K_v) \right\}.$$

(The Cyrillic letter Ш is pronounced "sha.")

Remark 4.1.1. The exact sequences $(*_v)$ require us to extend each $v \in M_K$ to $\bar{K}$, so the groups $S^{(\phi)}(E/K)$ and $\text{Ш}(E/K)$ might depend on this choice. However, in order to determine whether an element of $\mathrm{WC}(E/K)$ becomes trivial in $\mathrm{WC}(E/K_v)$, we must check whether the associated homogeneous space, which is a curve defined over K, has any points defined over K_v. This last problem is clearly independent of our choice of extension of v to $\bar{K}$, since v itself determines the embedding of K into K_v. Therefore $S^{(\phi)}(E/K)$ and $\text{Ш}(E/K)$ depend only on E and K.

Alternatively, one can check directly by working with cocycles that the cohomological definitions of $S^{(\phi)}$ and Ш do not depend on the extension of the $v \in M_K$ to $\bar{K}$. We leave this verification for the reader. (See also Exercise B.6.)

Remark 4.1.2. A good way to view $\text{III}(E/K)$ is as the group of homogeneous spaces for E/K that possess a K_v-rational point for every $v \in M_K$. Equivalently, the Shafarevich–Tate group $\text{III}(E/K)$ is the group of homogeneous spaces, modulo equivalence, that are everywhere locally trivial.

Theorem 4.2. Let $\phi : E/K \to E'/K$ be an isogeny of elliptic curves defined over K.

(a) *There is an exact sequence*

$$0 \longrightarrow E'(K)/\phi(E(K)) \longrightarrow S^{(\phi)}(E/K) \longrightarrow \text{III}(E/K)[\phi] \longrightarrow 0.$$

(b) *The Selmer group $S^{(\phi)}(E/K)$ is finite.*

PROOF. (a) This is immediate from the diagram $(**)$ and the definitions of the Selmer and Shafarevich–Tate groups.

(b) If we take $E = E'$ and $\phi = [m]$, then (a) and the finiteness of $S^{(m)}(E/K)$ imply the weak Mordell–Weil theorem. On the other hand, in order to prove that $S^{(\phi)}(E/K)$ is finite for a general map ϕ, we must essentially re-prove the weak Mordell–Weil theorem. The arguement goes as follows.

Let $\xi \in S^{(\phi)}(E/K)$, and let $v \in M_K$ be a finite place of K not dividing $m = \deg(\phi)$ and such that E/K has good reduction at v. We claim that ξ is unramified at v. (See (VIII §2) for the definition of an unramified cocycle.)

To check this, let $I_v \in G_v$ be the inertia group for v. Since $\xi \in S^{(\phi)}(E/K)$, we know that ξ is trivial in $\text{WC}(E/K_v)$. Hence from the sequence $(*_v)$ given earlier, there is a point $P \in E(\bar{K}_v)$ such that

$$\xi_\sigma = \{P^\sigma - P\} \qquad \text{for all } \sigma \in G_v.$$

(Note that $P^\sigma - P \in E[\phi]$.) In particular, this holds for all σ in the inertia group. But if $\sigma \in I_v$, then looking at the "reduction modulo v" map $E \to \tilde{E}_v$ yields

$$\widetilde{P^\sigma - P} = \tilde{P}^\sigma - \tilde{P} = \tilde{O},$$

since by definition inertia acts trivially on $\tilde{E}_v$. Thus $P^\sigma - P$ is in the kernel of reduction modulo v. But $P^\sigma - P$ is also in $E[\phi]$, which is contained in $E[m]$; and from (VIII.1.4) we know that $E(K)[m]$ injects into $\tilde{E}_v$. Therefore $P^\sigma = P$, and hence

$$\xi_\sigma = \{P^\sigma - P\} = 0 \qquad \text{for all } \sigma \in G_v.$$

This proves that every element in $S^{(\phi)}(E/K)$ is unramified at all but a fixed, finite set of places $v \in M_K$. The following lemma allows us to conclude that $S^{(\phi)}(E/K)$ is finite. $\qquad \square$

Lemma 4.3. Let M be a finite (abelian) $G_{\bar{K}/K}$-module, let $S \subset M_K$ be a finite set of places, and define

$$H^1(G_{\bar{K}/K}, M; S) = \{\xi \in H^1(G_{\bar{K}/K}, M) : \xi \text{ is unramified outside } S\}.$$

Then $H^1(G_{\bar{K}/K}, M; S)$ is finite.

PROOF. Since M is finite and $G_{\bar{K}/K}$ acts continuously on M, there is a subgroup of finite index in $G_{\bar{K}/K}$ that fixes every element of M. Using the inflation–restriction sequence (B.2.4), it suffices to prove the lemma with K replaced by a finite extension, so we may assume that the action of $G_{\bar{K}/K}$ on M is trivial. Then

$$H^1(G_{\bar{K}/K}, M; S) = \operatorname{Hom}(G_{\bar{K}/K}, M; S).$$

Let m be the exponent of M, i.e., the smallest positive integer such that $mx = 0$ for all $x \in M$, and let L/K be the maximal abelian extension of K having exponent m that is unramified outside of S. Since M has exponent m, the natural map

$$\operatorname{Hom}(G_{L/K}, M; S) \longrightarrow \operatorname{Hom}(G_{\bar{K}/K}, M; S)$$

is an isomorphism. But we know from (VIII.1.6) that L/K is a *finite* extension. Therefore $\operatorname{Hom}(G_{\bar{K}/K}, M; S)$ is finite. $\square$

We record as a corollary an important property of the Selmer group that was derived during the course of proving (X.4.2), where we use the fact (VII.7.2) that isogenous elliptic curves have the same set of primes of bad reduction.

Corollary 4.4. *Let $\phi : E/K \to E'/K$ be as in (X.4.2), and let $S \subset M_K$ be a finite set of places containing*

$$M_K^\infty \cup \{v \in M_K^0 : E \text{ has bad reduction at } v\} \cup \{v \in M_K^0 : v(\deg \phi) > 0\}.$$

Then

$$S^{(\phi)}(E/K) \subset H^1(G_{\bar{K}/K}, E[\phi]; S).$$

Remark 4.5. Certainly in theory, and often in practice, the Selmer group is effectively computable. This is true because the finite group $H^1(G_{\bar{K}/K}, E[\phi]; S)$ is effectively computable. Then, in order to determine whether a given element $\xi \in H^1(G_{\bar{K}/K}, E[\phi]; S)$ is in $S^{(\phi)}(E/K)$, we take the corresponding homogeneous spaces $\{C/K\} \in \operatorname{WC}(E/K)$ and check, for each of the finitely many $v \in S$, whether $C(K_v) \neq \emptyset$. This last problem may be reduced, using Hensel's lemma, to a finite amount of computation.

Example 4.5.1. We reformulate the example described in (X §1) in these terms, leaving some details to the reader. Let E/K be an elliptic curve with $E[m] \subset E(K)$, let $S \subset M_K$ be the usual set of places (X.4.4), and let $K(S, m)$ be as in (X.1.1c). We choose a basis for $E[m]$ and use it to identify $E[m]$ with $\boldsymbol{\mu}_m \times \boldsymbol{\mu}_m$ as $G_{\bar{K}/K}$-modules. Then

$$H^1(G_{\bar{K}/K}, E[m]; S) \cong K(S, m) \times K(S, m),$$

where this map uses the isomorphism $K^*/(K^*)^m \xrightarrow{\sim} H^1(G_{\bar{K}/K}, \boldsymbol{\mu}_m)$.

Restricting attention now to the case $m = 2$, the homogeneous space associated to a pair $(b_1, b_2) \in K(S, m) \times K(S, m)$ is the curve in $\mathbb{P}^3$ given by the equations (cf. (X.1.4))

$$C : b_1 z_1^2 - b_2 z_2^2 = (e_2 - e_1) z_0^2, \quad b_1 z_1^2 - b_1 b_2 z_3^2 = (e_3 - e_1) z_0^2.$$

For any given pair (b_1, b_2) and any absolute value $v \in S$, it is easy to check whether $C(K_v) \neq \emptyset$, and thus to calculate $S^{(2)}(E/K)$. For example, the conclusion of (X.1.5) may be summarized by stating that the curve

$$E : y^2 = x^3 - 12x^2 + 20x$$

satisfies

$$S^{(2)}(E/\mathbb{Q}) = (\mathbb{Z}/2\mathbb{Z})^2 \quad \text{and} \quad \text{III}(E/\mathbb{Q})[2] = 0.$$

The conclusion about III follows from the exact sequence (X.4.2a), since in (X.1.5) we proved that every element of $S^{(2)}(E/\mathbb{Q})$ is the image of a point in $E(\mathbb{Q})$.

Suppose that we have computed the Selmer group $S^{(\phi)}(E/K)$ for an isogeny ϕ. Each $\xi \in S^{(\phi)}(E/K)$ corresponds to a homogeneous space C_ξ/K that has a point defined over every local field K_v. Suppose further that we are lucky and can show that $\text{III}(E/K)[\phi] = 0$. This means that we are able to find a K-rational point on each C_ξ. It then follows from (X.4.2a) that $E'(K)/\phi(E(K)) \cong S^{(\phi)}(E/K)$, and all that remains is to explain how to find generators for $E'(K)/\phi(E(K))$ in terms of the points that we found in each $C_\xi(K)$. This is done in the next proposition.

Proposition 4.6. *Let* $\phi : E/K \to E'/K$ *be a K-isogeny, let ξ be a cocycle representing an element of $H^1(G_{\bar{K}/K}, E[\phi])$, and let C/K be a homogeneous space representing the image of ξ in* $\text{WC}(E/K)$. *Choose a $\bar{K}$-isomorphism $\theta : C \to E$ satisfying*

$$\theta^\sigma \circ \theta^{-1} = (\text{translation by } \xi_\sigma) \quad \text{for all } \sigma \in G_{\bar{K}/K}.$$

(a) *The map $\phi \circ \theta : C \to E'$ is defined over K.*
(b) *Suppose that there is a point $P \in C(K)$, so $\{C/K\}$ is trivial in $\text{WC}(E/K)$. Then the point $\phi \circ \theta(P) \in E'(K)$ maps to ξ via the connecting homomorphism $\delta : E'(K) \to H^1(G_{\bar{K}/K}, E[\phi])$.*

PROOF. (a) Let $\sigma \in G_{\bar{K}/K}$ and let $P \in C$. Then, since ϕ is defined over K and $\xi_\sigma \in E[\phi]$, we have

$$\left(\phi \circ \theta(P)\right)^\sigma = (\phi \circ \theta^\sigma)(P^\sigma) = \phi\big(\theta(P^\sigma) + \xi_\sigma\big) = \phi \circ \theta(P^\sigma).$$

Therefore $\phi \circ \theta$ is defined over K.
(b) This is just a matter of unwinding definitions. Thus

$$\delta\big(\phi \circ \theta(P)\big)_\sigma = \theta(P)^\sigma - \theta(P) = \theta(P^\sigma) + \xi_\sigma - \theta(P) = \xi_\sigma. \qquad \square$$

Remark 4.7. We have been working with arbitrary isogenies $\phi : E \to E'$, but in order to compute the Mordell–Weil group of E', we must find generators for $E'(K)/mE'(K)$ for some integer m; simply knowing $E'(K)/\phi(E(K))$ is not enough. The solution to this dilemma is to work with both ϕ and its dual

$\hat{\phi} : E' \to E$. Using the procedure described in this section, we compute both Selmer groups $S^{(\phi)}(E/K)$ and $S^{(\hat{\phi})}(E'/K)$, and with a little bit of luck, we find generators for the two quotient groups $E'(K)/\phi(E(K))$ and $E(K)/\hat{\phi}(E'(K))$. It is then a simple matter to compute generators for $E(K)/mE(K)$, where $m = \deg(\phi)$, using the elementary exact sequence (note that $\hat{\phi} \circ \phi = [m]$)

$$0 \longrightarrow \frac{E'(K)[\hat{\phi}]}{\phi(E(K)[m])} \longrightarrow \frac{E'(K)}{\phi(E(K))} \xrightarrow{\hat{\phi}} \frac{E(K)}{mE(K)} \longrightarrow \frac{E(K)}{\hat{\phi}(E'(K))} \longrightarrow 0.$$

Example 4.8. *Two-isogenies.* We illustrate the general theory by completely analyzing the case of isogenies of degree 2. Let $\phi : E \to E'$ be an isogeny of degree 2 defined over K. Then the kernel $E[\phi] = \{O, T\}$ is defined over K, so $T \in E(K)$. Moving this K-rational 2-torsion point to $(0,0)$, we can find a Weierstrass equation for E/K of the form

$$E : y^2 = x^3 + ax^2 + bx.$$

Let $S \subset M_K$ be the usual set of places (X.4.4). Identifying $E[\phi]$ with μ_2 (as $G_{\bar{K}/K}$-modules), we see that $K^*/(K^*)^2 \cong H^1(G_{\bar{K}/K}, E[\phi])$. Thus, using notation from (X.1.1c) and (X.4.3), we have

$$H^1(G_{\bar{K}/K}, E[\phi]; S) \cong K(S, 2).$$

More precisely, if $d \in K(S, 2)$, then tracing through the above identification shows that the corresponding cocycle is

$$\sigma \longmapsto \begin{cases} O & \text{if } \sqrt{d}^\sigma = \sqrt{d}, \\ T & \text{if } \sqrt{d}^\sigma = -\sqrt{d}. \end{cases}$$

The homogeneous space C_d/K associated to this cocycle was computed in (X.3.7); it is given by the equation

$$C_d : dw^2 = d^2 - 2adz^2 + (a^2 - 4b)z^4.$$

We can now compute the Selmer group $S^{(\phi)}$ by checking whether $C_d(K_v) = \emptyset$ for each of the finitely many $d \in K(S, 2)$ and $v \in S$.

The isogenous curve E'/K has Weierstrass equation

$$E' : Y^2 = X^3 - 2aX^2 + (a^2 - 4b)X,$$

and the isogeny $\phi : E \to E'$ is given by the formula (III.4.5)

$$\phi(x, y) = \left(\frac{y^2}{x^2}, \frac{y(b - x^2)}{x^2} \right).$$

In (X.3.7) we gave an isomorphism $\theta : C_d \to E$ defined over $K(\sqrt{d})$. Computing the composition $\phi \circ \theta$ yields the map

$$\theta \circ \phi : C_d \longrightarrow E', \qquad \theta \circ \phi(z,w) = \left(\frac{d}{z^2}, -\frac{dw}{z^3} \right),$$

described in (X.4.6). Finally, just as we did in (X.1.4) (see also Exercise 10.1), we can compute the connecting homomorphism

$$\delta : E'(K) \longrightarrow H^1(G_{\bar{K}/K}, E[\phi]) \cong K^*/(K^*)^2.$$

It is given by

$$\delta(O) = 1, \qquad \delta(0,0) = a^2 - 4b, \qquad \delta(X,Y) = X \quad \text{if } X \neq 0, \infty.$$

We summarize (X.4.8) in the next proposition.

Proposition 4.9. (Descent via Two-Isogeny) *Let E/K and E'/K be elliptic curves given by the equations*

$$E : y^2 = x^3 + ax^2 + bx \qquad and \qquad E' : Y^2 = x^3 - 2aX^2 + (a^2 - b)X,$$

and let

$$\phi : E \longrightarrow E', \qquad \phi(x,y) = \left(\frac{y^2}{x^2}, \frac{y(b - x^2)}{x^2} \right),$$

be the isogeny of degree 2 with kernel $E[\phi] = \{O, (0,0)\}$. Let

$$S = M_K^\infty \cup \{v \in M_K^0 : v(2) \neq 0 \text{ or } v(b) \neq 0 \text{ or } v(a^2 - 4b) \neq 0\}.$$

Further, for each $d \in K^*$, let C_d/K be the homogeneous space for E/K given by the equation

$$C_d : dw^2 = d^2 - 2adz^2 + (a^2 - 4b)z^4.$$

Then there is an exact sequence

$$0 \longrightarrow E'(K)/\phi(E(K)) \overset{\delta}{\longrightarrow} K(S,2) \longrightarrow WC(E/K)[\phi],$$
$$(X,Y) \longmapsto X, \qquad d \longmapsto \{C_d/K\},$$
$$O \longmapsto 1,$$
$$(0,0) \longmapsto a^2 - 4b.$$

The ϕ-Selmer group is

$$S^{(\phi)}(E/K) \cong \{d \in K(S,2) : C_d(K_v) \neq \emptyset \text{ for all } v \in S\}.$$

Finally, the map

$$\psi : C_d \longrightarrow E', \qquad \psi(z,w) = \left(\frac{d}{z^2}, -\frac{dw}{z^3} \right),$$

has the property that if $P \in C_d(K)$, then

$$\delta\big(\psi(P)\big) \equiv d \pmod{(K^*)^2}.$$

Remark 4.9.1. Note that since the isogenous curve E' in (X.4.9) has the same form as E, everything in (X.4.9) applies equally well to the dual isogeny $\hat{\phi} : E' \to E$. Then, using the exact sequence (X.4.7), we can try to compute $E(K)/2E(K)$.

Remark 4.9.2. If E/K is an elliptic curve that has a K-rational 2-torsion point, then (III.4.5) says that E automatically has an isogeny of degree 2 defined over K. Thus the procedure described in (X.4.8) may be applied to any elliptic curve satisfying $E(K)[2] \neq 0$. In particular, (X.4.9) in some sense subsumes (X.1.4), which described how to try to compute $E(K)/2E(K)$ when $E[2] \subset E(K)$.

Example 4.10. We use (X.4.9) to compute $E(\mathbb{Q})/2E(\mathbb{Q})$ for the elliptic curve

$$E : y^2 = x^3 - 6x^2 + 17x.$$

This equation has discriminant $\Delta = -147968 = -2^9 17^2$, so our set S is $\{\infty, 2, 17\}$ and we may identify $\mathbb{Q}(S, 2)$ with $\{\pm 1, \pm 2, \pm 17, \pm 34\}$. The curve that is 2-isogenous to E has equation

$$E : Y^2 = X^3 + 12X^2 - 32X,$$

and for $d \in \mathbb{Q}(S, 2)$, the corresponding homogeneous space is

$$C_d : dw^2 = d^2 + 12dz^2 - 32z^4.$$

From (X.4.9) we know that the point $(0, 0) \in E'(\mathbb{Q})$ maps to

$$\delta(0, 0) = -32 \equiv -2 \pmod{(\mathbb{Q}^*)^2},$$

so $-2 \in S^{(\phi)}(E/\mathbb{Q})$. It remains to check the other values of $d \in \mathbb{Q}(S, 2)$.

$\boxed{d = 2}$ $\qquad\qquad C_2 : 2w^2 = 4 + 24z^2 - 32z^4.$

Dividing by 2 and letting $z = Z/2$ gives the equation

$$w^2 = 2 + 3Z^2 - Z^4,$$

which by inspection has the rational point $(Z, w) = (1, 2)$. Then (X.4.9) tells us that the point $(z, w) = (\frac{1}{2}, 2) \in C_2(\mathbb{Q})$ maps to to $\psi(\frac{1}{2}, 2) = (8, -32) \in E'(\mathbb{Q})$. Further, as the theory predicts, we have $\delta(8, -32) = 8 \equiv 2 \pmod{(\mathbb{Q}^*)^2}$.

$\boxed{d = 17}$ $\qquad\qquad C_{17} : 17w^2 = 17^2 + 12 \cdot 17z^2 - 32z^4.$

Suppose that $C_{17}(\mathbb{Q}_{17}) \neq \emptyset$. Since $\operatorname{ord}_{17}(17w^2)$ is odd and $\operatorname{ord}_{17}(32z^4)$ is even, we see that necessarily $z, w \in \mathbb{Z}_{17}$. But then the equation for C_{17} implies first that $z \equiv 0 \pmod{17}$, then that $w \equiv 0 \pmod{17}$, and finally that $17^2 \equiv 0 \pmod{17^3}$. This contradiction shows that $C_{17}(\mathbb{Q}_{17}) = \emptyset$, and hence that $17 \notin S^{(\phi)}(E/\mathbb{Q})$.

We now know that

$$1, -2, 2 \in S^{(\phi)}(E/\mathbb{Q}) \qquad \text{and} \qquad 17 \notin S^{(\phi)}(E/\mathbb{Q}).$$

Since $S^{(\phi)}(E/\mathbb{Q})$ is a subgroup of $\mathbb{Q}(S, 2)$, we have $S^{(\phi)}(E/\mathbb{Q}) = \{\pm 1, \pm 2\}$. Further, we have shown that $E'(\mathbb{Q})$ surjects onto $S^{(\phi)}(E/\mathbb{Q})$, and hence from (X.4.2a) we see that $Ш(E/\mathbb{Q})[\phi] = 0$.

We now repeat the above computation with the roles of E and E' reversed. Thus for $d \in \mathbb{Q}(S, 2)$ we look at the homogeneous space

$$C'_d : dw^2 = d^2 - 24dz^2 + 272z^4.$$

As above, the point $(0, 0) \in E(\mathbb{Q})$ maps to $\delta(0, 0) = 272 \equiv 17 \pmod{(\mathbb{Q}^*)^2}$. Next, if $d < 0$, then clearly $C'_d(\mathbb{R}) = \emptyset$, so $d \notin S^{(\hat\phi)}(E'/\mathbb{Q})$. Finally, for $d = 2$, if we let $z = Z/2$, then C'_2 has the equation

$$2w^2 = 4 - 12Z^2 + 17Z^4.$$

If $C'_2(\mathbb{Q}_2) \neq \emptyset$, then necessarily $Z, w \in \mathbb{Z}_2$, and then the equation allows us to deduce successively

$$Z \equiv 0 \pmod{2}, \qquad w \equiv 0 \pmod{2}, \qquad 4 \equiv 0 \pmod{2^3}.$$

Therefore $C_2(\mathbb{Q}_2) = \emptyset$, and hence $2 \notin S^{(\hat\phi)}(E'/\mathbb{Q})$. Thus $S^{(\hat\phi)}(E'/\mathbb{Q}) = \{1, 17\}$ and $Ш(E'/\mathbb{Q})[\hat\phi] = 0$.

To recapitulate, we now know that

$$E'(\mathbb{Q})/\phi\big(E(\mathbb{Q})\big) \cong (\mathbb{Z}/2\mathbb{Z})^2 \qquad \text{and} \qquad E(\mathbb{Q})/\hat\phi\big(E'(\mathbb{Q})\big) \cong \mathbb{Z}/2\mathbb{Z},$$

the former being generated by $\{(0, 0), (8, -32)\}$ and the latter by $\{(0, 0)\}$. The exact sequence (X.4.7) then yields

$$E(\mathbb{Q})/2E(\mathbb{Q}) \cong (\mathbb{Z}/2\mathbb{Z})^2 \qquad \text{and} \qquad E'(\mathbb{Q})/2E'(\mathbb{Q}) \cong (\mathbb{Z}/2\mathbb{Z})^2,$$

and hence

$$E(\mathbb{Q}) \cong E'(\mathbb{Q}) \cong \mathbb{Z} \times \mathbb{Z}/2\mathbb{Z}.$$

Remark 4.11. In all of the examples up to this point, we have been lucky in the sense that for every locally trivial homogeneous space that has appeared, we were able to find (by inspection) a global rational point. Another way to say this is that we have yet to see a nontrivial element of the Shafarevich–Tate group. The first examples of such spaces are due to Lind [150] and independently, but a bit later, to Reichardt [207]. For example, they proved that the curve

$$2w^2 = 1 - 17z^4$$

has no $\mathbb{Q}$-rational points, while it is easy to check that it has a point defined over every $\mathbb{Q}_p$. Shortly thereafter, Selmer [225, 227] made an extensive study of the curves $ax^3 + by^3 + cz^3 = 0$, which are homogeneous spaces for the elliptic curves $x^3 + y^3 + dz^3 = 0$. He gave many examples of locally trivial, globally nontrivial homogeneous spaces, of which the simplest is

$$3x^3 + 4y^3 + 5z^3 = 0.$$

It is a difficult problem, in general, to divide the Selmer group into the piece coming from rational points on the elliptic curve and the piece giving nontrivial elements of the Shafarevich–Tate group. At present there is no algorithm known that is guaranteed to solve this problem. The procedure that we now describe often works in practice, although it tends to lead to fairly elaborate computations in algebraic number fields.

Recall that for each integer $m \geq 2$ there is an exact sequence (X.4.2a)

$$E(K) \xrightarrow{\delta} S^{(m)}(E/K) \longrightarrow \text{Ш}(E/K)[m] \longrightarrow 0,$$

where at least in theory, the finite group $S^{(m)}(E/K)$ is effectively computable; see (X.4.5). If we knew some way of computing $\text{Ш}(E/K)[m]$, then we could find generators for $E(K)/mE(K)$, and thence for $E(K)$. Unfortunately, a general procedure for computing $\text{Ш}(E/K)[m]$ is still being sought. However, for each integer $n \geq 1$ we can combine different versions of the above exact sequence to form a commutative diagram

$$
\begin{array}{ccccccc}
E(K) & \longrightarrow & S^{(m^n)}(E/K) & \longrightarrow & \text{Ш}(E/K)[m^n] & \longrightarrow & 0 \\
\downarrow{\scriptsize\begin{array}{l}\text{identity}\\\text{map}\end{array}} & & \downarrow & & \downarrow{\scriptsize\begin{array}{l}\text{multiplication}\\\text{by } m^{n-1}\end{array}} & & \\
E(K) & \longrightarrow & S^{(m)}(E/K) & \longrightarrow & \text{Ш}(E/K)[m] & \longrightarrow & 0
\end{array}
$$

Now at least in principle, the middle column of this diagram is effectively computable. This allows us to make the following refinement to the exact sequence in (X.4.2a).

Proposition 4.12. *Let E/K be an elliptic curve. For any integers $m \geq 2$ and $n \geq 1$, let $S^{(m,n)}(E/K)$ be the image of $S^{(m^n)}(E/K)$ in $S^{(m)}(E/K)$. Then there exists an exact sequence*

$$0 \longrightarrow E(K)/mE(K) \longrightarrow S^{(m,n)}(E/K) \longrightarrow m^{n-1}\text{Ш}(E/K)[m^n] \longrightarrow 0.$$

PROOF. This is immediate from the commutative diagram given above. $\qquad\square$

Now to find generators for $E(K)$, we can try the following procedure. Compute successively the *relative Selmer groups*

$$S^{(m)}(E/K) = S^{(m,1)}(E/K) \supset S^{(m,2)}(E/K) \supset S^{(m,3)}(E/K) \supset \cdots$$

and the *rational-point groups*

$$T_{(m,1)}(E/K) \subset T_{(m,2)}(E/K) \subset T_{(m,3)}(E/K) \subset \cdots,$$

where $T_{(m,r)}(E/K)$ is the subgroup of $S^{(m)}(E/K)$ generated by all of the points $P \in E(K)$ with height $h_x(P) \leq r$. Eventually, with sufficient perseverance, we hope to arrive at an equality

$$S^{(m,n)}(E/K) = T_{(m,r)}(E/K).$$

Once this happens, we know that $m^{n-1}\text{Ш}(E/K)[m^n] = 0$ and that the points with height $h_x(P) \leq r$ generate $E(K)/mE(K)$. The difficulty lies in the fact that as far as is currently known, there is nothing to prevent $\text{Ш}(E/K)$ from containing an element that is infinitely m-divisible, i.e., a nonzero element $\xi \in \text{Ш}(E/K)$ such that for every $n \geq 1$ there is an element $\xi_n \in \text{Ш}(E/K)$ satisfying $\xi = m^n \xi_n$. If such an element exists, then the above procedure never terminates. However, opposed to such a gloomy scenario is the following optimistic conjecture.

Conjecture 4.13. *Let E/K be an elliptic curve. Then $\text{Ш}(E/K)$ is finite.*

The finiteness of Ш has been proven for certain elliptic curves by Kolyvagin [130] and Rubin [215]. Note that the successful carrying out of the procedure described above shows only that the m-primary component of $\text{Ш}(E/K)$ is finite. This has, of course, been done in many cases. For example, in (X.4.10) we showed that $\text{Ш}(E/\mathbb{Q})[2] = 0$ for a particular elliptic curve.

We conclude this section with a beautiful result of Cassels, which says something interesting about the order of a group that is not known in general to be finite.

Theorem 4.14. ([38], [281]) *Let E/K be an elliptic curve. There exists an alternating bilinear pairing*

$$\Gamma : \text{Ш}(E/K) \times \text{Ш}(E/K) \longrightarrow \mathbb{Q}/\mathbb{Z}$$

whose kernel on each side is exactly the subgroup of divisible elements of $\text{Ш}(E/K)$. In other words, if $\Gamma(\alpha, \beta) = 0$ for all $\beta \in \text{Ш}(E/K)$, then for every integer $N \geq 1$ there exists an element $\alpha_N \in \text{Ш}(E/K)$ satisfying $N\alpha_N = \alpha$.

In particular, if $\text{Ш}(E/K)$ is finite, then its order is a perfect square, and the same is true of any p-primary component of $\text{Ш}(E/K)$. (See Exercise 10.20.)

X.5 Twisting—Elliptic Curves

As in (X §§2,3), we let K be an arbitrary (perfect) field and we let E/K be an elliptic curve. We saw in (X.2.2) that if we consider E merely to be a curve and ignore the base point O, then the twists of E/K correspond to the elements of the (pointed) cohomology set $H^1(G_{\bar{K}/K}, \text{Isom}(E))$. The group $\text{Isom}(E)$ has two natural subgroups, namely $\text{Aut}(E)$ and E, where we identify E with the set of translations $\{\tau_P\}$ in $\text{Isom}(E)$. We also observe that $\text{Aut}(E)$ acts naturally on E. The next proposition describes $\text{Isom}(E)$.

Proposition 5.1. *The map*

$$E \times \text{Aut}(E) \longrightarrow \text{Isom}(E), \qquad (P, \alpha) \longmapsto \tau_P \circ \alpha,$$

is a bijection of sets. It identifies $\text{Isom}(E)$ with the product of E and $\text{Aut}(E)$ twisted by the natural action of $\text{Aut}(E)$ on E. In other words, the group $\text{Isom}(E)$ is the set of ordered pairs $E \times \text{Aut}(E)$ with the group law

$$(P, \alpha) \cdot (Q, \beta) = (P + \alpha Q, \alpha \circ \beta).$$

PROOF. Let $\phi \in \mathrm{Isom}(E)$. Then $\tau_{-\phi(O)} \circ \phi \in \mathrm{Aut}(E)$, so writing

$$\phi = \tau_{\phi(O)} \circ \left(\tau_{-\phi(O)} \circ \phi \right)$$

shows that the map is surjective. On the other hand, if $\tau_P \circ \alpha = \tau_Q \circ \beta$, then evaluating at O gives $P = Q$, and then also $\alpha = \beta$. This proves injectivity. Finally, the twisted nature of the group law follows from the calculation

$$\tau_P \circ \alpha \circ \tau_Q \circ \beta = \tau_P \circ \tau_{\alpha Q} \circ \alpha \circ \beta. \qquad \square$$

We have already extensively studied those twists of E/K that arise from translations; these are the twists corresponding to elements of the group

$$H^1(G_{\bar{K}/K}, E) \cong \mathrm{WC}(E/K)$$

that we studied in (X §§3,4). We now look at the twists of E/K coming from isomorphisms of E as an *elliptic curve*, i.e., isomorphisms of the pair (E, O). In other words, we consider the twists of E corresponding to elements of the cohomology group $H^1(G_{\bar{K}/K}, \mathrm{Aut}(E))$. We start with a general proposition and then, for $\mathrm{char}(K) \neq 2, 3$, we derive explicit equations for the associated twists.

Remark 5.2. In the literature, the phrase "let C be a twist of E" generally means that C corresponds to an element of $H^1(G_{\bar{K}/K}, \mathrm{Aut}(E))$. More properly, such a C should be called a twist of the pair (E, O), since the group of isomorphisms of (E, O) with itself is the group we denote by $\mathrm{Aut}(E)$. However, one can generally resolve any ambiguity from context.

Proposition 5.3. *Let E/K be an elliptic curve.*
(a) *The natural inclusion $\mathrm{Aut}(E) \subset \mathrm{Isom}(E)$ induces an inclusion*

$$H^1(G_{\bar{K}/K}, \mathrm{Aut}(E)) \subset H^1(G_{\bar{K}/K}, \mathrm{Isom}(E)).$$

Identifying the latter set with $\mathrm{Twist}(E/K)$ via (X.2.2), we denote the former by $\mathrm{Twist}((E, O)/K)$.
(b) *Let $C/K \in \mathrm{Twist}((E, O)/K)$. Then $C(K) \neq \emptyset$, so C/K can be given the structure of an elliptic curve over K. N.B. The curve C/K is generally not K-isomorphic to E/K; cf. (X.3.3).*
(c) *Conversely, if E'/K is an elliptic curve that is isomorphic to E over $\bar{K}$, then E'/K represents an element of $\mathrm{Twist}((E, O)/K)$.*

PROOF. (a) Let $i : \mathrm{Aut}(E) \to \mathrm{Isom}(E)$ be the natural inclusion. From (X.5.1), there is a homomorphism $j : \mathrm{Isom}(E) \to \mathrm{Aut}(E)$ satisfying $j \circ i = 1$. It follows that the induced map

$$H^1(G_{\bar{K}/K}, \mathrm{Aut}(E)) \xrightarrow{\;i\;} H^1(G_{\bar{K}/K}, \mathrm{Isom}(E))$$

is one-to-one.

(b) Let $\phi : C \to E$ be an isomorphism defined over $\bar{K}$ such that the cocycle

$$\sigma \longmapsto \phi^\sigma \circ \phi^{-1}$$

represents the element of $H^1\big(G_{\bar{K}/K}, \operatorname{Aut}(E)\big)$ corresponding to C/K. Then we have $\phi^\sigma \circ \phi^{-1}(O) = O$, so

$$\phi^{-1}(O) = \phi^{-1}(O)^\sigma \qquad \text{for all } \sigma \in G_{\bar{K}/K}.$$

Hence $\phi^{-1}(O) \in C(K)$, so $\big(C, \phi^{-1}(O)\big)$ is an elliptic curve defined over K.

(c) Let $\phi : E' \to E$ be a $\bar{K}$-isomorphism of elliptic curves, so in particular, $\phi(O') = O$, where $O \in E(K)$ and $O' \in E'(K)$ are the respective zero points of E and E'. Then for any $\sigma \in G_{\bar{K}/K}$ we have

$$\phi^\sigma \circ \phi^{-1}(O) = \phi^\sigma(O') = \phi(O')^\sigma = O^\sigma = O.$$

Thus $\phi^\sigma \circ \phi^{-1} \in \operatorname{Aut}(E)$, so the cocycle corresponding to E'/K lies in the group $H^1\big(G_{\bar{K}/K}, \operatorname{Aut}(E)\big)$ as desired. $\qquad\square$

If the characteristic of K is not equal to 2 or 3, then the elements of the group $\operatorname{Twist}\big((E, O)/K\big)$ can be described quite explicitly.

Proposition 5.4. *Assume that* $\operatorname{char}(K) \neq 2, 3$, *and let*

$$n = \begin{cases} 2 & \text{if } j(E) \neq 0, 1728, \\ 4 & \text{if } j(E) = 1728, \\ 6 & \text{if } j(E) = 0. \end{cases}$$

Then $\operatorname{Twist}\big((E, O)/K\big)$ *is canonically isomorphic to* $K^*/(K^*)^n$.

More precisely, choose a Weierstrass equation

$$E : y^2 = x^3 + Ax + B$$

for E/K, *and let* $D \in K^*$. *Then the elliptic curve* $E_D \in \operatorname{Twist}\big((E, O)/K\big)$ *corresponding to* $D \pmod{(K^*)^n}$ *has Weierstrass equation*

(i) $E_D : y^2 = x^3 + D^2 Ax + D^3 B$ *if* $j(E) \neq 0, 1728$,

(ii) $E_D : y^2 = x^3 + DAx$ *if* $j(E) = 1728$ (*so* $B = 0$),

(iii) $E_D : y^2 = x^3 + DB$ *if* $j(E) = 0$ (*so* $A = 0$).

Corollary 5.4.1. *Define an equivalence relation on the set* $K \times K^*$ *by*

$$(j, D) \sim (j', D') \qquad \text{if} \quad j = j' \quad \text{and} \quad D/D' \in (K^*)^{n(j)},$$

where $n(j) = 2$ (*respectively* 4, *respectively* 6) *if* $j \neq 0, 1728$ (*respectively* $j = 1728$, *respectively* $j = 0$). *Then the K-isomorphism classes of elliptic curves* E/K *are in one-to-one correspondence with the elements of the quotient*

$$\frac{K \times K^*}{\sim}.$$

PROOF. From (III.10.2.) we have an isomorphism

$$\mathrm{Aut}(E) \cong \mu_n$$

of $G_{\bar{K}/K}$-modules. It follows from (B.2.5c) that

$$\mathrm{Twist}\big((E,O)/K\big) = H^1\big(G_{\bar{K}/K}, \mathrm{Aut}(E)\big) \cong H^1\big(G_{\bar{K}/K}, \mu_n\big) \cong K^*/(K^*)^n.$$

The calculation of an equation for the twist of E is straightforward. The case $j(E) \neq 0, 1728$ was done in (X.2.4). We do $j(E) = 1728$ here and leave $j(E) = 0$ for the reader.

Thus let $D \in K^*$, let $\delta \in \bar{K}$ be a fourth root of D, and define a cocycle

$$\xi : G_{\bar{K}/K} \longrightarrow \mu_4, \qquad \xi_\sigma = \delta^\sigma/\delta.$$

We also fix an isomorphism

$$[\;] : \mu_4 \longrightarrow \mathrm{Aut}(E), \qquad [\zeta](x,y) = (\zeta^2 x, \zeta y).$$

Then E_D corresponds to the cocycle $\sigma \mapsto [\xi_\sigma]$ in $H^1\big(G_{\bar{K}/K}, \mathrm{Aut}(E)\big)$.

The action of $G_{\bar{K}/K}$ on the twisted field $\bar{K}(E)_\xi$ is given by

$$\delta^\sigma = \xi_\sigma \delta, \qquad x^\sigma = \xi_\sigma^2 x, \qquad y^\sigma = \xi_\sigma y.$$

The subfield fixed by $G_{\bar{K}/K}$ thus contains the functions

$$X = \delta^{-2} x \qquad \text{and} \qquad Y = \delta^{-1} y,$$

and these functions satisfy the equation

$$Y^2 = DX^3 + AX.$$

This gives an equation for the twist E_D/K, and the substitution

$$(X,Y) = (D^{-1}X', D^{-1}Y')$$

puts it into the desired form.

The corollary follows by combining the proposition and (X.5.3c) with (III.1.4bc), which says that up to $\bar{K}$-isomorphism, the elliptic curves E/K are in one-to-one correspondence with their j-invariants $j(E) \in K$. $\square$

X.6   The Curve $Y^2 = X^3 + DX$

Many of the deepest theorems and conjectures in the arithmetic theory of elliptic curves have had as their testing ground one of the families of curves given in (X.5.4). To illustrate the theory that we have developed, let's see what we can say about the family of elliptic curves $E/\mathbb{Q}$ with j-invariant $j(E) = 1728$.

One such curve is given by the equation

$$y^2 = x^3 + x,$$

and then (X.5.3) and (X.5.4) tell us that every such curve has an equation of the form

$$E : y^2 = x^3 + Dx,$$

where D ranges over representatives for the cosets in $\mathbb{Q}^*/(\mathbb{Q}^*)^4$. Thus if we specify that D is a fourth-power-free integer, then D is uniquely determined by E. We observe that the equation for E has discriminant $\Delta(E) = -4D^3$, so E has good reduction at all primes not dividing $2D$, and the given Weierstrass equation is minimal at all odd primes.

Let p be a prime not dividing $2D$ and consider the reduced curve $\tilde{E}$ over the finite field $\mathbb{F}_p$. From (V.4.1) we find that $\tilde{E}$ is supersingular if and only if the coefficient of x^{p-1} in $(x^3 + Dx)^{(p-1)/2}$ is zero. In particular, if $p \equiv 3 \pmod 4$, then $\tilde{E}/\mathbb{F}_p$ is supersingular, and hence we conclude (see Exercise 5.10) that

$$\#\tilde{E}(\mathbb{F}_p) = p + 1 \qquad \text{for all } p \equiv 3 \pmod 4.$$

(See Exercise 10.17 for an elementary derivation of this result.)

Next we recall from (VII.3.5) that if $p \neq 2$ and if E has good reduction at p, then $E_{\text{tors}}(\mathbb{Q})$ injects into the reduction $\tilde{E}(\mathbb{F}_p)$. It follows from this discussion that $\#E_{\text{tors}}(\mathbb{Q})$ divides $p + 1$ for all but finitely many primes $p \equiv 3 \pmod 4$, and hence that $\#E_{\text{tors}}(\mathbb{Q})$ divides 4. Since $(0, 0) \in E(\mathbb{Q})[2]$, the possibilities for $E_{\text{tors}}(\mathbb{Q})$ are $\mathbb{Z}/2\mathbb{Z}$, $(\mathbb{Z}/2\mathbb{Z})^2$, and $\mathbb{Z}/4\mathbb{Z}$.

We have $E[2] \subset E(\mathbb{Q})$ if and only if the polynomial $x^3 + Dx$ factors completely over $\mathbb{Q}$, so if and only if $-D$ is a perfect square. Similarly, $E(\mathbb{Q})$ has a point of order 4 if and only if $(0, 0) \in 2E(\mathbb{Q})$. The duplication formula for E reads

$$x(2P) = \frac{(x^2 - D)^2}{4x^3 + 4Dx},$$

so we see that

$$(0, 0) = [2]\big(D^{1/2}, (4D^3)^{1/4}\big).$$

Hence assuming that D is a fourth-power-free integer, we conclude that

$$(0, 0) \in 2E(\mathbb{Q}) \quad \text{if and only if} \quad D = 4,$$

in which case $(0, 0) = [2](2, \pm 4)$.

Next, since $E(\mathbb{Q})$ contains a 2-torsion point, we can use (X.4.9) to try to calculate $E(\mathbb{Q})/2E(\mathbb{Q})$. The curve E is isogenous to the curve

$$E' : Y^2 = X^3 - 4DX$$

via the isogeny

$$\phi : E \longrightarrow E', \qquad \phi(x,y) = \left(\frac{y^2}{x^2}, \frac{y(D-x^2)}{x^2} \right).$$

The set $S \subset M_{\mathbb{Q}}$ consists of ∞ and the primes dividing $2D$, and for each $d \in \mathbb{Q}(S,2)$, the corresponding homogeneous space $C_d/\mathbb{Q} \in \mathrm{WC}(E/\mathbb{Q})$ is given by the equation

$$C_d : dw^2 = d^2 - 4Dz^4.$$

Similarly, working with the dual isogeny $\hat{\phi} : E' \to E$ leads to the homogeneous spaces $C'_d/\mathbb{Q} \in \mathrm{WC}(E'/\mathbb{Q})$ with equations

$$C'_d : dW^2 = d^2 + DZ^4.$$

(More precisely, using (X.4.9) leads to the equation $dW^2 = d^2 + 16DZ^4$, but we are free to replace Z with $Z/2$.)

Let $\nu(2D)$ denote the number of distinct primes dividing $2D$. The group $\mathbb{Q}(S,2)$ is generated by -1 and the primes dividing $2D$, so we have the estimate

$$\dim_2 E(\mathbb{Q})/2E(\mathbb{Q}) \le 2 + 2\nu(2D) - \dim_2 E'(\mathbb{Q})[\hat{\phi}] + \dim_2 \phi\big(E(\mathbb{Q})[2]\big).$$

Here $\dim_2$ denotes the dimension of an $\mathbb{F}_2$-vector space. We clearly have

$$E'(\mathbb{Q})[\hat{\phi}] \cong \mathbb{Z}/2\mathbb{Z}.$$

In order to deal with the other two terms, we consider two cases.

(1) $E(\mathbb{Q})[2] \cong \mathbb{Z}/2\mathbb{Z}$.
   Then $\phi\big(E(\mathbb{Q})[2]\big) \cong 0$ and $\dim_2 E(\mathbb{Q})/2E(\mathbb{Q}) = \mathrm{rank}\, E(\mathbb{Q}) + 1$.

(2) $E(\mathbb{Q})[2] \cong \mathbb{Z}/2\mathbb{Z} \times \mathbb{Z}/2\mathbb{Z}$.
   Then $\phi\big(E(\mathbb{Q})[2]\big) \cong \mathbb{Z}/2\mathbb{Z}$ and $\dim_2 E(\mathbb{Q})/2E(\mathbb{Q}) = \mathrm{rank}\, E(\mathbb{Q}) + 2$.

Substituting these values into the above inequality yields in both cases the estimate

$$\mathrm{rank}\, E(\mathbb{Q}) \le 2\nu(2D).$$

Notice that we have obtained this upper bound without having checked for local triviality of any of the homogeneous spaces C_d and C'_d. By inspection, if $d < 0$, then either $C_d(\mathbb{R}) = \emptyset$ or $C'_d(\mathbb{R}) = \emptyset$. Thus the upper bound my be decreased by 1, giving the small improvement

$$\mathrm{rank}\, E(\mathbb{Q}) \le 2\nu(2D) - 1.$$

The preceding discussion is summarized in the following proposition.

Proposition 6.1. *Let $D \in \mathbb{Z}$ be a fourth-power-free integer, and let E_D be the elliptic curve*

$$E_D : y^2 = x^3 + Dx.$$

(a)
$$E_{D,\text{tors}}(\mathbb{Q}) \cong \begin{cases} \mathbb{Z}/4\mathbb{Z} & \text{if } D = 4, \\ \mathbb{Z}/2\mathbb{Z} \times \mathbb{Z}/2\mathbb{Z} & \text{if } -D \text{ is a perfect square}, \\ \mathbb{Z}/2\mathbb{Z} & \text{otherwise}. \end{cases}$$

(b)
$$\text{rank } E(\mathbb{Q}) \le 2\nu(2D) - 1.$$

Remark 6.1.1. The estimate in (X.6.1b) cannot be improved in general. For example, the curve $E : y^2 = x^3 - 82x$ has

$$\text{rank } E(\mathbb{Q}) = 3,$$

while $\nu(-164) = \nu(2^4 \cdot 41) = 2$. See Exercise 10.18.

We now restrict attention to the special case that $D = p$ is an odd prime. The next proposition gives a complete description of the relevant Selmer groups and deduces corresponding upper bounds for the rank of $E(\mathbb{Q})$ and the dimension of $\text{III}(E/\mathbb{Q})[2]$.

Proposition 6.2. *Let p be an odd prime, let E_p be the elliptic curve*

$$E_p : y^2 = x^3 + px,$$

and let $\phi : E_p \to E'_p$ be the isogeny of degree 2 with kernel $E_p[\phi] = \{O, (0,0)\}$.
(a)
$$E_{p,\text{tors}}(\mathbb{Q}) \cong \mathbb{Z}/2\mathbb{Z}.$$

(b)
$$S^{(\hat\phi)}(E'_p/\mathbb{Q}) \cong \mathbb{Z}/2\mathbb{Z}.$$

$$S^{(\phi)}(E_p/\mathbb{Q}) \cong \begin{cases} \mathbb{Z}/2\mathbb{Z} & \text{if } p \equiv 7, 11 \ (\text{mod } 16), \\ (\mathbb{Z}/2\mathbb{Z})^2 & \text{if } p \equiv 3, 5, 13, 15 \ (\text{mod } 16), \\ (\mathbb{Z}/2\mathbb{Z})^3 & \text{if } p \equiv 1, 9 \ (\text{mod } 16). \end{cases}$$

(c)
$$\text{rank } E_p(\mathbb{Q}) + \dim_2 \text{III}(E_p/\mathbb{Q})[2] = \begin{cases} 0 & \text{if } p \equiv 7, 11 \ (\text{mod } 16), \\ 1 & \text{if } p \equiv 3, 5, 13, 15 \ (\text{mod } 16), \\ 2 & \text{if } p \equiv 1, 9 \ (\text{mod } 16). \end{cases}$$

PROOF. To ease notation, we let $E = E_p$ and $E' = E'_p$.
(a) This is a special case of (X.6.1a).
(b) As usual, we take representatives $\{\pm 1, \pm 2, \pm p, \pm 2p\}$ for the cosets in the finite group $\mathbb{Q}(S, 2)$. From (X.4.9) we know that the images of the 2-torsion points in the Selmer groups are given by

$$-p \in S^{(\phi)}(E/\mathbb{Q}) \quad \text{and} \quad p \in S^{(\hat\phi)}(E'/\mathbb{Q}).$$

Further, if $d < 0$, then $C_d(\mathbb{R}) = \emptyset$, so $d \notin S^{(\hat{\phi})}(E'/\mathbb{Q})$.

Next we consider the homogeneous space

$$C_2' : 2W^2 = 4 + pZ^4.$$

If $(Z, W) \in C_2'(\mathbb{Q}_2)$, then necessarily $Z, W \in \mathbb{Z}_2$, which allows us to conclude that $Z \equiv 0 \pmod 2$, so $W \equiv 0 \pmod 2$, and thus $0 \equiv 4 \pmod 8$. Therefore $C_2'(\mathbb{Q}_2) = \emptyset$, and hence $2 \notin S^{(\hat{\phi})}(E'/\mathbb{Q})$. We now know that

$$p \in S^{(\hat{\phi})}(E'/\mathbb{Q}) \qquad \text{and} \qquad -1, \pm 2, -p, -2p \notin S^{(\hat{\phi})}(E'/\mathbb{Q}).$$

It follows that $S^{(\hat{\phi})}(E'/\mathbb{Q}) = \{1, p\} \cong \mathbb{Z}/2\mathbb{Z}$.

It remains to calculate $S^{(\phi)}(E/\mathbb{Q})$, and from the form of the answer, it is clear that there will be many cases to be considered. The best approach is to look at the various $d \in \mathbb{Q}(S, 2)$ and check for which primes the homogeneous space is locally trivial. Note that (X.4.9) says that

$$d \in S^{(\phi)}(E/\mathbb{Q}) \qquad \text{if and only if} \qquad C_d(\mathbb{Q}_p) \neq \emptyset \quad \text{and} \quad C_d(\mathbb{Q}_2) \neq \emptyset,$$

i.e., it suffices to check whether C_d is locally trivial at the primes p and 2. We make frequent use of Hensel's lemma (Exercise 10.12), which gives a criterion for when a solution of an equation modulo q^n lifts to a solution in $\mathbb{Q}_q$.

$\boxed{d = -1}$ $\qquad\qquad\qquad C_{-1} : w^2 + 1 = 4pz^4.$

(i) If $(z, w) \in C_{-1}(\mathbb{Q}_p)$, then necessarily $z, w \in \mathbb{Z}_p$, so $w^2 \equiv -1 \pmod p$. Conversely, from Exercise 10.12 we see that any solution to the congruence $w^2 \equiv -1 \pmod p$ lifts to a point in $C_{-1}(\mathbb{Q}_p)$. Therefore

$$C_{-1}(\mathbb{Q}_p) \neq \emptyset \quad \Longleftrightarrow \quad p \equiv 1 \pmod 4.$$

(ii) From (i) we may assume that $p \equiv 1 \pmod 4$. If $p \equiv 1 \pmod 8$, then we let

$$(z, w) = (Z/4, W/8).$$

Our equation becomes $W^2 + 64 = pZ^4$, and the solution $(1, 1)$ to the congruence

$$W^2 + 64 \equiv pZ^4 \pmod 8$$

lifts to a point in $C_{-1}(\mathbb{Q}_2)$. Similarly, if $p \equiv 5 \pmod 8$, then we let

$$(z, w) = (Z/2, W/2)$$

and consider the solution $(Z, W) = (1, 1)$ to the congruence

$$W^2 + 4 = pZ^4 \pmod 8.$$

This solution lifts to a point in $C_{-1}(\mathbb{Q}_2)$. This shows that if $p \equiv 1 \pmod 4$, then $C_{-1}(\mathbb{Q}_2) \neq \emptyset$.

Combining (i) and (ii) yields

$$-1 \in S^{(\phi)}(E/\mathbb{Q}) \quad \Longleftrightarrow \quad p \equiv 1 \pmod{4}.$$

$\boxed{d = -2}$ $C_{-2} : w^2 + 2 = 2pz^4.$

(i) If $(z, w) \in C_{-2}(\mathbb{Q}_p)$, then $z, w \in \mathbb{Z}_p$ and $w^2 \equiv -2 \pmod{p}$. Conversely, a solution to $w^2 \equiv -2 \pmod{p}$ lifts to a point of $C_{-1}(\mathbb{Q}_p)$. Therefore

$$C_{-2}(\mathbb{Q}_p) \neq \emptyset \quad \Longleftrightarrow \quad p \equiv 1, 3 \pmod{8}.$$

(ii) If $(z, w) \in C_{-2}(\mathbb{Q}_2)$, then $z, w \in \mathbb{Z}_2$ and $w \equiv 0 \pmod{2}$. So after setting $(z, w) = (Z, 2W)$, we must check whether the equation

$$2W^2 + 1 = pZ^4$$

has any solutions $Z, W \in \mathbb{Z}_2$. From (i) we see that it suffices to consider primes $p \equiv 1, 3 \pmod{8}$. The congruence $2W^2 + 1 \equiv pZ^4 \pmod{16}$ has no solutions if $p \equiv 11 \pmod{16}$, so

$$p \equiv 11 \pmod{16} \quad \Longrightarrow \quad C_{-2}(\mathbb{Q}_2) = \emptyset.$$

On the other hand, if we can find solutions modulo $2^5 = 32$, then Exercise 10.12 says that they lift to points in $C_{-2}(\mathbb{Q}_2)$. The following table gives solutions (Z, W) to the congruence

$$2W^2 + 1 \equiv pZ^4 \pmod{32}$$

for each of the remaining values of p mod 32:

$p \bmod 32$	1	3	9	17	19	25
(Z, W)	$(1, 0)$	$(3, 11)$	$(1, 2)$	$(3, 0)$	$(1, 3)$	$(3, 2)$

Combining (i) and (ii), we have proven that

$$-2 \in S^{(\phi)}(E/\mathbb{Q}) \quad \Longleftrightarrow \quad p \equiv 1, 3, 9 \pmod{16}.$$

$\boxed{d = 2}$ $C_2 : w^2 = 2 - 2pz^4.$

This case is entirely similar to the case $d = -2$ that we just completed. A point $(z, w) \in C_2(\mathbb{Q}_p)$ has $z, w \in \mathbb{Z}_p$ and $w^2 \equiv 2 \pmod{p}$, and any such solutions lifts, so

$$C_2(\mathbb{Q}_p) \neq \emptyset \quad \Longleftrightarrow \quad p \equiv 1, 7 \pmod{8}.$$

Next, if $p \equiv 1 \pmod{8}$, then from above we have $-1, -2 \in S^{(\phi)}(E/\mathbb{Q})$, so certainly $2 \in S^{(\phi)}(E/\mathbb{Q})$. It remains to consider the case $p \equiv 7 \pmod{8}$.

A point $(z, w) \in C_2(\mathbb{Q}_2)$ satisfies $(z, w) = (Z, 2W)$ with $Z, W \in \mathbb{Z}_2$ and

$$2W^2 = 1 - pZ^4.$$

If $p \equiv 7 \pmod{16}$, then this equation has no solutions modulo 16. On the other hand, if $p \equiv 15 \pmod{16}$, then we have solutions

$$
\begin{aligned}
2 \cdot 3^2 &\equiv 1 - p \cdot 1^4 \pmod{32} &&\text{if } p \equiv 15 \pmod{32}, \\
2 \cdot 1^2 &\equiv 1 - p \cdot 1^4 \pmod{32} &&\text{if } p \equiv 31 \pmod{32},
\end{aligned}
$$

and these solutions lift to points in $C_2(\mathbb{Q}_2)$. Putting all of this together, we have shown that

$$2 \in S^{(\phi)}(E/\mathbb{Q}) \iff p \equiv 1, 9, 15 \pmod{16}.$$

We have now determined exactly which of the values -1, 2, and -2 are in $S^{(\phi)}(E/\mathbb{Q})$ in terms of the residue of p modulo 16. Since we also know that $-p \in S^{(\phi)}(E/\mathbb{Q})$, it is now a simple matter to reconstruct the table for $S^{(\phi)}(E/\mathbb{Q})$ given in (b). In fact, we obtain more information, namely a precise list of which elements of $\mathbb{Q}(S, 2)$ are in $S^{(\phi)}(E/\mathbb{Q})$.

(c) We use (X.4.7) and (X.4.2a) to compute

$$
\begin{aligned}
\dim_2 E'(\mathbb{Q})[\hat{\phi}]/\phi\big(E(\mathbb{Q})[2]\big) &+ \dim_2 E(\mathbb{Q})/2E(\mathbb{Q}) \\
&= \dim_2 E'(\mathbb{Q})/\phi\big(E(\mathbb{Q})\big) + \dim_2 E(\mathbb{Q})/\hat{\phi}\big(E(\mathbb{Q})\big) \\
&= \dim_2 S^{(\phi)}(E/\mathbb{Q}) - \dim_2 \text{Ш}(E/\mathbb{Q})[\phi] \\
&\quad + \dim_2 S^{(\hat{\phi})}(E'/\mathbb{Q}) - \dim_2 \text{Ш}(E'/\mathbb{Q})[\hat{\phi}].
\end{aligned}
$$

From (a) we see that

$$E'(\mathbb{Q})/\phi\big(E(\mathbb{Q})[2]\big) \cong \mathbb{Z}/2\mathbb{Z} \quad \text{and} \quad E(\mathbb{Q})/2E(\mathbb{Q}) \cong (\mathbb{Z}/2\mathbb{Z})^{1+\operatorname{rank} E(\mathbb{Q})}.$$

Further, since $E(\mathbb{Q})/\hat{\phi}\big(E'(\mathbb{Q})\big) \cong S^{(\hat{\phi})}(E'/\mathbb{Q}) \cong \mathbb{Z}/2\mathbb{Z}$ from (b), the exact sequence given in (X.4.2a) implies that $\text{Ш}(E'/\mathbb{Q})[\hat{\phi}] = 0$. Hence the exact sequence

$$0 \longrightarrow \text{Ш}(E/\mathbb{Q})[\phi] \longrightarrow \text{Ш}(E/\mathbb{Q})[2] \xrightarrow{\phi} \text{Ш}(E'/\mathbb{Q})[\hat{\phi}] = 0$$

gives

$$\dim \text{Ш}(E/\mathbb{Q})[2] = \dim_2 \text{Ш}(E/\mathbb{Q})[\phi],$$

and combining this with the above results yields

$$1 + \big(1 + \operatorname{rank} E(\mathbb{Q})\big) = \dim_2 S^{(\phi)}(E/\mathbb{Q}) + \dim_2 S^{(\hat{\phi})}(E'/\mathbb{Q}) - \dim_2 \text{Ш}(E/\mathbb{Q})[2].$$

Now (c) is immediate from the calculation of $S^{(\phi)}(E/\mathbb{Q})$ and $S^{(\hat{\phi})}(E'/\mathbb{Q})$ given in (b). $\qquad\square$

Corollary 6.2.1. *There are infinitely many elliptic curves* $E/\mathbb{Q}$ *satisfying*

$$\operatorname{rank} E(\mathbb{Q}) = 0 \quad \text{and} \quad \mathrm{III}(E/\mathbb{Q})[2] = 0.$$

PROOF. From (X.6.2), the elliptic curves $y^2 = x^3 + px$ with $p \equiv 7, 11 \pmod{16}$ have this property. $\qquad \square$

Remark 6.3. One of the consequences of (X.6.2) is that if p is a prime satisfying $p \equiv 5 \pmod 8$, then the elliptic curve

$$E_p : y^2 = x^3 + px$$

has rank at most one. Further, examining the proof of (X.6.2) shows that the group $E_p(\mathbb{Q})$ has rank 1 if and only if the homogeneous space

$$C_{-1} : w^2 + 1 = 4pz^4$$

has a $\mathbb{Q}$-rational point, and if there is such a point, then we can find a point of infinite order in $E(\mathbb{Q})$ by using the map

$$\hat{\phi} \circ \psi : C_1 \longrightarrow E, \qquad \hat{\phi} \circ \psi(z, w) = \left(\frac{w^2}{4z^2}, \frac{w(w^2 + 2)}{8z^3} \right);$$

cf. (X.4.9). Taking the first few primes $p \equiv 5 \pmod 8$, in each case we find points in $C_{-1}(\mathbb{Q})$, and these give points of infinite order in $E_p(\mathbb{Q})$ as listed in the following table:

p	5	13	29	37
(x, y)	$\left(\frac{1}{4}, \frac{9}{8} \right)$	$\left(\frac{9}{4}, \frac{51}{8} \right)$	$\left(\frac{25}{4}, \frac{165}{8} \right)$	$\left(\frac{22801}{900}, \frac{3540799}{27000} \right)$

Suppose that we knew, a priori, that the Shafarevich–Tate group $\mathrm{III}(E_p/\mathbb{Q})$ was finite, or even that its 2-primary component was finite. Then the existence of the Cassels pairing (X.4.14) implies that $\dim_2 \mathrm{III}(E_p/\mathbb{Q})[2]$ is even, and hence that $E_p(\mathbb{Q})$ has rank 1 for *all* primes $p \equiv 5 \pmod 8$. This also follows from a conjecture of Selmer [226] concerning the difference between the number of "first and second descents," and it is also a consequence of the conjecture of Birch and Swinnerton-Dyer (C.16.5). Bremner and Cassels [26, 27] have verified numerically that $\operatorname{rank} E_p(\mathbb{Q}) = 1$ for all such primes less than 20000, and Monsky [182] has shown that $\operatorname{rank} E_p(\mathbb{Q}) = 1$ for all primes $p \equiv 5 \pmod{16}$.

In order to give the reader an idea of the magnitude of the solutions that may occur, we mention that for $p = 877$, the Mordell–Weil group of the elliptic curve

$$y^2 = x^3 + 877x$$

is generated by the 2-torsion point $(0, 0)$ and the point

$$\left(\frac{612776083187947368101^2}{7884153586068300210^2}, \frac{256256267988926809388776834045513089648669153204356603464786949}{7884153586068300210^3} \right).$$

Similarly, if $p \equiv 3, 15 \pmod{16}$ and if the 2-primary component of $\mathrm{III}(E_p/\mathbb{Q})$ is finite, then (X.6.2) and (X.4.14) imply that $E_p(\mathbb{Q})$ has rank exactly one. The fact that the rank is one for any particular prime p may be verified numerically by searching for a point in $C_{-2}(\mathbb{Q})$ and $C_2(\mathbb{Q})$ respectively. See, for example, the tables in [20] and [54] and online at [53] and [274].

Remark 6.4. If $p \equiv 7, 11 \pmod{16}$, then (X.6.2c) says that $E_p(\mathbb{Q})$ consists of only two points, while if $p \equiv 3, 5, 13, 15 \pmod{16}$, then (X.6.2c) combined with the reasonable conjecture that $\text{III}(E_p/\mathbb{Q})[2^\infty]$ is finite tells us that $E_p(\mathbb{Q}) \cong \mathbb{Z}/2\mathbb{Z} \times \mathbb{Z}$. In the remaining case, namely $p \equiv 1 \pmod 8$, there are two possibilities. First, $E_p(\mathbb{Q})$ might have rank 2. This can certainly occur. For example, the curves

$$y^2 = x^3 + 73x \quad \text{and} \quad y^2 = x^3 + 89x$$

both have rank 2, independent points being given by

$$\left(\frac{9}{16}, \frac{411}{64}\right), (36, 222) \in E_{73}(\mathbb{Q}) \quad \text{and} \quad \left(\frac{25}{16}, \frac{765}{64}\right), \left(\frac{4}{9}, \frac{170}{27}\right) \in E_{89}(\mathbb{Q}).$$

Second, $E_p(\mathbb{Q})$ might have rank 0, in which case $\text{III}(E_p/\mathbb{Q})[2] \cong (\mathbb{Z}/2\mathbb{Z})^2$. (Note that rank $E_p(\mathbb{Q}) = 1$ is precluded if we assume that $\text{III}(E/\mathbb{Q})$ is finite.) The next proposition gives a fairly general condition under which the second possibility holds. It also provides our first examples of homogeneous spaces that are everywhere locally trivial, but have no global rational points.

Proposition 6.5. *Let* $p \equiv 1 \pmod 8$ *be a prime for which* 2 *is not a quartic residue.*
(a) *The curves*

$$w^2 + 1 = 4pz^4, \quad w^2 + 2 = 2pz^4, \quad w^2 + 2pz^4 = 2,$$

have points defined over every completion of $\mathbb{Q}$, *but they have no* $\mathbb{Q}$-*rational points.*
(b) *The elliptic curve*

$$E_p : y^2 = x^3 + px$$

satisfies

$$\text{rank } E_p(\mathbb{Q}) = 0 \quad \text{and} \quad \text{III}(E_p/\mathbb{Q})[2] \cong (\mathbb{Z}/2\mathbb{Z})^2.$$

Remark 6.5.1. Any prime $p \equiv 1 \pmod 8$ can be written as $p = A^2 + B^2$ with $A, B \in \mathbb{Z}$ satisfying $AB \equiv 0 \pmod 4$. A theorem of Gauss, which we prove later in this section (X.6.6), says that 2 is a quartic residue modulo p if and only if $AB \equiv 0 \pmod 8$. Thus, for example, 2 is a quartic nonresidue for the primes

$$17 = 1^2 + 4^2, \quad 41 = 5^2 + 4^2, \quad 97 = 9^2 + 4^2, \quad \text{and} \quad 193 = 7^2 + 12^2,$$

so these primes satisfy the conclusion of (X.6.5).

PROOF OF (X.6.5). During the course of proving (X.6.2b), we showed that the Selmer group $S^{(p)}(E_p/\mathbb{Q}) \subset \mathbb{Q}^*/(\mathbb{Q}^*)^2$ is given by $\{\pm 1, \pm 2, \pm p, \pm 2p\}$. Further, we showed that $-p$ is the image of the 2-torsion point $(0, 0) \in E_p(\mathbb{Q})$. Thus in order to show that $\text{III}(E_p/\mathbb{Q})[\phi]$ has order 4, it suffices to prove that the homogeneous spaces C_{-1}, C_2, and C_{-2} have no $\mathbb{Q}$-rational points. These are the three curves listed in (a), and so, once we prove that they have no $\mathbb{Q}$-rational points, all of (X.6.5) will follow from (X.6.2). Our proof is based on ideas of Lind and Mordell [150, 41]; see also [207, 184, 20].

Case I. $$C_{\pm 2} : w^2 = 2 - 2pz^4.$$

Suppose that $(z, w) \in C_{\pm 2}(\mathbb{Q})$. Writing z and w in lowest terms, we see that they necessarily have the form $(z, w) = (r/t, 2s/t^2)$, where $r, s, t \in \mathbb{Z}$ satisfy

$$\pm 2s^2 = t^4 - pr^4 \quad \text{and} \quad \gcd(r, s, t) = 1.$$

We write $(a|b)$ for the Legendre symbol. Let q be an odd prime dividing s. Then $(p|q) = 1$, so $(q|p) = 1$ by quadratic reciprocity. Since also $(2|p) = 1$, we see that $(s|p) = 1$, so $(s^2|p)_4 = 1$, i.e., s^2 is a quartic residue modulo p. Now the equation implies that $(\pm 2|p)_4 = 1$. But -1 is always a quartic residue for primes $p \equiv 1 \pmod 8$, while by assumption 2 is a quartic nonresidue modulo p. This contradiction proves that $C_{\pm 2}(\mathbb{Q}) = \emptyset$.

Case II. $$C_{-1} : -w^2 = 1 - 4pz^4.$$

Writing $(z, w) \in C_{-1}(\mathbb{Q})$ in (almost) lowest terms as $(z, w) = (r/2t, s/2t^2)$, we have

$$s^2 + 4t^4 = pr^4 \quad \text{with} \quad \gcd(r, t) = 1.$$

(We do not preclude the possibility that r is even.) Since $p \equiv 1 \pmod 4$, there are integers $A \equiv 1 \pmod 2$ and $B \equiv 0 \pmod 2$ such that

$$p = A^2 + B^2.$$

It is a simple matter to verify the identity

$$(pr^2 + 2Bt^2)^2 = p(Br^2 + 2t^2)^2 + A^2 s^2,$$

from which we obtain the factorization

$$(pr^2 + 2Bt^2 + As)(pr^2 + 2Bt^2 - As) = p(Br^2 + 2t^2)^2.$$

It is not difficult to check that $\gcd(pr^2 + 2Bt^2 + As, pr^2 + 2Bt^2 - As)$ is either a square or twice a square; up to multiplication by 2, it is a square divisor of $\gcd(A, s)^2$. Hence the above factorization implies that there are integers u and v satisfying

$$\begin{bmatrix} pr^2 + 2Bt^2 \pm As = pu^2 \\ pr^2 + 2Bt^2 \mp As = v^2 \\ Br^2 + 2t^2 = uv \end{bmatrix} \quad \text{or} \quad \begin{bmatrix} pr^2 + 2Bt^2 \pm As = 2pu^2 \\ pr^2 + 2Bt^2 \mp As = 2v^2 \\ Br^2 + 2t^2 = 2uv \end{bmatrix}.$$

Eliminating s from these equations, we obtain two systems of equations:

$$\boxed{\begin{array}{c} 2pr^2 + 4Bt^2 = pu^2 + v^2 \\ Br^2 + 2t^2 = uv \end{array}} \qquad \boxed{\begin{array}{c} pr^2 + 2Bt^2 = pu^2 + v^2 \\ Br^2 + 2t^2 = 2uv \end{array}}$$

We prove later (X.6.6) that the assumptions that 2 is a quartic nonresidue modulo p and that $p \equiv 1 \pmod 8$ imply that $B \equiv 4 \pmod 8$. Reducing each system of equations modulo 8, it is now a simple matter to verify that in both cases, any solution must satisfy $r \equiv t \equiv 0 \pmod 2$. This contradicts our initial assumption that $\gcd(r, t) = 1$, which completes the proof that $C_{-1}(\mathbb{Q}) = \emptyset$. $\square$

We close this section with the theorem of Gauss describing the quartic character of 2 that was used in the proof of (X.6.5). The proof that we give is due to Dirichlet [66]; see also [184].

Proposition 6.6. *Let p be a prime satisfying $p \equiv 1 \pmod 8$, and write p as a sum of two squares, $p = A^2 + B^2$. Then*

$$\left(\frac{2}{p}\right)_4 = (-1)^{AB/4}.$$

In other words, 2 is a quartic residue modulo p if and only if $AB \equiv 0 \pmod 8$.

PROOF. Using the fact that $A^2 + B^2 \equiv 0 \pmod p$, we compute

$$(A + B)^{(p-1)/2} \equiv (2AB)^{(p-1)/4} \pmod p$$
$$\equiv 2^{(p-1)/4}(-1)^{(p-1)/8} A^{(p-1)/2} \pmod p.$$

Switching A and B if necessary, we may assume that A is odd, and then the fact that $p \equiv 1 \pmod 4$ implies that

$$\left(\frac{A}{p}\right) = \left(\frac{p}{A}\right) = \left(\frac{B^2}{A}\right) = 1.$$

Hence

$$\left(\frac{A + B}{p}\right) = (-1)^{(p-1)/8}\left(\frac{2}{p}\right)_4.$$

Finally, we observe that

$$\left(\frac{A + B}{p}\right) = \left(\frac{p}{A + B}\right) = \left(\frac{2}{A + B}\right)\left(\frac{2p}{A + B}\right) = \left(\frac{2}{A + B}\right) = (-1)^{\frac{(A+B)^2 - 1}{8}},$$

since the identity

$$2p = (A + B)^2 + (A - B)^2 \quad \text{implies that} \quad \left(\frac{2p}{A + B}\right) = 1.$$

Substituting this above yields

$$\left(\frac{2}{p}\right)_4 = (-1)^{\frac{(A+B)^2 - 1}{8} - \frac{p-1}{8}} = (-1)^{\frac{AB}{4}}. \square$$

Exercises

10.1. Let $\phi : E/K \to E'/K$ be an isogeny of degree m of elliptic curves defined over an arbitrary (perfect) field K. Assume that $E[\hat{\phi}] \subset E(K)$. Generalize (X.1.1) as follows:

(a) Prove that there is a bilinear pairing

$$b : E'(K)/\phi\big(E(K)\big) \times E'[\hat{\phi}] \longrightarrow K(S,m)$$

defined by

$$e_\phi\big(\delta_\phi(P), T\big) = \delta_K\big(b(P,T)\big).$$

Here e_ϕ is the generalized Weil pairing (Exercise 3.15) and

$$\delta_\phi : E'(K) \to H^1\big(G_{\bar{K}/K}, E[\phi]\big) \qquad \text{and} \qquad \delta_K : K^* \to H^1\big(G_{\bar{K}/K}, \mu_m\big)$$

are the usual connecting homomorphisms.

(b) Prove that the pairing in (a) is nondegenerate on the left.

(c) For $T \in E[\hat{\phi}]$, let $f_T \in K(E')$ and $g_T \in K(E)$ be functions satisfying

$$\text{div}(f_T) = m(T) - m(O) \qquad \text{and} \qquad f_T \circ \phi = g_T^m.$$

Prove that

$$b(P,T) = f_T(P) \bmod (K^*)^m \qquad \text{for all } P \neq O, T.$$

(d) In particular, if $\deg(\phi) = 2$, so $E'[\hat{\phi}] = \{O, T\}$, then

$$b(P,T) = x(P) - x(T) \bmod (K^*)^2.$$

We thus recover part of (X.4.9).

10.2. Let K be an arbitrary (perfect) field, let E/K be an elliptic curve, and let C_1/K and C_2/K be homogeneous spaces for E/K.

(a) Prove that there exist a homogeneous space C_3/K for E/K and a morphism

$$\phi : C_1 \times C_2 \to C_3$$

defined over K such that for all $p_1 \in C_1$, $p_2 \in C_2$, and $P_1, P_2 \in E$,

$$\phi(p_1 + P_1, p_2 + P_2) = \phi(p_1, p_2) + P_1 + P_2.$$

(b) Prove that C_3 is uniquely determined, up to equivalence of homogeneous spaces, by C_1 and C_2.

(c) Prove that

$$\{C_1\} + \{C_2\} = \{C_3\},$$

the sum taking place in $\text{WC}(E/K)$.

10.3. Let C/K be a curve of genus one defined over an arbitrary (perfect) field.

(a) Prove that there exists an elliptic curve E/K such that C/K is a homogeneous space for E/K. (*Hint.* Use Exercise 3.22 to show that $C/K \in \text{Twist}(E/K)$. Then find an element $\{\xi\} \in H^1\big(G_{\bar{K}/K}, \text{Aut}(E)\big)$ such that C/K is the homogeneous space for the twist of E by ξ.)

(b) Prove that E is unique up to K-isomorphism.

10.4. Let K be an arbitrary (perfect) field and let E/K be an elliptic curve.

(a) Prove that there is a natural action of $\mathrm{Aut}_K(E)$ on $\mathrm{WC}(E/K)$ defined by letting an automoprhism $\alpha \in \mathrm{Aut}_K(E)$ act on $\{C/K, \mu\} \in \mathrm{WC}(E/K)$ via

$$\{C/K, \mu\}^\alpha = \{C/K, \mu \circ (1 \times \alpha)\}.$$

In other words, take the same curve, but define a new action of E on C by the rule

$$\mu^\alpha(p, P) = \mu(p, \alpha P).$$

(b) Conversely, if $\{C/K, \mu\}$ and $\{C/K, \mu'\}$ are elements of $\mathrm{WC}(E/K)$, prove that there exists an $\alpha \in \mathrm{Aut}_K(E)$ such that $\mu' = \mu \circ (1 \times \alpha)$.

(c) Conclude that for a given curve C/K of genus one, there are only finitely many inequivalent ways to make C/K into a homogeneous space. In particular, if C satisfies $j(C) \neq 0, 1728$, then there are at most two. (See also Exercise B.5.)

10.5. Let $\phi : E/K \to E'/K$ be a separable isogeny of elliptic curves defined over an arbitrary (perfect) field K, and let C/K be a homogeneous space for E/K. The finite group $E[\phi]$ acts on C, and we let $C' = C/E[\phi]$ be the quotient curve (cf. Exercise 3.13).

(a) Prove that C' is a curve of genus one defined over K.

(b) Prove that C'/K is a homogeneous space for E'/K and that the natural map

$$\phi : \mathrm{WC}(E/K) \to \mathrm{WC}(E'/K)$$

sends $\{C/K\}$ to $\{C'/K\}$.

(c) In particular, if $\{C/K\} \in \mathrm{WC}(E/K)[\phi]$, then C' is isomorphic to E' over K. Prove that this isomorphism can be chosen so that the natural projection $C \to C/E[\phi] \cong E'$ is the map $\phi \circ \theta$ defined in (X.4.6a).

10.6. *WC Over Finite Fields.* Let $\mathbb{F}_q$ be a finite field with q elements, let $C/\mathbb{F}_q$ be a curve of genus one, and pick any point of $C(\bar{\mathbb{F}}_q)$ as origin to make C into an elliptic curve. Let $\phi : C \to C$ be the qth-power Frobenius map on C.

(a) Prove that there are an endomorphism $f \in \mathrm{End}(C)$ and a point $P_0 \in C(\bar{\mathbb{F}}_q)$ satisfying $\phi(P) = f(P) + P_0$.

(b) Prove that f is inseparable, and conclude that there exists a point $P_1 \in C(\bar{\mathbb{F}}_q)$ satisfying $(1 - f)(P_1) = P_0$.

(c) Prove that $\phi(P_1) = P_1$, and hence that $P_1 \in C(\mathbb{F}_q)$.

(d) Let $E/\mathbb{F}_q$ be an elliptic curve. Prove that $\mathrm{WC}(E/\mathbb{F}_q) = 0$.

10.7. *WC Over $\mathbb{R}$.* Let $E/\mathbb{R}$ be an elliptic curve.

(a) Prove that

$$\mathrm{WC}(E/\mathbb{R}) = \begin{cases} \mathbb{Z}/2\mathbb{Z} & \text{if } \Delta(E/\mathbb{R}) > 0, \\ 0 & \text{if } \Delta(E/\mathbb{R}) < 0. \end{cases}$$

(b) Assuming that $\Delta(E/\mathbb{R}) > 0$, find an equation for a homogeneous space representing the nontrivial element of $\mathrm{WC}(E/\mathbb{R})$ in terms of a given Weierstrass equation for $E/\mathbb{R}$.

10.8. Let E/K be an elliptic curve, let $m \geq 2$ be an integer, and assume that $E[m] \subset E(K)$. Let $v \in M_K^0$ be a prime not dividing m. Prove that the restriction map

$$\mathrm{WC}(E/K)[m] \longrightarrow \mathrm{WC}(E/K_v)[m]$$

is surjective. (*Hint.* Show that the map on the $H^1(\,\cdot\,, E[m])$ groups is surjective.)

10.9. Let E/K be an elliptic curve, let $T \in E[m]$, and suppose that the field $L = K(T)$ has maximal degree, namely $[L : K] = m^2 - 1$. (Note that L/K is generally not a Galois extension.) Let e_m denote the Weil pairing and consider the chain of maps

$$\alpha : E(K) \xrightarrow{\delta} H^1\big(G_{\bar{K}/K}, E[m]\big) \xrightarrow{\text{res}} H^1\big(G_{\bar{K}/L}, E[m]\big) \to H^1\big(G_{\bar{K}/L}, \boldsymbol{\mu}_m\big) \cong L^*/(L^*)^m,$$
$$\xi_\sigma \mapsto e_m(\xi_\sigma, T).$$

(a) Let $f_T \in L(E)$ be as in (X.1.1d), i.e.,

$$\text{div}(f_T) = m(T) - m(O) \qquad \text{and} \qquad f_T \circ [m] \in \big(L(E)^*\big)^m.$$

Prove that

$$\alpha(P) = f_T(P) \bmod (L^*)^m.$$

(b) Prove that for all $P \in E(K)$,

$$N_{L/K}\big(\alpha(P)\big) \in (K^*)^m.$$

(c) Let $S \subset M_L$ be the set of places of L containing all archimedean places, all places dividing m, and all places at which E/L has bad reduction. Show that if $P \in E(K)$ and $v \in M_L$ with $v \notin S$, then

$$\text{ord}_v\big(\alpha(P)\big) \equiv 0 \pmod{m}.$$

(d) For $m = 2$, prove that the kernel of α is exactly $2E(K)$. Hence in this case there is an *injective* homomorphism from $E(K)/2E(K)$ into the group

$$\big\{a \in L^*/(L^*)^2 : N_{L/K}(a) \in (K^*)^2 \text{ and } \text{ord}_v(a) \equiv 0 \, (\text{mod } 2) \text{ for all } v \notin S\big\}$$

given by the map

$$P \longmapsto x(P) - x(T).$$

This map may often be used to compute $E(K)/2E(K)$. (*Hint.* Expand the quantity $x(P) - x(T) = r + sx(T) + tx(T)^2$ and use the resulting relations on $r, s, t \in K$ to show that P is in $2E(K)$.)

(e) Use (d) to compute $E(\mathbb{Q})/2E(\mathbb{Q})$ for the curve

$$E : y^2 + y = x^3 - x.$$

(*Hint.* Let $K/\mathbb{Q}$ be the totally real cubic field generated by a root of the polynomial $4x^3 - 4x + 1$. Start by showing that K has class number one and that every totally positive unit in K is a square.)

10.10. Let C/K be a curve of genus one, and suppose that $C(K_v) \neq \emptyset$ for all $v \in M_K$. Prove that the map

$$\text{Div}_K(C) \longrightarrow \text{Pic}_K(C)$$

is surjective. (*Hint.* Take Galois cohomology of the exact sequence

$$1 \longrightarrow \bar{K}^* \longrightarrow \bar{K}(C)^* \longrightarrow \text{Div}(C) \longrightarrow \text{Pic}(C) \longrightarrow 0.$$

Use Noether's generalization of Hilbert's Theorem 90,

$$H^1\big(G_{\bar{K}/K}, \bar{K}(C)^*\big) = 0,$$

and the (cohomological version) of the Brauer–Hasse–Noether theorem [288, §9.6], which says that an element of $H^2(G_{\bar{K}/K}, \bar{K}^*)$ is trivial if and only if it is trivial in $H^2(G_{\bar{K}_v/K_v}, \bar{K}_v^*)$ for every $v \in M_K$.)

10.11. *Index and Period in WC.* Let K be an arbitrary (perfect) field, let E/K be an elliptic curve, and let C/K be a homogeneous space for E/K. The *period of C/K* is defined to be the exact order of $\{C/K\}$ in $WC(E/K)$, and the *index of C/K* is the smallest degree of an extension L/K such that $C(L) \neq \emptyset$. So for example, (X.3.3) says that the period is equal to 1 if and only if the index is equal to 1.

(a) Prove that the period may also be characterized as the smallest integer $m \geq 1$ such that there exists a point $p \in C$ satisfying

$$p^{\sigma} - p \in E[m] \qquad \text{for every } \sigma \in G_{\bar{K}/K}.$$

(b) Prove that the index may also be characterized as the smallest degree among the *positive* divisors in $\mathrm{Div}_K(C)$.

(c) Prove that the period divides the index.

(d) Prove that the period and the index are divisible by the same set of primes.

(e) * Give an example with $K = \mathbb{Q}$ showing that the period may be strictly smaller than the index.

(f) Prove that if K is a number field and if C/K represents an element of $\text{III}(E/K)$, then the period and the index are equal. (*Hint.* Use (a), (b), (c), and Exercise 10.10.)

(g) * Let $K/\mathbb{Q}_p$ be a finite extension. Prove that the period and the index are equal.

10.12. *Hensel's Lemma.* The following version of Hensel's lemma is often useful for proving that a homogeneous space is locally trivial. Let R be a ring that is complete with respect to a discrete valuation v.

(a) Let $f(T) \in R[T]$ be a polynomial and $a_0 \in R$ a value satisfying

$$v\big(f(a_0)\big) > 2v\big(f'(a_0)\big).$$

Define a sequence of elements $a_n \in R$ recursively by

$$a_{n+1} = a_n - \frac{f(a_n)}{f'(a_n)} \qquad \text{for } n = 1, 2, \ldots.$$

Prove that a_n converges to an element $a \in R$ satisfying

$$f(a) = 0 \qquad \text{and} \qquad v(a - a_0) \geq v\left(\frac{f(a_0)}{f'(a_0)^2}\right) > 0.$$

(b) More generally, let $F(X_1, \ldots, X_N) \in R[X_1, \ldots, X_N]$, and suppose that there are an index $1 \leq i \leq N$ and a point $(a_1, \ldots, a_N) \in R^N$ satisfying

$$v\big(F(a_1, \ldots, a_N)\big) > 2v\left(\frac{\partial F}{\partial X_i}(a_1, \ldots, a_N)\right).$$

Prove that F has a root in R^N.

(c) Show that the curve

$$3X^3 + 4Y^3 + 5Z^3 = 0$$

in $\mathbb{P}^2$ has a point defined over $\mathbb{Q}_p$ for every prime p.

10.13. Use (X.1.4) to compute $E(\mathbb{Q})/2E(\mathbb{Q})$ for each of the following elliptic curves.

(a) $E : y^2 = x(x - 1)(x + 3)$.

(b) $E : y^2 = x(x - 12)(x - 36)$.

10.14. Use (X.4.9) to compute $E(\mathbb{Q})/2E(\mathbb{Q})$ for each of the following elliptic curves.
(a) $E : y^2 = x^3 + 6x^2 + x$.
(b) $E : y^2 = x^3 + 14x^2 + x$.
(c) $E : y^2 = x^3 + 9x^2 - x$.

10.15. Let E/K be an elliptic curve, let $\xi \in H^1\big(G_{\bar{K}/K}, \mathrm{Aut}(E)\big)$, and let E_ξ be the twist of E corresponding to ξ. Let $v \in M_K$ be a finite place at which E has good reduction. Prove that E_ξ has good reduction at v if and only if ξ is unramified at v. (See (VIII §2) for the definition of an unramified cocycle. *Hint.* If the residue characteristic is not 2 or 3, you can use explicit Weierstrass equations. In general, use the criterion of Néron–Ogg–Shafarevich (VII.7.1).)

10.16. Let E/K be an elliptic curve, let $D \in K^*$ be such that $L = K(\sqrt{D})$ is a quadratic extension, and let E_D/K be the twist of E/K given by (X.5.4). Prove that

$$\mathrm{rank}\, E(L) = \mathrm{rank}\, E(K) + \mathrm{rank}\, E_D(K).$$

10.17. Let $p \equiv 3 \pmod 4$ be a prime and let $D \in \mathbb{F}_p^*$.
(a) Show directly that the equation

$$C : v^2 = u^4 - 4D$$

has $p - 1$ solutions $(u, v) \in \mathbb{F}_p \times \mathbb{F}_p$. (*Hint.* Since $p \equiv 3 \pmod 4$, the map $u^2 \mapsto u^4$ is an automorphism of $(\mathbb{F}_p^*)^2$.)
(b) Let $E/\mathbb{F}_p$ be the elliptic curve

$$E : y^2 = x^3 + Dx.$$

Use the map

$$\phi : C \longrightarrow E, \qquad \phi(u, v) = \left(\frac{u^2 + v}{2}, \frac{u(u^2 + v)}{2} \right),$$

to prove that

$$\#E(\mathbb{F}_p) = p + 1.$$

10.18. Let p be an odd prime. Do a computation analogous to (X.6.2) to determine the Selmer groups and a bound for the ranks of the following families of elliptic curves $E/\mathbb{Q}$.
(a) $E : y^2 = x^3 - 2px$. (The curve with $p = 41$ has rank 3.)
(b) $E : y^2 = x^3 + p^2 x$.

10.19. Let $E/\mathbb{Q}$ be an elliptic curve with $j(E) = 0$.
(a) Prove that there is a unique sixth-power-free integer D such that E is given by the Weierstrass equation

$$E : y^2 = x^3 + D.$$

(b) Let $p \equiv 2 \pmod 3$ be a prime not dividing $6D$. Prove that

$$\#E(\mathbb{F}_p) = p + 1.$$

(c) Prove that $\#E(\mathbb{Q})_{\mathrm{tors}}$ divides 6.

(d) More precisely, prove that the following list gives a complete description of $E_{\text{tors}}(\mathbb{Q})$:

$$E_{\text{tors}}(\mathbb{Q}) \cong \begin{cases} \mathbb{Z}/6\mathbb{Z} & \text{if } D = 1, \\ \mathbb{Z}/3\mathbb{Z} & \text{if } D \neq 1 \text{ is a cube, or if } D = -432, \\ \mathbb{Z}/2\mathbb{Z} & \text{if } D \neq 1 \text{ is a square,} \\ 1 & \text{otherwise.} \end{cases}$$

10.20. Let A be a finite abelian group, and suppose that there exists a bilinear, alternating, nondegenerate pairing

$$\Gamma : A \times A \longrightarrow \mathbb{Q}/\mathbb{Z}.$$

Prove that $\#A$ is a perfect square.

10.21. Let E/K be an elliptic curve defined over a field of characteristic not equal to 2 or 3, fix a Weierstrass equation for E/K, and let c_4 and c_6 be the usual quantities (III §1) associated to the equation. Assuming that $j(E) \neq 0, 1728$, we define

$$\gamma(E/K) = -c_4/c_6 \in K^*/(K^*)^2.$$

(a) Prove that $\gamma(E/K)$ is well-defined as an element of $K^*/(K^*)^2$, independent of the choice of Weierstrass equation for E/K.
(b) Let E'/K be another elliptic curve with $j(E') \neq 0, 1728$. Prove that E and E' are isomorphic over K if and only if $j(E) = j(E')$ and $\gamma(E/K) = \gamma(E'/K)$.
(c) If $j(E) = j(E') \neq 0, 1728$, prove that E and E' are isomorphic over the field

$$K\left(\sqrt{\frac{\gamma(E/K)}{\gamma(E'/K)}}\right).$$

10.22. Let E/K be an elliptic curve over an arbitrary (perfect) field, let L/K be a finite Galois extension, and define a trace map

$$T_{L/K} : E(L) \longrightarrow E(K), \qquad P \longmapsto \sum_{\sigma \in G_{L/K}} P^\sigma.$$

(a) Prove that $T_{L/K}$ is a homomorphism.
(b) If K is a finite field, prove that $T_{L/K} : E(L) \to E(K)$ is surjective.
(c) Assume that $[L : K] = 2$ and that $\text{char}(K) \neq 2$, and write $L = K(\sqrt{D})$. Fix a Weierstrass equation for E/K of the form

$$E : y^2 = x^3 + ax^2 + bx + c,$$

and let E_D be the quadratic twist of E given by the equation (cf. (X.5.4))

$$E_D : y^2 = x^3 + Dax^2 + D^2bx + D^3c.$$

(i) Prove that the kernel of $T_{L/K} : E(L) \to E(K)$ is isomorphic to $E_D(K)$.
(ii) Prove that the image of $T_{L/K} : E(L) \to E(K)$ contains $2E(K)$.
(iii) Deduce that there are an exact sequence

$$0 \longrightarrow E_D(K) \longrightarrow E(L) \xrightarrow{\ T_{L/K}\ } E(K) \longrightarrow V \longrightarrow 0$$

and a surjective homomorphism $E(K)/2E(K) \twoheadrightarrow V$.

(iv) Suppose further that K is a number field. Re-prove Exercise 10.16, i.e.,

$$\operatorname{rank} E(L) = \operatorname{rank} E(K) + \operatorname{rank} E_D(K).$$

10.23. Let $a \equiv 1 \pmod 4$ be an integer with the property that $p = a^2 + 64$ is prime. (It is conjectured, but not known, that there exist infinitely many such primes.) Let $E_a/\mathbb{Q}$ be the elliptic curve

$$E_a : y^2 = x^3 + ax^2 - 16x.$$

These are known as Neumann–Setzer curves.

(a) Prove that $\mathcal{D}_{E_a/\mathbb{Q}} = (p)$. More precisely, prove that E_a has split multiplicative reduction at p and good reduction at all other primes. (N.B. The given Weierstrass equation is not minimal.)

(b) Perform a two-descent (X.4.9) and prove that

$$E(\mathbb{Q}) \cong \mathbb{Z}/2\mathbb{Z} \qquad \text{and} \qquad \text{Ш}(E/\mathbb{Q})[2] = 0.$$

(c) * Let $E/\mathbb{Q}$ be an elliptic curve with the following two properties: (i) $E(\mathbb{Q})$ contains a 2-torsion point. (ii) E has multiplicative reduction at a single prime $p > 17$ and good reduction at all other primes. Prove that p has the form $p = a^2 + 64$ and that E is either isomorphic or 2-isogenous to the curve E_a.

10.24. Let E/K be an elliptic curve and let $m \ge 1$. This exercise describes the *Tate pairing*

$$\langle\,\cdot\,,\,\cdot\,\rangle_{\text{Tate}} : E(K)/mE(K) \times \text{WC}(E/K)[m] \longrightarrow \text{Br}(K),$$

where $\text{Br}(K) = H^2(G_{\bar{K}/K}, \bar{K}^*)$ is the Brauer group of K. (This exercise assumes that the reader is familiar with higher cohomology groups; see for example [9, 233].)

Let $P \in E(K)/mE(K)$ and let $C \in \text{WC}(E/K)[m]$. We use the Kummer sequence (VIII §2)

$$0 \longrightarrow E(K)/mE(K) \xrightarrow{\ \delta\ } H^1\big(G_{\bar{K}/K}, E[m]\big) \xrightarrow{\ \epsilon\ } \text{WC}(E/K)[m] \longrightarrow 0$$

to push forward the point P by δ and to pull back the homogeneous space C by ϵ to obtain 1-cocycles

$$\delta_P : G_{\bar{K}/K} \longrightarrow E[m] \qquad \text{and} \qquad \xi_C : G_{\bar{K}/K} \longrightarrow E[m].$$

We use δ_P and ξ_C to define a map

$$\lambda_{P,C} : G_{\bar{K}/K} \times G_{\bar{K}/K} \longrightarrow \mu_m, \qquad \lambda_{P,C}(\sigma,\tau) = e_m\big(\delta_P(\sigma), \xi_C(\tau)\big),$$

where e_m is the Weil pairing.

(a) Prove that the map $\lambda_{P,C}$ is a 2-cocycle.

(b) Prove that changing either δ_P or ξ_C by a 1-coboundary has the effect of changing $\lambda_{P,C}$ by a 2-coboundary.

(c) Prove that pulling back C to some other element ξ'_C changes $\lambda_{P,C}$ by a 2-coboundary.

(d) Conclude that

$$\langle P, \xi \rangle_{\text{Tate}} = \text{cohomology class of } \lambda_{P,C}$$

gives a well-defined pairing

$$\langle\,\cdot\,,\,\cdot\,\rangle_{\text{Tate}} : E(K)/mE(K) \times \text{WC}(E/K)[m] \longrightarrow \text{Br}(K).$$

(e) Prove that the pairing in (d) is bilinear.

(f) * Let $K/\mathbb{Q}_p$ be a finite extension. A basic result in local class field theory says that $\text{Br}(K) \cong \mathbb{Q}/\mathbb{Z}$; see [233, XII §3, Theorem 2]. Prove in this case that the Tate pairing is nondegenerate.

Chapter XI

Algorithmic Aspects of Elliptic Curves

The burgeoning field of computational number theory asks for practical algorithms to compute solutions to arithmetic problems. For example, the Mordell–Weil theorem (VIII.6.7) says that the group of rational points on an elliptic curve is finitely generated, and although we still lack an effective algorithm that is guaranteed to find a set of generators, there are algorithms that often work well in practice. Similarly, Siegel's theorem (IX.3.2.1) says that an elliptic curve has only finitely many S-integral points, but it took 50 years from Siegel's proof of finiteness to Baker's theorem giving an effective bound for the height of the largest solution (IX §5). And Baker's theorem is only the beginning of the story, since it leads to estimates that, although effective, are not practical without the introduction of significant additional ideas.

A full introduction to the computational theory of elliptic curves would require (at least) a book of its own, so in this single chapter we touch on only a few of the many algorithms in the theory. We decided to concentrate on aspects that are especially useful for applications to cryptography, not because these are intrinsically more interesting than other computational problems, but because they form a satisfying whole and because they tie in with many of the other topics covered in this book.

The theme of this chapter is thus that of computations on elliptic curves over (large) finite fields. We describe fast algorithms for computing multiples of points, for determining the number of points in $E(\mathbb{F}_q)$, and for computing the Weil pairing. We briefly survey some cryptographic constructions based on the difficulty of solving the elliptic curve discrete logarithm (ECDLP), and we describe algorithms to solve the ECDLP. We explain how elliptic curves can be used to factor large numbers. In the final section we define and analyze the Tate–Lichtenbaum pairing, which is frequently used in cryptography because it is easier to compute than the Weil pairing.

Lack of space precludes our covering computational problems over global fields, although these are also extremely interesting. In particular, we do not cover

J.H. Silverman, *The Arithmetic of Elliptic Curves, Second Edition*, Graduate Texts in Mathematics 106, DOI 10.1007/978-0-387-09494-6_XI,
© Springer Science+Business Media, LLC 2009

algorithms related to modular aspects of elliptic curves, including in particular the computation of L-series, nor do we discuss more advanced algorithmic methods for computing Mordell–Weil groups or for finding integer points on elliptic curves. We do not discuss methods used to find curves of high rank over $\mathbb{Q}$, nor how to find precise estimates for $|\hat{h} - h|$. For an introduction to these and other algorithmic topics, see for example [50, 54, 55, 58, 67, 76, 86, 171, 188, 219, 265, 315].

There are two other topics that would fit naturally into this chapter, but were omitted due to lack of space. The first is efficient implementation of elliptic curve addition, which includes issues of affine versus projective coordinates and the choice of different sorts of equations to minimize the number of field additions, multiplications, and inversions. See for example [16], [22, IV.1], [51, §§13.2, 13.3], or [71]. The second topic is the use of elliptic curves to prove that a number is prime; see [10], [22, §IX.3], [50, §9.2], [51, §25.2.2], or [97].

Finally, while on the topic of elliptic curve algorithms, we mention the free computer packages Pari [202] and Sage [275], both of which contain extensive libraries of algorithms for doing computations on elliptic curves. In particular, Sage includes Cremona's mwrank package, which (attempts to) compute the Mordell–Weil group of elliptic curves over $\mathbb{Q}$. There are also extensive online tables of elliptic curves of various types, e.g., of small conductor, and of modular forms associated to elliptic curves. See for example [53] and [274].

XI.1   Double-and-Add Algorithms

Let E/K be an elliptic curve and let $P \in E(K)$ be a point on E. Suppose that we need to compute $[n]P$ for some large value of n. An obvious way to do this is to compute successively

$$P, \quad [2]P = P + P, \quad [3]P = [2]P + P, \quad \ldots \quad [n]P = [n-1]P + P.$$

This naive algorithm takes $n - 1$ steps, where a "step" consists of adding two points.

If n is large, the naive algorithm is completely useless. All practical applications of elliptic curves over large finite fields rely on the following exponential improvement.

Double-and-Add Algorithm 1.1. *Let E/K be an elliptic curve, let $P \in E(K)$, and let $n \geq 2$ be an integer. The algorithm described in Figure 11.1 computes $[n]P$ using no more than $\log_2(n)$ point doublings and no more than $\log_2(n)$ point additions.*

PROOF. During the ith iteration of the loop, the value of Q is $[2^i]P$. Since R is incremented by Q if and only if $\epsilon_i = 1$, the final value of R is

$$\sum_{i \text{ with } \epsilon_i = 1} [2^i]P = \sum_{i=0}^{t} [\epsilon_i 2^i]P = \left[\sum_{i=0}^{i} \epsilon_i 2^i \right] P = [n]P.$$

Each iteration of the loop requires one point duplication and at most one point addition, and since $t \leq \log_2 n$, the running time of the algorithm is as stated. $\square$

(1) Write the binary expansion of n as

$$n = \epsilon_0 + \epsilon_1 \cdot 2 + \epsilon_2 \cdot 2^2 + \epsilon_3 \cdot 2^3 + \cdots + \epsilon_t \cdot 2^t$$

$$\text{with } \epsilon_0, \ldots, \epsilon_t \in \{0, 1\} \text{ and } \epsilon_t = 1.$$

(2) Set $Q = P$ and $R = \begin{cases} O & \text{if } \epsilon_0 = 0, \\ P & \text{if } \epsilon_0 = 1. \end{cases}$

(3) Loop $i = 1, 2, \ldots, t$.

(4)   Set $Q = [2]Q$.

(5)   If $\epsilon_i = 1$, set $R = R + Q$.

(6) End Loop

(7) Return R, which is equal to $[n]P$.

Figure 11.1: The double-and-add algorithm.

Remark 1.2. The double-and-add algorithm (XI.1.1) is not unique to elliptic curves; it is applicable to any group. Thus if G is a group and $g \in G$, we use the binary expansion $n = \sum \epsilon_i 2^i$ to compute g^n as $g^n = \prod (g^{2^i})^{\epsilon_i}$. This requires at most $\log_2 n$ group squarings and at most $\log_2 n$ group multiplications. When the group law in G is written multiplicatively, for example for $G = \mathbb{F}_q^*$, the double-and-add algorithm is instead called the *square-and-multiply* algorithm.

Remark 1.3. The double-and-add algorithm is most often applied to a finite group such as $E(\mathbb{F}_q)$ or $\mathbb{F}_q^*$, rather than to an infinite group such as $E(\mathbb{Q})$. To see why, note that if $P \in E(\mathbb{Q})$, then the theory of canonical heights (VIII §9) says that it takes $O(n^2)$ bits to write down the coordinates of $[n]P$. Thus it is not feasible to compute $[n]P$ for, say, $n > 2^{80}$. On the other hand, the double-and-add algorithm allows us to easily compute $[n]P$ in $E(\mathbb{F}_q)$ when, say, q and n are as large as 2^{1000}. Of course, when we say that the computation is easy, we mean on a computer, not with paper and pencil!

Remark 1.4. The average running time of the double-and-add algorithm to compute $[n]P$ is $\log_2 n$ doublings and $\frac{1}{2} \log_2 n$ additions, since the binary expansion of a random integer n has an equal number of 1's and 0's. We can reduce the average running time by using a *ternary expansion of n*,

$$n = \epsilon_0 + \epsilon_1 \cdot 2 + \epsilon_2 \cdot 2^2 + \epsilon_3 \cdot 2^3 + \cdots + \epsilon_t \cdot 2^t$$

$$\text{with } \epsilon_1, \ldots, \epsilon_t \in \{-1, 0, 1\} \text{ and } \epsilon_t = \pm 1.$$

The only changes in (XI.1.1) are in step (2), where we set $R = -P$ if $\epsilon_0 = -1$, and in step (5), where we set $R = R \pm Q$ if $\epsilon_i = \pm 1$. It is not hard to show that every integer has a unique ternary expansion in which no two consecutive coefficients are nonzero; see Exercise 11.2.

There are two complementary reasons why ternary expansions are advantageous for computing $[n]P$ in $E(\mathbb{F}_q)$. First, ternary expansions tend to have significantly fewer nonzero ϵ_i, so the number of point additions is reduced. Second, and equally important, the negation operation in $E(\mathbb{F}_q)$ is computationally trivial, so subtraction is no more difficult than addition. This is in marked contrast to $\mathbb{F}_q^*$, where group negation (inversion) is much slower than group addition (multiplication).

Remark 1.5. Koblitz has suggested using the Frobenius map to further speed the computation of $[n]P$. The idea is to use an elliptic curve $E/\mathbb{F}_p$ with p small and to take a point $P \in E(\mathbb{F}_{p^r})$. Then we replace the doubling map with the easier-to-compute Frobenius map. As a practical matter, Koblitz's idea works especially well for $p = 2$, so for concreteness we illustrate using the curve $E/\mathbb{F}_2$ given by the equation

$$E : y^2 + xy = x^3 + 1.$$

We have $E(\mathbb{F}_2) = \mathbb{Z}/4\mathbb{Z}$, so $E/\mathbb{F}_2$ is ordinary and

$$p + 1 - \#E(\mathbb{F}_p) = 2 + 1 - 4 = -1.$$

We use (V.2.3.1) to deduce that the Frobenius map

$$\tau : E(\mathbb{F}_{2^r}) \longrightarrow E(\mathbb{F}_{2^r}), \qquad (x, y) \longmapsto (x^2, y^2),$$

satisfies

$$\tau^2 + \tau + 2 = 0.$$

Using this relation, it is easy to write any integer n in the form

$$n = \epsilon_0 + \epsilon_1 \tau + \epsilon_2 \tau^2 + \cdots + \epsilon_t \tau^t \qquad \text{with } \epsilon_0, \ldots, \epsilon_t \in \{-1, 0, 1\},$$

where $t \approx 2\log_2(n)$ and at most one-third of the ϵ_i are nonzero. (With somewhat more work, the length of the expansion can be reduced to $t \approx \log_2(n)$; see Exercise 11.3.) Then $[n]P$ can be computed via

$$[n]P = \epsilon_0 P + \epsilon_1 \tau(P) + \epsilon_2 \tau^2(P) + \cdots + \epsilon_t \tau^t(P).$$

This is generally faster than using the binary or ternary expansion of n, because the Frobenius map on E is far easier to compute than the duplication map.

Remark 1.6. There are many variants of the basic double-and-add method that are used to make it more efficient in various situations. See for example [22, Chapter IV], [51, §9], and Exercise 11.4,

XI.2　Lenstra's Elliptic Curve Factorization Algorithm

Factorization of large numbers has been studied since antiquity, but the subject acquired added significance with the invention of public key cryptography, and in particular the development of the RSA cryptosystem, whose security depends on the difficulty of the factorization problem. Public key cryptography in general, and RSA in particular, are described in many books; see for example [116, 169, 277]. In this section we focus on the factorization problem itself.

The modern theory of factorization, by which we mean factorization algorithms that take less than exponential time,[1] dates back only to the 1920s. The fastest factorization algorithm currently known is the *number field sieve*, which factors an integer N in approximately

$$\exp\left(c\sqrt[3]{(\log N)(\log\log N)^2}\right) \text{ steps.}$$

Before the invention of the number field sieve, the fastest factorization method was the *quadratic sieve*, whose running time is approximately

$$\exp\left(c\sqrt{(\log N)(\log\log N)}\right) \text{ steps.}$$

(Notice that the cube root has been replaced by a square root. However, due to the different values of the constants, the quadratic sieve is actually the faster of the two algorithms for factoring numbers up to about 10^{100}.)

In this section we describe a factorization method due to Hendrik Lenstra that uses elliptic curves and has a running time comparable to the quadratic sieve. However, Lenstra's algorithm has one useful characteristic that differentiates it from sieve methods. If p is the smallest prime factor of N, then the running time of Lenstra's algorithm is actually

$$\exp\left(c\sqrt{(\log p)(\log\log p)}\right) \text{ steps.}$$

Thus Lenstra's algorithm is especially good at finding prime factors of N that are significantly smaller than $\sqrt{N}$. However, we note that the moduli used for RSA have the form $N = pq$ with primes $p \approx q$, so sieve algorithms are more efficient than Lenstra's algorithm for factoring such numbers.

The prototype for Lenstra's work is an earlier factorization algorithm, due to Pollard, which we briefly describe. Pollard's algorithm is good at factoring numbers N that have a prime factor p such that $p - 1$ is a product of small primes. Numbers that are a product of small primes are called *smooth numbers*.[2]

Pollard's $p - 1$ Algorithm 2.1. *Suppose that N is a composite number that has a prime factor p such that $p - 1$ factors into primes as*

$$p - 1 = q_1^{e_1} q_2^{e_2} \cdots q_t^{e_t}.$$

---

[1]The running time of an algorithm is measured as a function of the number of bits of the input and output. Thus an algorithm that factors an integer N in time $O(N^c)$ for some $c > 0$ takes *exponential time*, since the number of bits of the input is $\log_2(N)$. Similarly, a *polynomial-time algorithm* is one that runs in time $O\big((\log_2 N)^c\big)$, and a a *subexponential-time algorithm* is one that runs faster than $O(N^\epsilon)$ for every $\epsilon > 0$.

[2]More precisely, a number m is said to be *B-smooth* if every prime p dividing m satisfies $p \le B$. An important theorem of Canfield, Erdős, and Pomerance [34] gives an estimate for the number of B-smooth numbers less than a given bound.

(1)   Choose a base value $2 \leq a < N$ and set $A = a$.

(2)   Loop $i = 1, 2, \ldots, L$.

(3)   Replace A with $A^i \bmod N$.

(4)   Compute $F = \gcd(A - 1, N)$.

(5)   If $1 < F < N$, then return F, which is a nontrivial factor of N.

(6)   If $F = N$, go to step (1) and choose a new value of a.

(7)   End Loop

Figure 11.2: Pollard's $p - 1$ algorithm.

Let L be the quantity

$$L = \max_{1 \leq j \leq t} e_j q_j.$$

Then for most base values, the algorithm described in Figure 11.2 finds a nontrivial factor of N for some value of $i \leq L$ in the main loop (steps (2)–(7)).

PROOF. During the ith iteration of the loop, the value of A is $a^{i!} \bmod N$. The definition of L ensures that $q_j^{e_j}$ divides $L!$ for each $1 \leq j \leq t$, so $p - 1 \mid L!$. (See also Exercise 11.6.) It follows from Fermat's little theorem that $a^{L!} \equiv 1 \pmod{p}$, so the value of F in step (4) is divisible by p. Since it is unlikely that $a^{L!} \equiv 1 \pmod{N}$, we obtain a nontrivial factor of N. $\square$

Example 2.2. We use Pollard's algorithm to factor $N = 71384665949740607$. Using the base $a = 2$, we find on the 33rd iteration of the loop that

$$2^{33!} \equiv 58248995050016779 \quad (\bmod\ 71384665949740607),$$
$$\gcd(58248995050016778, 71384665949740607) = 228266501.$$

Thus

$$N = 71384665949740607 = 228266501 \cdot 312725107,$$

and one can check that both factors are prime.

Pollard's algorithm works well for this N because the prime $p = 228266501$ satisfies

$$p - 1 = 228266500 = 2^2 \cdot 5^3 \cdot 7^3 \cdot 11^3,$$

so $p - 1$ divides $33!$.

Pollard's algorithm is a valuable factorization tool, but it applies only to special sorts of numbers, namely those divisible by a prime p such that $p - 1$ is smooth. The significance of $p - 1$ lies in the fact that the multiplicative group $\mathbb{F}_p^*$ has order $p - 1$, so N and $a^{L!} - 1$ share a common factor of p as soon as $L!$ is divisible by $p - 1$. Lenstra's brilliant innovation was to observe that if one replaces the multiplicative group $\mathbb{F}_p^*$ by the points $E(\mathbb{F}_p)$ of an elliptic curve, then the group order varies as E varies. This allows the elliptic curve algorithm to factor a much larger set of numbers.

We first state the algorithm and then explain the various steps in more detail.

(0)   Choose a loop bound L.

(1)   Choose an elliptic curve $E \bmod N$ and a point $P \in E(\mathbb{Z}/N\mathbb{Z})$.

(2)   Set $Q = P$.

(3)   Loop $i = 2, 3, \ldots, L$.

(4)   Replace Q with $[i]Q$, working in $E(\mathbb{Z}/N\mathbb{Z})$.

(5)   If, during the computation of $[i](Q)$, you need the inverse of an
           element $a \in \mathbb{Z}/N\mathbb{Z}$ and that inverse does not exist, then
           $\gcd(a, N)$ is (probably) a nontrivial factor of N.

(6)   End i Loop

(7)   Go to step (1) and choose a new curve and point.

Figure 11.3: Lenstra's elliptic curve factorization algorithm.

Lenstra's Elliptic Curve Factorization Algorithm 2.3. *Let N be a positive integer to be factored, and consider the algorithm described in Figure 11.3. Suppose that N has a prime divisor p such that the loop bound L chosen in step (0) and the elliptic curve E chosen in step (1) satisfy*

$$\#E(\mathbb{F}_p) = q_1^{e_1} q_2^{e_2} \cdots q_t^{e_t} \qquad \text{and} \qquad L \geq \max\{e_1 q_1, \ldots, e_t q_t\}.$$

(Here $q_1, \ldots, q_t$ are distinct primes.) Then with high probability, the algorithm described in Figure 11.3 factors the integer N. (See (XI.2.4.5) for advice on how to choose the loop bound L.)

We now use a series of remarks to discuss various aspects of Lenstra's algorithm.

Remark 2.4.1. We have not heretofore worked with elliptic curves over rings such as $\mathbb{Z}/N\mathbb{Z}$ when N is composite. The fancy way to do this is via the theory of group schemes [266, Chapter IV], but for our purposes it suffices to take $A, B \in \mathbb{Z}/N\mathbb{Z}$ and use a Weierstrass equation

$$E : y^2 = x^3 + Ax + B \qquad \text{with} \quad \Delta = -16(4A^3 + 27B^2) \in (\mathbb{Z}/N\mathbb{Z})^*.$$

The choice of an elliptic curve modulo N in step (1) is thus simply a choice of $A, B \in \mathbb{Z}/N\mathbb{Z}$. (If we are unlucky and $\Delta \notin (\mathbb{Z}/N\mathbb{Z})^*$, then with high probability, $\gcd(\Delta, N)$ is a nontrivial factor of N.)

Remark 2.4.2. Having chosen an elliptic curve modulo N, it is not clear how to efficiently choose a point on that curve, since taking square roots modulo an unfactored N is a hard problem. The trick is to first choose A and the point $P = (x_0, y_0)$, and then set $B = y_0^2 - x_0^3 - Ax_0$.

Remark 2.4.3. Steps (4) and (5) require some explanation. The double-and-add method (XI.1.1) for computing $[i]Q$ involves additions $R_1 + R_2$ and duplications $[2]R$. We perform these operations using the standard formulas from (III.2.3), always working modulo N. For example, to add $R_1 = (x_1, y_1)$ and $R_2 = (x_2, y_2)$, we must compute

$$x(R_1 + R_2) \equiv \left(\frac{y_2 - y_1}{x_2 - x_1}\right)^2 + a_1 \left(\frac{y_2 - y_1}{x_2 - x_1}\right)^2 - a_2 - x_1 - x_2 \pmod{N}.$$

This works fine if the quantity $x_2 - x_1$ is invertible modulo N. However, if it is not invertible and we are unable to compute $R_1 + R_2 \bmod N$, then (Eureka!)

$$\gcd(x_2 - x_1, N) > 1,$$

and there is a good chance that $\gcd(x_2 - x_1, N)$ is a nontrivial factor of N. Thus in computing $[i]Q$ modulo N in step (4), either the computation works, or else we have (probably) factored N.

Remark 2.4.4. If we successfully complete L iterations of the i loop (steps (3)–(6)), the final value of Q is

$$Q = [L!]P \quad \text{in} \quad E(\mathbb{Z}/N\mathbb{Z}).$$

When does this computation fail?

Let p be a prime divisor of N, and let $n = n_p(E) = \#E(\mathbb{F}_p)$. We can use the reduction modulo p map $E(\mathbb{Z}/N\mathbb{Z}) \to E(\mathbb{Z}/p\mathbb{Z})$ to send the point P to the group $E(\mathbb{F}_p)$. Then, when we use the double-and-add method to compute $[n]P = O$ in $E(\mathbb{F}_p)$, at some stage we get a "zero in the denominator." Hence if $n_p(E) \mid L!$, then step (5) is executed and we (probably) find a nontrivial factor of N.

Remark 2.4.5. Notice the analogy with Pollard's algorithm (XI.2.1), where the relevant condition for success was $\#\mathbb{F}_p^* \mid L!$. The advantage of Lenstra's algorithm is that

$$n_p(E) = \#E(\mathbb{F}_p) = p + 1 - a_p(E)$$

varies as we choose different elliptic curves. (Here $a_p(E)$ is the trace of Frobenius (V.2.6).) Lenstra's algorithm succeeds if we manage to choose an elliptic curve E such that for some prime factor p of N, the order of the group $E(\mathbb{F}_p)$ is a smooth number.

In order to optimally implement Lenstra's algorithm, we need to decide how large to choose L, since this determines when we give up on a particular elliptic curve and choose a new one. We know from (V.1.1) that $a_p(E)$ satisfies $|a_p(E)| \leq 2\sqrt{p}$, and it is not unreasonable to assume that the $a_p(E)$ values are more or less equidistributed in this range as E varies. (See (C.21.4) for a more precise statement.) So the probability of success for a chosen E depends on the distribution of smooth numbers in an interval around p. Using [34] as a heuristic to estimate the number of smooth numbers in short intervals, one can show that the optimal choice for L is approximately $\exp\big(c_1\sqrt{(\log p)(\log \log p)}\big)$, and that with this choice of L, the expected number of elliptic curves used before a factor of N is found is L^{c_2}. (Here c_1 and c_2 are small absolute constants.) Thus, as noted above, Lenstra's algorithm has the same qualitative running time as the quadratic sieve for numbers that are a product of two large primes, but it is generally much faster for numbers that have a comparatively small prime factor.

Remark 2.4.6. There are many implementation tricks that are used to make Lenstra's algorithm more efficient. We mention in particular the use of several elliptic curves in parallel to save on mod N inversions, and the use of so-called Stage 2 computations, which are also used for Pollard's algorithm. For details see for example [50, §8.8].

Example 2.5. We use Lenstra's algorithm to factor $N = 6887$. We randomly select $P = (1512, 3166)$ and $A = 14$, and we set

$$B \equiv 3166^2 - 1512^3 - 14 \cdot 1512 \equiv 19 \pmod{6887},$$

so P is a mod N point on the elliptic curve

$$E : Y^2 = X^3 + 14X + 19.$$

We compute successively (always working modulo 6887)

$$[2]P \equiv (3466, 2996),$$
$$[3!]P = [3]([2]P) \equiv (3067, 396) ,$$
$$[4!]P = [4]([3!]P) \equiv (6507, 2654),$$
$$[5!]P = [5]([4!]P) \equiv (2783, 6278),$$
$$[6!]P = [6]([5!]P) \equiv (6141, 5581).$$

These values are not, themselves, of any intrinsic interest. To ease notation, we let $Q = [6!]P = (6141, 5581)$. We use the double-and-add algorithm (XI.1.1) to compute $[7]Q = [7!]P$. Thus

$$Q \equiv (6141, 5881), \quad [2]Q \equiv (5380, 174), \quad [4]Q \equiv [2]([2]Q) \equiv (203, 2038),$$

and

$$\begin{aligned}
7Q &\equiv (Q + [2]Q) + [4]Q \\
&\equiv ((6141, 5581) + (5380, 174)) + (203, 2038) \\
&\equiv (984, 589) + (203, 2038) \\
&\equiv ???
\end{aligned}$$

When we attempt to perform the final addition, we need the inverse of $203 - 984$ modulo 6887, but

$$\gcd(203 - 984, 6887) = \gcd(-781, 6887) = 71.$$

Thus $71 \mid 6887$, and we find the factorization $6887 = 71 \cdot 97$.

It turns out that in $E(\mathbb{F}_{71})$, the point P satisfies $[63]P \equiv \mathcal{O} \pmod{71}$, while in $E(\mathbb{F}_{97})$, the point P satisfies $[107]P \equiv \mathcal{O} \pmod{97}$. The reason that we succeed in factoring 6887 using $[7!]P$, but not with a smaller multiple of P, is due to the fact that 7! is the smallest factorial that is divisible by 63 (and because 7! is not divisible by 107).

XI.3   Counting the Number of Points in $E(\mathbb{F}_q)$

Let $E/\mathbb{F}_q$ be an elliptic curve defined over a finite field. Hasse's theorem (V.1.1) says that

$$\#E(\mathbb{F}_q) = q + 1 - a_q \quad \text{with} \quad |a_q| \leq 2\sqrt{q}.$$

For many applications, especially in cryptography, it is important to have an efficient way to compute the number of points in $E(\mathbb{F}_q)$. For simplicity, we assume that q is odd and that E is given by a Weierstrass equation of the form

$$E : y^2 = f(x) = 4x^3 + b_2x^2 + 2b_4x + b_6,$$

but with minor modifications, everything that we do also works in characteristic 2.

     A straightforward, but not very efficient, method to find $\#E(\mathbb{F}_q)$ is to compute the sum (cf. (V.1.3))

$$a_q = \sum_{x \in \mathbb{F}_q} \left(\frac{f(x)}{q} \right).$$

Each Legendre symbol $\left(\frac{f(x)}{q}\right)$ can be computed by quadratic reciprocity in $O(\log q)$ steps, so this explicit formula takes $O(q \log q)$ steps, making it an exponential-time algorithm. (See also Exercise 11.14 for an algorithm that computes $\#E(\mathbb{F}_q)$ in $O(\sqrt{q})$ steps.)

     In this section we describe Schoof's algorithm [223], which computes $\#E(\mathbb{F}_q)$ in polynomial time, i.e., it computes $\#E(\mathbb{F}_q)$ in $O\big((\log q)^c\big)$ steps, where c is fixed, independent of q. The idea is to compute the value of a_q modulo ℓ for a lot of small primes ℓ and then use the Chinese remainder theorem to reconstruct a_q.

     Let

$$\tau : E(\bar{\mathbb{F}}_q) \longrightarrow E(\bar{\mathbb{F}}_q), \qquad (x, y) \longmapsto (x^q, y^q),$$

be the q-power Frobenius map, so (V.2.3.1b) tells us that

$$\tau^2 - a_q\tau + q = 0 \qquad \text{in } \mathrm{End}(E).$$

In particular, if $P \in E(\mathbb{F}_q)[\ell]$, then

$$\tau^2(P) - [a_q]\tau(P) + [q]P = O,$$

so if we write $P = (x, y)$ (we assume that $P \neq O$), then

$$(x^{q^2}, y^{q^2}) - [a_q](x^q, y^q) + [q](x, y) = O.$$

A key observation is that since the point $P = (x, y)$ is assumed to have order ℓ, we have

$$[a_q](x^q, y^q) = [n_\ell](x^q, y^q), \quad \text{where } n_\ell \equiv a_q \pmod{\ell} \text{ with } 0 \leq n_\ell < \ell.$$

Similarly, we can compute $[q](x, y)$ by first reducing q modulo ℓ.

Of course, we don't know the value of n_ℓ, so for each integer n between 0 and ℓ we compute $[n](x^q, y^q)$ for a point $(x, y) \in E[\ell] \smallsetminus \{O\}$ and check to see whether it satisfies

$$[n](x^q, y^q) = (x^{q^2}, y^{q^2}) + [q](x, y).$$

However, the individual points in $E[\ell]$ tend to be defined over fairly large extension fields of $\mathbb{F}_q$, so we instead work with all of the ℓ-torsion points simultaneously. To do this, we use the division polynomial (see Exercise 3.7)

$$\psi_\ell(x) \in \mathbb{F}_q[x],$$

whose roots are the x-coordinates of the nonzero ℓ-torsion points of E. (For simplicity, we assume that $\ell \neq 2$.) This division polynomial has degree $\frac{1}{2}(\ell^2 - 1)$ and is easily computed using the recurrence described in Exercise 3.7. We now perform all computations in the quotient ring

$$R_\ell = \frac{\mathbb{F}_q[x, y]}{\left(\psi_\ell(x), y^2 - f(x)\right)}.$$

Thus anytime we have a nonlinear power of y, we replace y^2 with $f(x)$, and anytime we have a power x^d with $d \geq \frac{1}{2}(\ell^2 - 1)$, we divide by $\psi_\ell(x)$ and take the remainder. In this way we never have to work with polynomials of degree greater than $\frac{1}{2}(\ell^2 - 3)$.

Our goal is to compute the value of $a_q \bmod \ell$ for enough primes ℓ to determine a_q. Hasse's theorem (V.1.1) says that $|a_q| \leq 2\sqrt{q}$, so it suffices to use all primes $\ell \leq \ell_{\max}$ such that

$$\prod_{\ell \leq \ell_{\max}} \ell \geq 4\sqrt{q}.$$

The preceding discussion shows that the following algorithm computes $\#E(\mathbb{F}_q)$. The subsequent proof estimates how long the computation takes.

Schoof's Algorithm 3.1. *Let $E/\mathbb{F}_q$ be an elliptic curve. The algorithm described in Figure 11.4 is a polynomial-time algorithm to compute $\#E(\mathbb{F}_q)$; more precisely, it computes $\#E(\mathbb{F}_q)$ in $O\left((\log q)^8\right)$ steps.*

PROOF. We prove that the running time of Schoof's algorithm is $O\left((\log q)^8\right)$. We begin by verifying three claims.

(a) *The largest prime ℓ used by the algorithm satisfies $\ell \leq O(\log q)$.*

The prime number theorem is equivalent to the statement [4, Theorem 4.4(9)]

$$\lim_{X \to \infty} \frac{1}{X} \sum_{\substack{\ell \leq X \\ \ell \text{ prime}}} \log \ell = 1.$$

Hence $\prod_{\ell \leq X} \ell \approx e^X$, so in order to make the product larger than $4\sqrt{q}$, it suffices to take $X \approx \frac{1}{2}\log(16q)$.

(1)   Set $A = 1$ and $\ell = 3$.

(2)   Loop while $A < 4\sqrt{q}$.

(3)   Loop $n = 0, 1, 2, \ldots, \ell - 1$.

(4)   Working in the ring R_ℓ, if
$$(x^{q^2}, y^{q^2}) + [q](x, y) = [n](x^q, y^q),$$
then break out of the n loop.

(5)   End n Loop

(6)   Set $A = \ell \cdot A$

(7)   Set $n_\ell = n$

(8)   Replace ℓ by the next largest prime.

(9)   End A Loop

(10) Use the Chinese remainder theorem to find an integer a
satisfying $a \equiv n_\ell \pmod{\ell}$ for all of the stored values of n_ℓ.

(11) Return the value $\#E(\mathbb{F}_q) = q + 1 - a$.

Figure 11.4: Schoof's algorithm.

(b) *Multiplication in the ring R_ℓ can be done in $O(\ell^4 (\log q)^2)$ bit operations.*[3]

Elements of the ring R_ℓ are polynomials of degree $O(\ell^2)$. Multiplication of two such polynomials and reduction modulo $\psi_\ell(x)$ takes $O(\ell^4)$ elementary operations (additions and multiplications) in the field $\mathbb{F}_q$. Similarly, multiplication in $\mathbb{F}_q$ takes $O((\log q)^2)$ bit operations. So basic operations in R_ℓ take $O(\ell^4 (\log q)^2)$ bit operations.

(c) *It takes $O(\log q)$ ring operations in R_ℓ to reduce $x^q, y^q, x^{q^2}, y^{q^2}$ in the ring R_ℓ.*

In general, the square-and-multiply algorithm (XI.1.2) allows us to compute powers x^n and y^n using $O(\log n)$ multiplications in R_ℓ. We note that this computation is done only once, and then the points

$$(x^{q^2}, y^{q^2}) + [q \bmod \ell](x, y) \qquad \text{and} \qquad (x^q, y^q)$$

are computed and stored for use in step (4) of Schoof's algorithm.

We now use (a), (b), and (c) to estimate the running time of Schoof's algorithm. From (a), we need to use only primes ℓ that are less than $O(\log q)$. There are $O(\log q / \log \log q)$ such primes, so that is how many times the A-loop, steps (2)–(9), is executed. Then, each time we go through the A-loop, the n loop, steps (3)–(5), is executed $\ell = O(\log q)$ times.

---

[3]A bit operation is a basic computer operation on one or two bits. Examples of bit operations include addition, multiplication, and, or, xor, and complement. Fancier multiplication methods based on fast Fourier transforms or Karatsuba multiplication can be used to reduce multiplication in R_ℓ to $O((\ell^2 \log q)^{1+\epsilon})$ bit operations, at the cost of a larger big-O constant.

Further, since $\ell = O(\log q)$, claim (b) says that basic operations in R_ℓ take $O((\log q)^6)$ bit operations. The value of $[n](x^q, y^q)$ in step (4) can be computed in $O(1)$ operations in R_ℓ from the previous value $[n-1](x^q, y^q)$, or we can be inefficient and compute it in $O(\log n) = O(\log \ell) = O(\log \log q)$ R_ℓ-operations using the double-and-add algorithm (XI.1.1).

Hence the total number of bit operations required by Schoof's algorithm is

$$\overbrace{O(\log q)}^{A \text{ loop}} \cdot \overbrace{O(\log q)}^{n \text{ loop}} \cdot \overbrace{O((\log q)^6)}^{\substack{\text{bit operations per} \\ R_\ell \text{ operation}}} = O((\log q)^8) \text{ bit operations.}$$

This completes the proof that Schoof's algorithm computes $\#E(\mathbb{F}_q)$ in polynomial time. $\qquad\square$

The most time-consuming part of Schoof's algorithm consists of computations in the ring R_ℓ, which is an extension of $\mathbb{F}_q$ of degree $2\ell^2$. So even though the bound for ℓ is linear in $\log q$, if q is reasonably large, then the bound for ℓ and the $\mathbb{F}_q$-dimension of the ring R_ℓ are large.

Example 3.2. Let $q \approx 2^{256}$, which is a typical size used in cryptographic applications. We have

$$\prod_{\ell \le 103} \ell \approx 2^{133.14} > 4\sqrt{q} = 2^{130},$$

so the largest prime ℓ required by Schoof's algorithm is $\ell = 103$. An element of $\mathbb{F}_q[x]/(\psi_\ell(x))$ is represented by an $\mathbb{F}_q$-vector of dimension $103^2 \approx 2^{13.4}$, and each element of $\mathbb{F}_q$ is a 256-bit number, so elements of $\mathbb{F}_q[x]/(\psi_\ell(x))$ are approximately 2^{22} bits, which is more than 16 KB. Although modern computers are quite capable of working with rings whose elements are 16 KB, extensive calculations in such rings take nontrivial amounts of time.

The SEA Algorithm 3.3. Suppose that the ℓ-division polynomial $\psi_\ell(x)$ factors in $\mathbb{F}_q[x]$, say $f_\ell(x) \mid \psi_\ell(x)$ with $\deg f_\ell = \ell$. Then we can obtain significant savings in Schoof's algorithm by working in the smaller ring

$$R'_\ell = \frac{\mathbb{F}_q[x, y]}{(f_\ell(x), y^2 - f(x))}.$$

Multiplication in the ring R'_ℓ takes $O(\ell^2 (\log q)^2)$ bit operations, as compared to $O(\ell^4 (\log q)^2)$ bit operations for multiplication in R_ℓ. For the example described in (XI.3.2), this amounts to a potential speedup on the order of 10^4.

The division polynomial $\psi_\ell(x)$ has an $\mathbb{F}_q[x]$-factor of degree ℓ if and only if the group of ℓ-torsion points $E[\ell]$ has a cyclic subgroup $C \subset E[\ell]$ of order ℓ that is defined over $\mathbb{F}_q$, i.e., such that C is a $G_{\bar{\mathbb{F}}_q/\mathbb{F}_q}$-invariant subgroup. Equivalently, from (III.4.12), $\psi_\ell(x)$ factors in this way if and only if there is an isogeny $E \to E'$ of degree ℓ defined over $\mathbb{F}_q$.

The idea of using factors of $\psi_\ell(x)$ was proposed and developed by Schoof, Elkies, and Atkin and is known as the *SEA algorithm*. Efficient computation of a factor $f_\ell(x)$ uses the modular polynomial [266, II.6.3]. For a description of the SEA algorithm, see for example [22, Chapter 7], [79], or [224].

XI.4   Elliptic Curve Cryptography

Public key cryptography was invented by Diffie and Hellman[4] in 1976 [65], although they were not able to find a practical method to implement their idea. The first practical public key cryptosystem was devised the following year by Rivest, Shamir, and Adleman [209]. The famed RSA cryptosystem bases its security on the difficulty of factoring large numbers. However, Diffie and Hellman did describe a key exchange algorithm whose security relies on the discrete logarithm problem in $\mathbb{F}_q^*$, and subsequently ElGamal created a public key cryptosystem based on the same underlying problem. Koblitz [128] and Miller [176] suggested replacing the finite field $\mathbb{F}_q$ with an elliptic curve E, with the hope that the discrete logarithm problem in the elliptic curve group $E(\mathbb{F}_q)$ might be harder to solve than the discrete logarithm problem in the multiplicative group $\mathbb{F}_q^*$. Their intuition led to the creation of elliptic curve cryptography.

The subject of cryptography is vast, and although cryptography is not the focus of this book, we take the opportunity in this section and in (XI §7) to briefly indicate some of the ways in which elliptic curves are applied. This material is meant merely to whet the reader's appetite, so be aware that we ignore many of the subtleties inherent in the subject. Readers desiring more information on the mathematical aspects of cryptography may consult any of the numerous volumes on the subject, including for example [116], [169], or [277]. For books devoted to the use of elliptic (and hyperelliptic) curves in cryptography, see for example [22] or [51].

Public key cryptosystems rely on what are known as *one-way trapdoor functions*. These are easy-to-compute injective functions $f : A \to B$ with the property that f^{-1} is very hard to compute in general, but f^{-1} becomes quite easy to compute if someone possesses an extra piece of information k. Thus, if Alice[5] knows the value of k, then Bob can send her a message $a \in A$ by sending her the quantity $b = f(a)$. Alice easily recovers $a = f^{-1}(b)$, since she knows k, while Eve, who does not know k, is unable to compute $f^{-1}(b)$.

It is not clear that one-way trapdoor functions exist, and indeed, it is still an open problem to prove their existence. However, a number of hard mathematical problems have been proposed as the bases for one-way trapdoor functions, including in particular the discrete logarithm problem.

Definition. Let G be group, and let $x, y \in G$ be elements such that y is in the subgroup generated by x. The *discrete logarithm problem* (DLP) is the problem of determining an integer $m \geq 1$ such that

$$x^m = y.$$

---

[4]The concept of public key cryptography was actually originally described by James Ellis in 1969, but his discovery was classified by the British government and not declassified until after his death in 1997. Two other British government employees, Williamson and Cocks, are the original inventors of the Diffie–Hellman key exchange algorithm and the RSA public key cryptosystem, respectively, but their discoveries were also classified.

[5]In cryptography it is customary to personalize the participants. Typically, Alice and Bob want to communicate, while Eve, the eavesdropper, intercepts and tries to read their messages.

Example 4.1. Each group G has its own discrete logarithm problem. In the next section we describe a collision algorithm (XI.5.2) that takes $O(\sqrt{\#G})$ steps to solve the DLP in virtually any group G. However, this square root estimate is only an upper bound for the computational complexity of the DLP. It turns out that the DLP is significantly easier in some groups than it is in others. We mention three examples in increasing order of difficulty.

(a) *The Additive Group $\mathbb{F}_q^+$.* The DLP for the additive group of a finite field $\mathbb{F}_q$ asks for a solution m to the linear equation $xm = y$ for given $x, y \in \mathbb{F}_q$. To solve this equation, we need only find the multiplicative inverse of x in $\mathbb{F}_q$, which takes $O(\log q)$ steps using the Euclidean algorithm. Thus the DLP in $\mathbb{F}_q^+$ is a very easy problem.

(b) *The Multiplicative Group $\mathbb{F}_q^*$.* The DLP for the multiplicative group of a finite field $\mathbb{F}_q$ asks for a solution m to the exponential equation $x^m = y$ for given $x, y \in \mathbb{F}_q^*$. As already noted, the DLP in any group of order $O(q)$ can be solved in $O(\sqrt{q})$ steps, but there are algorithms for the DLP in $\mathbb{F}_q^*$ that are much faster, taking fewer than $O(q^\epsilon)$ steps for every $\epsilon > 0$. These go by the general name of *index calculus* methods, and they solve the DLP in $\mathbb{F}_q^*$ in

$$\exp\left(c\sqrt[3]{(\log q)(\log\log q)^2}\right) \text{ steps,}$$

where c is a small absolute constant. Thus the index calculus is a subexponential algorithm. We note that it is not a coincidence that the running time of the number field sieve and the index calculus have the same form, since both rely on the distribution of smooth numbers. For further information about the index calculus, see for example [116, §3.8] or [277, §6.2.4].

(c) *An Elliptic Curve $E(\mathbb{F}_q)$.* The elliptic curve discrete logarithm problem, which is abbreviated ECDLP, asks for a solution m to the equation $[m]P = Q$ for given points $P, Q \in E(\mathbb{F}_q)$. Despite extensive research since the mid-1980s, the fastest known algorithms to solve the ECDLP on general curves are collision algorithms taking $O(\sqrt{q})$ steps. Thus the best known algorithms to solve the ECDLP in $E(\mathbb{F}_q)$ take exponential time, i.e., the running time is exponential in $\log q$. This fact is the primary attraction for using elliptic curves in cryptography.

There is a key exchange system based on the DLP that is due to Diffie and Hellman and a public key cryptosystem based on the DLP that is due to ElGamal. These systems work, mutatis mutandis, for any group and are typically applied to (subgroups) of either $\mathbb{F}_q^*$ or $E(\mathbb{F}_q)$. In keeping with the primary subject of this book, we describe these systems in terms of elliptic curves. As already noted, the primary advantage of using elliptic curves is that at present, it is much harder to solve the ECDLP in $E(\mathbb{F}_q)$ than it is to solve the DLP in $\mathbb{F}_q^*$. This means that *elliptic curve cryptography* has key and message sizes that are 5 to 10 times smaller than those for other systems, including RSA and $\mathbb{F}_q^*$-based DLP systems.

The first algorithm that we describe allows Alice and Bob to securely exchange a piece of information whose value neither one of them knows in advance. We discuss later (XI.4.3.3) why this might be useful.

Diffie–Hellman Key Exchange 4.2. *The following procedure allows Alice and Bob to securely exchange the value of a point on an elliptic curve, although neither of them initially knows the value of the point*:

(1) Alice and Bob agree on a finite field $\mathbb{F}_q$, an elliptic curve $E/\mathbb{F}_q$, and a point $P \in E(\mathbb{F}_q)$.

(2) Alice selects a secret integer a and computes the point $A = [a]P \in E(\mathbb{F}_q)$.

(3) Bob selects a secret integer b and computes the point $B = [b]P \in E(\mathbb{F}_q)$.

(4) Alice and Bob exchange the values of A and B over a possibly insecure communication line.

(5) Alice computes $[a]B$ and Bob computes $[b]A$. They have now shared the value of the point $[ab]P$.

We briefly mention a few of the many issues that must be addressed before this rough outline of Diffie–Hellman key exchange becomes a usable system.

Remark 4.3.1. Typically, the finite field $\mathbb{F}_q$, the elliptic curve $E/\mathbb{F}_q$, and the point $P \in E(\mathbb{F}_q)$ are preselected and published by a standards body. See (XI.4.7).

Remark 4.3.2. It is essential that the order of P be divisible by a large prime, because a Chinese remainder algorithm due to Pohlig and Hellman [205] shows that the solution time of the ECDLP depends only on the largest prime dividing the order of P. For this and other reasons, it is generally advisable to use a point P of prime order. For details of the Pohlig–Hellman algorithm, see [116, § 2.9], [277, §5.1.1], or Exercise 11.9.

Remark 4.3.3. Diffie–Hellman key exchange allows Alice and Bob to exchange a piece of data that neither knows in advance. This may not seem very useful. However, it is useful, because they can use this "random" piece of data as the secret key for a private key cryptosystem such as the advanced encryption standard (AES).

Remark 4.3.4. Alice and Bob's adversary Eve knows the values of P, $A = [a]P$, and $B = [b]P$, so if Eve can solve the ECDLP, then she can find a (or b) and recover Alice and Bob's secret value. However, in principle, Eve does not need to find a or b. What Eve needs to do is to solve the following problem:

> *Elliptic Curve Diffie–Hellman Problem*
>
> Given three points P, $[a]P$, and $[b]P$
> in $E(\mathbb{F}_q)$, compute the point $[ab]P$.

At present, the only way known to solve the elliptic curve Diffie-Hellman problem is to solve the associated elliptic curve discrete logarithm problem, i.e., no one knows how to compute $[ab]P$ from P, $[a]P$, and $[b]P$ without knowing one of a or b.

Remark 4.3.5. There is clearly no need for Alice and Bob to exchange both the x and y coordinates of a point, since the x-coordinate alone determines y up to ± 1. Thus given Bob's x value, Alice can determine $\pm y$ by computing a square root in $\mathbb{F}_q$. (See Exercise 11.8.) It thus suffices for Bob to send the value of x and one additional bit that specifies which square root to take for y. In cryptographic circles, the idea of sending the x-coordinate plus one extra bit is known as *point compression*.

Diffie–Hellman key exchange allows Alice and Bob to exchange a random bit string, but a true public key cryptosystem such as RSA allows Bob to send a specific message to Alice. A public key cryptosystem based on the discrete logarithm problem in $\mathbb{F}_q^*$ was proposed in 1985 by ElGamal [74]. Here is an elliptic curve version.

ElGamal Public Key Cryptosystem 4.4. *The following procedure allows Bob to securely send a message to Alice without any previous communication.*

(1) Alice and Bob agree on a finite field $\mathbb{F}_q$, an elliptic curve $E/\mathbb{F}_q$, and a point $P \in E(\mathbb{F}_q)$.

(2) Alice selects a secret integer a and computes the point $A = [a]P \in E(\mathbb{F}_q)$.

(3) Alice publishes the point A. This is her *public key*. The secret multiplier a is her *private key*.

(4) Bob chooses a *plaintext* (i.e., a message) $M \in E(\mathbb{F}_q)$ and a random integer k. He computes the two points

$$B_1 = [k]P \in E(\mathbb{F}_q) \qquad \text{and} \qquad B_2 = M + [k]A \in E(\mathbb{F}_q).$$

(5) Bob sends the *ciphertext* (B_1, B_2) to Alice over a potentially insecure communication line.

(6) Alice uses her secret key a to compute $B_2 - [a]B_1 \in E(\mathbb{F}_q)$. This value is equal to Bob's plaintext M.

There is much to say about the ElGamal cryptosystem, but we content ourselves with a few brief remarks.

Remark 4.5.1. It is easy to verify that Alice recovers Bob's plaintext. Thus

$$B_2 - [a]B_1 = \big(M + [k]A\big) - [a][k]P = M + [k][a]P - [a][k]P = M.$$

Notice how Bob's random integer k disappears from the calculation.

Remark 4.5.2. Just as with Diffie–Hellman key exchange, the field $\mathbb{F}_q$, curve $E/\mathbb{F}_q$, and point $P \in E(\mathbb{F}_q)$ are typically chosen from a list published by some trusted authority; see (XI.4.7). We also note that Alice may choose her private key (step 2) and publish her public key (step 3) without knowing who is planning to send messages to her, nor when those messages will be sent.

Remark 4.5.3. An ElGamal plaintext is a point $M \in E(\mathbb{F}_q)$, while an ElGamal ciphertext is a pair of points $B_1, B_2 \in E(\mathbb{F}_q)$. Thus even with point compression (XI.4.3.5), Bob has to send two bits of information to Alice for every one bit in his message. We say that ElGamal has 2-*to*-1 *message expansion*. This is less efficient than the RSA cryptosystem, whose plaintexts and ciphertexts are the same size.

Remark 4.5.4. In practice, there is no natural way to assign a message written in, say, English, to a point $M \in E(\mathbb{F}_p)$. A variant of the ElGamal system due to Menezes and Vanstone uses the coordinates of a point in E as a mask for the actual message; see Exercise 11.10. We also note that if ElGamal is used in the raw state described in (XI.4.4), then it is subject to various sorts of attacks. All practical secure implementations of modern public key cryptosystems include some sort of internal message structure that allows Alice to verify that Bob's message was properly encrypted.

An example of such a method is the Integrated Encryption Scheme (IES) due to Abdalla, Bellare, and Rogaway [1]. We briefly describe the elliptic curve variant of IES in Exercise 11.11.

Remark 4.5.5. As with Diffie–Hellman key exchange, the ElGamal cryptosystem can be broken by solving the Diffie–Hellman problem (XI.4.3.4). Thus Eve knows $A = [a]P$ and $B_1 = [k]P$, so if she can solve the Diffie–Hellman problem, then she can compute $[ak]P = [k]A$. Since she also knows B_2, she is then able to compute $B_2 - [k]A = M$.

A public key cryptosystem allows Bob and Alice to exchange information. A *digital signature scheme* has a different purpose. It allows Alice to use a private key to sign a digital document, e.g., a computer file, in such a way that Bob can use Alice's public key to verify the validity of the signature. There are a number of practical digital signature algorithms; see for example [116, 169, 277]. We describe one such algorithm that uses elliptic curves.

Elliptic Curve Digital Signature Algorithm (ECDSA) 4.6. *The following procedure allows Alice to sign a digital document and Bob to verify that the signature is valid*:

(1) Alice and Bob agree on a finite field $\mathbb{F}_p$, an elliptic curve $E/\mathbb{F}_p$, and a point $P \in E(\mathbb{F}_p)$ of (prime) order N.

(2) Alice selects a secret integer a and computes the point $A = [a]P \in E(\mathbb{F}_p)$.

(3) Alice publishes the point A. This is her *public verification key*. The secret multiplier a is her *private signing key*.

(4) Alice chooses a digital document $d \bmod N$ to sign.[6] She also chooses a random integer $k \bmod N$. Alice computes $[k]P$ and sets

$$s_1 \equiv x([k]P) \pmod{N} \qquad \text{and} \qquad s_2 \equiv (d + as_1)k^{-1} \pmod{N}.$$

(Here $x([k]P)$ is only in $\mathbb{F}_p$, but we choose an integer representative between 0 and $p - 1$.) Alice publishes the signature (s_1, s_2) for the document d.

(5) Bob computes

$$v_1 \equiv ds_2^{-1} \pmod{N} \qquad \text{and} \qquad v_2 \equiv s_1 s_2^{-1} \pmod{N}.$$

He then computes $[v_1]P + [v_2]A \in E(\mathbb{F}_p)$ and verifies that

$$x([v_1]P + [v_2]A) \equiv s_1 \pmod{N}.$$

PROOF. We need to check that if Alice follows the procedure described in step (4), then Bob's verification in step (5) works. The point that Bob computes in step (5) is

---

[6]In practice, Alice applies a *hash function* to her actual document in order to obtain an integer $d \bmod N$. This allows her to sign long documents and prevents various types of attacks. For information about hash functions and their use in cryptography, see for example [51, §§1.6.3, 24.2.5], [116, §8.1], or [169, Chapter 9].

$$[v_1]P + [v_2]A = [ds_2^{-1}]P + [s_1 s_2^{-1}][a]P \quad \text{using the values of } s_1, s_2, \text{ and } A,$$
$$= [s_2^{-1}(d + as_1)]P$$
$$= [k]P \quad \text{using the value of } s_2.$$

Hence

$$x([v_1]P + [v_2]A) = x([k]P) \equiv s_1 \pmod{N}$$

by definition of s_1. $\qquad\qquad\square$

Remark 4.7. Before using elliptic curves to exchange keys or messages or to sign documents, Alice and Bob need to choose a finite field $\mathbb{F}_q$, an elliptic curve $E/\mathbb{F}_q$, and a point $P \in E(\mathbb{F}_q)$ having large prime order. This selection process can be time consuming, but there is no need for every Alice and every Bob to choose their own individual fields, curves, and points. The only secret personal information utilized by Alice and Bob consists of the multipliers they use to form multiples of P. In order to make Alice and Bob's life easier, the United States National Institute of Standards and Technology (NIST) published a list [191] of fifteen fields, curves, and points for Alice and Bob to use. For each of five different security levels, NIST gives one curve $E/\mathbb{F}_p$ with p a large prime, one curve $E/\mathbb{F}_{2^k}$, and one Koblitz curve $E/\mathbb{F}_2$ as in (XI.1.5) with a point $P \in E(\mathbb{F}_{2^k})$.

XI.5 Solving the Elliptic Curve Discrete Logarithm Problem: The General Case

Recall that the discrete logarithm problem (DLP) for elements x and y in a group G asks for an integer m such that $x^m = y$. In this section we discuss the best known algorithms for solving the DLP in arbitrary groups. The following rough criteria will be used to describe the complexity of an algorithm.

Definition. We will say that a discrete logarithm algorithm *takes T steps and requires S storage* if, on average, the algorithm needs to compute T group operations and needs to store the values of S group elements. (We will ignore the time it takes to sort or compare lists of elements, since the time for such operations is generally logarithmically small compared to the number of group operations.)

Example 5.1. Let $x \in G$ be an element of order n. The *naive algorithm* for solving the DLP is to compute $x, x^2, x^3, \ldots$ until y is found. This method takes $O(n)$ steps and requires $O(1)$ storage.

We now describe a general algorithm that "square-roots" the number of steps required to solve the DLP compared to the naive algorithm, albeit at the cost of using a significant amount of storage. Algorithms of this type are called *collision algorithms*, because they depend on the fact that it is easier to find collisions (elements that are common to two subsets) than it is to find specific elements in a set. This phenomenon is also known as the *birthday paradox*; see Exercise 11.5.

Proposition 5.2. (Shanks's Babystep–Giantstep Algorithm) *Let G be a group, let $x, y \in G$, and let n be the order of x. Then the following algorithm solves the DLP in $O(\sqrt{n})$ steps with $O(\sqrt{n})$ storage:*

(1) Let $N = \lceil \sqrt{n} \rceil$ be the "ceiling" of n, i.e., N is the smallest integer that is greater than or equal to $\sqrt{n}$.

(2) Make a list of the elements (these are the babysteps)

$$x, \, x^2, \, x^3, \, \ldots, \, x^N.$$

(3) Let $z = (x^N)^{-1}$ and make a list of the elements (these are the giantsteps)

$$yz, \, yz^2, \, yz^3, \, \ldots, \, yz^N.$$

(4) Look for a match between the lists in steps (2) and (3). If there is a match, say $x^i = yz^j$, then $y = x^{i+jN}$; otherwise y is not a power of x.

PROOF. Suppose that y is equal to a power of x, say $y = x^m$ with $0 \le m < n$. We write $m = jN + i$ with $0 \le i < N$, so

$$0 \le j = (m - i)/N \le N, \qquad \text{since } m \le n \text{ and } N \ge \sqrt{n}.$$

It follows that x^i is in the first list and $yz^j = yx^{-jN}$ is in the second list, so there is a match $x^i = yz^j$. Hence $y = x^i z^{-j} = x^i (x^{-N})^{-j} = x^{i+jN}$.

We note that there are many ways to check for matches in step (4). For example, we can sort the elements $x, x^2, \ldots, x^N$ in step (2) in $O(N \log N)$ steps, and then it takes $O(\log N)$ steps to check whether any particular element yz^i from step (3) is in the list. So Shanks's algorithm really takes $O(\sqrt{n}(\log n)^2)$ steps (or a bit less, using fancier sorting algorithms), but as noted earlier, we will ignore the log factors. $\square$

The babystep–giantstep algorithm (XI.5) requires a considerable amount of storage. An alternative collision algorithm, due to Pollard, takes approximately the same number of steps and reduces the storage to essentially nothing. Pollard's algorithm and its variants, which are the most practical methods currently known for solving the ECDLP, rely on the following collision theorem that describes the likelihood of finding terms satisfying $x_{2i} = x_i$ in an iterative sequence $x_0, x_1, x_2, \ldots$.

Theorem 5.3. *Let S be a finite set containing N elements, and let $f : S \to S$ be a function. Starting with an initial value $x_0 \in S$, define a sequence of points $x_0, x_1, x_2, \ldots$ by*

$$x_i = f(x_{i-1}) = \underbrace{f \circ f \circ \cdots \circ f}_{i \text{ iterations of } f}(x_0).$$

Let T be the tail length and let L be the loop length of the orbit

$$x_0, x_1, x_2, x_3, \ldots$$

of x, as illustrated in Figure 11.5. *Formally,*

$T = $ *largest integer such that x_{T-1} appears only once in the sequence $(x_i)_{i \ge 0}$,*

$L = $ *smallest integer such that $x_{T+L} = x_T$.*

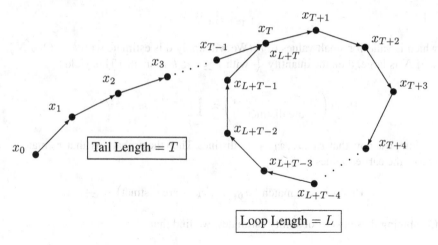

Figure 11.5: The orbit of x_0 in Pollard's ρ algorithm.

(a) *There exists an index $1 \le i < T + L$ such that $x_{2i} = x_i$.*

(b) *If $f : S \to S$ and its iterates are "sufficiently random" at mixing the elements of S, then the expected value of $T + L$ is $\sqrt{\pi N / 2}$.*

Remark 5.3.1. The shape of the path in Figure 11.5 explains why (XI.5.3) is called the "ρ algorithm."

PROOF. (a) It is clear from Figure 11.5 that for $j > i$ we have

$$x_j = x_i \quad \text{if and only if} \quad i \ge T \quad \text{and} \quad j \equiv i \;(\mathrm{mod}\; L).$$

Hence $x_{2i} = x_i$ if and only if $i \ge T$ and $L \mid i$. The first such i lies between T and $T + L - 1$.

(b) We sketch the proof, which is an exercise in discrete probability theory, and leave the reader to fill in error-estimate details.

If k points $x_0, \ldots, x_{k-1}$ are chosen randomly from S, then the probability that they are distinct is

$$\mathrm{Prob}\begin{pmatrix} x_0, x_1, \ldots, x_{k-1} \\ \text{are distinct} \end{pmatrix} = \prod_{i=1}^{k-1} \mathrm{Prob}\begin{pmatrix} x_i \ne x_j \text{ for} & x_0, x_1, \ldots, x_{i-1} \\ \text{all } 0 \le j < i & \text{are distinct} \end{pmatrix}$$

$$= \prod_{i=1}^{k-1} \left(\frac{N - i}{N} \right)$$

$$= \prod_{i=1}^{k-1} \left(1 - \frac{i}{N} \right).$$

We approximate this last product using the estimate

$$1 - t \approx e^{-t},$$

which is valid for small values of t. (We will apply this estimate with $k = O(\sqrt{N})$, so if N is large, then the quantity $\frac{i}{N}$ with $1 \le i < k$ is small.) This yields

$$\text{Prob}\left(\begin{matrix} x_0, x_1, \ldots, x_{k-1} \\ \text{are distinct} \end{matrix}\right) \approx \prod_{i=1}^{k-1} e^{-i/N} \approx e^{-k^2/2N}.$$

Suppose now that $x_0, \ldots, x_{k-1}$ are distinct. Then the probability that x_k matches one of the earlier values is

$$\text{Prob}\big(x_k \text{ is a match} \mid x_0, \ldots, x_{k-1} \text{ are distinct}\big) = \frac{k}{N}.$$

Combining these two probability estimates, we find that

$$\text{Prob}\big(x_k \text{ is the first match}\big)$$
$$= \text{Prob}\big(x_k \text{ is a match AND } x_0, \ldots, x_{k-1} \text{ are distinct}\big)$$
$$= \text{Prob}\big(x_k \text{ is a match} \mid x_0, \ldots, x_{k-1} \text{ are distinct}\big)$$
$$\qquad\qquad\qquad\qquad\qquad\qquad \cdot \text{Prob}\big(x_0, \ldots, x_{k-1} \text{ are distinct}\big)$$
$$\approx \frac{k}{N} \cdot e^{-k^2/2N}.$$

Hence the expected number of steps before finding the first match is

$$\sum_{k \ge 1} k \cdot \text{Prob}\big(x_k \text{ is the first match}\big)$$

$$\approx \sum_{k \ge 1} \frac{k^2}{N} \cdot e^{-k^2/2N}$$

$$= \sum_{k \ge 1} \phi(k/\sqrt{N}) \qquad\qquad \text{letting } \phi(t) = t^2 e^{-t^2/2},$$

$$\approx \sqrt{N} \cdot \int_0^\infty t^2 e^{-t^2/2} \, dt \qquad \text{using } \sum_{k=1}^\infty \phi(k/n) \approx n \int_0^\infty \phi(t) \, dt,$$

$$= \sqrt{\pi N/2}.$$

(The square of the integral in the last step may be evaluated via the usual polar coordinates trick.) □

Pollard's algorithm to solve the discrete logarithm problem in a group G uses a self-map $f : G \to G$ that is easy to compute, yet whose iterates mix up the elements of G in a random fashion.

Algorithm 5.4. (Pollard's ρ algorithm)
Let G be a group and let $x, y \in G$. Our goal is to compute an integer m satisfying $x^m = y$. We will use (XI.5.3) to find integers i, j, k, ℓ such that

$$x^i y^j = x^k y^\ell.$$

Then $x^{i-k} = y^{j-\ell}$, and assuming that $j - \ell$ is relatively prime to the order n of x,[7] we can solve for y as a power of x.

It is not clear how to define a function $f : G \to G$ that is complicated enough to provide good mixing, yet simple enough to keep track of its orbits. Pollard [206] suggests splitting G into a disjoint union of three sets of approximately equal size,

$$G = A \cup B \cup C,$$

and using the function

$$f(z) = \begin{cases} xz & \text{if } z \in A, \\ z^2 & \text{if } z \in B, \\ yz & \text{if } z \in C. \end{cases}$$

(See Exercise 11.13 for an elliptic curve example.) Such functions work reasonably well in practice, although more complicated functions with better mixing properties are known [293, 294].

Consider the outcome when we repeatedly apply f to the initial point $z_0 = 1$. After i iterations we arrive at a point

$$z_i = \underbrace{f \circ f \circ \cdots \circ f}_{i \text{ iterations of } f}(1) = x^{\alpha_i} y^{\beta_i}$$

for certain integers α_i and β_i. It is difficult to predict, a priori, the values of α_i and β_i, but we can compute them at the same time that we compute $z_1, z_2, \ldots$ by starting with $\alpha_0 = \beta_0 = 0$ and using the iterative formulas

$$\alpha_{i+1} = \begin{cases} \alpha_i + 1 & \text{if } z_i \in A, \\ 2\alpha_i & \text{if } z_i \in B, \\ \alpha_i & \text{if } z_i \in C, \end{cases} \qquad \beta_{i+1} = \begin{cases} \beta_i & \text{if } z_i \in A, \\ 2\beta_i & \text{if } z_i \in B, \\ \beta_i + 1 & \text{if } z_i \in C. \end{cases}$$

Note that we need only keep track of α_i and β_i modulo n, since $x^n = 1$. This keeps the values of α_i and β_i at a manageable size.

In a similar fashion we compute the sequence of points

$$w_0 = 1 \qquad \text{and} \qquad w_{i+1} = f\big(f(w_i)\big).$$

Then

$$w_i = z_{2i} = x^{\gamma_i} y^{\delta_i},$$

where γ_i and δ_i can be computed from γ_{i-1} and δ_{i-1} using two repetitions of the recurrences for α_i and β_i. Of course, the first time we use $w_i = z_{2i}$ to determine which case to apply, and the second time we use $f(w_i) = z_{2i+1}$ to decide. See Exercise 11.12.

---

[7]In practical applications, the element x usually has prime order (XI.4.3.2), in which case $j - \ell$ is almost certainly prime to n. The general case is discussed later in this section.

We now compute (z_1, w_1), (z_2, w_2), (z_3, w_3),... until we find a pair whose coordinates are the same. Note that each successive (z_i, w_i) may be computed solely in terms of the previous (z_{i-1}, w_{i-1}), so we never need to store more than a few numbers. Assuming that A, B, and C are sufficiently good at mixing the elements of G, our analysis in (XI.5.3) says that we will find a match

$$z_i = w_i = z_{2i}$$

in $O(\sqrt{n})$ steps. The equality $z_i = w_i$ implies that

$$x^{\alpha_i - \gamma_i} = y^{\delta_i - \beta_i}.$$

If $\gcd(\delta_i - \beta_i, n) = 1$, as is typically the case in applications where n is prime, then

$$m \equiv (\alpha_i - \gamma_i)(\delta_i - \beta_i)^{-1} \pmod{n}$$

solves $x^m = y$.

In general, if $d = \gcd(\delta_i - \beta_i, n) > 1$, then we can express y^d as a power of x, say $y^d = x^e$. Then y is equal to one of the elements $x^{(e+nu)/d}$ with $0 \le u < d$. So if d is not too large, this solves the DLP, and if d is large, we can rerun Pollard's algorithm to find another relation between x and y.

Remark 5.4.1. The proof of (XI.5.3) and its subsequent application to Pollard's ρ algorithm (XI.5.4) rely on two heuristic assumptions. First, they assume that iterations of certain self-maps behave as if they were random mixing maps. Second, they assume that the resulting collision is nondegenerate, e.g., in the notation of (XI.5.4), if $n \mid \delta_i - \beta_i$, then the algorithm yields no information. For a proof that the running time of Pollard's algorithm is $O(\sqrt{n})$, see [126, 174], and for a rigorous analysis of the nondegeneracy assumption, see [175].

Remark 5.4.2. Shanks's and Pollard's algorithms (XI.5.2), (XI.5.4) apply to (virtually) any group and show that the DLP in a cyclic group G of order n can be solved in $O(\sqrt{n})$ steps. Now imagine that you are given a black box that performs the group operations in G. This means that you may feed any two group elements x_1 and x_2 into the box and it will compute for you the value of their product $x_1 x_2$, but you have no knowledge of how the computation is performed. In this situation Shoup [253] has shown that any algorithm that solves the DLP in G takes on average at least $O(\sqrt{n})$ steps. Thus despite the fact that the group law on an elliptic curve is far from being a black box, the best known algorithms to solve the ECDLP are qualitatively no better than a black box algorithm.

XI.6 Solving the Elliptic Curve Discrete Logarithm Problem: Special Cases

The fastest known algorithms that solve the ECDLP on all elliptic curves are collision algorithms such as (XI.4.1) and (XI.4.2). However, not all elliptic curves are created equal. Menezes, Okamoto, and Vanstone [168] suggested using the Weil pairing to reduce the ECDLP to an easier DLP in the multiplicative group of a finite field. (An alternative reduction using the Tate–Lichtenbaum pairing was suggested by Frey and Rück [91].)

Definition. Let $\mathbb{F}_q$ be a finite field, and let $N \geq 1$ be an integer. The *embedding degree of N in* $\mathbb{F}_q$ is the smallest integer $d \geq 1$ such that

$$\mu_N \subset \mathbb{F}_{q^d}^*.$$

Since $\mathbb{F}_{q^d}^*$ is a cyclic group of order $q^d - 1$, this is equivalent to d being the smallest integer satisfying

$$q^d \equiv 1 \pmod{N}.$$

Proposition 6.1. (MOV Algorithm [168]) *Let $E/\mathbb{F}_q$ be an elliptic curve, let $P, Q \in E(\mathbb{F}_q)$ be points of prime order N, and let d be the embedding degree of N in $\mathbb{F}_q$. Assume that $\gcd(q - 1, N) = 1$. Then there is a polynomial-time algorithm that reduces the ECDLP for P and Q to the DLP in $\mathbb{F}_{q^d}^*$.*

PROOF. We are looking for an integer m such that $Q = [m]P$. We choose a point $T \in E[N](\bar{\mathbb{F}}_q)$ such that P and T generate $E[N]$. Then the value of the Weil pairing $e_N(P, T)$ is a primitive N^{th} root of unity (III.8.1.1), so by definition of embedding degree we have $e_N(P, T) \in \mathbb{F}_{q^d}^*$. Linearity of the Weil pairing (III.8.1) gives

$$e_N(Q, T) = e_N([m]P, T) = e_N(P, T)^m.$$

We know the values of points P, Q, and T, so if we can solve the discrete logarithm problem

$$e_N(Q, T) = e_N(P, T)^m$$

in $\mathbb{F}_{q^d}^*$, then we recover the value of m, which also solves the ECDLP for P and Q.

The running time of the MOV algorithm is determined by how long it takes to find the point T and how long it takes to compute the Weil pairing values $e_N(Q, T)$ and $e_N(P, T)$. The Weil pairing computations are not a problem, since they may be computed using Miller's (linear-time) algorithm as described in (XI §8). A key fact, which we prove below (XI.6.2), is that $E[N] \subset E(\mathbb{F}_{q^d})$, where d is the embedding degree. Thus all computations may be done in the field $\mathbb{F}_{q^d}$. To construct an appropriate point $T \in E(\mathbb{F}_{q^d})$, we randomly choose points T in $E(\mathbb{F}_{q^d})$ of order N until we find one such that $e_N(P, T)$ is a primitive N^{th} root of unity.[8] $\square$

Lemma 6.2. *Let $E/\mathbb{F}_q$ be an elliptic curve, let $N \geq 1$ be an integer satisfying $\gcd(q-1, N) = 1$, let d be the embedding degree of N in $\mathbb{F}_q$, and suppose that $E(\mathbb{F}_q)$ contains a point of exact order N. Then $E[N] \subset E(\mathbb{F}_{q^d})$.*

---

[8]We first compute $n = \#E(\mathbb{F}_{q^d})$, which takes polynomial time (XI.3.1). Then, since N is typically a large prime, most points $S \in E(\mathbb{F}_{q^d})$ have the property that $T = [n/N^2]S$ has order N.

PROOF. Let $P \in E(\mathbb{F}_q)$ be the given point of exact order N defined over $\mathbb{F}_q$, and choose a point $T \in E[N]$ such that $\{P, T\}$ is a basis for $E[N]$. Let $\phi \in G_{\bar{\mathbb{F}}_q/\mathbb{F}_q}$ be the q-power Frobenius map. Since $P \in E(\mathbb{F}_q)$, we have

$$P^\phi = P \quad \text{and} \quad T^\phi = [a]P + [b]T \quad \text{for some } a, b \in \mathbb{Z}/N\mathbb{Z}.$$

Using basic properties of the Weil pairing (III.8.1), we find that

$$e_N(P,T)^q = e_N(P,T)^\phi = e_N(P^\phi, T^\phi)$$
$$= e_N(P, [a]P + [b]T) = e_N(P,P)^a e_N(P,T)^b = e_N(P,T)^b.$$

Since $e_N(P,T)$ is a primitive N^{th} root of unity, cf. (III.8.1.1), this implies that

$$b = q \bmod N.$$

Thus $T^\phi = [a]P + [q]T$. Applying ϕ repeatedly to T and using the fact that ϕ fixes P gives

$$T^{\phi^d} = \left[a(1 + q + q^2 + \cdots + q^{d-1})\right]P + [q^d]T.$$

By definition of embedding degree, we have $q^d \equiv 1 \pmod N$, so $[q^d]T = T$. Further, the assumption that $\gcd(q-1, N) = 1$ implies that

$$1 + q + q^2 + \cdots + q^{d-1} \equiv 0 \pmod N,$$

so $[1 + q + q^2 + \cdots + q^{d-1}]P = O$. Therefore $T^{\phi^d} = T$, which proves that $T \in E(\mathbb{F}_{q^d})$. $\qquad\square$

Remark 6.3. Under plausible assumptions, Balasubramanian and Koblitz [13] show that for most elliptic curves $E/\mathbb{F}_q$, if N is a large prime divisor of $\#E(\mathbb{F}_q)$, then the embedding degree of N in $\mathbb{F}_q$ is proportional to N. Hence for a randomly chosen $E/\mathbb{F}_q$, the MOV algorithm reduces the ECDLP in $E(\mathbb{F}_q)$ to a much harder DLP in $\mathbb{F}_{q^d}$. However, there are special cases for which the embedding degree is small, as in the following example. (See also (XI.9.8).)

Example 6.4. Let $p \geq 5$ be prime, and let $E/\mathbb{F}_p$ be a supersingular elliptic curve. We can compute embedding degrees for E using Exercise 5.15, which implies that

$$\#E(\mathbb{F}_p) = p + 1.$$

Suppose that $P \in E(\mathbb{F}_p)$ is a point of exact order N. Then N divides $\#E(\mathbb{F}_p)$, so $p \equiv -1 \pmod N$. Hence $p^2 \equiv 1 \pmod N$, so N has embedding degree 2 in $\mathbb{F}_p$. Thus the ECDLP on a supersingular curve over $\mathbb{F}_p$ can be reduced to solving the DLP in $\mathbb{F}_{p^2}^*$, for which there are subexponential algorithms. This militates against using supersingular curves in most cryptographic settings. However, we will see later (XI §7) that there are cryptographic applications that make use of the Weil pairing and low embedding degrees. For these applications, supersingular curves may be used; we simply must ensure that it is computationally infeasible to solve the associated DLP in $\mathbb{F}_{p^2}^*$.

The MOV algorithm (XI.6.1) shows that the ECDLP on an elliptic curve with low embedding degree may be reduced to a potentially easier DLP in the multiplicative group of a finite field. We next describe a special situation in which the ECDLP is reduced to an essentially trivial additive DLP

Proposition 6.5. (Semaev [228], Satoh–Araki [218], Smart [269]) *Let $p \geq 3$ and let $E/\mathbb{F}_p$ be an elliptic curve satisfying*

$$\#E(\mathbb{F}_p) = p.$$

(Such curves are called anomalous.*) The following algorithm solves the ECDLP in $E(\mathbb{F}_p)$.*

(1) Let $P, Q \in E(\mathbb{F}_p)$ be nonzero points satisfying $Q = [m]P$, where the integer m is not known.

(2) Choose an elliptic curve $E'/\mathbb{Q}_p$ whose reduction modulo p is $E/\mathbb{F}_p$.

(3) Use Hensel's lemma to lift the points P, Q to points $P', Q' \in E'(\mathbb{Q}_p)$.

(4) The points $[p]P'$ and $[p]Q'$ are in the formal group $E'_1(\mathbb{Q}_p)$. Let

$$\log_E : E'_1(\mathbb{Q}_p) \longrightarrow \hat{\mathbb{G}}_a(p\mathbb{Z}_p) \cong p\mathbb{Z}_p^+$$

be the formal logarithm map (IV §5, IV.6.4), and compute

$$pa = \log_E\big([p]P'\big) \in p\mathbb{Z}_p \quad \text{and} \quad pb = \log_E\big([p]Q'\big) \in p\mathbb{Z}_p.$$

(5) Then $m \equiv a^{-1}b \pmod{p}$.

PROOF. Using the fact that $\#E(\mathbb{F}_p) = p$, we have

$$\widetilde{[p]P'} = [p]P = O \quad \text{and} \quad \widetilde{[p]Q'} = [p]Q = O \qquad \text{in } E(\mathbb{F}_p),$$

so $[p]P'$ and $[p]Q'$ are in the kernel of reduction modulo p. Hence they are in the formal group $E'_1(\mathbb{Q}_p)$. Similarly, if we let $R' = Q' - [m]P'$, then the reduction of R' modulo p is

$$\tilde{R}' = \tilde{Q}' - [m]\tilde{P}' = Q - [m]P = O \qquad \text{in } E(\mathbb{F}_p),$$

so $R' \in E'_1(\mathbb{Q}_p)$. We now compute

$$
\begin{aligned}
\log_E\big([p]Q'\big) &= \log_E\big([p]([m]P' + R')\big) && \text{since } R' = Q' - [m]P', \\
&= m\log_E\big([p]P'\big) + p\log_E(R') && \text{valid since } [p]P', R' \in E_1(\mathbb{Q}_p), \\
&\equiv m\log_E\big([p]P'\big) \pmod{p^2} && \text{since } \log_E(R') \in p\mathbb{Z}_p.
\end{aligned}
$$

Substituting $\log_E\big([p]P'\big) = pa$ and $\log_E\big([p]Q'\big) = pb$ as in step (4) of the algorithm gives $pb \equiv mpa \pmod{p^2}$, so $m \equiv a^{-1}b \pmod{p}$. $\square$

Remark 6.6. The algorithm described in (XI.6.5) may seem impractical, since it requires lifting points in $E(\mathbb{F}_p)$ to points in $E'(\mathbb{Q}_p)$. However, an examination of the proof shows that we need only lift points modulo p^2 and compute formal logarithms in $\hat{\mathbb{G}}_a(p\mathbb{Z}_p)/\hat{\mathbb{G}}_a(p^2\mathbb{Z}_p) \cong p\mathbb{Z}_p/p^2\mathbb{Z}_p$. Thus we work on an elliptic curve over the ring $\mathbb{Z}/p^2\mathbb{Z}$, or in fancier language, on an elliptic scheme; see (XI.2.4.1).

Example 6.7. We work over the field $\mathbb{F}_{127}$ and consider the supersingular curve and points

$$E : y^2 = x^3 + 19x + 112, \quad P = (106, 72) \in E(\mathbb{F}_{127}), \quad Q = (12, 121) \in E(\mathbb{F}_{127}).$$

We take the same equation to be our lift of E to $\mathbb{Z}/127^2\mathbb{Z}$, and we lift the points P and Q to

$$P' = (106, 13026) \in E(\mathbb{Z}/127^2\mathbb{Z}) \quad \text{and} \quad Q' = (12, 5201) \in E(\mathbb{Z}/127^2\mathbb{Z}).$$

We now want to compute multiples of P' and Q' working modulo 127^2. In order to avoid noninvertible denominators, it is convenient to make the change of variables $z = -x/y$ and $w = -1/y$. This has the effect of moving O to $(z, w) = (0, 0)$; cf. (IV §1). The equation for E now reads

$$E : w = z^3 + 19zw^2 + 112w^3.$$

We use the double-and-add algorithm, working modulo 127^2 with (z, w)-coordinates, to compute

$$[127]P' = (12319, 0) \in E(\mathbb{Z}/127^2\mathbb{Z}) \quad \text{and} \quad [127]Q' = (2159, 0) \in E(\mathbb{Z}/127^2\mathbb{Z}).$$

The elliptic logarithm for $y^2 = x^3 + Ax + B$ starts $\log_E(z) = z + \frac{2}{5}Az^5 + \cdots$, so since we are working modulo 127^2, it suffices to use $\log_E(z) \approx z$. Thus

$$\log_E\big([127]P'\big) \equiv 12319 \equiv 97 \cdot 127 \pmod{127^2},$$
$$\log_E\big([127]Q'\big) \equiv 2159 \equiv 17 \cdot 127 \pmod{127^2}.$$

Finally, we compute $m \equiv 97^{-1} \cdot 17 \equiv 46 \pmod{127}$, which is the desired discrete logarithm. We can check our answer by verifying that $[46]P = Q$ in $E(\mathbb{F}_{127})$.

XI.7   Pairing-Based Cryptography

The Diffie–Hellman key exchange algorithm allows two people to exchange an unspecified piece of data. It was a long-standing problem to find a method that allows three people to perform a similar exchange. Joux found a solution using the Weil pairing

$$e_N : E[N] \times E[N] \longrightarrow \mu_N,$$

which we recall (III §8) is nondegenerate, bilinear, and *alternating*. For cryptographic applications we need a pairing that is nondegenerate on a cyclic subgroup of $E[N]$, but unfortunately the Weil pairing is trivial on such subgroups. One way around this difficulty is to use a curve that admits an isogeny

$$\phi : E \longrightarrow E$$

such that $E[N]$ has a basis of the form $\{T, \phi(T)\}$. In the cryptographic literature, the map ϕ is called a *distortion map*; see [51, §24.2.1b] or [116, §5.9.2]. Using a distortion map, the modified Weil pairing

$$\langle \cdot, \cdot \rangle : E[N] \times E[N] \longrightarrow \mu_N, \qquad \langle P, Q \rangle = e_N\big(P, \phi(Q)\big)$$

has the property that $\langle T, T \rangle$ is a primitive N^{th} root of unity.

Example 7.1. Let E be the elliptic curve $y^2 = x^3 + x$ having complex multiplication by $\mathbb{Z}[i]$, and let ϕ be the isogeny

$$\phi : E \longrightarrow E, \qquad \phi(x, y) = [i](x, y) = (-x, iy).$$

Then ϕ is a distortion map on $E[N]$ for all integers N satisfying $N \equiv 3 \pmod 4$. To see this, let $T \in E[N]$ be a point of exact order N, and suppose that some linear combination of T and $\phi(T)$ is zero. Then

$$\begin{aligned}
[a]T + [b]\phi(T) = O \quad &\Longleftrightarrow \quad [a + bi](T) = O \\
&\Longrightarrow \quad [a^2 + b^2](T) = O \\
&\Longrightarrow \quad a^2 + b^2 \equiv 0 \pmod N \\
&\Longrightarrow \quad a \equiv b \equiv 0 \pmod N,
\end{aligned}$$

where the last line follows from the assumption that $N \equiv 3 \pmod 4$. (See Exercise 3.26 for another example of a distortion map.)

An alternative to the modified Weil pairing is the Tate–Lichtenbaum pairing

$$\tau : E(K)/NE(K) \times E(K)[N] \longrightarrow K^*/(K^*)^N,$$

which we discuss in (XI §9). If K is a finite (or local) field, then under appropriate conditions the Tate–Lichtenbaum pairing is nondegenerate, in which case we can define a nondegenerate pairing by

$$\langle \cdot, \cdot \rangle : E(\mathbb{F}_q)[N] \times E(\mathbb{F}_q)[N] \longrightarrow \mu_N, \qquad \langle P, Q \rangle = \tau(P, Q)^{(q-1)/N}.$$

From a practical perspective, the primary advantage of the Tate–Lichtenbaum pairing over the Weil pairing is that the former can be computed in roughly half the time that it takes to compute the latter (XI.9.3.2). In any case, there is a double-and-add algorithm for both pairings, so they are both easy to compute; see (XI §8) for details.

Tripartite Diffie–Hellman Key Exchange 7.2. (Joux [120]) *The following procedure allows Alice, Bob, and Carl to securely exchange a a piece of information whose value none of them knows in advance*:

(1) Alice, Bob, and Carl agree on a finite field $\mathbb{F}_q$, an elliptic curve $E/\mathbb{F}_q$, a prime N, and a point $T \in E(\mathbb{F}_q)[N]$ such that there is a bilinear pairing

$$\langle \cdot, \cdot \rangle : E(\mathbb{F}_q)[N] \times E(\mathbb{F}_q)[N] \longrightarrow \mu_N$$

with the property that $\langle T, T \rangle$ is a primitive N^{th} root of unity.

(2) Alice, Bob, and Carl choose secret integers a, b, and c, respectively, and they compute

Alice computes this	Bob computes this	Carl computes this
$A = [a]T,$	$B = [b]T,$	$C = [c]T.$

(3) Alice, Bob, and Carl publish the values of A, B, and C.

(4) Alice, Bob, and Carl compute, respectively,

Alice computes this   Bob computes this   Carl computes this

$$\langle B, C \rangle^a, \qquad \langle A, C \rangle^b, \qquad \langle A, B \rangle^c.$$

(5) Alice, Bob, and Carl have now shared the value $\langle T, T \rangle^{abc}$.

Remark 7.3. If Eve can solve the ECDLP, then she can certainly break tripartite Diffie–Hellman key exchange, since for example she can recover Alice's secret multiplier a from the publicly available points T and $A = [a]T$. However, Eve can also break the system if she can solve the discrete logarithm problem in the multiplicative group $\mathbb{F}_q^*$. Thus Eve knows the values of T, A, and B, so if she can solve the DLP

$$\langle T, T \rangle^m = \langle A, B \rangle$$

in $\mathbb{F}_q^*$, she recovers the value $m = ab$. From this she obtains Alice, Bob, and Carl's shared value by computing $\langle T, C \rangle^{ab}$.

The DLP in $\mathbb{F}_q^*$ can be solved in subexponential time (XI.4.1b), so it is currently significantly easier to solve the DLP in $\mathbb{F}_q^*$ than it is to solve the ECDLP in $E(\mathbb{F}_q)$. Thus tripartite Diffie–Hellman key exchange and other pairing-based cryptographic algorithms require that q be sufficiently large to preclude the solution of the DLP in $\mathbb{F}_q^*$.

Another cryptographic application of pairings on elliptic curves is a digital signature scheme that has extremely short signatures, as described in the following result of Boneh, Lynn, and Shacham [24]; see also [51, §24.1.3].

Theorem 7.4. *The following procedure allows Alice to sign a digital document and Bob to verify that the signature is valid.*

(1) Alice and Bob agree on a finite field $\mathbb{F}_q$, an elliptic curve $E/\mathbb{F}_q$, a prime N, and a point $T \in E(\mathbb{F}_q)[N]$ such that there is a bilinear pairing

$$\langle \,\cdot\, , \,\cdot\, \rangle : E(\mathbb{F}_q)[N] \times E(\mathbb{F}_q)[N] \longrightarrow \mu_N$$

with the property that $\langle T, T \rangle$ is a primitive N^{th} root of unity.

(2) Alice selects a secret integer a and computes the point $A = [a]T \in E(\mathbb{F}_q)$.

(3) Alice publishes the point A. This is her *public verification key*. The secret multiplier a is her *private signing key*.

(4) Alice chooses a digital document $D \in E(\mathbb{F}_q)$ to sign.[9] She computes and publishes the signature

$$S = [a]D.$$

(5) Bob accepts the signature as valid if the two quantities

$$\langle A, D \rangle \quad \text{and} \quad \langle T, S \rangle$$

are equal.

---

[9] As with ECDSA (XI.4.6), the point D is really a hash of the actual document. See [51, §24.2.5].

PROOF. Assuming that Alice has constructed A and S as in steps (2) and (4), bilinearity of the pairing yields

$$\langle A, D \rangle = \langle [a]T, D \rangle = \langle T, D \rangle^a \quad \text{and} \quad \langle T, S \rangle = \langle T, [a]D \rangle = \langle T, D \rangle^a,$$

so Bob accepts the signature. □

Remark 7.5.1. The reason that (XI.7.4) is called a short signature scheme is because the signature consists of only a single point in $E(\mathbb{F}_q)$, so with point compression (XI.4.3.5), a single number in $\mathbb{F}_q$. Thus (XI.7.4) gives signatures that are half the size of those produced by ECDSA (XI.4.6). (The use of hyperelliptic curves allows the size of q to be further reduced; see [51].)

Remark 7.5.2. Recall (XI.4.3.4) that the elliptic curve Diffie–Hellman problem asks for the value of $[ab]P$, given the three points P, $[a]P$, and $[b]P$. Potentially easier is the following decision version of this problem:

> *Decision Diffie–Hellman Problem*
> Given four points P, $[a]P$, $[b]P$, and Q in $E(\mathbb{F}_q)$, determine whether Q is equal to $[ab]P$.

The short signature scheme (XI.7.4) uses the fact that if E is an elliptic curve with a nondegenerate bilinear pairing, then the decision Diffie–Hellman problem is easy to solve, since $Q = [ab]P$ if and only if $\langle [a]P, [b]P \rangle = \langle Q, P \rangle$.

Remark 7.6. There are a number of other cryptographic constructions that use bilinear pairings on elliptic curves and that depend for their security on the difficulty of solving both the ECDLP in $E(\mathbb{F}_q)$ and the DLP in $\mathbb{F}_q^*$. We mention in particular *ID-based cryptography*, in which Alice may use an arbitrary character string as her public key. For example, her public key could be her email address. To send a message, Bob combines Alice's public key with a universal public key available from some trusted authority. The trusted authority also provides Alice with a personal private key that goes with her ID-based public key. The idea of ID-based cryptosystems was proposed by Shamir [245] in 1985, and a practical system using elliptic curves and pairings was devised by Boneh and Franklin [23] in 2001. For further details, see for example [51, §24.1.2] or [116, §5.10.2].

XI.8 Computing the Weil Pairing

The abstract definition of the Weil pairing requires functions having specified divisors. In this section we describe a double-and-add algorithm due to Victor Miller that computes such functions in linear time. Miller's algorithm makes pairings practical for use in applications such as cryptography.

Theorem 8.1. *Let E be an elliptic curve given by a Weierstrass equation*

$$E : y^2 + a_1 xy + a_3 y = x^3 + a_2 x^2 + a_4 x + a_6,$$

and let $P = (x_P, y_P)$ and $Q = (x_Q, y_Q)$ be nonzero points on E.

(1) Set $T = P$ and $f = 1$
(2) Loop $i = t - 1$ down to 0
(3)   Set $f = f^2 \cdot h_{T,T}$
(4)   Set $T = 2T$
(5)   If $\epsilon_i = 1$
(6)   Set $f = f \cdot h_{T,P}$
(7)   Set $T = T + P$
(8)   End If
(9) End i Loop
(10) Return the value f

Figure 11.6: Miller's algorithm.

(a) *Let λ be the slope of the line connecting P and Q, or the slope of the tangent line to E at P if $P = Q$. (If the line is vertical, set $\lambda = \infty$.) Define a function $h_{P,Q}$ on E as follows:*

$$h_{P,Q} = \begin{cases} \dfrac{y - y_P - \lambda(x - x_P)}{x + x_P + x_Q - \lambda^2 - a_1\lambda + a_2} & \text{if } \lambda \neq \infty, \\ x - x_P & \text{if } \lambda = \infty. \end{cases}$$

Then

$$\mathrm{div}(h_{P,Q}) = (P) + (Q) - (P + Q) - (O).$$

(b) **Miller's algorithm.** *Let $N \geq 1$ and write the binary expansion of N as*

$$N = \epsilon_0 + \epsilon_1 \cdot 2 + \epsilon_2 \cdot 2^2 + \cdots + \epsilon_t \cdot 2^t \qquad \text{with } \epsilon_i \in \{0, 1\} \text{ and } \epsilon_t \neq 0.$$

The algorithm described in Figure 11.6 returns a function f_P whose divisor satisfies

$$\mathrm{div}(f_P) = N(P) - ([N]P) - (N - 1)(O),$$

where the functions $h_{T,T}$ and $h_{T,P}$ used by the algorithm are as defined in (a). *In particular, if $P \in E[N]$, then $\mathrm{div}(f_P) = N(P) - N(O)$.*

PROOF. (a) Suppose first that $\lambda \neq \infty$, and let $y = \lambda x + \nu$ be the line through P and Q, or the tangent line at P if $P = Q$. This line intersects E at the three points P, Q, and $-P - Q$, so

$$\mathrm{div}(y - \lambda x - \nu) = (P) + (Q) + (-P - Q) - 3(O).$$

Vertical lines intersect E at points and their negatives, so

$$\mathrm{div}(x - x_{P+Q}) = (P + Q) + (-P - Q) - 2(O).$$

It follows that

$$h_{P,Q} = \frac{y - \lambda x - \nu}{x - x_{P+Q}}$$

has the divisor stated in (a). Finally, the addition formula (III.2.3d) tells us that $x_{P+Q} = \lambda^2 + a_1\lambda - a_2 - x_P - x_Q$, and we can eliminate ν from the numerator of $h_{P,Q}$ using $y_P = \lambda x_P + \nu$.

If $\lambda = \infty$, then $P+Q = O$, so we need $h_{P,Q}$ to have divisor $(P)+(-P)-2(O)$. The function $x - x_P$ has this divisor.

(b) This is a standard double-and-add algorithm, similar to (XI.1.1). The key to analyzing the algorithm comes from (a), which tells us that the functions $h_{T,T}$ and $h_{T,P}$ used in steps (3) and (6) have divisors

$$\begin{aligned}
\operatorname{div}(h_{T,T}) &= 2(T) - (2T) - (O), \\
\operatorname{div}(h_{T,P}) &= (T) + (P) - (T + P) - (O).
\end{aligned}$$

We consider the effect of executing the i loop, steps (2)–(9), for a given value of i. At the start of the loop the variables T and f have initial values T_i^{start} and f_i^{start}, and at the end of (one execution of) the loop they have final values T_i^{end} and f_i^{end}. We start with T. During the loop, the value of T is doubled and then, if $\epsilon_i = 1$, the value is incremented by P. This gives the relation

$$T_i^{\text{end}} = 2T_i^{\text{start}} + \epsilon_i P.$$

Similarly, the value of f is squared, multiplied by $h_{T,T}$, and then, if $\epsilon_i = 1$, it is multiplied by $h_{2T,P}$. (Note that the value of T has been doubled in step (4) before it is used in step (6).) This yields

$$f_i^{\text{end}} = (f_i^{\text{start}})^2 \cdot h_{T_i^{\text{start}}, T_i^{\text{start}}} \cdot h_{2T_i^{\text{start}}, P}^{\epsilon_i}.$$

Hence the divisors of f_i^{start} and f_i^{end} are related by

$$\begin{aligned}
\operatorname{div}(f_i^{\text{end}}) &= 2\operatorname{div}(f_i^{\text{start}}) + \operatorname{div}(h_{T_i^{\text{start}}, T_i^{\text{start}}}) + \epsilon_i \operatorname{div}(h_{2T_i^{\text{start}}, P}) \\
&= 2\operatorname{div}(f_i^{\text{start}}) + \big(2(T_i^{\text{start}}) - (2T_i^{\text{start}}) - (O)\big) \\
&\quad + \epsilon_i\big((2T_i^{\text{start}}) + (P) - (2T_i^{\text{start}} + P) - (O)\big) \\
&= 2\operatorname{div}(f_i^{\text{start}}) + 2(T_i^{\text{start}}) - (2T_i^{\text{start}} + \epsilon_i P) + \epsilon_i(P) - (1 + \epsilon_i)(O) \\
&\qquad\qquad\qquad\qquad\qquad\qquad\qquad\qquad \text{since } \epsilon_i \in \{0, 1\}, \\
&= 2\operatorname{div}(f_i^{\text{start}}) + 2(T_i^{\text{start}}) - (T_i^{\text{end}}) + \epsilon_i(P) - (1 + \epsilon_i)(O).
\end{aligned}$$

Of course, the final values of T and f after a given iteration of the i loop are the initial values for the next iteration, i.e., $T_i^{\text{end}} = T_{i-1}^{\text{start}}$ and $f_i^{\text{end}} = f_{i-1}^{\text{start}}$. (Note that the i loop decrements from $t-1$ to 0.) This allows us to rewrite the recurrences for T and f as

$$\begin{aligned}
T_{i-1}^{\text{start}} - 2T_i^{\text{start}} &= \epsilon_i P, \\
\operatorname{div}(f_{i-1}^{\text{start}}) - 2\operatorname{div}(f_i^{\text{start}}) &= 2(T_i^{\text{start}}) - (T_{i-1}^{\text{start}}) + \epsilon_i(P) - (1 + \epsilon_i)(O).
\end{aligned}$$

These formulas are designed to telescope when they are summed. For example, when the algorithm terminates, the final value of T is

$$
\begin{aligned}
T_0^{\text{end}} &= \epsilon_0 P + 2T_0^{\text{start}} \\
&= \epsilon_0 P + \left[\sum_{i=1}^{t-1} 2^i (T_{i-1}^{\text{start}} - 2T_i^{\text{start}}) \right] + 2^t T_{t-1}^{\text{start}} \\
&= \epsilon_0 P + \sum_{i=1}^{t-1} 2^i \epsilon_i P + 2^t T_{t-1}^{\text{start}} \qquad \text{using the recurrence for } T_i^{\text{start}}, \\
&= \sum_{i=0}^{t} 2^i \epsilon_i P \qquad\qquad\qquad\quad \text{since } T_{t-1}^{\text{start}} = P \text{ and } \epsilon_t = 1, \\
&= NP. \qquad\qquad\qquad\qquad\quad\; \text{since } N = \sum \epsilon_i 2^i.
\end{aligned}
$$

Finally, we compute the divisor of the function f returned by Miller's algorithm:

$$
\begin{aligned}
&\text{div}(f_0^{\text{end}}) \\
&= 2\,\text{div}(f_0^{\text{start}}) + 2(T_0^{\text{start}}) - (T_0^{\text{end}}) + \epsilon_0(P) - (1+\epsilon_0)(O) \\
&= \left[\sum_{i=1}^{t-1} 2^i (\text{div}(f_{i-1}^{\text{start}}) - 2\,\text{div}(f_i^{\text{start}})) \right] + 2(T_0^{\text{start}}) - (NP) + \epsilon_0(P) - (1+\epsilon_0)(O) \\
&\qquad\qquad\qquad\qquad\qquad\qquad\qquad\quad \text{since } f_{t-1}^{\text{start}} = 1 \text{ and } T_0^{\text{end}} = NP, \\
&= \left[\sum_{i=1}^{t-1} 2^i \big(2(T_i^{\text{start}}) - (T_{i-1}^{\text{start}}) + \epsilon_i(P) - (1+\epsilon_i)(O)\big) \right] + 2(T_0^{\text{start}}) \\
&\qquad\qquad\qquad\qquad\qquad\qquad - (NP) + \epsilon_0(P) - (1+\epsilon_0)(O) \\
&= 2^t(T_{t-1}^{\text{start}}) + \sum_{i=0}^{t-1} 2^i \epsilon_i(P) - \sum_{i=0}^{t-1} 2^i(1+\epsilon_i)(O) - (NP) \\
&= N(P) - (N-1)(O) - (NP) \qquad \text{since } T_{t-1}^{\text{start}} = P, \epsilon_t = 1, \text{ and } N = \sum \epsilon_i 2^i.
\end{aligned}
$$

This completes the proof that the function returned by Miller's algorithm has the stated divisor. $\qquad\qquad\qquad\qquad\qquad\qquad\qquad\qquad\qquad\qquad\qquad\qquad$ $\square$

Remark 8.2. Let $P \in E[N](K)$. Miller's algorithm (XI.8.1b) tells us how to compute a function $f_P \in K(E)$ with divisor $N(P) - N(O)$. Further, if $R \in E$ is any point, we can use Miller's algorithm to directly evaluate $f_P(R)$ by evaluating the functions $h_{T,T}(R)$ and $h_{T,P}(R)$ in steps (3) and (6). This allows us to compute the Weil pairing $e_N(P, Q)$ via the alternative definition of $e_N(P, Q)$ from Exercise 3.16. We choose any point $S \in E$ not in the subgroup generated by P and Q, and then

$$
e_N(P, Q) = \frac{f_P(Q+S)}{f_P(S)} \bigg/ \frac{f_Q(P-S)}{f_Q(-S)}.
$$

The right-hand side of this formula can be computed using four applications of Miller's algorithm (XI.8.1b). For added efficiency, the two values $f_P(Q + S)$ and $f_P(S)$ of f_P may be computed simultaneously, and similarly for $f_Q(P - S)$ and $f_Q(-S)$.

Example 8.3. Let $E/\mathbb{F}_{631}$ be the elliptic curve

$$y^2 = x^3 + 30x + 34.$$

We have $E(\mathbb{F}_{631}) \cong \mathbb{Z}/5\mathbb{Z} \times \mathbb{Z}/130\mathbb{Z}$, and it is easy to check that the points $P = (36, 60)$ and $Q = (121, 387)$ generate $E(\mathbb{F}_{631})[5]$. In order to compute the Weil pairing with Miller's algorithm, we use the auxiliary point $S = (0, 36) \in E(\mathbb{F}_{631})$. The point S has order 10, so it is not in the subgroup spanned by P and Q. Miller's algorithm gives

$$f_P(Q + S) = 103, \quad f_P(S) = 219, \quad f_Q(P - S) = 284, \quad f_Q(-S) = 204.$$

Hence

$$e_5(P, Q) = \frac{103}{219} \Big/ \frac{284}{204} = 242 \in \mathbb{F}_{631}.$$

We check that $(242)^5 = 1$, so $e_5(P, Q)$ is indeed a fifth root of unity in $\mathbb{F}_{631}$.

Remark 8.4. As an alternative to the Weil pairing, cryptographers often use the Tate–Lichtenbaum pairing described in (XI §9). Miller's algorithm (XI.8.1b) can also be used to compute the Tate–Lichtenbaum pairing (XI.9.3.2).

Remark 8.5. Another linear-time algorithm to compute the Weil and Tate–Lichtenbaum pairings, due to Shipsey and Stange, makes use of elliptic divisibility sequences (Exercises 3.34–3.36) and elliptic nets. See [252, 270, 271] for details.

XI.9 The Tate–Lichtenbaum Pairing

The Weil pairing is often used to define other pairings on elliptic curves. In this section we describe the Tate–Lichtenbaum pairing, which has both theoretical and cryptographic applications. See (C §17) and Exercise 10.24 for other instances of pairings on elliptic curves.

Definition. Let E/K be an elliptic curve, and let $N \geq 1$ be an integer that is prime to $p = \operatorname{char}(K)$ if $p > 0$. The *Tate–Lichtenbaum pairing*

$$\tau : \frac{E(K)}{NE(K)} \times E(K)[N] \longrightarrow \frac{K^*}{(K^*)^N}$$

is defined as follows. Let $P \in E(K)$ and $T \in E(K)[N]$. Choose a point $Q \in E(\bar{K})$ satisfying $[N]Q = P$. The map

$$G_{\bar{K}/K} \longrightarrow \mu_N, \qquad \sigma \longrightarrow e_N(Q^\sigma - Q, T),$$

is a 1-cocycle (see below), so it represents an element of $H^1(G_{\bar{K}/K}, \mu_N)$. Hilbert's Theorem 90 (B.2.5c) says that the connecting homomorphism

$$K^*/(K^*)^N \longrightarrow H^1(G_{\bar{K}/K}, \mu_N)$$

is an isomorphism; hence there exists an element $\alpha \in K^*$, unique up to N^{th} powers, with the property that

$$e_N(Q^\sigma - Q, T) = \sqrt[N]{\alpha}^\sigma / \sqrt[N]{\alpha} \qquad \text{for all } \sigma \in G_{\bar{K}/K}.$$

The value of the Tate–Lichtenbaum pairing is then

$$\tau(P, T) = \alpha \bmod (K^*)^N.$$

Proposition 9.1. *The Tate–Lichtenbaum pairing is a well-defined bilinear pairing.*

PROOF. Let $\xi(\sigma) = e_N(Q^\sigma - Q, T)$ be the given map $\xi : G_{\bar{K}/K} \to \mu_N$. We use basic properties of the Weil pairing (III.8.1ad) and the assumption that $T \in E(K)[N]$ to verify that ξ is a 1-cocycle:

$$\begin{aligned}
\xi(\sigma\tau) &= e_N(Q^{\sigma\tau} - Q, T) \\
&= e_N(Q^{\sigma\tau} - Q^\tau + Q^\tau - Q, T) \\
&= e_N(Q^\sigma - Q, T)^\tau e_N(Q^\tau - Q, T) \\
&= \xi(\sigma)^\tau \xi(\tau).
\end{aligned}$$

Next we show that $\xi(\sigma)$ depends only on P modulo $NE(K)$. If we replace P by $P + NR$ for some $R \in E(K)$, then Q is replaced by $Q + R$. Since R is defined over K, we have

$$(Q + R)^\sigma - (Q + R) = Q^\sigma - Q \qquad \text{for all } \sigma \in G_{\bar{K}/K},$$

so the value $\xi(\sigma)$ does not change.

Suppose that we replace Q by some other point Q' satisfying $[N]Q' = P$. Then the difference $S = Q' - Q$ is in $E[N]$, so

$$\begin{aligned}
e_N(Q'^\sigma - Q', T) &= e_N\big((Q + S)^\sigma - (Q + S), T\big) \\
&= e_N(Q^\sigma - Q, T) e_N(S^\sigma - S, T) \\
&= e_N(Q^\sigma - Q, T) \frac{e_N(S, T)^\sigma}{e_N(S, T)}.
\end{aligned}$$

(Note that $e_N(S, T)$ is well-defined, since $S \in E[N]$, while $e_N(Q, T)$ is not defined, since in general $Q \notin E[N]$.) Thus the $G_{\bar{K}/K}$-to-μ_N cocycle coming from Q' differs from the $G_{\bar{K}/K}$-to-μ_N cocycle coming from Q by the $G_{\bar{K}/K}$-to-μ_N coboundary $\sigma \mapsto e_N(S, T)^\sigma / e_N(S, T)$, so they represent the same cohomology class.

This completes the proof that the Tate–Lichtenbaum pairing is well-defined. The bilinearity is immediate from the bilinearity of the Weil pairing. $\qquad\square$

The Weil pairing is defined by evaluating certain functions at certain points. We can do the same for the Tate–Lichtenbaum pairing.

Proposition 9.2. *Let* $T \in E(K)[N]$ *and choose a function* $f \in K(E)$ *satisfying*

$$\operatorname{div}(f) = N(T) - N(O) \quad \text{and} \quad f \circ [N] \in (K(E)^*)^N.$$

Then for all $P \in E(K) \smallsetminus \{O, T\}$ *we have*

$$\tau(P, T) = f(P) \bmod (K^*)^N.$$

Remark 9.3.1. Although (XI.9.2) excludes the case of $\tau(T, T)$, we can use bilinearity to compute $\tau(T, T)$ as

$$\tau(T, T) = \frac{\tau(T + Q, T)}{\tau(Q, T)} = \frac{f(T + Q)}{f(Q)}.$$

More generally,

$$\tau(P, T) = \frac{\tau(P + Q, T)}{\tau(Q, T)} = \frac{f(P + Q)}{f(Q)}$$

for any $Q \in E$ such that f is defined and nonzero at both $P + Q$ and Q, i.e., any point $Q \notin \{O, T, -P, T - P\}$. Notice that with this formulation, there is no need to choose f to satisfy $f \circ [N] \in (K(E)^*)^N$, since the divisor relation $\operatorname{div}(f) = N(T) - N(O)$ determines f up to multiplication by a constant, and taking the ratio eliminates the dependence on the constant.

Remark 9.3.2. Miller's algorithm (XI.8.1b) can be used to efficiently compute the Tate–Lichtenbaum pairing, since it gives a linear-time algorithm to compute the value of f. Comparing the formula for the Tate–Lichtenbaum pairing (XI.9.3.1) to the formula for the Weil pairing (XI.8.2), we see that τ requires two values of f, while e_N requires four values of f. Thus the former is twice as efficient as the latter, which is why the Tate–Lichtenbaum pairing is often preferred for real-world cryptographic applications.

PROOF OF (XI.9.2). As explained in the construction of the Weil e_N-pairing (III §8), there are functions $f, g \in \bar{K}(E)$ satisfying

$$\operatorname{div}(f) = N(T) - N(O) \quad \text{and} \quad f \circ [N] = g^N,$$

and (II.5.8) implies that we may choose f and g in $K(E)$, since their divisors are $G_{\bar{K}/K}$-invariant. From the definition of e_N we have

$$e_N(Q^\sigma - Q, T) = \frac{g(X + Q^\sigma - Q)}{g(X)} \quad \text{for } X \in E.$$

In particular, setting $X = Q$ gives

$$e_N(Q^\sigma - Q, T) = \frac{g(Q^\sigma)}{g(Q)} = \frac{g(Q)^\sigma}{g(Q)}.$$

Comparing this formula with the definition of the Tate–Lichtenbaum pairing yields

$$\tau(P,T) = g(Q)^N = f \circ [N](Q) = f(P) \quad (\text{mod } (K^*)^N). \qquad \square$$

The Tate–Lichtenbaum pairing has many applications, both theoretical and practical. In cryptography, it is used on elliptic curves over finite fields; see (XI §7). The following result provides an important nondegeneracy criterion in this situation. For applications of the Tate–Lichtenbaum pairing over local and global fields, see for example [149, 177, 281, 286].

Theorem 9.4. *Let $E/\mathbb{F}_q$ be an elliptic curve defined over a finite field, let $N \geq 1$, let $T \in E(\mathbb{F}_q)[N]$ be a point of exact order N, and make the following assumptions:*

(i) *$\mu_N \subset \mathbb{F}_q$, or equivalently, $q \equiv 1 \pmod{N}$.*

(ii) *$E(\mathbb{F}_q)[N^2] = \mathbb{Z}T$, i.e., the only rational N^2-torsion points are the multiples of T.*

Then the Tate–Lichtenbaum pairing is a perfect pairing, and $\tau(T,T)^{(q-1)/N}$ is a primitive N^{th} root of unity in $\mathbb{F}_q^*$.

PROOF. We begin by proving that the Tate–Lichtenbaum pairing is nondegenerate on the left. Let $\phi \in G_{\bar{\mathbb{F}}_q/\mathbb{F}_q}$ be the q-power Frobenius map. Choose another N-torsion point T' so that T and T' generate $E[N]$. We know that $T^\phi = T$, since $T \in E(\mathbb{F}_q)$, and we write

$$T'^\phi = [a]T + [b]T' \qquad \text{for some } a, b \in \mathbb{Z}/N\mathbb{Z}.$$

We use basic properties of the Weil pairing (II.8.1) to compute

$$\begin{aligned}
e_N(T,T')^q = e_N(T,T')^\phi &= e_N(T^\phi, T'^\phi) \\
&= e_N(T, [a]T + [b]T') \\
&= e_N(T,T)^a e_N(T,T')^b \\
&= e_N(T,T')^b.
\end{aligned}$$

Since $e_N(T,T')$ is a primitive N^{th} root of unity (cf. (III.8.1.1)), this implies that

$$b \equiv q \pmod{N}.$$

But our assumption that $\mu_N \subset \mathbb{F}_q$ tells us that $q \equiv 1 \pmod{N}$, so the action of Frobenius on T' is given by

$$T'^\phi = [a]T + T' \qquad \text{for some } a \in \mathbb{Z}/N\mathbb{Z}.$$

We claim that $a \in (\mathbb{Z}/N\mathbb{Z})^*$. To see this, let $d = N/\gcd(a, N)$. Then

$$([d]T')^\phi = [da]T + [d]T' = [d]T',$$

so $[d]T' \in E(\mathbb{F}_q)[N]$. But by assumption, the point T generates $E(\mathbb{F}_q)[N]$, while T and T' generate all of $E[N]$. Hence $[d]T' = O$, which implies that $d = N$ and $\gcd(a, N) = 1$.

Suppose now that $P \in E(\mathbb{F}_q)$ satisfies $\tau(P, T) = 1$. What this really means is that in the definition of the Tate–Lichtenbaum pairing, we have $\alpha \in (\mathbb{F}_q^*)^N$. Thus $\sqrt[N]{\alpha} \in \mathbb{F}_q^*$, which implies that $e_N(Q^\phi - Q, T) = 1$. So if we write

$$Q^\phi - Q = [A]T + [B]T',$$

then again using properties of the Weil pairing we find that

$$1 = e_N(Q^\phi - Q, T) = e_N([A]T + [B]T', T) = e_N(T', T)^B.$$

This implies that $B \equiv 0 \pmod{N}$, so $Q^\phi - Q = [A]T$. Now consider the point

$$Q' = Q - [a^{-1}A]T',$$

where a^{-1} is the inverse of a modulo N. Then

$$Q'^\phi = \left(Q + [A]T\right) - [a^{-1}A]\left([a]T + T'\right) = Q - [a^{-1}A]T' = Q'.$$

Thus $Q' \in E(\mathbb{F}_q)$, so

$$P = [N]Q = [N]Q' \in NE(\mathbb{F}_q).$$

This proves nondegeneracy of the Tate–Lichtenbaum pairing on the left.

We now consider the value of $\tau(T, T)$. Let k denote the order of $\tau(T, T)$ in $\mathbb{F}_q^*/(\mathbb{F}_q^*)^N$. Bilinearity of the Tate–Lichtenbaum pairing implies that

$$\tau([k]T, T) = \tau(T, T)^k = 1,$$

and then the nondegeneracy that we already proved implies that $[k]T \in NE(\mathbb{F}_q)$. So we can write $[k]T = [N]S$ for some $S \in E(\mathbb{F}_q)$. The point T has exact order N, so $[k]T$ has exact order N/k, and hence S has exact order N^2/k. By assumption, the only $\mathbb{F}_q$-rational points in $E[N^2]$ are the multiples of T, all of which have order dividing N. Therefore $k = N$, which proves that $\tau(T, T)$ is an element of exact order N in $\mathbb{F}_q^*/(\mathbb{F}_q^*)^N$. It follows immediately that $\tau(T, T)^{(q-1)/N}$ is a primitive N^{th} root of unity.

It also proves nondegeneracy on the right. To see this, let $Q \in E(\mathbb{F}_q)[N]$ satisfy $\tau(P, Q) = 1$ for all $P \in E(\mathbb{F}_q)$. Then $Q = [n]T$ for some integer n, so taking $P = T$ yields

$$1 = \tau(P, Q) = \tau(T, [n]T) = \tau(T, T)^n.$$

We know that $\tau(T, T)$ has exact order N, so $N \mid n$, which shows that $Q = [n]T = O$. $\qquad\square$

Remark 9.5. In the situation of (XI.9.4), the natural map

$$E(\mathbb{F}_q)[N] \to E(\mathbb{F}_q)/NE(\mathbb{F}_q)$$

is an isomorphism of cyclic groups of order N, so we can use the Tate–Lichtenbaum pairing to define a nondegenerate symmetric bilinear pairing

$$E(\mathbb{F}_q)[N] \times E(\mathbb{F}_q)[N] \longrightarrow \mu_N, \qquad (P,Q) \longmapsto \tau(P,Q)^{(q-1)/N}.$$

This is the pairing that is typically used for cryptographic applications.

Remark 9.6. Suppose that we randomly choose an elliptic curve $E/\mathbb{F}_q$, compute the order of the group $E(\mathbb{F}_q)$ (e.g., using (XI §3)), and find that there is a large prime N dividing $\#E(\mathbb{F}_q)$. This almost puts us into the situation to apply (XI.9.4), but we may need to extend the field $\mathbb{F}_q$ in order to ensure that it contains μ_N. Let $d \geq 1$ be the embedding degree (XI §6) of N in $\mathbb{F}_q$, i.e., d is the smallest integer such that $\mu_N \subset \mathbb{F}_{q^d}^*$, or equivalently, such that $q^d \equiv 1 \pmod{N}$. How large should we expect d to be?

If we write

$$\#E(\mathbb{F}_q) = q + 1 - a$$

as usual, then the assumption that $\#E(\mathbb{F}_q)$ is divisible by N implies that

$$q + 1 - a = \#E(\mathbb{F}_q) \equiv 0 \pmod{N}.$$

Similarly, since we have chosen d to satisfy $\mu_N \subset \mathbb{F}_{q^d}^*$, we have

$$q^d - 1 = \#\mathbb{F}_{q^d}^* \equiv 0 \pmod{N}.$$

Hence

$$(a-1)^d \equiv 1 \pmod{N}.$$

We know from (V.1.1) that $|a| \leq 2\sqrt{q}$, but within this allowed range, the value of a is more-or-less randomly distributed; see (C.21.4) for a precise statement. The expected order of a randomly chosen element of a randomly chosen cyclic group is a constant multiple of the order of the group (see Exercise 11.17), so if we choose $E/\mathbb{F}_q$ randomly, the embedding degree d is almost certain to be too large for practical applications. (See [13] for a more detailed analysis.)

Remark 9.7. *Supersingular Elliptic Curves.* Let $E/\mathbb{F}_p$ be supersingular elliptic curve with $p \geq 5$ prime, and suppose that $E(\mathbb{F}_p)$ contains a point of prime order N. We have seen (XI.6.4) that the embedding degree of N in $\mathbb{F}_p$ is 2, i.e., $\mu_N \subset \mathbb{F}_{p^2}$, but unfortunately condition (ii) of (XI.9.4) is never true in this situation; see Exercise 11.18. In general, we know from (V.3.1) that every supersingular elliptic curve is isomorphic to a curve defined over $\mathbb{F}_{p^2}$, and supersingular curves always have small embedding degree, so it may be possible to apply (XI.9.4) to a supersingular curve defined over $\mathbb{F}_{p^2}$. As an alternative, one can use a distortion map as in (XI §7); see [51, §24.2.1b] or [116, §5.9.2].

Remark 9.8. *Pairing-Friendly Elliptic Curves.* In general we would like to construct an elliptic curve $E/\mathbb{F}_q$ such that $E(\mathbb{F}_q)$ contains a point of large prime order N and such that the embedding degree d of N in $\mathbb{F}_q$ is not too large. These are called *pairing-friendly elliptic curves.* The exact constraints on the parameters q, N, and d depend on the desired security level, but in any case it is important to balance the difficulty of solving the ECDLP in a subgroup of $E(\mathbb{F}_q)$ of order N against the difficulty of solving the DLP in $\mathbb{F}_{q^d}^*$. For the former, only exponential-time algorithms are known, while there are subexponential algorithms for the latter. See the discussion in (XI.4.1). Further, for computational efficiency we should choose q to be as small as possible.

For example, current algorithms to solve the ECDLP when $N \approx 2^{160}$ take about the same amount of time as current algorithms to solve the DLP when $q^d \approx 2^{1024}$. So for this security level, the embedding degree should be $d \approx 6.4 \log q / \log N$. Typically $\sqrt{q} < N < q$, so d should be roughly between 6 and 12.

Atkin and Morain [10, 183] devised a method using the theory of complex multiplication to find elliptic curves with points of large order and small embedding degree. Their idea is to fix positive integers D and d and to search for integers a, N, and p satisfying the following four conditions:

(1) N and p are primes.

(2) $N \mid p + 1 - a$.

(3) $N \mid p^d - 1$.

(4)   The equation $Dy^2 = 4p - a^2$ has a solution $y \in \mathbb{Z}$.

If we can find values for a, N, and p satisfying (1)–(4) and if the class number of the quadratic field $\mathbb{Q}(\sqrt{-D})$ is not too large, say less than 10^5, then Atkin and Morain's CM method yields an elliptic curve $E/\mathbb{F}_p$ with $N \mid E(\mathbb{F}_p)$ and $\boldsymbol{\mu}_N \subset \mathbb{F}_{p^d}$. For a description of the CM method, see for example [22, Chapter VIII] or [51, §18.1], and for various algorithms that have been devised to find (a, N, p) satisfying (1)–(4), see for example [29, 69, 87, 181].

Exercises

11.1. Use the double-and-add algorithm (XI.1.1) to compute $[n]P$ in $E(\mathbb{F}_p)$ for each of the following curves and points.

(a) $E : Y^2 = X^3 + 143X + 367$, $p = 613$, $P = (195, 9)$, $n = 23$.

(b) $E : Y^2 = X^3 + 1541X + 1335$, $p = 3221$, $P = (2898, 439)$, $n = 3211$.

11.2. Let n be a positive integer.
(a) Prove that n has a unique ternary expansion

$$n = \epsilon_0 + \epsilon_1 \cdot 2 + \epsilon_2 \cdot 2^2 + \epsilon_3 \cdot 2^3 + \cdots + \epsilon_t \cdot 2^t, \qquad \epsilon_0, \ldots, \epsilon_t \in \{-1, 0, 1\},$$

with the property that no two consecutive ϵ_i are nonzero. Such an expansion is called *nonadjacent form* (NAF). (*Hint.* For existence, start with the binary expansion of n and replace consecutive nonzero terms $2^i + 2^{i+1} + \cdots + 2^{i+j-1} + 0 \cdot 2^{i+j}$ with $-2^i + 2^{i+j}$.)

(b) If we assume that the expansion in (a) has $\epsilon_t \neq 0$, prove that $t \leq \log_2(2n)$.

(c) Prove that most positive integers have a ternary expansion as in (a) with approximately one-third of the ϵ_i being nonzero.

(d) Convert your proof in (a) into an algorithm and find a ternary expansion for each of the following numbers. Compare the number of nonzero terms in the ternary expansion with the number of nonzero terms in the binary expansion.

(i) 349.   (ii) 9337.   (iii) 38728.   (iv) 8379483273489.

11.3. Let τ represent a quantity satisfying $\tau^2 + \tau + 2 = 0$.

(a) Prove that every positive integer n can be written in the form

$$n = \epsilon_0 + \epsilon_1 \tau + \epsilon_2 \tau^2 + \cdots + \epsilon_t \tau^t, \qquad \epsilon_0, \ldots, \epsilon_t \in \{-1, 0, 1\},$$

with $t \leq 2\lceil \log_2 n \rceil + 1$ and at most one-third of the ϵ_i nonzero. (*Hint.* Repeatedly write integers as $2a + b$ and replace the 2 with $-\tau - \tau^2$.)

(b) More generally, prove that (a) is true for any $n \in \mathbb{Z}[\tau]$, where the upper bound for t is approximately $\log_2 N_{\mathbb{Z}[\tau]/\mathbb{Z}}(n)$.

(c) Let $E/\mathbb{F}_2$ be the curve in (XI.1.5), let $\tau(x, y) = (x^2, y^2)$ be the Frobenius map, let $P \in E(\mathbb{F}_{2^r})$, and let n be a positive integer. Prove that there is an element

$$\nu = \epsilon_0 + \epsilon_1 \tau + \epsilon_2 \tau^2 + \cdots + \epsilon_t \tau^t, \qquad \epsilon_0, \ldots, \epsilon_t \in \{-1, 0, 1\},$$

with $[\nu]P = [n]P$ and such that t is (approximately) bounded above by $\log_2 n$. (*Hint.* "Divide" n by $\tau^r - 1$ in $\mathbb{Z}[\tau]$ to find a remainder ν whose norm is approximately bounded by 2^r. Then use (b) and the fact that $\tau^r(P) = P$.)

(d) Devise an algorithm implementing your results in this exercise, and use your algorithm to compute a τ-adic expansion for each of the following values of n.

(i) $n = 931$ (ii) $n = 32755$ (iii) $n = 82793729188$

11.4. The double-and-add algorithm described in (XI.1.1) reads the bits of n from right to left, where we view n as a binary number such as 1001101. Prove that the following left-to-right version also computes nP.

(1) Write the binary expansion of n as $\sum_{i=0}^{t} \epsilon_i \cdot 2^i$.

(2) Set $Q = O$.

(3) Loop $i = t, t - 1, \ldots, 2, 1, 0$.

(4)   Set $Q = [2]Q$.

(5)   If $\epsilon_i = 1$, set $Q = Q + P$.

(6) End Loop

(7) Return Q.

11.5. Let S be a set containing n elements.

(a) Suppose that we select an element of S at random, note its value, return the element to the set, and repeat the process m times. Find a formula for the probability that some element has been selected at least twice. (If this happens, we say that there has been a collision.)

(b) If $n = 365$ and $m = 50$, what is the probability of a collision? (This is the probability that among 50 people in a room, at least two have the same birthday. The surprising answer is the origin of the name "birthday paradox.") How many people are required to have a 50% chance that two share a common birthday?

(c) Suppose that n is large. Give a good approximation for the probability of a collision if $m = c\sqrt{n}$, where c is a small constant, say $1 \leq c \leq 10$. What value of c gives a 50% chance of a collision? What value of c gives a $1 - 10^{-6}$ probability of a collision?

11.6. Pollard's algorithm (XI.2.1) says that if N has a prime factor p with $p - 1 = \prod q_j^{e_j}$, then it suffices to take $L = \max e_j q_j$ in order to (probably) factor N. Show that it is enough that L satisfy $L \geq \sum_{t \geq 1} \lfloor L/q^t \rfloor$. Give a similar statement for the elliptic curve factorization algorithm (XI.2.3).

11.7. Implement Lenstra's elliptic curve factorization algorithm (XI.2.3) and use it to factor N using the given elliptic curve E and point P.

(a) $N = 589$, $\qquad E : Y^2 = X^3 + 4X + 9$, $\qquad P = (2, 5)$.

(b) $N = 26167$, $\qquad E : Y^2 = X^3 + 4X + 128$, $\qquad P = (2, 12)$.

(c) $N = 1386493$, $\qquad E : Y^2 = X^3 + 3X - 3$, $\qquad P = (1, 1)$.

(d) $N = 28102844557$, $\qquad E : Y^2 = X^3 + 18X - 453$, $\qquad P = (7, 4)$.

11.8. We noted (XI.4.3.5) that it suffices for Bob to send Alice the x-coordinate of his point $P \in E(\mathbb{F}_q)$, together with one extra bit that specifies which of the two possible y-coordinates to use. However, this means that Alice needs to compute a square root in $\mathbb{F}_q$.

(a) Suppose that $q \equiv 3 \pmod 4$, and let $a \in \mathbb{F}_q$ be an element that is a square. Prove that $b = a^{(q+1)/4}$ is a square root of a.

(b) Suppose that q is prime and satisfies $q \equiv 5 \pmod 8$. Let $a \in \mathbb{F}_q$ be an element that is a square, and let

$$b = \begin{cases} a^{(q+3)/8} & \text{if } a^{(q-1)/4} = 1, \\ 2a(4a)^{(q-5)/8} & \text{if } a^{(q-1)/4} = -1. \end{cases}$$

Prove that $b^2 = a$.

11.9. Let G be a group, and suppose that you know an algorithm that takes $T(n)$ steps to solve any discrete logarithm problem $h = g^m$ in G if the element g has order n.

(a) Let $g \in G$ have order n and suppose that n factors as $n = n_1 n_2 \cdots n_t$ with $\gcd(n_i, n_j) = 1$ for all $i \neq j$. Find an algorithm that solves $h = g^m$ in (approximately) $\sum T(n_i)$ steps. (*Hint.* Solve $(g^{n/n_i})^m = h^{n/n_i}$ for each i and combine the solutions using the Chinese remainder theorem.)

(b) Let $g \in G$ have order n with $n = \ell^k$ a prime power. Find an algorithm that solves $h = g^m$ in (approximately) $kT(\ell)$ steps.

11.10. This exercise describes the Menezes–Vanstone variant of the ElGamal cryptosystem.

1. Alice and Bob agree on a finite field $\mathbb{F}_q$, an elliptic curve $E/\mathbb{F}_q$, and a point $P \in E(\mathbb{F}_q)$.
2. Alice selects a secret integer a and computes the point $A = [a]P \in E(\mathbb{F}_q)$.
3. Alice publishes the point A. This is her *public key*. The secret multiplier a is her *private key*.
4. Bob chooses a *plaintext* $(m_1, m_2) \in \mathbb{F}_q^2$ and a random integer k. He computes the two points $B_1 = [k]P$ and $B_2 = [k]A$.
5. Bob writes B_2 as $(x, y) \in E(\mathbb{F}_q)$, sets $c_1 = xm_1$ and $c_2 = ym_2$, and sends Alice the ciphertext (B_1, c_1, c_2).

(a) Explain how Alice can use the ciphertext (B_1, c_1, c_2) and her secret multiplier a to recover Bob's plaintext (m_1, m_2).

(b) What is the message expansion (XI.4.5.3) of MV-ElGamal?

(c) Explain how Eve can break MV-ElGamal if she can solve the Diffie–Hellman problem (XI.4.3.4).

11.11. This exercise describes the Elliptic Curve Integrated Encryption Scheme (ECIES). It combines the discrete logarithm problem with several other cryptographic constructions, including a hash function that we denote by $\mathcal{H}$, a message authentication code that we denote by $\mathcal{M}$, and a private key cryptosystem that we denote by $\mathcal{P}$.[10]

1. Alice and Bob agree on a finite field $\mathbb{F}_q$, an elliptic curve $E/\mathbb{F}_q$, and a point $P \in E(\mathbb{F}_q)$.
2. Alice selects a secret integer a and computes the point $A = [a]P \in E(\mathbb{F}_q)$.
3. Alice publishes the point A. This is her *public key*. The multiplier a is her *private key*.
4. Bob chooses a *plaintext* m and a random number k.
5. Bob computes $[k]A$ and uses the hash function to compute $\mathcal{H}\big(x\big([k]A\big)\big)$. He breaks this value into two pieces, say b_1 and b_2, which he uses as keys.
6. Bob uses the private key cryptosystem and the MAC to compute the two values

$$c = \mathcal{P}(b_1; m) \qquad \text{and} \qquad d = \mathcal{M}(b_2; c).$$

7. Bob computes $B = [k]P$.
8. Bob sends Alice the triple (B, c, d).

(a) Explain how Alice can recover the value of $x\big([k]A\big)$ that Bob used in step (5). This allows Alice to use the hash function to compute b_1 and b_2.

(b) Explain how Alice can then recover the message m.

(c) Explain how Alice can check the validity of the ciphertext c by recomputing $\mathcal{M}(b_2; c)$ and verifying that it agrees with the value of d sent by Bob.

(d) Explain why it is difficult for Eve to find a triple (B, c, d) that Alice accepts as valid unless she knows the plaintext m that corresponds to c via Alice's decryption process.

11.12. In the description of Pollard's ρ method in (XI §5), we gave an algorithm for computing the coefficients α_i and β_i in the expression $R_i = [\alpha_i]P + [\beta_i]Q$. Give a similar algorithm, in the form of two tables, for the coefficients γ_i and δ_i of the point

$$S_i = R_{2i} = [\gamma_i]P + [\delta_i]Q.$$

In other words, give the values of γ_{i+1} and δ_{i+1} in terms of γ_i and δ_i depending on whether the values of x_{S_i} and $x_{f(S_i)}$ are in A, B, or C.

11.13. Let $E/\mathbb{F}_p$ be an elliptic curve defined over a field of prime order. As described in (XI.5.4), one way to define a mixing function $f : E(\mathbb{F}_p) \to E(\mathbb{F}_p)$ for use in Pollard's ρ algorithm (XI.5.3) is to write $E(\mathbb{F}_p)$ as a disjoint union of three sets A, B, C. For example, we might take

$$A = \big\{P \in E(\mathbb{F}_p) : 0 \le x(P) < \tfrac{1}{3}p\big\},$$
$$B = \big\{P \in E(\mathbb{F}_p) : \tfrac{1}{3}p \le x(P) < \tfrac{2}{3}p\big\},$$
$$C = \big\{P \in E(\mathbb{F}_p) : \tfrac{2}{3}p \le x(P) < p\big\}.$$

---

[10]Informally, a hash function is an easy-to-compute, hard-to-invert function; a message authentication code (MAC) is a hash function that requires a secret key; and a private key cryptosystem is a one-to-one function that is easy to compute in both directions if one knows the secret key, but hard to compute otherwise. For precise definitions and examples, see [169].

Using this choice of A, B, and C, write a computer program implementing Pollard's ρ algorithm and use it to solve the following discrete logarithm problems, i.e., find a value of m satisfying $Q = [m]P$.

(a) $p = 541$, $E : y^2 = x^3 + 442x + 211$, $P = (238, 345)$,
      $Q = (180, 148)$.

(b) $p = 7919$, $E : y^2 = x^3 + 1356x + 1654$, $P = (6007, 296)$,
      $Q = (2821, 6396)$.

(c) $p = 104729$, $E : y^2 = x^3 + 25780x + 74070$, $P = (6588, 76182)$,
      $Q = (14624, 59879)$.

11.14. Let G be an abelian group whose order is bounded by a known quantity, say $\#G \le n$, and let $x \in G$.

(a) Adapt Shanks's babystep–giantstep algorithm (XI.5.2) to find the order of x in time $O(\sqrt{n})$ and space $O(\sqrt{n})$.

(b) Adapt Pollard's ρ algorithm (XI.5.4) to find the order of x in time $O(\sqrt{n})$ while using only space $O(1)$. (*Hint.* A direct adaptation of (XI.5.4) does not work, since the exponents $\alpha_i, \ldots, \delta_i$ cannot be reduced by the unknown order of the group. Instead write G as a disjoint union $G = A_1 \cup \cdots \cup A_t$, choose several random exponents $e_1, \ldots, e_t$ between 2 and n, and define $f : G \to G$ by $f(z) = g^{e_j} z$ if $z \in A_j$. Show that a match $z_{2i} = z_i$ is likely to be found with exponents $\alpha_i, \cdots, \delta_i$ that are $O(n\sqrt{n})$.)

(c) Explain how to use an algorithm that finds the order of elements in G to determine the order of the group G.

11.15. Working over the field $\mathbb{F}_{137}$, consider the curve and points

$$E : y^2 = x^3 + 86x + 98, \quad P = (56, 85) \in E(\mathbb{F}_{137}), \quad Q = (54, 86) \in E(\mathbb{F}_{137}).$$

(a) Verify that E is anomalous, i.e., $\#E(\mathbb{F}_{137}) = 137$.

(b) Lift P and Q to points $P' = (56, —)$ and $Q' = (54, —)$ in $E(\mathbb{Z}/137^2\mathbb{Z})$.

(c) Compute the elliptic logarithms of $[137]P'$ and $[137]Q'$ modulo 137^2.

(d) As in (XI.6.5) and (XI.6.7), use the results from (c) to solve the discrete logarithm problem, i.e., find an integer m such that $Q = [m]P$ in $E(\mathbb{F}_{137})$.

11.16. Let $E/\mathbb{F}_{631}$ be the elliptic curve $y^2 = x^3 + 30x + 34$ from (XI.8.3).

(a) The points $P' = (617, 5)$ and $Q' = (121, 244)$ are in $E(\mathbb{F}_{631})[5]$. Use Miller's algorithm to compute $e_5(P', Q')$.

(b) Let $P = (36, 60)$ and $Q = (121, 387)$ be the points from (XI.8.3). Express P' and Q' as linear combinations of P and Q, and use linearity of e_N to express $e_5(P', Q')$ as a power of $e_5(P, Q)$.

(c) Verify that the value of $e_5(P', Q')$ from (a) and the value of $e_5(P, Q)$ from (XI.8.3) are consistent with the relation that you found in (b).

11.17. (a) Let C_m be a cyclic group of order m. Prove that the average order of an element of C_m is

$$A(C_m) = \frac{1}{m} \sum_{d \mid m} d\phi(d).$$

(b) Prove that

$$\frac{1}{X} \sum_{m \leq X} A(C_m) = \frac{\zeta(3)}{2\zeta(2)} X + O(\log X),$$

where $\zeta(s)$ is the Riemann zeta function.

(c) Deduce that the expected order of a randomly chosen element in a randomly chosen cyclic group is proportional to the order of the group.

11.18. Let $p \geq 5$, let $E/\mathbb{F}_p$ be a supersingular elliptic curve, and let N be a prime such that $E(\mathbb{F}_p)$ contains a point of order N.

(a) Prove that $N^2 \mid \#E(\mathbb{F}_{p^2})$. (*Hint.* Use Exercise 5.15.)

(b) Deduce that one of the following statements is true:

(i) $E[N] \subset E(\mathbb{F}_{p^2})$.

(ii) $E(\mathbb{F}_{p^2})$ contains a point of order N^2.

Appendix A

Elliptic Curves in Characteristics 2 and 3

In this appendix we prove some of the results for elliptic curves in characteristics 2 and 3 that were omitted in the main body of the text. To simplify the computations, we begin by giving normal forms for the Weierstrass equations of such curves.

Proposition 1.1. *Let E/K be a curve given by a Weierstrass equation. Then, under the boxed assumptions, there is a substitution*

$$x = u^2 x' + r, \qquad y = u^3 y' + u^2 s x' + t, \qquad \text{with } u \in K^* \text{ and } r, s, t \in K,$$

that transforms the given Weierstrass equation into a Weierstrass equation of the indicated form.

(a) $\boxed{\text{char } K \neq 2, 3}$

$$y^2 = x^3 + a_4 x + a_6, \qquad \Delta = -16(4a_4^3 + 27a_6^2), \qquad j = 1728 \frac{4a_4^3}{4a_4^3 + 27a_6^2}.$$

(b) $\boxed{\text{char } K = 3 \quad \text{and} \quad j(E) \neq 0}$

$$y^2 = x^3 + a_2 x^2 + a_6, \qquad \Delta = -a_2^3 a_6, \qquad j = -a_2^3/a_6.$$

$\boxed{\text{char } K = 3 \quad \text{and} \quad j(E) = 0}$

$$y^2 = x^3 + a_4 x + a_6, \qquad \Delta = -a_4^3, \qquad j = 0.$$

(c) $\boxed{\text{char } K = 2 \quad \text{and} \quad j(E) \neq 0}$

$$y^2 + xy = x^3 + a_2 x^2 + a_6, \qquad \Delta = a_6, \qquad j = 1/a_6.$$

$\boxed{\text{char } K = 2 \quad \text{and} \quad j(E) = 0}$

$$y^2 + a_3 y = x^3 + a_4 x + a_6, \qquad \Delta = a_3^3, \qquad j = 0.$$

J.H. Silverman, *The Arithmetic of Elliptic Curves, Second Edition*, Graduate Texts in Mathematics 106, DOI 10.1007/978-0-387-09494-6_A,
© Springer Science+Business Media, LLC 2009

PROOF. (a) See (III §1).

(b) We start with a general Weierstrass equation and complete the square on the left. This gives an equation of the form

$$y^2 = x^3 + a_2 x^2 + a_4 x + a_6$$

with invariants

$$\Delta = a_2^2 a_4^2 - a_2^3 a_6 - a_4^3, \qquad j = a_2^3/\Delta.$$

(Remember that char $K = 3$.) If $j = 0$, then $a_2 = 0$, so the equation is already in the right shape. On the other hand, if $j \neq 0$, then $a_2 \neq 0$ and the substitution $x = x' + a_4/a_2$ eliminates the linear term.

(c) Again starting with a general Weierstrass equation

$$y^2 + a_1 xy + a_3 y = x^3 + a_2 x^2 + a_4 x + a_6,$$

an easy computation (in characteristic 2) yields

$$j = a_1^{12}/\Delta.$$

If $j \neq 0$, so $a_1 \neq 0$, then the substitution

$$x = a_1^2 x' + a_3/a_1, \qquad y = a_1^3 y' + (a_1^2 a_4 + a_3^2)/a_1^3,$$

gives an equation of the desired form. Finally, if $j = a_1 = 0$, then the substitution

$$x = x' + a_2, \qquad y = y',$$

has the desired effect.

Note that there is no deep theory involved in finding these substitutions. One merely looks at the transformation formulas given in Table 3.1 on page 45, sets various coefficients equal to 0 or 1, and chooses appropriate values for u, r, s, and t. □

It is now a simple matter to complete the proofs of (III.1.4) and (III.10.1), parts of which we restate here.

Proposition 1.2. (a) *A curve given by a Weierstrass equation is nonsingular if and only if the discriminant of the equation is nonzero.*

(b) *Two elliptic curves E/K and E'/K are isomorphic over $\bar{K}$ if and only if they have the same j-invariant.*

(c) *Let E/K be an elliptic curve. Then $\mathrm{Aut}(E)$ is a finite group of order:*

$$
\begin{array}{ll}
2 & \textit{if } j(E) \neq 0, 1728, \\
4 & \textit{if } j(E) = 1728 \textit{ and char } K \neq 2, 3, \\
6 & \textit{if } j(E) = 0 \textit{ and char } K \neq 2, 3, \\
12 & \textit{if } j(E) = 0 = 1728 \textit{ and char } K = 3, \\
24 & \textit{if } j(E) = 0 = 1728 \textit{ and char } K = 2.
\end{array}
$$

(See also Exercise A.1.)

PROOF. (a) We already proved most of this result in (III.1.4a). All that remains is to show that if $\operatorname{char}(K) = 2$ and $\Delta = 0$, then the curve is singular. But this is immediate from the normal forms given in (A.1.1c).

(b, c) Again referring to the proofs of (III.1.4b) and (III.10.1), we need only deal with the cases that $\operatorname{char}(K) = 2$ or 3. We use the normal forms given in (A.1.1bc) and consider four cases.

Case I. char $K = 3$ *and* $j(E) \neq 0$. In this case E and E' have Weierstrass equations of the form
$$y^2 = x^3 + a_2 x^2 + a_6.$$

The only substitutions preserving this type of equation are
$$x = u^2 x' \quad \text{and} \quad y = u^3 y'.$$

Since $j(E) = j(E')$, we have $a_2^3 a_6 = a_2'^3 a_6' \neq 0$, so taking $u^2 = a_2/a_2'$ gives an isomorphism from E to E'. Further, if $E = E'$, then we must have $u^2 = 1$, so $\operatorname{Aut}(E) \cong \{\pm 1\}$.

Case II. char $K = 3$ *and* $j(E) = 0$. In this case E and E' have Weierstrass equations of the form
$$y^2 = x^3 + a_4 x + a_6.$$

The substitutions preserving this form look like
$$x = u^2 x' + r \quad \text{and} \quad y = u^3 y'.$$

Note that we have $a_4, a_4' \neq 0$. An isomorphism from E to E' is obtained by choosing u and r to satisfy
$$u^4 = a_4'/a_4 \quad \text{and} \quad r^3 + a_4 r + a_6 - u^6 a_6'.$$

Further, if $E = E'$, then automorphisms of E have
$$u^4 = 1 \quad \text{and} \quad r^3 + a_4 r + (1 - u^2)a_6 = 0.$$

Since $a_4 \neq 0$, there are exactly 12 such pairs (u, r) making up $\operatorname{Aut}(E)$.

Case III. char $K = 2$ *and* $j(E) \neq 0$. In this case E and E' are given by equations of the form
$$y^2 + xy = x^3 + a_2 x^2 + a_6.$$

The substitutions preserving this form look like
$$x = x', \quad y = y' + sx'.$$

Since $j(E) = j(E')$, we have $a_6 = a_6' \neq 0$, so isomorphisms from E to E' come from taking s to be a root of the equation
$$s^2 + s + a_2 + a_2' = 0.$$

Similarly, the automorphisms of E are obtained by taking $s \in \{0, 1\}$.

Case IV. char $K = 2$ *and* $j(E) = 0$. Here E and E' have equations of the form

$$y^2 + a_3 y = x^3 + a_4 x + a_6,$$

and allowable substitutions look like

$$x = u^2 x' + s^2, \qquad y = u^3 y' + u^2 s x' + t.$$

By assumption, $a_3, a_3' \neq 0$, so in order to map E to E', we must choose u, s, and t to satisfy the equations

$$u^3 = a_3 / a_3', \qquad s^4 + a_3 s + a_4 - u^4 a_4' = 0,$$
$$t^2 + a_3 t + s^6 + a_4 s^2 + a_6 - u^6 a_6' = 0.$$

Finally, the automorphism group of E is given by the set of triples (u, s, t) satisfying the equations

$$u^3 = 1, \qquad s^4 + a_3 s + (1 - u) a_4 = 0, \qquad t^2 + a_3 t + s^6 + a_4 s^2 = 0.$$

Since $a_3 \neq 0$, we see that $\text{Aut}(E)$ has order 24. $\qquad \square$

The next proposition gives a normal form for Weierstrass equations that is similar to Legendre form, but valid in characteristic 2. This will allow us to easily complete the proofs of (VII.5.4c) and (VII.5.5).

Proposition 1.3. (Deuring Normal Form) *Let E/K be an elliptic curve over a field with* char $K \neq 3$. *Then E has a Weierstrass equation over $\bar{K}$ of the form*

$$E_\alpha : y^2 + \alpha x y + y = x^3, \qquad \alpha \in \bar{K}, \ \alpha^3 \neq 27.$$

This Weierstrass equation has discriminant and j-invariant

$$\Delta = \alpha^3 - 27 \qquad and \qquad j = \frac{\alpha^3 (\alpha - 24)^3}{\alpha^3 - 27}.$$

PROOF. The computation of Δ and j for E_α is an exercise. In order to show that E has an equation of the form E_α, we could find appropriate substitutions. However, using (A.1.2b), a quicker route is available. Given an elliptic curve E/K, we let $\alpha \in \bar{K}$ be a solution to the equation

$$\alpha^3 (\alpha^3 - 24)^3 - (\alpha^3 - 27) j(E) = 0.$$

Since char $K \neq 3$, we see that $\alpha^3 \neq 27$, so E_α is an elliptic curve with the same j-invariant as E. It follows from (A.1.2b) that E and E' are isomorphic over $\bar{K}$. $\qquad \square$

Corollary 1.4. *Let E/K be an elliptic curve defined over a local field, i.e., the field K comes equipped with a discrete valuation.*

(a) *There exists a finite extension K'/K such that E has either good or split multiplicative reduction over K'.*

(b) *E has potential good reduction if and only if its j-invariant is integral.*

PROOF. Let R be the ring of integers of K, let $\mathcal{M}$ be its maximal ideal, and let $k = R/\mathcal{M}$ be its residue field. From the proofs of (VII.5.4c) and (VII.5.5), we are left to deal with the case that char $K = 2$. In particular, we may assume that char $K \neq 3$. Replacing K by a finite extension, we choose an equation for E in Deuring normal form,

$$E_\alpha : y^2 + \alpha xy + y = x^3, \qquad \alpha^3 \neq 27.$$

This equation has

$$c_4 = \alpha(\alpha^3 - 24) \qquad \text{and} \qquad \Delta = \alpha^3 - 27.$$

(a) We consider three cases.

Case I. $\alpha \in R$ and $\alpha^3 \not\equiv 27$ (mod $\mathcal{M}$). Then $\Delta \not\equiv 0$ (mod $\mathcal{M}$), so the given equation has good reduction.

Case II. $\alpha \in R$ and $\alpha^3 \equiv 27$ (mod $\mathcal{M}$). Then

$$\Delta \equiv 0 \ (\text{mod } \mathcal{M}) \qquad \text{and} \qquad c_4^3 \equiv 3^6 \not\equiv 0 \ (\text{mod } \mathcal{M}),$$

so (VII.5.1b) tells us that the given equation for E has multiplicative reduction. To obtain split multiplicative reduction then requires at most a quadratic extension of K.

Case III. $\alpha \notin R$. Let π be a uniformizer for R and choose an integer $r \geq 1$ such that $\pi^r \alpha \in R^*$. Then the substitution $x = \pi^{-2r} x'$, $y = \pi^{-3r} y'$, gives an equation of the form

$$y'^2 + \beta x'y' + \pi^{3r} y' = x'^3 \qquad \text{with } \beta = \pi^r \alpha \in R^*.$$

This equation has

$$c_4' = \beta(\beta^3 - 24\pi^{3r}) \equiv \beta^4 \not\equiv 0 \quad (\text{mod } \mathcal{M}),$$
$$\Delta = \pi^{9r}(\beta^3 - 27\pi^{3r}) \equiv 0 \quad (\text{mod } \mathcal{M}),$$

so again from (VII.5.1b), the equation has multiplicative reduction. Further, the reduced curve is $y(y + \beta x) \equiv x^3$ (mod $\mathcal{M}$), so the reduction is split multiplicative.

(b) Suppose first that $j(E)$ is integral. Since $j(E)$ and α are related by

$$\alpha^3(\alpha^3 - 24)^3 - (\alpha^3 - 27)j(E) = 0,$$

the integrality of $j(E)$ implies that α is integral. Further, since the characteristic of k is different from 3, the equation implies that $\alpha^3 \not\equiv 27$ (mod $\mathcal{M}$). Thus the Deuring normal equation has integral coefficients and good reduction.

Conversely, suppose that E has potential good reduction. Replacing K by a finite extension, we can find a Weierstrass equation for E with integral coefficients and discriminant $\Delta \in R^*$. Then $c_4 \in R$ and $j(E) = c_4^3/\Delta \in R$. $\qquad \square$

Exercises

A.1. Let E/K be an elliptic curve with $j(E) = 0$. Strengthen (A.1.2) by showing that the automorphism group of E may be described as follows:

(a) If $\operatorname{char}(K) = 3$, then $\operatorname{Aut}(E)$ is the twisted product of C_4 and C_3, where C_n denotes a cyclic group of order n. Here C_3 is a normal subgroup of $\operatorname{Aut}(E)$ and the twisting is via the natural action of C_4 on C_3.

(b) If $\operatorname{char}(K) = 2$, then $\operatorname{Aut}(E)$ is the twisted product of C_3 and the quaternion group. The quaternion group is a normal subgroup of $\operatorname{Aut}(E)$, and if we write the quaternions as $\{\pm 1, \pm i \pm j, \pm k\}$, then a generator of C_3 acts by permuting i, j, and k.

A.2. Let K be a field of characteristic 2, and let E/K be a curve with $j(E) \neq 0$ given by a Weierstrass equation

$$y^2 + xy = x^3 + a_2 x^2 + a_6.$$

Let $\xi \in H^1\big(G_{\bar{K}/K}, \operatorname{Aut}(E)\big) = \operatorname{Hom}(G_{\bar{K}/K}, \mathbb{Z}/2\mathbb{Z})$, and let L/K be the quadratic extension corresponding to the character ξ. Suppose that L/K is the Artin–Schreier extension generated by a root of the polynomial

$$t^2 - t - D = 0 \qquad \text{for some } D \in K.$$

Prove that the twist of E by ξ as described in (X §5) is given by the equation

$$y^2 + xy = x^3 + (a_2 + D)x^2 + a_6.$$

A.3. Let E/K be an elliptic curve with Weierstrass coordinate functions x and y. Show that the differential dx is holomorphic if and only if $\operatorname{char}(K) = 2$ and $j(E) = 0$.

A.4. Let E/K and E'/K be elliptic curves over a *not necessarily perfect* field K. Suppose that $j(E) = j(E')$. Prove that E and E' are isomorphic over a *separable* extension L of K whose degree divides 24. If $j(E) \neq 0, 1728$, prove that L may be chosen to have degree 2.

Appendix B

Group Cohomology (H^0 and H^1)

In this appendix we discuss the basic properties of group cohomology that are used in Chapter VIII §2 and Chapter X. Since only H^0 and H^1 are needed in this book, we have restricted attention to these two groups. The reader desiring to learn more about group cohomology might consult [9], [105], [238], or [233].

B.1 Cohomology of Finite Groups

Let G be a finite group, and let M be an abelian group on which G acts. We denote the action of $\sigma \in G$ on $m \in M$ by $m \mapsto m^\sigma$. Then M is a (*right*) *G-module* if the action of G on M satisfies

$$m^1 = m, \qquad (m + m')^\sigma = m^\sigma + m'^\sigma, \qquad (m^\sigma)^\tau = m^{\sigma\tau}.$$

Let M and N be G-modules. A *G-module homomorphism* is a homomorphism

$$\phi : M \longrightarrow N$$

commuting with the action of G, i.e.,

$$\phi(m^\sigma) = \phi(m)^\sigma \qquad \text{for all } m \in M \text{ and all } \sigma \in G.$$

For a given G-module M, we are often interested in calculating the largest submodule of M on which G acts trivially.

Definition. The 0th *cohomology group of the G-module* M, which is denoted by M^G or $H^0(G, M)$, is the set

$$H^0(G, M) = \{m \in M : m^\sigma = m \text{ for all } \sigma \in G\}.$$

It is the submodule of M consisting of all *G-invariant elements*, i.e., elements that are fixed by G.

J.H. Silverman, *The Arithmetic of Elliptic Curves, Second Edition*, Graduate Texts in Mathematics 106, DOI 10.1007/978-0-387-09494-6_B,
© Springer Science+Business Media, LLC 2009

Let
$$0 \longrightarrow P \xrightarrow{\phi} M \xrightarrow{\psi} N \longrightarrow 0$$
be an *exact sequence of G-modules*, i.e., the maps ϕ and ψ are G-module homomorphisms with ϕ injective, ψ surjective, and $\mathrm{Image}(\phi) = \mathrm{Kernel}(\psi)$. It is easy to check that taking G-invariants gives an exact sequence
$$0 \longrightarrow P^G \xrightarrow{\phi} M^G \xrightarrow{\psi} N^G,$$
but the map on the right may not be surjective. In order to measure the lack of surjectivity, we make the following definitions.

Definition. Let M be a G-module. The *group of 1-cochains (from G to M)* is defined by
$$C^1(G, M) = \{\text{maps } \xi : G \to M\}.$$
The *group of 1-cocycles (from G to M)* is given by
$$Z^1(G, M) = \{\xi \in C^1(G, M) : \xi_{\sigma\tau} = \xi_\sigma^\tau + \xi_\tau \text{ for all } \sigma, \tau \in G\}.$$
The *group of 1-coboundaries (from G to M)* is defined by
$$B^1(G, M) = \left\{\xi \in C^1(G, M) : \begin{array}{l} \text{there exists an } m \in M \text{ such that} \\ \xi_\sigma = m^\sigma - m \text{ for all } \sigma \in G \end{array}\right\}.$$

One easily checks that $B^1(G, M) \subset Z^1(G, M)$. The 1st *cohomology group* of the G-module M is the quotient group
$$H^1(G, M) = \frac{Z^1(G, M)}{B^1(G, M)}.$$

In other words, $H^1(G, M)$ is the group of 1-cocycles $\xi : G \to M$ modulo the equivalence relation that two cocycles are identified if their difference has the form $\sigma \mapsto m^\sigma - m$ for some $m \in M$.

Remark 1.1. Notice that if the action of G on M is trivial, then
$$H^0(G, M) = M \qquad \text{and} \qquad H^1(G, M) = \mathrm{Hom}(G, M).$$

These both follow immediately from the definitions; for the latter, the 1-cocycles are homomorphisms and all of the 1-coboundaries are 0.

Proposition 1.2. *Let*
$$0 \longrightarrow P \xrightarrow{\phi} M \xrightarrow{\psi} N \longrightarrow 0$$
be an exact sequence of G-modules. *Then there is a long exact sequence*
$$0 \longrightarrow H^0(G, P) \longrightarrow H^0(G, M) \longrightarrow H^0(G, N)$$
$$\xrightarrow{\ \delta\ }$$
$$\longrightarrow H^1(G, P) \longrightarrow H^1(G, M) \longrightarrow H^1(G, N),$$

where the connecting homomorphism δ *is defined as follows:*

Let $n \in H^0(G, N) = N^G$. *Choose an* $m \in M$ *such that* $\psi(m) = n$ *and define a cochain* $\xi \in C^1(G, M)$ *by*

$$\xi_\sigma = m^\sigma - m.$$

Then the values of ξ *are in* P, *so* $\xi \in Z^1(G, P)$, *and we define* $\delta(n)$ *to be the cohomology class in* $H^1(G, P)$ *of the 1-cocycle* ξ.

PROOF. Everything follows from a straightforward, but tedious, diagram chase that we leave to the reader (Exercise B.1). Or see any of the references listed at the beginning of this appendix. □

Suppose now that H is a subgroup of G. Then any G-module is automatically an H-module. Further, if $\xi : G \to M$ is a 1-cochain, then by restricting the domain of ξ to H, we obtain an H-to-M cochain. It is clear that this process takes cocycles to cocycles and coboundaries to coboundaries, so in this way we obtain a *restriction homomorphism*

$$\mathrm{Res} : H^1(G, M) \longrightarrow H^1(H, M).$$

Suppose further that H is a normal subgroup of G. Then the submodule M^H of M consisting of elements fixed by H has a natural structure as a G/H-module. Let $\xi : G/H \to M^H$ be a 1-cochain from G/H to M^H. Then composing with the projection $G \to G/H$ and with the inclusion $M^H \subset M$ gives a G-to-M cochain

$$G \longrightarrow G/H \xrightarrow{\xi} M^H \subset M.$$

Again it is easy to see that if ξ is a cocycle or coboundary, then the new G-to-M cochain has the same property. This gives an *inflation homomorphism*

$$\mathrm{Inf} : H^1(G/H, M^H) \longrightarrow H^1(G, M).$$

Proposition 1.3. (Inflation–Restriction Sequence) *Let M be a G-module and let H be a normal subgroup of G. Then the following sequence is exact:*

$$0 \longrightarrow H^1(G/H, M^H) \xrightarrow{\mathrm{Inf}} H^1(G, M) \xrightarrow{\mathrm{Res}} H^1(H, M).$$

PROOF. From the definitions it is clear that $\mathrm{Res} \circ \mathrm{Inf} = 0$.

Next let $\xi : G/H \to M^H$ be a 1-cocycle with $\mathrm{Inf}\{\xi\} = 0$, where we use braces $\{\cdot\}$ to indicate the cohomology class of a cocycle. Thus there is an $m \in M$ such that $\xi_\sigma = m^\sigma - m$ for all $\sigma \in G$. But ξ depends only on $\sigma \pmod{H}$, so

$$m^\sigma - m = m^{\tau\sigma} - m \qquad \text{for all } \tau \in H.$$

Taking $\sigma = 1$, we find that $m^\tau - m = 0$ for all $\tau \in H$, so $m \in M^H$, and hence ξ is a G/H-to-M^H coboundary.

Finally, suppose that $\xi : G \to M$ is a 1-cocycle with $\mathrm{Res}\{\xi\} = 0$. Thus there is an $m \in M$ such that

$$\xi_\tau = m^\tau - m \qquad \text{for all } \tau \in H.$$

Subtracting the G-to-M coboundary $\sigma \mapsto m^\sigma - m$ from ξ, we may assume that $\xi_\tau = 0$ for all $\tau \in H$. Then the coboundary condition applied to $\sigma \in G$ and $\tau \in H$ yields

$$\xi_{\tau\sigma} - \xi_\tau^\sigma + \xi_\sigma = \xi_\sigma.$$

Thus ξ_σ depends only on the class of σ in G/H. Since H is a normal subgroup, there is a $\tau' \in H$ such that $\sigma\tau = \tau'\sigma$. Using the cocycle condition again, together with the fact that ξ is a map on G/H, gives

$$\xi_\sigma = \xi_{\tau'\sigma} = \xi_{\sigma\tau} = \xi_\sigma^\tau + \xi_\tau = \xi_\sigma^\tau.$$

This proves that ξ gives a map from G/H to M^H, and hence $\{\xi\} \in H^1(G/H, M^H)$. $\qquad\square$

B.2   Galois Cohomology

Let K be a perfect field, let $\bar{K}$ be an algebraic closure of K, and let $G_{\bar{K}/K}$ be the Galois group of $\bar{K}$ over K. We recall that $G_{\bar{K}/K}$ is the inverse limit of $G_{L/K}$ as L varies over all finite Galois extensions of K. Thus $G_{\bar{K}/K}$ is a profinite group, i.e., an inverse limit of finite groups. As such, it comes equipped with a topology in which a basis of open sets around the identity consists of the collection of normal subgroups having finite index in $G_{\bar{K}/K}$. These are the subgroups that are kernels of maps $G_{\bar{K}/K} \to G_{L/K}$ for finite Galois extensions L/K.

Definition. A (*discrete*) $G_{\bar{K}/K}$-*module* is an abelian group on which $G_{\bar{K}/K}$ acts such that the action is continuous for the profinite topology on $G_{\bar{K}/K}$ and the discrete topology on M. Equivalently, the action of $G_{\bar{K}/K}$ on M has the property that for all $m \in M$, the stabilizer of m,

$$\{\sigma \in G_{\bar{K}/K} : m^\sigma = m\},$$

is a subgroup of finite index in $G_{\bar{K}/K}$. Since all of our $G_{\bar{K}/K}$-modules will be discrete, we will normally just refer to them as $G_{\bar{K}/K}$-modules.

Example 2.1.1. Both $\bar{K}^+$ and $\bar{K}^*$ are $G_{\bar{K}/K}$-modules under the natural action of $G_{\bar{K}/K}$. This is true, because for any $x \in \bar{K}$, the extension $K(x)/K$ is finite, so the stabilizer of x has finite index in $G_{\bar{K}/K}$.

Example 2.1.2. In general, let $\mathcal{D}/K$ be any algebraic group. Then $\mathcal{D} = \mathcal{D}(\bar{K})$ is a $G_{\bar{K}/K}$-module, since again the coordinates of any point of $\mathcal{D}$ generate a finite extension of K.

The 0th cohomology group of a $G_{\bar{K}/K}$-module is defined just as in the case of finite groups.

Definition. The 0th *cohomology group of the* $G_{\bar{K}/K}$-*module* M is the group of $G_{\bar{K}/K}$-invariant elements of M,

$$M^{G_{\bar{K}/K}} = H^0(G_{\bar{K}/K}, M) = \{m \in M : m^\sigma = m \text{ for all } \sigma \in G_{\bar{K}/K}\}.$$

We could define H^1 for $G_{\bar{K}/K}$ as we did for finite groups, but instead we use the fact that the group $G_{\bar{K}/K}$ is profinite and the module is discrete to put some restrictions on the allowable cocycles.

Definition. Let M be a $G_{\bar{K}/K}$-module. A map $\xi : G_{\bar{K}/K} \to M$ is *continuous* if it is continuous for the profinite topology on $G_{\bar{K}/K}$ and the discrete topology on M. Equivalently, for each $m \in M$, the set $\xi^{-1}(m)$ is a union of cosets of subgroups of finite index in $G_{\bar{K}/K}$. The *group of continuous 1-cocycles from $G_{\bar{K}/K}$ to M*, denoted by $Z^1_{\mathrm{cont}}(G_{\bar{K}/K}, M)$, is the group of continuous maps $\xi : G_{\bar{K}/K} \to M$ satisfying the cocycle condition

$$\xi_{\sigma\tau} = \xi_\sigma^\tau + \xi_\tau.$$

We observe that $Z^1_{\mathrm{cont}}(G_{\bar{K}/K}, M)$ is a subgroup of the full group of 1-cocycles $Z^1(G_{\bar{K}/K}, M)$. Further, since M has the discrete topology, every coboundary

$$\sigma \mapsto m^\sigma - m$$

is automatically continuous. The 1st *cohomology group of the $G_{\bar{K}/K}$-module M* is the quotient group

$$H^1(G_{\bar{K}/K}, M) = \frac{Z^1_{\mathrm{cont}}(G_{\bar{K}/K}, M)}{B^1(G_{\bar{K}/K}, M)}.$$

Remark 2.2. Just as in the case of finite groups, if $G_{\bar{K}/K}$ acts trivially on M, we have

$$H^0(G_{\bar{K}/K}, M) = M \qquad \text{and} \qquad H^1(G_{\bar{K}/K}, M) = \mathrm{Hom}_{\mathrm{cont}}(G_{\bar{K}/K}, M),$$

where $\mathrm{Hom}_{\mathrm{cont}}$ denotes the group of continuous homomorphisms.

The fundamental exact sequences (B.1.2) and (B.1.3) carry over word for word from finite groups to profinite groups.

Proposition 2.3. *Let*

$$0 \longrightarrow P \overset{\phi}{\longrightarrow} M \overset{\psi}{\longrightarrow} N \longrightarrow 0$$

be an exact sequence of $G_{\bar{K}/K}$-modules. Then there is a long exact sequence

$$0 \longrightarrow H^0(G_{\bar{K}/K}, P) \longrightarrow H^0(G_{\bar{K}/K}, M) \longrightarrow H^0(G_{\bar{K}/K}, N)$$
$$\overset{\delta}{}$$
$$\longrightarrow H^1(G_{\bar{K}/K}, P) \longrightarrow H^1(G_{\bar{K}/K}, M) \longrightarrow H^1(G_{\bar{K}/K}, N),$$

where the connecting homomorphism δ is defined as in (B.1.2).

Let M be a $G_{\bar{K}/K}$-module and let L/K be a finite Galois extension. Then $G_{\bar{K}/L}$ is a subgroup of finite index in $G_{\bar{K}/K}$, so M is naturally a $G_{\bar{K}/L}$-module. This leads to a *restriction map* on cohomology,

$$\text{Res} : H^1(G_{\bar{K}/K}, M) \longrightarrow H^1(G_{\bar{K}/L}, M).$$

Further, $G_{\bar{K}/L}$ is a normal subgroup of $G_{\bar{K}/K}$, and the quotient $G_{\bar{K}/K}/G_{\bar{K}/L}$ is the finite group $G_{L/K}$. The submodule of invariants $M^{G_{\bar{K}/L}}$ has a natural structure as a $G_{L/K}$-module. Then any 1-cocycle $\xi : G_{L/K} \to M^{G_{\bar{K}/L}}$ becomes a 1-cocycle for $G_{\bar{K}/K}$ via the composition

$$G_{\bar{K}/K} \longrightarrow G_{L/K} \xrightarrow{\xi} M^{G_{\bar{K}/L}} \subset M.$$

This gives an *inflation map*

$$\text{Inf} : H^1(G_{L/K}, M^{G_{\bar{K}/L}}) \longrightarrow H^1(G_{\bar{K}/K}, M).$$

Proposition 2.4. (Inflation–Restriction Sequence) *With notation as above, there is an exact sequence*

$$0 \longrightarrow H^1(G_{L/K}, M^{G_{\bar{K}/L}}) \xrightarrow{\text{Inf}} H^1(G_{\bar{K}/K}, M) \xrightarrow{\text{Res}} H^1(G_{\bar{K}/L}, M).$$

PROOF. Virtually identical to the proof of (B.1.3). □

The next proposition describes some fundamental facts about the cohomology of the additive and multiplicative groups of a field.

Proposition 2.5. *Let K be a field.*
(a) $H^1(G_{\bar{K}/K}, \bar{K}^+) = 0$.
(b) $H^1(G_{\bar{K}/K}, \bar{K}^*) = 0$. *This is* Hilbert's Theorem 90.
(c) *Assume that either* $\text{char}(K) = 0$ *or that* $\text{char}(K)$ *does not divide m. Then*

$$H^1(G_{\bar{K}/K}, \boldsymbol{\mu}_m) \cong K^*/(K^*)^m.$$

PROOF. (a) [233, Chapter X, Proposition 1].
(b) [233, Chapter X, Proposition 2].
(c) Consider the following exact sequence of $G_{\bar{K}/K}$-modules:

$$1 \longrightarrow \boldsymbol{\mu}_m \longrightarrow \bar{K}^* \xrightarrow{z \to z^m} \bar{K}^* \longrightarrow 1.$$

Applying (B.2.3) yields the long exact sequence

$$\longrightarrow K^* \xrightarrow{z \to z^m} K^* \xrightarrow{\delta} H^1(G_{\bar{K}/K}, \boldsymbol{\mu}_m) \longrightarrow H^1(G_{\bar{K}/K}, \bar{K}^*) \longrightarrow .$$

From (b) we know that $H^1(G_{\bar{K}/K}, \bar{K}^*) = 0$, which gives the desired result. □

B.3   Nonabelian Cohomology

We again start with a finite group G and a group M on which G acts, but we no longer require that M be abelian. (To emphasize the possible noncommutativity of M, we write the group law on M multiplicatively.) As always, the 0th *cohomology group of M* is the subgroup of G-invariant elements,

$$H^0(G, M) = M^G = \{m \in M : m^\sigma = m \text{ for all } \sigma \in G\}.$$

We further define the *set of 1-cocycles of G into M* to be the set of maps

$$\xi : G \longrightarrow M \qquad \text{satisfying} \quad \xi_{\sigma\tau} = (\xi_\sigma)^\tau \xi_\tau \quad \text{for all } \sigma, \tau \in G.$$

We emphasize that the set of 1-cocycles does not, in general, form a group, since the noncommutativity of M may prevent the product of two cocycles from being a cocycle.

Two 1-cocycles ξ and ζ are said to be *cohomologous* if there is an $m \in M$ such that

$$m^\sigma \xi_\sigma = \zeta_\sigma m \qquad \text{for all } \sigma \in G.$$

It is easy to check that this defines an equivalence relation on the set of 1-cocycles. The 1st *cohomology set of M*, denoted by $H^1(G, M)$, is the set of 1-cocycles modulo this relation. We observe that $H^1(G, M)$ has a distinguished element, namely the equivalence class of the identity cocycle. Thus $H^1(G, M)$ is a *pointed set*, i.e., a set with a distinguished element.

Continuing as in (B §2), we say that the Galois group $G_{\bar{K}/K}$ acts *discretely* on a (possibly nonabelian) group M if the stabilizer of any element of M is a subgroup of finite index in $G_{\bar{K}/K}$. We again define a *continuous 1-cocycle from $G_{\bar{K}/K}$ to M* to be a map $\xi : M \to G_{\bar{K}/K}$ that satisfies the cocycle condition and is continuous for the profinite topology on $G_{\bar{K}/K}$ and the discrete topology on M. Two cocycles ξ and ζ are *cohomologous* if $m^\sigma \xi_\sigma = \zeta_\sigma m$ for some $m \in M$, and the 0th *cohomology group* and the 1st *cohomology group* of M are defined as usual by

$$H^0(G_{\bar{K}/K}, M) = M^{G_{\bar{K}/K}} = \{m \in M : m^\sigma = m \text{ for all } \sigma \in G_{\bar{K}/K}\},$$

$$H^1(G_{\bar{K}/K}, M) = \frac{\text{set of continuous 1-cocycles from } G_{\bar{K}/K} \text{ to } M}{\text{equivalence of cohomologous 1-cocycles}}.$$

Example 3.1. If $\mathcal{D}/K$ is any algebraic group, then there is a natural action of $G_{\bar{K}/K}$ on $\mathcal{D} = \mathcal{D}(\bar{K})$, and as explained earlier (B.2.1.2), this action is discrete. It is clear that

$$H^0(G_{\bar{K}/K}, \mathcal{D}) = \mathcal{D}(K)$$

is the subgroup of K-rational points of $\mathcal{D}$. The structure of $H^1(G_{\bar{K}/K}, \mathcal{D})$ is harder to describe, but for the general linear group there is the following generalization of Hilbert's Theorem 90.

Proposition 3.2. *For all integers $n \geq 1$,*

$$H^1\big(G_{\bar{K}/K}, \mathrm{GL}_n(\bar{K})\big) = \{1\}.$$

PROOF. [233, Chapter X, Proposition 3].   □

Exercises

B.1. Prove that the sequence in (B.1.2) is exact.

B.2. Let G be a finite group and let M be a G-module.
(a) If G has order n, prove that every element of $H^1(G, M)$ is killed by n.
(b) If M is finitely generated as a G-module, prove that $H^1(G, M)$ is finite.

B.3. Let G be a finite group, let M be a G-module, and let H be a normal subgroup of G.
(a) Show that there is a natural action of G/H on $H^1(H, M)$.
(b) Prove that the image of the restriction map $\mathrm{Res} : H^1(G, M) \to H^1(H, M)$ lies in the subgroup of $H^1(H, M)$ fixed by G/H. This allows (B.1.3) to be refined to

$$0 \longrightarrow H^1(G/H, M^H) \xrightarrow{\ \mathrm{Inf}\ } H^1(G, M) \xrightarrow{\ \mathrm{Res}\ } H^1(H, M)^{G/H}.$$

This exact sequence is a piece of the Serre–Hochschild spectral sequence for group cohomology [115].

B.4. Let M be a (discrete) $G_{\bar{K}/K}$-module. For any tower of fields $F/L/K$, there is an inflation map

$$H^1(G_{L/K}, M^{G_{\bar{K}/L}}) \longrightarrow H^1(G_{F/K}, M^{G_{\bar{K}/F}}).$$

Prove that these inflation maps fit together to form a direct system and that there is an isomorphism

$$H^1(G_{\bar{K}/K}, M) \cong \varinjlim H^1(G_{L/K}, M^{G_{\bar{K}/L}}),$$

where the direct limit is over all finite Galois extensions L/K. This provides an alternative definition for the cohomology of $G_{\bar{K}/K}$-modules.

B.5. Let G be a finite group, and let E and A be groups on which G acts. Assume that E is abelian and that A acts on E in a manner compatible with the action of G. In other words, assume that $(\alpha x)^\sigma = \alpha^\sigma x^\sigma$ for all $\alpha \in A$, $x \in E$, and $\sigma \in G$. The *twisted product of E and A*, denoted by $E \ltimes A$, is the group whose underlying set is $E \times A$ and whose group law is given by

$$(x, \alpha) \star (y, \beta) = \big(x(\alpha y), \alpha\beta\big).$$

Notice that G acts on $E \ltimes A$ via $(x, \alpha)^\sigma = (x^\sigma, \alpha^\sigma)$.
(a) Prove that there are exact sequences

$$1 \longrightarrow E \longrightarrow E \ltimes A \longrightarrow A \longrightarrow 1$$

and

$$1 \longrightarrow E^G \longrightarrow E \ltimes A^G \longrightarrow A^G \longrightarrow 1.$$

(b) Any $\alpha \in A^G$ gives a G-isomorphism $\alpha : E \to E$, and so induces an automorphism of $H^1(G, E)$. Show that two elements $\xi_1, \xi_2 \in H^1(G, E)$ have the same image under the natural map $H^1(G, E) \to H^1(G, E \ltimes A)$ if and only if there is an $\alpha \in A^G$ such that $\alpha\xi_1 = \xi_2$.

B.6. Let G be a finite group, let M be a G-module, and let H_1 and H_2 be subgroups of G. Suppose further that H_1 and H_2 are conjugate, i.e., $H_1 = \sigma H_2 \sigma^{-1}$ for some $\sigma \in G$. Prove that the restriction maps

$$\mathrm{Res} : H^1(G, M) \longrightarrow H^1(H_1, M) \quad \text{and} \quad \mathrm{Res} : H^1(G, M) \longrightarrow H^1(H_2, M)$$

have the same kernel.

Appendix C

Further Topics: An Overview

In this volume we have tried to give an essentially self-contained introduction to the basic theory of the arithmetic of elliptic curves. Unfortunately, due to limitations of time and space, many important topics have had to be omitted. This appendix contains a *very brief* introduction to some of the material that could not be included in the main body of the text. Further details may be found in the companion volume [266] and in the references listed at the end of each section.

The first ten topics covered in this appendix were originally supposed to form Chapters XI through XX of this book, so they have been numbered as sections 11 through 20. An additional section has been added for the second edition. The contents of Appendix C are as follows:

C.11 Complex Multiplication

The Kronecker–Weber theorem says that the maximal abelian extension $\mathbb{Q}^{\mathrm{ab}}$ of $\mathbb{Q}$ is generated by roots of unity, and so the class field theory of $\mathbb{Q}$ is given explicitly by an isomorphism

$$G_{\mathbb{Q}^{\mathrm{ab}}/\mathbb{Q}} \cong \prod_p \mathbb{Z}_p^*.$$

J.H. Silverman, *The Arithmetic of Elliptic Curves, Second Edition*, Graduate Texts in Mathematics 106, DOI 10.1007/978-0-387-09494-6_C,
© Springer Science+Business Media, LLC 2009

The theory of complex multiplication provides a similar description for the abelian extensions of imaginary quadratic fields.

Let $\mathcal{K}/\mathbb{Q}$ be an imaginary quadratic field, let $\mathcal{R} \subset \mathcal{K}$ be the ring of integers of $\mathcal{K}$, and let $\mathcal{Cl}(\mathcal{R})$ be the ideal class group of $\mathcal{R}$. If we fix an embedding $\mathcal{K} \subset \mathbb{C}$, then each ideal Λ of $\mathcal{R}$ is a lattice $\Lambda \subset \mathbb{C}$, so we may consider the elliptic curve $\mathbb{C}/\Lambda$. From (VI.4.1) we have

$$\mathrm{End}(\mathbb{C}/\Lambda) \cong \{\alpha \in \mathbb{C} : \alpha\Lambda \subset \Lambda\} = \mathcal{R}.$$

Further, (VI.4.1.1) says that up to isomorphism, the elliptic curve $\mathbb{C}/\Lambda$ depends only on the ideal class $\{\Lambda\} \in \mathcal{Cl}(\mathcal{R})$.

Conversely, suppose that $E/\mathbb{C}$ satisfies $\mathrm{End}(E) \cong \mathcal{R}$. Then (VI.5.11) implies that $E(\mathbb{C}) \cong \mathbb{C}/\Lambda$ for a unique ideal class $\{\Lambda\} \in \mathcal{Cl}(\mathcal{R})$. We have proven the following result.

Proposition 11.1. *With notation as above, there is a one-to-one correspondence between ideal classes in $\mathcal{Cl}(\mathcal{R})$ and isomorphism classes of elliptic curves $E/\mathbb{C}$ with $\mathrm{End}(E) \cong \mathcal{R}$.*

Corollary 11.1.1. (a) *There are only finitely many isomorphism classes of elliptic curves $E/\mathbb{C}$ with $\mathrm{End}(E) \cong \mathcal{R}$.*
(b) *Let $E/\mathbb{C}$ be an elliptic curve with $\mathrm{End}(E) \cong \mathcal{R}$. Then $j(E)$ is algebraic over $\mathbb{Q}$.*

PROOF. (a) Clear from (C.11.1), since $\mathcal{Cl}(\mathcal{R})$ is finite.
(b) Let $\sigma \in \mathrm{Aut}(\mathbb{C}/\mathbb{Q})$. Then $\mathrm{End}(E^\sigma) \cong \mathrm{End}(E) \cong \mathcal{R}$. It follows from (a) that $\{E^\sigma : \sigma \in \mathrm{Aut}(\mathbb{C}/\mathbb{Q})\}$ contains only finitely many isomorphism classes of elliptic curves. Since $j(E^\sigma) = j(E)^\sigma$, the set $\{j(E)^\sigma : \sigma \in \mathrm{Aut}(\mathbb{C}/\mathbb{Q})\}$ is finite. It follows that $j(E)$ is algebraic over $\mathbb{Q}$. $\square$

Actually, we can say quite a bit more about the j-invariant of an elliptic curve having complex multiplication. For $\{\Lambda\} \in \mathcal{Cl}(\mathcal{R})$, we denote the j-invariant of $\mathbb{C}/\Lambda$ by $j(\Lambda)$.

Theorem 11.2. (Weber, Fueter) *Let $\{\Lambda\} \in \mathcal{Cl}(\mathcal{R})$.*
(a) $j(\Lambda)$ *is an algebraic integer.*
(b) $[\mathcal{K}(j(\Lambda)) : \mathcal{K}] = [\mathbb{Q}(j(\Lambda)) : \mathbb{Q}].$
(c) *The field $\mathcal{H} = \mathcal{K}(j(\Lambda))$ is the maximal unramified abelian extension of $\mathcal{K}$, i.e., $\mathcal{H}$ is the Hilbert class field of $\mathcal{K}$.*
(d) *Let $\{\Lambda_1\}, \ldots, \{\Lambda_h\}$ be a complete set of representatives for $\mathcal{Cl}(\mathcal{R})$. Then $j(\Lambda_1), \ldots, j(\Lambda_h)$ is a complete set of $G_{\bar{\mathcal{K}}/\mathcal{K}}$ conjugates for $j(\Lambda)$.*

PROOF. (a) The original proof of the integrality of $j(\Lambda)$ uses the theory of modular functions; see, for example, [140, Chapter 5, Theorem 4], [249, §4.6], or [266, II §6]. An algebraic proof that generalizes to higher dimensions can be based on the criterion of Néron–Ogg–Shafarevich; see [239, Theorem 6], [266, II §6], and Exercise 7.10. There is also a proof that uses Tate curves (C §14); see [266, V.6.3].
(b), (c), (d) See [140, Chapter 10, Theorem 1], [230], [249, Theorem 5.7], or [266, II.4.3]. $\square$

Example 11.3.1. Suppose that $E/\mathbb{Q}$ is an elliptic curve with complex multiplication, and suppose that $\text{End}(E)$ is the full ring of integers $\mathcal{R}$ in the field $\mathcal{K} = \text{End}(E) \otimes \mathbb{Q}$. (Note that (VI.5.5) tells us that $\mathcal{K}$ is imaginary quadratic.) Since $j(E) \in \mathbb{Q}$, it follows from (C.11.2c) that

$$\mathcal{H} = \mathcal{K}(j(E)) = \mathcal{K},$$

and thus that $\mathcal{K}$ has class number one.

Conversely, if $\mathcal{K}/\mathbb{Q}$ is an imaginary quadratic field with class number one, then (C.11.2bc) implies that

$$j(\Lambda) \in \mathbb{Q} \quad \text{for all } \{\Lambda\} \in \mathcal{C\ell}(\mathcal{R}).$$

For example, this is true for $\Lambda = \mathcal{R}$. Hence $\mathbb{C}/\Lambda$ is (analytically) isomorphic to an elliptic curve $E/\mathbb{Q}$ satisfying $j(E) = j(\Lambda)$ and $\text{End}(E) \cong \mathcal{R}$.

Baker, Heegner, and Stark have shown that there are exactly nine imaginary quadratic fields whose ring of integers has class number one, namely the fields $\mathbb{Q}(\sqrt{-d})$ with $d \in \{1, 2, 3, 7, 11, 19, 43, 67, 163\}$. Hence there are only 9 possible j-invariants for elliptic curves E defined over $\mathbb{Q}$ for which $\text{End}(E)$ is the full ring of integers in $\text{End}(E) \otimes \mathbb{Q}$.

Example 11.3.2. If we relax the requirement that $\text{End}(E)$ be the full ring of integers of $\mathcal{K}$ and allow $\text{End}(E)$ to be an arbitrary order in $\mathcal{K}$, then $\text{End}(E)$ has the form $\mathbb{Z} + f\mathcal{R}$ for some integer $f \in \mathbb{Z}$; see Exercise 3.20. One can show in this case that

$$[\mathcal{K}(j(E)) : \mathcal{K}] = \#\mathcal{C\ell}(\mathbb{Z} + f\mathcal{R}),$$

where $\mathcal{C\ell}(\mathbb{Z} + f\mathcal{R})$ is the group of rank-1 projective $(\mathbb{Z} + f\mathcal{R})$-modules. In particular, if $j(E) \in \mathbb{Q}$, then $\mathcal{C\ell}(\mathbb{Z} + f\mathcal{R}) = \{1\}$. It turns out that there are only four such orders having $f \geq 2$, namely

$$\mathbb{Q}(\sqrt{-1}), \mathbb{Q}(\sqrt{-3}), \mathbb{Q}(\sqrt{-7}) \text{ with } f = 2 \text{ and } \mathbb{Q}(\sqrt{-3}) \text{ with } f = 3.$$

Combining this with (C.11.3.1), we see that up to isomorphism over $\bar{\mathbb{Q}}$, there are exactly 13 elliptic curves $E/\mathbb{Q}$ having complex multiplication. Of course, each $\bar{\mathbb{Q}}$-isomorphism class contains infinitely many $\mathbb{Q}$-isomorphism classes (X.5.4). For example, we studied the family of elliptic curves $E/\mathbb{Q}$ having $\text{End}(E) \cong \mathbb{Z}[\sqrt{-1}]$ in (X §6).

Returning now to the situation in (C.11.2), let $\{\Lambda\} \in \mathcal{C\ell}(\mathcal{R})$. Then (C.11.2) tells us that the Galois group $G_{\mathcal{H}/\mathcal{K}}$ acts on $\mathcal{K}(j(\Lambda))$. This action can be described quite precisely in terms of the Artin map.

Theorem 11.4. (Hasse) *Let* $\{\Lambda\} \in \mathcal{C\ell}(\mathcal{R})$, *and let* $\mathcal{H} = \mathcal{K}(j(\Lambda))$ *be as in* (C.11.2). *For each prime ideal* $\mathfrak{p}$ *of* $\mathcal{R}$, *let* $\text{Frob}(\mathfrak{p}) \in G_{\mathcal{H}/\mathcal{K}}$ *be the Frobenius element corresponding to* $\mathfrak{p}$. *Suppose that there is an elliptic curve with* j-*invariant* $j(\Lambda)$ *defined over* $\mathcal{H}$ *that has good reduction at all primes of* $\mathcal{H}$ *lying over* $\mathfrak{p}$. *Then*

$$j(\Lambda)^{\text{Frob}(\mathfrak{p})} = j(\Lambda \cdot \mathfrak{p}^{-1}),$$

where $\Lambda \cdot \mathfrak{p}^{-1}$ *is the usual product of fractional ideals in* $\mathcal{K}$.

PROOF. See [140, Chapter 10, Theorem 1], [230], [249, Theorem 5.7], or [266, II.4.3]. □

Suppose now that E/K is an elliptic curve with complex multiplication over K, i.e., $\text{End}_K(E) \neq \mathbb{Z}$. Then the fact that $G_{\bar{K}/K}$ and $\text{End}_K(E)$ commute with one another in their action on the Tate module $T_\ell(E)$ implies that the action of $G_{\bar{K}/K}$ is abelian. (This is essentially Schur's lemma; see Exercise 3.24.) Thus the field $K(E_{\text{tors}})$ obtained by adjoining to K the coordinates of all of the torsion points of E is an abelian extension of K.

We return now to the situation that $\{\Lambda\} \in \mathcal{Cl}(\mathcal{R})$ and $\mathcal{H} = \mathcal{K}(j(\Lambda))$, and we let $E/\mathcal{H}$ be an elliptic curve with j-invariant $j(\Lambda)$. Then $\mathcal{H}(E_{\text{tors}})$ is an abelian extension of $\mathcal{H}$, but it is not generally an abelian extension of $\mathcal{K}$. However, it turns out that $\mathcal{H}(E_{\text{tors}})$ contains $\mathcal{K}^{\text{ab}}$ and that $\mathcal{H}(E_{\text{tors}})/\mathcal{K}^{\text{ab}}$ is an abelian extension whose Galois group is (usually) a product of groups of order 2. To create $\mathcal{K}^{\text{ab}}$ itself, we instead adjoin (essentially) the x-coordinates of the torsion points.

More precisely, for any elliptic curve E/K, we define a *Weber function on E/K* to be a morphism defined over K of the form

$$\phi_E : E \longrightarrow E/\text{Aut}(E) \cong \mathbb{P}^1.$$

(See Exercise 3.13 for the definition of the quotient curve $E/\text{Aut}(E)$.) Classically, if E is given by a Weierstrass equation

$$E : y^2 = 4x^3 - g_2 x - g_3, \qquad g_2, g_3 \in \mathbb{C},$$

with discriminant $\Delta = g_2^3 - 27g_3^2$, then one defines the *Weber function* quite explicitly by the formula

$$\phi_E(P) = \begin{cases} (g_2 g_3/\Delta)x(P) & \text{if } j(E) \neq 0, 1728, \\ (g_2^2/\Delta)x(P)^2 & \text{if } j(E) = 1728, \\ (g_3/\Delta)x(P)^3 & \text{if } j(E) = 0. \end{cases}$$

Note that although g_2 and g_3 are in $\mathbb{C}$, the map $\phi_E : E \to \mathbb{P}^1$ is independent of the choice of Weierstrass equation for E, and thus ϕ_E is defined over any field of definition for E.

Theorem 11.5. *Let $\mathcal{K}$ be an imaginary quadratic field, let $\mathcal{R} \subset \mathcal{K}$ be its ring of integers, and let $E/\mathbb{C}$ be an elliptic curve with $\text{End}(E) \cong \mathcal{R}$.*
(a) The maximal unramified abelian extension of $\mathcal{K}$ is $\mathcal{K}(j(E))$.
(b) The maximal abelian extension $\mathcal{K}^{\text{ab}}$ of $\mathcal{K}$ is given by

$$\mathcal{K}^{\text{ab}} = \mathcal{K}(j(E); \phi_E(T), T \in E_{\text{tors}}),$$

i.e., $\mathcal{K}^{\text{ab}}$ is the field obtained by adjoining to $\mathcal{K}$ the j-invariant of E and the value of a Weber function at all of the torsion points of E.

PROOF. (a) This is a restatement of (C.11.2c).

(b) See [140, Chapter 10, Theorem 2], [230], [249, Corollary 5.6], or [266, II.5.7].

□

Remark 11.6. Let $\{\Lambda\}$ be an ideal class of $\mathcal{R}$, for example $\Lambda = \mathcal{R}$. Then, in (C.11.5), we may take E to be the elliptic curve $E(\mathbb{C}) \cong \mathbb{C}/\Lambda$ given by the Weierstrass equation

$$E : y^2 = 4x^3 - g_2(\Lambda)x - g_3(\Lambda).$$

(See (VI §3) for infinite series expansions of $g_2(\Lambda)$ and $g_3(\Lambda)$.) Then the Weber function

$$\phi_\Lambda : \mathbb{C}/\Lambda \longrightarrow \mathbb{C}$$

is given analytically by

$$\phi_\Lambda(z) = \begin{cases} (g_2(\Lambda)g_3(\Lambda)/\Delta)\wp(z,\Lambda) & \text{if } j(\Lambda) \neq 0, 1728, \\ (g_2(\Lambda)^2/\Delta)\wp(z,\Lambda)^2 & \text{if } j(\Lambda) = 1728, \\ (g_3(\Lambda)/\Delta)\wp(z,\Lambda)^3 & \text{if } j(\Lambda) = 0. \end{cases}$$

Since (C.11.5) says that $\mathcal{K}^{ab}$ is generated by $j(\Lambda)$ and $\phi_\Lambda(t)$ for $t \in \mathbb{Q}\Lambda \subset \mathbb{C}$, we see that $\mathcal{K}^{ab}$ is given explicitly by the values of an analytic function evaluated at the points of finite order on the complex torus $\mathbb{C}/\Lambda$. Notice the similarity with the situation over $\mathbb{Q}$, where $\mathbb{Q}^{ab}$ is generated by the values of the analytic function $\phi(z) = e^{2\pi i z}$ at the points of finite order on the cylinder $\mathbb{C}/\mathbb{Z}$.

Remark 11.7. Just as in (C.11.4), one can use the Artin map to describe the action of $G_{\mathcal{K}^{ab}/\mathcal{K}}$ on the elements $\phi_E(T)$ that generate $\mathcal{K}^{ab}/\mathcal{K}$. See, for example, [140, Chapter 10, Lemma 1 and Theorem 3], [249, Theorem 5.4], or [266, II.8.2].

References. [140], [230], [249], [266]. For generalizations to abelian varieties, see [137], [239], [250].

C.12 Modular Functions

We have seen (VI.5.1.1) that every elliptic curve $E/\mathbb{C}$ is analytically isomorphic to a complex torus $\mathbb{C}/\Lambda$, where E uniquely determines the lattice $\Lambda \subset \mathbb{C}$ up to homothety. Associated to the lattice Λ are the Eisenstein series $G_{2k}(\Lambda)$, the discriminant $\Delta(\Lambda)$, and the j-invariant $j(\Lambda)$. One easily verifies the homogeneity properties (Exercise 6.6)

$$G_{2k}(\alpha\Lambda) = \alpha^{-2k}G_{2k}(\Lambda), \qquad \Delta(\alpha\Lambda) = \alpha^{-12}\Delta(\Lambda), \qquad j(\alpha\Lambda) = j(\Lambda).$$

These functions have as their domain the space of lattices. Using homogeneity, it is enough to study them on the space of lattices modulo homothety. In order to do this, we set the following notation:

$$\mathbb{H} = \{\tau \in \mathbb{C} : \mathrm{Im}(\tau) > 0\},$$
$$\Lambda_\tau = \mathbb{Z} + \mathbb{Z}\tau \quad \text{for } \tau \in \mathbb{H},$$
$$G_{2k}(\tau) = G_{2k}(\Lambda_\tau), \qquad \Delta(\tau) = \Delta(\Lambda_\tau), \qquad j(\tau) = j(\Lambda_\tau).$$

It is clear that every lattice Λ is homothetic to Λ_τ for some $\tau \in \mathbb{H}$. To describe when two values of τ give the same lattice, we observe that the group

$$\mathrm{SL}_2(\mathbb{Z}) = \left\{ \begin{pmatrix} a & b \\ c & d \end{pmatrix} : a, b, c, d \in \mathbb{Z} \text{ and } ad - bc = 1 \right\}$$

acts on $\mathbb{H}$ by linear fractional transformations,

$$\gamma = \begin{pmatrix} a & b \\ c & d \end{pmatrix} : \mathbb{H} \longrightarrow \mathbb{H}, \qquad \gamma(\tau) = \frac{a\tau + b}{c\tau + d}.$$

The next proposition describes some properties of this action.

Proposition 12.1. (a) *The group* $\mathrm{SL}_2(\mathbb{Z})$ *acts properly discontinuously on* $\mathbb{H}$.
(b) *The region*

$$\mathcal{F} = \left\{ \tau \in \mathbb{H} : |\mathrm{Re}(\tau)| \leq \frac{1}{2} \text{ and } |\tau| \geq 1 \right\}$$

is a fundamental domain for $\mathbb{H}/\mathrm{SL}_2(\mathbb{Z})$. *More precisely, the natural map* $\mathcal{F} \to \mathbb{H}/\mathrm{SL}_2(\mathbb{Z})$ *is surjective and its restriction to the interior of* $\mathcal{F}$ *is injective.*

(c) *Let*

$$S = \begin{pmatrix} 0 & -1 \\ 1 & 0 \end{pmatrix} \qquad \text{and} \qquad T = \begin{pmatrix} 1 & 1 \\ 0 & 1 \end{pmatrix}.$$

They satisfy $S^2 = -1$ *and* $(ST)^3 = -1$. *The modular group*

$$\mathrm{PSL}_2(\mathbb{Z}) = \mathrm{SL}_2(\mathbb{Z})/\pm 1$$

is the free product of the cyclic groups of orders 2 and 3 generated by S *and* ST. *In particular,* S *and* T *generate* $\mathrm{PSL}_2(\mathbb{Z})$.

PROOF. See [5, Theorems 2.1 and 2.3], [232, VII §1], or [266, I.1.5, I.1.6]. □

Corollary 12.1.1. *Every lattice* $\Lambda \subset \mathbb{C}$ *is homothetic to a lattice* Λ_τ *for some* $\tau \in \mathcal{F}$.

Figure 3.1 illustrates the fundamental domain $\mathcal{F}$ and its translates under various elements of $\mathrm{SL}_2(\mathbb{Z})$.

Remark 12.2. Any two bases $\{\omega_1, \omega_2\}$ and $\{\omega_1', \omega_2'\}$ for a lattice Λ are related by a change-of-basis formula

$$\omega_1' = a\omega_1 + b\omega_2, \qquad \omega_2' = c\omega_1 + d\omega_2,$$

with $a, b, c, d \in \mathbb{Z}$ and $ad - bc = \pm 1$. If we use homotheties to replace these bases by bases of the form $\{1, \tau\}$ and $\{1, \tau'\}$ with $\tau, \tau' \in \mathbb{H}$, then the above change-of-basis action on the ω values becomes exactly the linear fractional action of $\mathrm{SL}_2(\mathbb{Z})$ on the τ values described earlier.

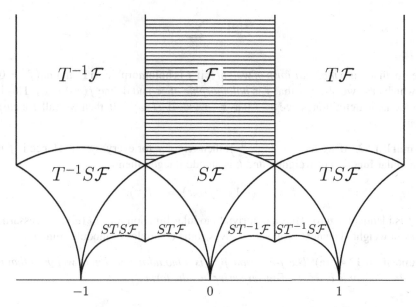

Figure 3.1: The fundamental domain $\mathcal{F}$ and some of its $\mathrm{PSL}_2(\mathbb{Z})$-translates.

The function $G_{2k}(\Lambda)$ depends only on the lattice Λ_τ and not on any particular choice of basis. However, if $\Lambda_\tau = \Lambda_{\tau'}$ for some $\tau, \tau' \in \mathcal{H}$, then $G_{2k}(\tau)$ and $G_{2k}(\tau')$ may not be equal. Tracing through the definitions, we find that

$$G_{2k}(\gamma\tau) = (c\tau + d)^{2k}G_{2k}(\tau) \qquad \text{for} \quad \begin{pmatrix} a & b \\ c & d \end{pmatrix} \in \mathrm{SL}_2(\mathbb{Z}).$$

Notice that if $c = 0$, then $a = d = \pm 1$, so $G_{2k}(\gamma\tau) = G_{2k}(\tau)$. In other words

$$G_{2k}(T^n\tau) = G_{2k}(\tau + n) = G_{2k}(\tau) \qquad \text{for all } n \in \mathbb{Z}.$$

This means that G_{2k} has a Fourier expansion

$$G_{2k}(\tau) = \sum_{n=-\infty}^{\infty} c(n)q^n,$$

where we write $q = e^{2\pi i \tau}$.

Definition. A meromorphic function f on $\mathbb{H}$ is called a *modular function of weight k* (*for* $\mathrm{SL}_2(\mathbb{Z})$) if it satisfies:

(i) $f(\tau) = (c\tau + d)^{-k}f(\gamma\tau)$ for all $\begin{pmatrix} a & b \\ c & d \end{pmatrix} \in \mathrm{SL}_2(\mathbb{Z})$.

(ii) There is an integer $n_0 = n_0(f)$ such that the Fourier expansion of f in the variable $q = e^{2\pi i \tau}$ has the form

$$f(\tau) = \sum_{n=n_0}^{\infty} c(n)q^n.$$

We say that f is a *modular form of weight k* if f is holomorphic on $\mathbb{H}$ and $n_0(f) = 0$, in which case we also say that f is *holomorphic at ∞* and define $f(\infty) = c_0$. This is a reasonable definition, since $q \to 0$ as $\tau \to i\infty$. If $f(\infty) = 0$, then we call f a *cusp form*.

Remark 12.3. If we let $\alpha = \left(\begin{smallmatrix} -1 & 0 \\ 0 & -1 \end{smallmatrix} \right)$, then $\alpha\tau = \tau$ for every $\tau \in \mathbb{H}$. Hence if f is a modular function of weight k and k is an odd integer, then

$$f(\tau) = (-1)^{-k} f(\alpha\tau) = -f(\tau),$$

so f is identically zero. Thus a nontrivial modular function for $\mathrm{SL}_2(\mathbb{Z})$ is necessarily of even weight. The next proposition gives some examples of modular functions.

Proposition 12.4. (a) *The j-function $j(\tau)$ is a modular function of weight 0 that is holomorphic on $\mathbb{H}$. Its Fourier series has the form*

$$j(\tau) = \frac{1}{q} + 744 + \sum_{n=1}^{\infty} c(n)q^n \qquad \text{with } c(n) \in \mathbb{Z}.$$

(b) *The Eisenstein series $G_{2k}(\tau)$ is a modular form of weight $2k$. Its Fourier series is*

$$G_{2k}(\tau) = 2\zeta(2k) + 2\frac{(2\pi i)^{2k}}{(2k-1)!} \sum_{n=1}^{\infty} \sigma_{2k-1}(n)q^n.$$

Here $\zeta(s) = \sum n^{-s}$ is the Riemann zeta function and $\sigma_\alpha(n)$ is the sum of divisors function $\sigma_\alpha(n) = \sum_{d|n} d^\alpha$.

(c) *The discriminant function $\Delta(\tau)$ is a cusp form of weight 12. Its Fourier series has the form*

$$\Delta(\tau) = (2\pi)^{12} \sum_{n=1}^{\infty} \tau(n)q^n \qquad \text{with } \tau(1) = 1 \text{ and } \tau(n) \in \mathbb{Z}.$$

The integer-valued function $n \mapsto \tau(n)$ is called the Ramanujan τ-function.

PROOF. See [5, Theorems 1.18, 1.19, 1.20], [232, VII Propositions 4, 5, 8], or [266, I.7.1, I.7.4]. $\qquad\qquad\qquad\qquad\qquad\qquad\qquad\qquad\qquad\qquad\qquad\qquad\qquad$ $\square$

Remark 12.4.1. The Fourier coefficients of $j(\tau)$ and $\Delta(\tau)$ have many interesting congruence properties. For example,

$$\tau(n) \equiv \sigma_{11}(n) \pmod{691} \qquad \text{for all } n \geq 1,$$

a result due to Ramanujan. We do not pursue this topic, but see, for example, [5, Chapter 4] or [232, VII §§3.3, 4.5].

Remark 12.4.2. The Fourier series for the Eisenstein series $G_{2k}(\tau)$ is often rewritten using the identity

$$\sum_{n=1}^{\infty} \sigma_\alpha(n) q^n = \sum_{n=1}^{\infty} \frac{n^\alpha q^n}{1 - q^n}.$$

It commonly appears in the literature in both forms.

The discriminant function $\Delta(\tau)$ also has a beautiful product expansion.

Theorem 12.5. (a) (Jacobi)

$$\Delta(\tau) = (2\pi)^{12} q \prod_{n=1}^{\infty} (1 - q^n)^{24}.$$

(b) *The* Dedekind η-*function is defined by the product*

$$\eta(\tau) = q^{1/24} \prod_{n=1}^{\infty} (1 - q^n),$$

where we set $q^{1/24} = e^{\pi i \tau / 12}$. Then $\Delta(\tau) = (2\pi)^{12} \eta(\tau)^{24}$, and the η-function satisfies the transformation formulas

$$\eta(\tau + 1) = e^{\pi i / 12} \eta(\tau) \qquad and \qquad \eta(-1/\tau) = (-i\tau)^{1/2} \eta(\tau),$$

where we take the branch of the square root that is real and positive on the positive real axis.

PROOF. See [5, Theorems 3.1, 3.3] or [232, VII Theorem 6]. □

Remark 12.5.1. Since the maps $\tau \mapsto \tau + 1$ and $\tau \mapsto -1/\tau$ generate the action of $\mathrm{SL}_2(\mathbb{Z})$ on $\mathbb{H}$ (C.12.1c), it is not hard to check that

$$\eta(\gamma\tau) = \epsilon \{-i(c\tau + d)\}^{1/2} \eta(\tau) \qquad \text{for all} \quad \begin{pmatrix} a & b \\ c & d \end{pmatrix} \in \mathrm{SL}_2(\mathbb{Z}),$$

where $\epsilon = \epsilon(a, b, c, d)$ satisfies $\epsilon^{24} = 1$. The exact value of $\epsilon(a, b, c, d)$ is complicated, but it can be expressed in terms of Dedekind sums and satisfies a beautiful reciprocity law. See [5, Chapter 3, especially Theorem 3.4] for details.

Elliptic functions such as the Weierstrass $\wp$-function may be treated as functions of two variables, the second variable being the lattice. We define

$$\wp(z; \tau) = \wp(z; \Lambda_\tau), \qquad \wp'(z; \tau) = \wp'(z; \Lambda_\tau), \qquad \sigma(z; \tau) = \sigma(z; \Lambda_\tau).$$

The q-expansions of these functions are given in the next proposition.

Proposition 12.6. *Let $q = e^{2\pi i \tau}$ and $u = e^{2\pi i z}$. Then*

$$(2\pi i)^{-2}\wp(z;\tau) = \sum_{n=-\infty}^{\infty} \frac{q^n u}{(1-q^n u)^2} + \frac{1}{12} - 2\sum_{n=1}^{\infty} \frac{q^n}{(1-q^n)^2},$$

$$(2\pi i)^{-3}\wp'(z;\tau) = \sum_{n=-\infty}^{\infty} \frac{q^n(1+q^n u^2)}{(1-q^n u)^3},$$

$$2\pi i\sigma(z;\tau) = e^{\eta z^2/2}(u^{1/2} - u^{-1/2}) \prod_{n=1}^{\infty} \frac{(1-q^n u)(1-q^n u^{-1})}{(1-q^n)^2}.$$

In this last formula, $\eta = \eta(1)$ is one of the quasiperiods associated to the lattice $\mathbb{Z} + \mathbb{Z}\tau$; see Exercise 6.6b.

PROOF. See [140, Chapter 18, §2], [210, II §5], or [266, I.6.2, I.6.4].   □

Remark 12.7. An elliptic curve $E/\mathbb{C}$ is analytically isomorphic to a torus $\mathbb{C}/\Lambda$, and we can choose the lattice to be of the form $\Lambda = \Lambda_\tau = \mathbb{Z} + \mathbb{Z}\tau$. Consider the exponential map $\exp(2\pi i \cdot) : \mathbb{C} \to \mathbb{C}^*$. The image of Λ_τ under this map is the cyclic group $q^{\mathbb{Z}} = \{q^n : n \in \mathbb{Z}\}$ in $\mathbb{C}^*$. Thus composing with the exponential map, we obtain an analytic isomorphism

$$\mathbb{C}^*/q^{\mathbb{Z}} \overset{\sim}{\longrightarrow} E(\mathbb{C}).$$

If we let u be the parameter on $\mathbb{C}^*$, i.e., $u = e^{2\pi i z}$, then this map is $[\wp, \wp', 1]$, where $\wp$ and $\wp'$ are given in terms of u by (C.12.6).

We now describe the field of modular functions and the algebra of modular forms.

Definition. Let

$$M_k = \{\text{modular forms of weight } 2k\},$$
$$M_{k,0} = \{\text{cusp forms of weight } 2k\}.$$

We observe that M_k and $M_{k,0}$ are $\mathbb{C}$-vector spaces. Further, if $f \in M_k$ and $g \in M_{k'}$, then $fg \in M_{k+k'}$. Thus the ring

$$M = \sum_{k=0}^{\infty} M_k$$

has a natural structure as a graded $\mathbb{C}$-algebra.

Theorem 12.8. (a) $j(\tau)$ is a modular function of weight 0. Every modular function of weight 0 is a rational function of $j(\tau)$.
(b) The map
$$\mathbb{C}[X, Y] \longrightarrow M, \qquad P(X, Y) \longmapsto P(G_4, G_6),$$

is an isomorphism of graded $\mathbb{C}$-algebras, where we assign weights $\mathrm{wt}(X) = 4$ and $\mathrm{wt}(Y) = 6$. In particular, every modular form is a polynomial in G_4 and G_6.

(c) *The dimension of the weight-2k part of M is*

$$\dim_{\mathbb{C}} M_k = \begin{cases} 0 & \text{if } k < 0, \\ [k/6] & \text{if } k \equiv 1 \pmod{6} \text{ and } k \geq 0, \\ [k/6] + 1 & \text{if } k \not\equiv 1 \pmod{6} \text{ and } k \geq 0. \end{cases}$$

(d) *Multiplication by $\Delta(\tau)$ defines an isomorphism from M_{k-6} to $M_{k,0}$.*

PROOF. See [5, Theorem 2.8, §§6.4, 6.5], [232, VII, §§3.2, 3.3], or [266, I.3.10, I.4.2, and exercise 1.10]. $\qquad\square$

The study of the spaces M_k and $M_{k,0}$ is facilitated by the existence of certain linear operators. For each integer $n \geq 1$, we define the *Hecke operator* $T(n)$ on the space M_k of modular forms of weight $2k$ by the formula

$$\bigl(T(n)f\bigr)(\tau) = n^{2k-1} \sum_{d|n} d^{-2k} \sum_{b=0}^{d-1} f\left(\frac{n\tau + bd}{d^2}\right).$$

(For a more intrinsic definition that explains the origin of this formula, see [5, §6.8], [232, VII §5.1], [249, Chapter 3], or [266, I §9].)

Proposition 12.9. (a) *If f is a modular form (respectively a cusp form) of weight $2k$, then $T(n)f$ is also a modular form (respectively a cusp form) of weight $2k$. In other words, $T(n)$ induces linear maps*

$$T(n) : M_k \longrightarrow M_k \qquad \text{and} \qquad T(n) : M_{k,0} \longrightarrow M_{k,0}.$$

(b) *For all integers m and n,*

$$T(m)T(n) = T(n)T(m).$$

(c) *If m and n are relatively prime, then*

$$T(mn) = T(m)T(n).$$

(d) *For all primes p and all integers $r \geq 1$,*

$$T(p^{r+1}) = T(p^r)T(p) - p^{2k-1}T(p^{r-1}).$$

PROOF. See [5, Theorems 6.11, 6.13], [232, VII §§5.1, 5.3], or [266, I.10.2, I.10.6]. $\qquad\square$

Application 12.10. Of particular interest are those modular forms that are *simultaneous eigenfunctions* for every Hecke operator $T(n)$. In other words,

$$T(n)f = \lambda(n)f \qquad \text{for all } n = 1, 2, \ldots,$$

where $\lambda(1), \lambda(2), \ldots$ are constants. If this occurs, then it is not hard to show that the Fourier expansion $f = \sum c(n)q^n$ of f satisfies

$$c(n) = c(1)\lambda(n) \qquad \text{for all } n = 1, 2, \ldots.$$

See [5, Theorems 6.14, 6.15], [232, VII §5.4], or [266, I.10.5]. In particular, if f is not constant, then $c(1) \neq 0$ and f is uniquely determined by $c(1)$ and the sequence of eigenvalues $(\lambda(n))_{n \geq 1}$.

Example 12.10.1. Consider the vector space $M_{6,0}$ of cusp forms of weight 12. We see from (C.12.8c) and (C.12.4c) that $M_{6,0}$ has dimension one and is generated by the discriminant function

$$\Delta = (2\pi)^{12} \prod_{n=1}^{\infty} (1 - q^n)^{24} = (2\pi)^{12} \sum_{n=1}^{\infty} \tau(n) q^n.$$

Since $T(n)\Delta$ is also in $M_{6,0}$, it follows that $T(n)\Delta$ is a multiple of Δ. Using (C.12.10), we conclude that

$$T(n)\Delta = \tau(n)\Delta \qquad \text{for all } n = 1, 2, \ldots.$$

(Note that $\tau(1) = 1$.) Now the identities (C.12.9cd) satisfied by the Hecke operators $T(n)$ lead to analogous formulas for the Ramanujan function:

$$\tau(mn) = \tau(m)\tau(n) \qquad\qquad \text{if } \gcd(m, n) = 1,$$
$$\tau(p^{r+1}) = \tau(p^r)\tau(p) - p^{12}\tau(p^{r-1}) \qquad \text{for } p \text{ prime and } r \geq 1.$$

These beautiful identities were conjectured by Ramanujan and proved by Mordell, who invented what are now known as Hecke operators for his proof. There is also a deep estimate for the size of the Ramanujan function that was demonstrated by Deligne as a (highly nontrivial) consequence of his proof of the Weil conjectures:

$$|\tau(p)| \leq 2p^{11/2} \qquad \text{for } p \text{ prime.}$$

Since $j(\tau)$ is a modular function of weight 0 (C.12.4a), it defines a function on the quotient space $\mathbb{H}/\operatorname{SL}_2(\mathbb{Z})$. This quotient has a natural structure as a Riemann surface, and one can show that $j(\tau)$ defines a holomorphic function; see [249, §§1.3, 1.4, 1.5] or [266, I.4.1].

Proposition 12.11. *The map*

$$j : \mathbb{H}/\operatorname{SL}_2(\mathbb{Z}) \longrightarrow \mathbb{C}$$

is a complex analytic isomorphism of (open) Riemann surfaces.

PROOF. See [232, VII Proposition 5] or [266, I.4.1]. □

Corollary 12.11.1. (Uniformization Theorem) *Let $E/\mathbb{C}$ be an elliptic curve. Then there exist a lattice $\Lambda \subset \mathbb{C}$ and a complex analytic isomorphism $\mathbb{C}/\Lambda \to E(\mathbb{C})$.*

PROOF. Let J be the j-invariant of E. From (C.12.11), there is a $\tau \in \mathbb{H}$ such that $j(\tau) = J$. Then the elliptic curve

$$E_\tau : y^2 = 4x^3 - g_2(\tau)x - g_3(\tau)$$

has j-invariant J, so $E_\tau \cong E$ from (III.1.4b). On the other hand, (VI.3.6b) says that there is a complex analytic isomorphism $\mathbb{C}/(\mathbb{Z} + \mathbb{Z}\tau) \to E_\tau(\mathbb{C})$, which gives the desired result. $\square$

From (C.12.11) we see that the Riemann surface $\mathbb{H}/\operatorname{SL}_2(\mathbb{Z})$ is not compact. Its natural compactification is $\mathbb{P}^1(\mathbb{C})$, obtained by adjoining a single extra point at infinity. However, with a view toward eventual generalizations, we take a different approach. We start by defining the extended upper half-plane to be the set

$$\mathbb{H}^* = \mathbb{H} \cup \mathbb{P}^1(\mathbb{Q}).$$

Here one should think of the points $[x, 1] \in \mathbb{P}^1(\mathbb{Q})$ as the usual copy of $\mathbb{Q}$ in $\mathbb{C}$, together with the point $[1, 0] \in \mathbb{P}^1(\mathbb{Q})$ at infinity. We note that $\operatorname{SL}_2(\mathbb{Z})$ acts on $\mathbb{P}^1(\mathbb{Q})$ in the standard manner,

$$\gamma : [x, y] \longmapsto [ax + by, cx + dy].$$

The quotient space $\mathbb{H}^*/\operatorname{SL}_2(\mathbb{Z})$ can be given the structure of a Riemann surface, and one can show that the j-function defines a complex analytic isomorphism

$$j : \mathbb{H}^*/\operatorname{SL}_2(\mathbb{C}) \overset{\sim}{\longrightarrow} \mathbb{P}^1(\mathbb{C}).$$

See [249, §§1.3, 1.4, 1.5] or [266, §§I.2, I.3, I.4] for details. Since $\operatorname{SL}_2(\mathbb{Z})$ acts transitively on $\mathbb{P}^1(\mathbb{Q})$, the net effect has been to add a single point, called a *cusp*, to $\mathbb{H}/\operatorname{SL}_2(\mathbb{Z})$.

Congruence Subgroups

In studying modular functions for $\operatorname{SL}_2(\mathbb{Z})$, one soon discovers the need to deal with functions that are modular only for certain subgroups of $\operatorname{SL}_2(\mathbb{Z})$.

Definition. For each integer $N \geq 1$, we define subgroups of $\operatorname{SL}_2(\mathbb{Z})$ as follows:

$$\Gamma_0(N) = \left\{ \begin{pmatrix} a & b \\ c & d \end{pmatrix} \in \operatorname{SL}_2(\mathbb{Z}) : c \equiv 0 \pmod{N} \right\},$$

$$\Gamma_1(N) = \left\{ \begin{pmatrix} a & b \\ c & d \end{pmatrix} \in \operatorname{SL}_2(\mathbb{Z}) : c \equiv 0 \pmod{N} \text{ and } a \equiv d \equiv 1 \pmod{N} \right\},$$

$$\Gamma(N) = \left\{ \begin{pmatrix} a & b \\ c & d \end{pmatrix} \in \operatorname{SL}_2(\mathbb{Z}) : b \equiv c \equiv 0 \pmod{N} \text{ and } a \equiv d \equiv 1 \pmod{N} \right\}.$$

More generally, a *congruence subgroup of* $\operatorname{SL}_2(\mathbb{Z})$ is a subgroup $\Gamma \subset \operatorname{SL}_2(\mathbb{Z})$ such that Γ contains $\Gamma(N)$ for some $N \geq 1$.

If Γ is a congruence subgroup of $SL_2(\mathbb{Z})$, then Γ acts on $\mathbb{H}^*$ and we can form the quotient space $\mathbb{H}^*/\Gamma$. This space has a natural structure as a Riemann surface [249, §§1.3, 1.5]. The action of Γ on $\mathbb{P}^1(\mathbb{Q}) \subset \mathbb{H}^*$ has finitely many orbits; the images of these orbits in $\mathbb{H}^*/\Gamma$ are called the *cusps of* Γ.

Example 12.12. If p is prime, then $\mathbb{H}^*/\Gamma_0(p)$ has two cusps, represented by the points $[1, 0]$ and $[0, 1]$ in $\mathbb{P}^1(\mathbb{Q})$.

Definition. Let Γ be a congruence subgroup of $SL_2(\mathbb{Z})$. A meromorphic function f on $\mathbb{H}$ is called a *modular function of weight k for* Γ if it satisfies the following two conditions:

(i) $f(\tau) = (c\tau + d)^{-k} f(\gamma\tau)$　　for all $\gamma \in \begin{pmatrix} a & b \\ c & d \end{pmatrix} \in \Gamma$.

(ii) f is meromorphic at each of the cusps of $\mathbb{H}^*/\Gamma$. (See [249, §2.1] for the precise definition.)

A modular function is called a *modular form* if it is holomorphic on $\mathbb{H}$ and at each of the cusps of $\mathbb{H}^*/\Gamma$, and it is a *cusp form* if it is a modular form that vanishes at every cusp.

Example 12.13. Let $\eta(\tau)$ be the Dedekind η-function (C.12.5b). Then the function

$$f(\tau) = \eta(\tau)^2 \eta(11\tau)^2$$

is a cusp form of weight 2 for the group $\Gamma_0(11)$.

If $f(\tau)$ is a modular form of weight 2 for Γ, then the differential form $f(\tau)\, d\tau$ on $\mathbb{H}$ is invariant under the action of Γ. This follows easily from the identity

$$d\left(\frac{a\tau + b}{c\tau + d}\right) = \frac{d\tau}{(c\tau + d)^2}.$$

(We follow standard practice of using the letter d as both a variable and to indicate differentiation.) If, further, f is a cusp form, then one can check that $f(\tau)\, d\tau$ is holomorphic at each of the cusps of Γ, and hence $f(\tau)\, d\tau$ defines a holomorphic 1-form on the quotient space $\mathbb{H}^*/\Gamma$.

Proposition 12.14. *Let Γ be a congruence subgroup of $SL_2(\mathbb{Z})$. There is a natural isomorphism between the space of weight-2 cusp forms for Γ and the space of holomorphic 1-forms on the Riemann surface $\mathbb{H}^*/\Gamma$.*

PROOF. See [249, §2.4].　　　　　　　　　　　　　　　　　　　　　□

Remark 12.15. It is not difficult to calculate the genus of $\mathbb{H}^*/\Gamma$ and thereby, using (C.12.14), find the dimension of the space of weight-2 cusp forms for Γ. For example, if p is prime with $p \equiv 11 \pmod{12}$, then the genus of $\mathbb{H}^*/\Gamma_0(p)$ is $\frac{1}{12}(p+1)$. For a general formula, see [249, Propositions 1.40, 1.43].

The Hecke operators defined above also act on the space of modular forms for congruence subgroups.

Proposition 12.16. *Let Γ be a congruence subgroup of $\mathrm{SL}_2(\mathbb{Z})$, say $\Gamma \supset \Gamma(N)$, and let $f(\tau)$ be a modular form of weight $2k$ for Γ. Then for each integer $n \geq 1$ satisfying $\gcd(n, N) = 1$, the function $T(n)f$ is again a modular form of weight $2k$ for Γ. Further, if f is a cusp form, then so is $T(n)f$.*

PROOF. See [249, Proposition 3.37]. $\square$

Remark 12.17. Just as in the case of the full modular group $\mathrm{SL}_2(\mathbb{Z})$, it is worth studying modular forms for Γ that are simultaneous eigenfunctions for all of the Hecke operators. For example, the Riemann surface $\mathbb{H}^*/\Gamma_0(11)$ has genus one (C.12.15), so it follows from (C.12.14) that the space of cusp forms of weight 2 for $\Gamma_0(11)$ has dimension one. Therefore the function

$$f(\tau) = \eta(\tau)^2 \eta(11\tau)^2$$

from (C.12.13) is an eigenfunction of $T(n)$ for every n satisfying $\gcd(n, 11) = 1$.

References. [3, Chapter 7], [5, Chapters 2, 3, 6], [21], [129], [140], [197], [210, I §§3,4], [232, Chapter VII], [249, Chapters 1, 2, 3], [266, Chapter I].

C.13   Modular Curves

Let Γ be a subgroup of $\mathrm{SL}_2(\mathbb{Z})$. If $\Gamma = \mathrm{SL}_2(\mathbb{Z})$, then we have seen in (C §12) that the points of the Riemann surface $\mathbb{H}/\Gamma$ are in one-to-one correspondence with the isomorphism classes of elliptic curves defined over $\mathbb{C}$. This correspondence associates to the point $\tau \pmod{\Gamma}$ of $\mathbb{H}/\Gamma$ the elliptic curve $E_\tau \cong \mathbb{C}/(\mathbb{Z} + \mathbb{Z}\tau)$. In this section we describe a similar interpretation for the points of $\mathbb{H}/\Gamma$ when Γ is a more general congruence subgroup of $\mathrm{SL}_2(\mathbb{Z})$.

For example, consider the subgroup $\Gamma_1(N)$, which we recall consists of all matrices

$$\begin{pmatrix} a & b \\ c & d \end{pmatrix} \quad \text{satisfying } c \equiv 0 \pmod{N} \text{ and } a \equiv d \equiv 1 \pmod{N}.$$

Since $\Gamma_1(N) \subset \mathrm{SL}_2(\mathbb{Z})$, we can associate an elliptic curve E_τ to each $\tau \in \mathbb{H}/\Gamma_1(N)$. This is nothing more than the natural map $\mathbb{H}/\Gamma_1(N) \to \mathbb{H}/\mathrm{SL}_2(\mathbb{Z})$. But a point of $\mathbb{H}/\Gamma_1(N)$ contains additional information. Consider the point $T_\tau \in E_\tau$ corresponding to $1/N \in \mathbb{C}/(\mathbb{Z} + \mathbb{Z}\tau)$, so in particular $T_\tau \in E_\tau[N]$. Then, for any $\gamma \in \mathrm{SL}_2(\mathbb{Z})$, the isomorphism

$$f : \frac{\mathbb{C}}{\mathbb{Z} + \mathbb{Z}\tau} \longrightarrow \frac{\mathbb{C}}{\mathbb{Z} + \mathbb{Z}\gamma(\tau)}, \qquad z \longmapsto \frac{z}{c\tau + d},$$

maps $1/N$ to $1/(N(c\tau + d))$, where $\gamma = \begin{pmatrix} a & b \\ c & d \end{pmatrix}$. If we further assume that γ is in $\Gamma_1(N)$, then

$$\frac{1}{N} - \frac{1}{N(c\tau + d)} = \frac{\dfrac{c}{N}\tau + \dfrac{d-1}{N}}{c\tau + d} \in f(\mathbb{Z} + \mathbb{Z}\tau) = \mathbb{Z} + \mathbb{Z}\gamma(\tau).$$

Thus the point $1/N \in \mathbb{C}/(\mathbb{Z} + \mathbb{Z}\tau)$ remains fixed when the basis for the lattice is changed by an element of $\Gamma_1(N)$. Hence a point in $\mathbb{H}/\Gamma_1(N)$ gives an elliptic curve $E_\tau/\mathbb{C}$ together with a specified point $T_\tau \in E_\tau$ of exact order N. Further, given any elliptic curve $E/\mathbb{C}$ and any point $T \in E$ of exact order N, there are a point $\tau \in \mathbb{H}/\Gamma_1(N)$ and an isomorphism $E_\tau \to E$ such that $T_\tau \to T$. Using fancier terminology, we say that the Riemann surface $\mathbb{H}/\Gamma_1(N)$ is a *moduli space* for the *moduli problem* of determining equivalence classes of pairs (E, T), where E is an elliptic curve defined over $\mathbb{C}$ and $T \in E$ is a point of exact order N. Two pairs (E, T) and (E', T') are deemed to be equivalent if there is an isomorphism $E \cong E'$ that takes T to T'.

Similarly, if $\gamma \in \Gamma_0(N)$ and $\tau \in \mathbb{H}/\Gamma_0(N)$, then one easily checks that the subgroup

$$\left\{ \frac{1}{N}, \frac{2}{N}, \ldots, \frac{N-1}{N} \right\} \subset \frac{\mathbb{C}}{\mathbb{Z} + \mathbb{Z}\tau}$$

remains invariant under the action of γ. Thus $\mathbb{H}/\Gamma_0(N)$ is a moduli space for the problem of determining (equivalence classes of) pairs (E, C), where E is an elliptic curve and $C \subset E$ is a cyclic subgroup of exact order N. Note that from (III.4.12), there is a one-to-one correspondence between finite subgroups $\Phi \subset E$ and isogenies $\phi : E \to E_1$ given by the association $\Phi \leftrightarrow \ker \phi$. Thus the points of $\mathbb{H}/\Gamma_0(N)$ may also be viewed as classifying triples (E, E', ϕ), where $\phi : E \to E'$ is an isogeny whose kernel is cyclic of order N.

Finally, we consider the moduli problem associated to the congruence subgroup $\Gamma(N)$. If $\gamma \in \Gamma(n)$ and $\tau \in \mathbb{H}/\Gamma(N)$, then as above one checks that the points $1/N$ and τ/N in $\mathbb{C}/(\mathbb{Z} + \mathbb{Z}\tau)$ remain invariant under the action of γ. Thus associated to a point of $\mathbb{H}/\Gamma(N)$ is an elliptic curve $\mathbb{C}/(\mathbb{Z} + \mathbb{Z}\tau)$, together with a basis $\{1/N, \tau/N\}$ for the group of N-torsion points. However, a point of $\mathbb{H}/\Gamma(N)$ contains one further piece of information. Recall (III §8) that there is a pairing e_N on the group of N-torsion points of an elliptic curve. Then one can check that

$$e_N \left(\frac{1}{N}, \frac{\tau}{N} \right) = e^{2\pi i/N}.$$

Thus not only do we get a basis for the N-torsion, but the two points forming that basis are mapped by the Weil pairing to a specific primitive N^{th} root of unity.

For arithmetic applications, it is important to understand when an elliptic curve $E/\mathbb{C}$ or a point $T \in E(\mathbb{C})$ is defined over a subfield of $\mathbb{C}$, for example, over a number field. To illustrate, we note that although the Riemann surface $\mathbb{H}/\operatorname{SL}_2(\mathbb{C})$ classifies elliptic curves only over $\mathbb{C}$, we have a complex analytic isomorphism (C.12.11)

$$j : \mathbb{H}/\operatorname{SL}_2(\mathbb{Z}) \longrightarrow \mathbb{A}^1(\mathbb{C}),$$

where $\mathbb{A}^1$ is a variety defined over $\mathbb{Q}$. Further, the elliptic curve E_τ corresponding to $\tau \in \mathbb{H}/\operatorname{SL}_2(\mathbb{Z})$ is isomorphic, over $\mathbb{C}$, to an elliptic curve *defined over* $\mathbb{Q}(j(\tau))$. There is a general theory that deals with fields of definition for the spaces $\mathbb{H}/\Gamma$ and their associated moduli problems, but we content ourselves with the following description for the quotient spaces associated to the three families of congruence subgroups $\Gamma_0(N)$, $\Gamma_1(N)$, and $\Gamma(N)$.

Theorem 13.1. *Let $N \geq 1$ be an integer.*

(a) *There exist a smooth projective curve $X_0(N)/\mathbb{Q}$ and a complex analytic isomorphism*

$$j_{N,0} : \mathbb{H}^* / \Gamma_0(N) \longrightarrow X_0(N)(\mathbb{C})$$

such that the following holds:

Let $\tau \in \mathbb{H}/\Gamma_0(N)$ and let $K = \mathbb{Q}(j_{N,0}(\tau))$. We have seen that τ corresponds to an equivalence class of pairs (E, C), where E is an elliptic curve and $C \subset E$ is a cyclic subgroup of order N. Then this equivalence class contains a pair such that both E and C are defined over K, i.e., E is an elliptic curve defined over K and $C \subset E(\bar{K})$ is $G_{\bar{K}/K}$-invariant.

(b) *There exists a smooth projective curve $X_1(N)/\mathbb{Q}$ and a complex analytic isomorphism*

$$j_{N,1} : \mathbb{H}^* / \Gamma_1(N) \longrightarrow X_1(N)(\mathbb{C})$$

such that the following holds:

Let $\tau \in \mathbb{H}/\Gamma_1(N)$ and let $K = \mathbb{Q}(j_{N,1}(\tau))$. We have seen that τ corresponds to an equivalence class of pairs (E, T), where E is an elliptic curve and $T \in E$ is a point of exact order N. Then this equivalence class contains a pair such that E is defined over K and $T \in E(K)$.

(c) *Fix a primitive N^{th} root of unity $\zeta \in \mathbb{C}$. There exist a smooth projective curve $X(N)/\mathbb{Q}$ and a complex analytic isomorphism*

$$j_N : \mathbb{H}^* / \Gamma(N) \longrightarrow X(N)(\mathbb{C})$$

such that the following holds:

Let $\tau \in \mathbb{H}/\Gamma(N)$ and let $K = \mathbb{Q}(\zeta, j_N(\tau))$. We have seen that τ corresponds to an equivalence class of triples (E, T_1, T_2), where E is an elliptic curve and (T_1, T_2) are generators for $E[N]$ satisfying $e_N(T_1, T_2) = \zeta$, where e_N is the Weil pairing (III §8). Then this equivalence class contains a triple such that E is defined over K and $T_1, T_2 \in E(K)$.

PROOF. See [249, §6.7]. $\qquad\qquad\qquad\qquad\qquad\qquad\qquad\qquad\qquad\qquad$ □

Remark 13.2. If Γ is any congruence subgroup of $\mathrm{SL}_2(\mathbb{Z})$, then there are a smooth projective curve $X(\Gamma)$ defined over some number field $K(\Gamma)$ and a complex analytic isomorphism $j_\Gamma : \mathbb{H}^* / \Gamma \to X(\Gamma)(\mathbb{C})$. See [249, §6.7] for details.

Definition. With notation as in (C.13.2), the curve $X(\Gamma)$ is called a *modular curve*. The *set of cusps* of $X(\Gamma)$ consists of the finite set of points $j_\Gamma(\mathbb{P}^1(\mathbb{Q})/\Gamma)$, i.e., the cusps of $X(\Gamma)$ are images under j_Γ of the cusps of Γ. We denote the complement of the set of cusps of $X(\Gamma)$ by $Y(\Gamma)$, so $Y(\Gamma)$ is a smooth affine curve.

Notation. For the congruence subgroups $\Gamma_0(N)$, $\Gamma_1(N)$, and $\Gamma(N)$ considered in (C.13.1), the curve $Y(\Gamma)$ is denoted, respectively, by $Y_0(N)$, $Y_1(N)$, and $Y(N)$.

Example 13.3. Let N be an odd prime. Then $X_0(N)$ has two cusps, both of which are rational over $\mathbb{Q}$, i.e., they are in $X_0(N)(\mathbb{Q})$. Similarly, the curve $X_1(N)$ has $N-1$ cusps, but now only half of the cusps are in $X_1(N)(\mathbb{Q})$. The other $\frac{1}{2}(N-1)$ cusps of $X_1(N)$ are defined over the maximal real subfield of $\mathbb{Q}(\zeta_N)$.

Example 13.4. The curve $X_1(7)$ is isomorphic to $\mathbb{P}^1$. To make this precise, we associate to each point $[t, 1] \in \mathbb{P}^1$ the pair (E_t, P_t), where E_t is the curve (defined over $\mathbb{Q}(t)$) given by the equation

$$E_t : y^2 + (1 + t - t^2)xy + (t^2 - t^3)y = x^3 + (t^2 - t^3)x^2,$$

and P_t is the point $P_t = (0, 0)$. We observe that E_t is an elliptic curve if and only if its discriminant

$$\Delta(t) = t^7(t - 1)^7(t^3 - 8t^2 + 5t + 1)$$

does not vanish. If $\Delta(t) \neq 0$, then one easily checks using the addition law that $[7]P_t = O$. The curve $X_1(7)$ has six cusps, corresponding to the values

$$t = 0, \quad t = 1, \quad t = \infty, \quad \text{and the roots of } t^3 - 8t^2 + 5t + 1.$$

The reader may verify that the latter three cusps generate the maximal real subfield of $\mathbb{Q}(\zeta_7)$, thereby verifying (C.13.3) for $N = 7$.

Remark 13.5. To illustrate one application of modular curves, we use them to rephrase the uniform boundedness theorem (VIII.7.5.1) for the torsion on elliptic curves. That theorem says that for any integer d, there is a bound $N(d)$ such that if $K/\mathbb{Q}$ is a number field of degree d and E/K is an elliptic curve, then $E(K)$ contains no torsion points of order $N(d)$ or greater. We observe that if $E(K)$ contains a point P of order N, then the pair (E, P) corresponds to a noncuspidal point in $X_1(N)(K)$, i.e., it corresponds to a K-rational point on the modular curve $X_1(N)$. Thus (VIII.7.5.1) is equivalent to the statement that if $N \geq N(d)$, then the set of rational points $X_1(N)(K)$ consists entirely of cusps for all number fields $K/\mathbb{Q}$ of degree d. The question of K-rational torsion points on elliptic curves is thus transformed into the question of K-rational points on modular curves. This idea provides the starting point for the results of Mazur (VIII.7.5), Merel (VIII.7.5.1), and Manin (VIII.7.6).

Some modular curves are themselves elliptic curves. For example, the curve $X_0(11)$ has genus one, and since it has two cusps defined over $\mathbb{Q}$ (C.13.3), we can use one of the cusps to make $X_0(11)$ into an elliptic curve. An elliptic curve such as $X_0(11)$ that is also a modular curve has a lot of additional structure, and one can use that extra information to study the arithmetic of $X_0(11)$. Unfortunately, the genus of $X_0(N)$ grows (somewhat irregularly) with N, so there are only finitely many curves $X_0(N)$ of any given genus. However, for a given elliptic curve $E/\mathbb{Q}$, one might ask whether there exists a finite map $\phi : X_0(N) \to E$ defined over $\mathbb{Q}$ for some modular curve $X_0(N)$. If this happens, then we say that the elliptic curve is *modular* and we call ϕ a *modular parametrization*. Such elliptic curves have a very rich structure that can be used to study their arithmetic properties, so the following theorem provides an extremely powerful tool for studying the arithmetic of elliptic curves defined over $\mathbb{Q}$.

Theorem 13.6. (Modularity Theorem, Wiles et. al. [28, 291, 311]) *Every elliptic curve defined over $\mathbb{Q}$ is modular, i.e., if $E/\mathbb{Q}$ is an elliptic curve, then there exist an integer N and a surjective morphism $\phi : X_0(N) \to E$ defined over $\mathbb{Q}$. More precisely, the integer N may be taken to be the conductor of $E/\mathbb{Q}$; see (C §16).*

We have credited the modularity theorem to Wiles et. al., since the history of the proof is somewhat complicated and involves a number of people, although certainly the most important new ideas were due to Wiles. The original announcement of a proof of (C.13.6) was made by Wiles in 1993, but a detailed examination of the proof revealed a serious gap in one part of the argument. Wiles then worked with Richard Taylor to devise an alternative approach that filled the gap and completed the proof. Wiles' main argument [311] and the additional Taylor–Wiles step [291] appeared in 1995.

However, these articles did not prove the full modularity conjecture. Instead, they proved modularity for all *semistable* curves over $\mathbb{Q}$, i.e., for elliptic curves $E/\mathbb{Q}$ having no additive reduction. (The semistable case sufficed to prove Fermat's last theorem.) A number of mathematicians then worked to extend and generalize Wiles' groundbreaking ideas, culminating in a proof of the full modularity conjecture by Breuil, Conrad, Harris, and Taylor [28] in 2001.

Remark 13.6.1. The history of the formulation of the modularity conjecture is also interesting, and not without controversy. A qualitative version was originally suggested by Taniyama and a more precise version formulated by Shimura. Weil's converse theorem [308] provided significant evidence for the validity of the conjecture. So at various times and in various publications, the conjecture had varying combinations of the names Shimura, Taniyama, and Weil attached to it.

Remark 3.6.2. An important motivation for Wiles' pursuit of the modularity conjecture was earlier work of Frey, Serre, and Ribet showing that the modularity conjecture implies Fermat's last theorem. Frey [88] noted that if $a^p + b^p = c^p$ with $abc \neq 0$ and $p \geq 3$ prime, then the minimal discriminant of the elliptic curve

$$E : y^2 = x(x + a^p)(x - b^p)$$

is essentially a $(2p)^{\text{th}}$ power, cf. (VIII.11.2) and (VIII.11.3), and he suggested that such a curve could not be modular. Serre [235] refined Frey's observation and showed that the nonmodularity of E follows from a precise level-lowering statement for modular forms, a statement that was subsequently proven by Ribet [208]. Thus Ribet proved that the Frey curve, built from a putative solution to the Fermat equation, is not modular, so Fermat's last theorem follows from Wiles' proof of the modularity conjecture for semistable elliptic curves. (The Frey curve is semistable.)

References. [21], [28], [124], [165], [166], [249], [291], [308], [311].

C.14 Tate Curves

For this section and the next, we let K be a local field that is complete with respect to a discrete valuation v. Recall that for elliptic curves over $\mathbb{C}$, the existence of a lattice $\Lambda \subset \mathbb{C}$ and a uniformization $E(\mathbb{C}) \cong \mathbb{C}/\Lambda$ provides a powerful tool for the study of $E(\mathbb{C})$. If we attempt to mimic this construction for K, we are immediately stymied, since the additive group of a local field has no nontrivial discrete subgroups. However, if Λ is normalized as $\mathbb{Z} + \mathbb{Z}\tau$, then applying the exponential map $e^{2\pi i \cdot}$ to $\mathbb{C}/\Lambda$ gives a new isomorphism $E(\mathbb{C}) \cong \mathbb{C}^*/q^{\mathbb{Z}}$. Here $q = e^{2\pi i \tau}$, and $q^{\mathbb{Z}}$ is the subgroup of $\mathbb{C}^*$ generated by q, cf. (C.12.7). The analogous situation for K now looks more promising, since the multiplicative group K^* has lots of discrete subgroups, namely those of the form $q^{\mathbb{Z}}$ with $|q|_v \neq 1$. Further, all of the classical q-expansions for the various elliptic and modular functions as described in (C §12) converge in the v-adic setting provided that q is chosen to satisfy $|q|_v < 1$.

For example, consider the elliptic curve

$$E_q : y^2 + xy = x^3 + a_4 x + a_6,$$

whose coefficients are given by the following power series, considered for the moment to be formal power series in the ring $\mathbb{Z}[\![q]\!]$:

$$a_4 = -5 \sum_{n \geq 1} \frac{n^3 q^n}{1 - q^n}, \qquad a_6 = -\frac{1}{12} \sum_{n \geq 1} \frac{(7n^5 + 5n^3)q^n}{1 - q^n}.$$

The curve E_q is called the *Tate curve*. Its discriminant and j-invariant are given by the familiar formulas from the complex case (C §12),

$$\Delta = q \prod_{n \geq 1} (1 - q^n)^{24}, \qquad j = \frac{1}{q} + 744 + 196884q + \cdots.$$

Note that except for its leading term, j is in $\mathbb{Z}[\![q]\!]$. Further, there is a point on the elliptic curve E_q with coordinates in the power series ring $\mathbb{Z}[\![q, u]\!]$ defined by

$$x = x(u, q) = \sum_{n \in \mathbb{Z}} \frac{q^n u}{(1 - q^n u)^2} - 2 \sum_{n \geq 1} \frac{q^n}{(1 - q^n)^2},$$

$$y = y(u, q) = \sum_{n \in \mathbb{Z}} \frac{q^n u^2}{(1 - q^n u)^3} + \sum_{n \geq 1} \frac{q^n}{(1 - q^n)^2}.$$

We now need merely observe that all of these formulas make sense if q and u are taken to be elements of K^* with $|q|_v < 1$ and $u \notin q^{\mathbb{Z}}$. Then all of the power series converge in K, since K is complete with respect to the absolute value v. We thus obtain a v-adic analytic uniformization

$$\phi : K^*/q^{\mathbb{Z}} \longrightarrow E_q(K), \qquad u \longmapsto \big(x(u, q), y(u, q)\big).$$

(Of course, we also set $\phi(1) = O$.) More generally, the power series $x(u, q)$ and $y(u, q)$ converge for any $u \in \bar{K}$ and thus induce a map

$$\phi : \bar{K}/q^{\mathbb{Z}} \longrightarrow E_q(\bar{K}).$$

(Note that although $\bar{K}$ is not v-adically complete, the convergence of the power series $\phi(u)$ is taking place in the finite extension $K(u)$. Alternatively, we could work in the v-adic completion of $\bar{K}$, which turns out to be algebraically closed.)

The Tate parametrization is arithmetically useful because the action of $G_{\bar{K}/K}$ on $\bar{K}$ is v-adically continuous, so this action commutes with the convergence of power series. Hence ϕ is an isomorphism of $G_{\bar{K}/K}$-modules, which allows us to use ϕ to make arithmetic deductions. In this respect the nonarchimedean uniformization is more useful than the corresponding uniformization over $\mathbb{C}$.

The uniformization theorem (VI.5.1.1) combined with the exponential map says that every elliptic curve over $\mathbb{C}$ is analytically isomorphic to $\mathbb{C}^*/q^{\mathbb{Z}}$ for some $q \in \mathbb{C}^*$ with $|q| < 1$. It is clear that this cannot be true over K, since examining the power series for $j = j(q)$, we see that if $|q|_v < 1$, then $|j(q)|_v > 1$. Thus every curve E_q has nonintegral j-invariant. More precisely, the reduction $\tilde{E}_q$ of E_q modulo v is given by the Weierstrass equation

$$\tilde{E}_q : y^2 + xy = x^3,$$

so E_q has split multiplicative reduction at v.

Theorem 14.1. (Tate) *Let K be a field that is complete with respect to a discrete valuation v.*

(a) *For every $q \in K^*$ with $|q|_v < 1$, the map*

$$\phi : \bar{K}^*/q^{\mathbb{Z}} \longrightarrow E_q(\bar{K})$$

as described above is an isomorphism of $G_{\bar{K}/K}$-modules.

(b) *For every $j_0 \in K^*$ with $|j_0| > 1$ there is a $q \in K^*$ with $|q|_v < 1$ such that the elliptic curve E_q/K has j-invariant j_0. The curve E_q is characterized up to isomorphism over K by $j(E_q) = j$ and the fact that it has split multiplicative reduction at v.*

(c) *Let R be the ring of integers of K. Then the isomorphism $E_q(K) \cong K^*/q^{\mathbb{Z}}$ induces identifications*

$$(E_q)_0(K) \cong R^* \qquad and \qquad (E_q)_1(K) \cong \{u \in R^* : u \equiv 1 \pmod{v}\}.$$

(d) *Let E/K be an elliptic curve with nonintegral j-invariant that does not have split multiplicative reduction. From (b) there is a value of $q \in K^*$ satisfying $j(E) = j(E_q)$. Then there is a unique quadratic extension L/K such that E is isomorphic to E_q over L. Further,*

$$E(K) \cong \frac{\{u \in L^* : \mathrm{Norm}_{L/K}(u) \in q^{\mathbb{Z}}\}}{q^{\mathbb{Z}}}.$$

The extension L/K is unramified if and only if E has (nonsplit) multiplicative reduction, in which case the residue field extension of L/K is generated by the slopes of the tangent lines to the node of the reduced curve $\tilde{E}$ of E modulo v.

PROOF. This result was originally proven by Tate in 1959, although not published until many years later [285]. Other accounts may be found in [210, II §5], [212], and [266, V.5.3, V.5.4]. □

References. [140, Chapter 15], [210], [212], [266, Chapter V], [285].

C.15   Néron Models and Tate's Algorithm

As in the last section, we let K be a field that is complete with respect to a discrete valuation v, and we let R be the ring of integers of K and k the residue field of K. Let E/K be an elliptic curve and choose a minimal Weierstrass equation for E at v. Suppose that we now consider this equation as defining a scheme E of $\mathrm{Spec}(R)$. The resulting scheme may not be regular (i.e., nonsingular), since if E has bad reduction at v, then the singular point on the special fiber $\tilde{E}$ of E may be a singular point of the scheme. By resolving the singularity, we obtain a scheme $\mathcal{C}/\mathrm{Spec}(R)$ whose generic fiber is E/K and whose special fiber is a union of curves (with multiplicities) over k.

Theorem 15.1. (Kodaira, Néron) *Let E/K be as above.*

(a) *There is a regular projective two-dimensional scheme $\mathcal{C}/\mathrm{Spec}(R)$ whose generic fiber $\mathcal{C} \times_{\mathrm{Spec}(R)} \mathrm{Spec}(K)$ is isomorphic over K to E/K. Suppose further that $\mathcal{C}$ is minimal, i.e., the map $\mathcal{C} \to \mathrm{Spec}(R)$ cannot be factored as*

$$\mathcal{C} \to \mathcal{C}' \to \mathrm{Spec}(R)$$

in such a way that

$$\mathcal{C} \times_{\mathrm{Spec}(R)} \mathrm{Spec}(K) \to \mathcal{C}' \times_{\mathrm{Spec}(R)} \mathrm{Spec}(K)$$

is an isomorphism. Then $\mathcal{C}$ is unique.

(b) *Let $\mathcal{E} \subset \mathcal{C}$ be the subscheme of $\mathcal{C}$ obtained by discarding all of the singular points of the special fiber $\tilde{\mathcal{C}} = \mathcal{C} \times_{\mathrm{Spec}(R)} \mathrm{Spec}(k)$. In other words, we discard all multiple fibral components, all intersections of fibral components, and all singular points of fibral components. (Note that these are not singular points of $\mathcal{C}$ itself, since $\mathcal{C}$ is regular.) Then $\mathcal{E}$ is a group scheme over $\mathrm{Spec}(R)$ whose generic fiber $\mathcal{E} \times_{\mathrm{Spec}(R)} \mathrm{Spec}(K)$ is isomorphic, as a group variety, to E/K. The scheme $\mathcal{E}$ is called the Néron minimal model of E/K.*

(c) *The natural map $\mathcal{E}(R) \to E(K)$ is an isomorphism, i.e., every section on the generic fiber $\mathrm{Spec}(K) \to E$ extends to a section $\mathrm{Spec}(R) \to \mathcal{E}$.*

(d) *Let $\tilde{\mathcal{E}} = \mathcal{E} \times_{\mathrm{Spec}(R)} \mathrm{Spec}(k)$ be the special fiber of $\mathcal{E}$. Then $\tilde{\mathcal{E}}$ is an algebraic group over k. Let $\tilde{\mathcal{E}}^0/k$ be its identity component, so $\tilde{\mathcal{E}}$ is an extension of $\tilde{\mathcal{E}}^0$ by a finite group. Note that there is a reduction map $\mathcal{E}(R) \to \tilde{\mathcal{E}}(k)$. Then, with the identification $\mathcal{E}(R) \cong E(K)$ from (c), we have*

(i) $$\tilde{\mathcal{E}}^0(k) \cong \tilde{E}_{\mathrm{ns}}(k) \cong E_0(K)/E_1(K).$$

(ii) $$\tilde{\mathcal{E}}(k)/\tilde{\mathcal{E}}^0(k) \cong E(K)/E_0(K).$$

PROOF. See [193] or [266, §§IV.5, IV.6].   □

Remark 15.1.1. In some sources, the scheme $\mathcal{C}$ is called the Néron minimal model of E/K. However, for abelian varieties of higher dimension, the Néron minimal model always refers to a group scheme analogous to our $\mathcal{E}$, and there is no natural analogue of $\mathcal{C}$. The ambiguity for elliptic curves results because the minimal model of E, considered as a *curve*, is $\mathcal{C}$, while the minimal model of E, considered as a group variety, is $\mathcal{E}$.

Notice that (C.15.1d(ii)) gives a description of $E(K)/E_0(K)$ in terms of the group of components of a certain algebraic group $\tilde{\mathcal{E}}/k$. It turns out that there are only a handful of possibilities for $\tilde{\mathcal{E}}$. More precisely, we can write down all of the possibilities for the special fiber $\tilde{\mathcal{C}} = \mathcal{C} \times_{\mathrm{Spec}(R)} \mathrm{Spec}(k)$, and then we obtain $\tilde{\mathcal{E}}$ by discarding all components of multiplicity greater than one, all points where components intersect, and all singular points of components. The results are described in the following theorem.

Theorem 15.2. (Kodaira, Néron) *With notation as in* (C.15.1), *all of the possibilities for the special fiber $\tilde{\mathcal{C}}$ and the group of components $\tilde{\mathcal{E}}(k)/\tilde{\mathcal{E}}^0(k)$ are given in Table 15.1. (Some of the components of the special fiber may only be defined over a finite extension of k.)*

Other than the case I_0, *each of the pictured components is a rational curve, i.e., a copy of $\mathbb{P}^1$. Further, the minimal discriminant Δ_v of E at v and the exponent of the conductor f_v (VIII §11, C §16) are as listed in the table.*

PROOF. See [193] or [266, §§IV.8, IV.9]. $\qquad\square$

Corollary 15.2.1. *The group $E(K)/E_0(K)$ is finite. If E has split multiplicative reduction, then it is cyclic of order $-\operatorname{ord}_v\bigl(j(E)\bigr)$. In all other cases it has order at most 4.*

PROOF. The first statement follows from (C.15.1d). For the second, if E has split multiplicative reduction, then (C.14.1ac) implies that

$$E(K) = K^*/q^{\mathbb{Z}} \qquad \text{and} \qquad E(K)/E_0(K) \cong \mathbb{Z}/\operatorname{ord}_v(q)\mathbb{Z}.$$

Further, the series for $j(E)$ shows that $\operatorname{ord}_v(q) = -\operatorname{ord}_v\bigl(j(E)\bigr)$. Next, if E has nonsplit multiplicative reduction, then it is easy to check using (C.14.1d) that the quotient $E(K)/E_0(K)$ has order 1 or 2. Finally, if E has additive reduction, then the result follows by inspection of Table 15.1. $\qquad\square$

Remark 15.3. If k does not have characteristic 2 or 3, then everything about E, i.e., reduction type, exponent of conductor, and the group $E(K)/E_0(K)$, may be read off from Table 15.1 once we have a minimal Weierstrass equation for E. (It may also be necessary to take a finite *unramified* extension of K to get the full group of components.) Further, a given Weierstrass equation is minimal if and only if either $\operatorname{ord}_v(\Delta) < 12$ or $\operatorname{ord}_v(c_4) < 4$; see Exercise 7.1a. So in this case it is easy to check for minimality. In general, for k of arbitrary characteristic, there is a straightforward (but somewhat lengthy) algorithm due to Tate [283] that computes the special fiber $\tilde{\mathcal{C}}$, after which one can read off the other information from the corresponding column in Table 15.1.

References. [193], [283], [266, Chapter IV].

Kodaira symbol	I₀	I_n $(n \geq 1)$	II	III	IV	I_0^*	I_n^* $(n \geq 1)$	IV*	III*	II*
Special fiber $\tilde{C}$ (the numbers indicate multiplicities)	(fiber diagram)	(fiber diagram)	(fiber diagram)	(fiber diagram)	(fiber diagram)	(fiber diagram)	(fiber diagram)	(fiber diagram)	(fiber diagram)	(fiber diagram)
m = number of irreducible components	1	n	1	2	3	5	$5+n$	7	8	9
$E(K)/E_0(K)$ $\cong \tilde{\mathcal{E}}(k)/\tilde{\mathcal{E}}^0(k)$	(0)	$\dfrac{\mathbb{Z}}{n\mathbb{Z}}$	(0)	$\dfrac{\mathbb{Z}}{2\mathbb{Z}}$	$\dfrac{\mathbb{Z}}{3\mathbb{Z}}$	$\dfrac{\mathbb{Z}}{2\mathbb{Z}} \times \dfrac{\mathbb{Z}}{2\mathbb{Z}}$	$\dfrac{\mathbb{Z}}{2\mathbb{Z}} \times \dfrac{\mathbb{Z}}{2\mathbb{Z}}$ n even / $\dfrac{\mathbb{Z}}{4\mathbb{Z}}$ n odd	$\dfrac{\mathbb{Z}}{3\mathbb{Z}}$	$\dfrac{\mathbb{Z}}{2\mathbb{Z}}$	(0)
$\tilde{\mathcal{E}}^0(k)$	$\tilde{E}(k)$	k^*	k^+	k^+	k^+	k^+	k^+	k^+	k^+	k^+
Entries below this line valid only for char$(k) = p$ as indicated										
char$(k) = p$			$p \neq 2,3$	$p \neq 2$	$p \neq 3$	$p \neq 2$	$p \neq 2$	$p \neq 3$	$p \neq 2$	$p \neq 2,3$
ord$_v(\Delta_v)$	0	n	2	3	4	6	$6+n$	8	9	10
f_v = exponent of conductor $= \operatorname{ord}_v(\Delta_v)+1-m$	0	1	2	2	2	2	2	2	2	2
behavior of j	$v(j) \geq 0$	$\operatorname{ord}_v(j) = -n$	$\tilde{j} = 0$	$\tilde{j} = 1728$	$\tilde{j} = 0$	$v(j) \geq 0$	$\operatorname{ord}_v(j) = -n$	$\tilde{j} = 0$	$\tilde{j} = 1728$	$\tilde{j} = 0$

Table 15.1: A table of reduction types

C.16 *L-Series*

The *L*-series of an elliptic curve is a generating function that records information about the reduction of the curve modulo every prime. Known results are fragmentary, but conjecturally such *L*-series contain a large amount of information concerning the set of global points on the curve, which is somewhat surprising in view of the failure of the Hasse principle for curves of genus one. Further, there are intimate relations connecting *L*-series on elliptic curves defined over $\mathbb{Q}$ and the theory of modular forms. In this section we describe these connections and state some fundamental conjectures.

Let E/K be an elliptic curve and let $v \in M_K$ be a finite place at which E has good reduction. We denote the residue field of K at v by k_v, the reduction of E at v by $\tilde{E}_v$, and we let $q_v = \#k_v$ be the norm of the prime ideal corresponding to v. We recall from (V §2) that the *zeta function of $\tilde{E}_v/k_v$* is the power series

$$Z(\tilde{E}_v/k_v; T) = \exp\left(\sum_{n=1}^{\infty} \#\tilde{E}_v(k_{v,n}) \frac{T^n}{n}\right),$$

where $k_{v,n}$ is the unique extension of k_v of degree n. We proved (V.2.4) that $Z(\tilde{E}_v/k_v; T)$ is a rational function,

$$Z(\tilde{E}_v/k_v; T) = \frac{L_v(T)}{(1-T)(1-q_v T)},$$

where

$$L_v(T) = 1 - a_v T + q_v T^2 \in \mathbb{Z}[T] \qquad \text{and} \qquad a_v = q_v + 1 - \#\tilde{E}_v(k_v).$$

We extend the definition of $L_v(T)$ to the case that E has bad reduction by setting

$$L_v(T) = \begin{cases} 1 - T & \text{if } E \text{ has split multiplicative reduction at } v, \\ 1 + T & \text{if } E \text{ has nonsplit multiplicative reduction at } v, \\ 1 & \text{if } E \text{ has additive reduction at } v. \end{cases}$$

Then in all cases we have the relation

$$L_v(1/q_v) = \#\tilde{E}_{\mathrm{ns}}(k_v)/q_v.$$

Definition. The *L*-series of E/K is defined by the Euler product

$$L_{E/K}(s) = \prod_{v \in M_K^0} L_v(q_v^{-s})^{-1}.$$

The product defining $L_{E/K}(s)$ converges and gives an analytic function for all $\mathrm{Re}(s) > \frac{3}{2}$. This is easy to prove using the fact (V.2.4) that $|a_v| \le 2\sqrt{q_v}$. It is conjectured that far more is true.

Conjecture 16.1. *The L-series $L_{E/K}(s)$ has an analytic continuation to the entire complex plane and satisfies a functional equation relating its values at s and $2 - s$.*

Work of Deuring [61, 62, 63, 64] and Weil [306] showed that (C.16.1) is true for elliptic curves having complex multiplication, in which case $L_{E/K}$ is equal to a (product of) Hecke L-series with Grössencharacter. Eichler [72] and Shimura [246, 249] proved that (C.16.1) is true for elliptic curves $E/\mathbb{Q}$ that have a modular parametrization (C §13), so Wiles' theorem (C.13.6) implies that (C.16.1) is true for all elliptic curves over $\mathbb{Q}$.

The *conductor of E/K* is an integral ideal of K that encodes the primes of bad reduction and is the same for isogenous elliptic curves. For any finite place $v \in M_K$, we define the *exponent of the conductor of E at v* by

$$
f_v = \begin{cases} 0 & \text{if } E \text{ has good reduction at } v, \\ 1 & \text{if } E \text{ has multiplicative reduction at } v, \\ 2 + \delta_v & \text{if } E \text{ has additive reduction at } v, \end{cases}
$$

where δ_v is a measure of the "wild ramification" in the action of the inertia group on $T_\ell(E)$; see [200], [239], or [266, IV §10]. In particular, if $\text{char}(k_v) \neq 2,3$, then $\delta_v = 0$. In all cases the exponent f_v my be computed using the following resul:

Proposition 16.2. (Ogg–Saito formula) *Let m_v be the number of irreducible components (ignoring multiplicities) on the special fiber of the minimal (complete) Néron model of E at v (C §15), and let $\mathcal{D}_{E/K}$ be the minimal discriminant of E/K. Then*

$$
f_v = \text{ord}_v(\mathcal{D}_{E/K}) + 1 - m_v.
$$

PROOF. This was proven by Ogg [200] except in the case that $\text{char}(K_v) = 0$ and $\text{char}(k_v) = 2$. A proof in all characteristics for curves of arbitrary genus was given by Saito [217]. $\qquad\qquad\qquad\qquad\qquad\qquad\qquad\qquad\qquad\qquad\qquad\quad\Box$

Definition. The *conductor of E/K* is the integral ideal of K defined by

$$
N_{E/K} = \prod_{v \in M_K^0} \mathfrak{p}_v^{f_v}.
$$

In order to simplify the exposition in the rest of this section, we henceforth restrict attention to the case $K = \mathbb{Q}$. In this case we may take $N_E = N_{E/\mathbb{Q}}$ to be a positive integer. We define a new function

$$
\xi_E(s) = N_E^{s/2}(2\pi)^{-s}\Gamma(s)L_E(s),
$$

where $\Gamma(s) = \int_0^\infty t^{s-1}e^{-t}\,dt$ is the gamma function. Then (C.16.1) has the following more precise formulation, but since we have assumed that $K = \mathbb{Q}$, it is a theorem, rather than a conjecture.

Theorem 16.3. *Let $E/\mathbb{Q}$ be an elliptic curve. Then the function $\xi_E(s)$ has an analytic continuation to the entire complex plane and satisfies the functional equation*

$$\xi_E(s) = w\xi_E(2 - s) \qquad \text{for some } w = \pm 1.$$

The quantity w is called the *sign of the functional equation*. Its parity determines whether the order of vanishing of $L_{E/\mathbb{Q}}(s)$ at $s = 1$ is odd or even.

The modularity of elliptic curves over $\mathbb{Q}$ and its implication for L-series are described in the next result, which combines Wiles' modularity theorem (C.13.6) with earlier work of Carayol [35], Eichler [72], Shimura [246, 249], and others, who showed that modular elliptic curves have the indicated properties.

Theorem 16.4. *Let $E/\mathbb{Q}$ be an elliptic curve of conductor N, let $L_E(s) = \sum c_n n^{-s}$ be its L-series, and let $f_E(\tau) = \sum c_n e^{2\pi i n \tau}$ be the inverse Mellin transform of L_E.*
(a) *$f_E(\tau)$ is a weight-2 cusp form for the congruence subgroup $\Gamma_0(N)$ of $\mathrm{SL}_2(\mathbb{Z})$.*
(b) *For each prime $p \nmid N$, let $T(p)$ be the associated Hecke operator, and define an operator W by $(Wf)(\tau) = f(-1/N\tau)$. Then*

$$T(p)f_E = c_p f_E \qquad \text{and} \qquad W f_E = w f_E,$$

where $w = \pm 1$ is the sign of the functional equation (C.16.3).
(c) *Let ω be an invariant differential on $E/\mathbb{Q}$. There exists a finite morphism $\phi : X_0(N) \to E$ defined over $\mathbb{Q}$ such that $\phi^*(\omega)$ is a multiple of the differential form on $X_0(N)$ represented by $f(\tau)\,d\tau$.*

Another important conjecture concerning the L-series of elliptic curves involves their behavior around $s = 1$. Before stating the conjecture, we set the following notation:

$E/\mathbb{Q}$ an elliptic curve.

ω the invariant differential $dx/(2y + a_1 x + a_3)$ on a global minimal Weierstrass equation for $E/\mathbb{Q}$; see (VIII §8).

Ω $= \int_{E(\mathbb{R})} |\omega|$, which equals either the real period or twice the real period, depending on whether $E(\mathbb{R})$ is connected.

$Ш(E/\mathbb{Q})$ the Shafarevich–Tate group of $E/\mathbb{Q}$; see (X §4).

$R(E/\mathbb{Q})$ the elliptic regulator of $E(\mathbb{Q})/E_{\text{tors}}(\mathbb{Q})$, computed using the canonical height pairing; see (VIII §9).[1]

c_p $= \#E(\mathbb{Q}_p)/E_0(\mathbb{Q}_p)$. In particular, $c_p = 1$ if E has good reduction at p. See (C §15) and (VII §6) for a geometric description of c_p. (Birch and Swinnerton-Dyer originally called these "fudge factors." Their presence was explained by Tate [283].)

---

[1]Many sources use an alternative definition of the height pairing that is twice our value, which has the effect of eliminating the factor of 2^r in (C.16.5b).

Conjecture 16.5. (Birch and Swinnerton-Dyer) *Let $E/\mathbb{Q}$ be an elliptic curve.*
(a) $L_E(s)$ *has a zero at $s = 1$ of order equal to the rank of $E(\mathbb{Q})$.*[2]
(b) *Let $r = \operatorname{rank} E(\mathbb{Q})$. Then with notation as above,*

$$\lim_{s \to 1} \frac{L_E(s)}{(s-1)^r} = \frac{2^r \Omega \#\text{Ш}(E/\mathbb{Q}) R(E/\mathbb{Q}) \prod_p c_p}{\#E_{\text{tors}}(\mathbb{Q})^2}.$$

As described by Tate in 1974, "this remarkable conjecture relates the behavior of a function L at a point where it is not at present known to be defined to the order of a group Ш which is not known to be finite!" Progress since that time (C.16.1) has shown that $L_E(s)$ is defined at $s = 1$, but there is still no general proof that Ш is always finite. However, there is a great deal of evidence supporting (C.16.5), of which we mention the following.

Evidence 16.5.1. Conjecture C.16.5 has been checked numerically in a large number of cases. Since in general Ш is not known to be finite, what this means is that (C.16.5b) is used to compute a hypothetical value for Ш. In all cases this has turned out to be the square of an integer, as it should because of Cassels' pairing (X.4.14), and the value agrees with the calculated value of the 2 and/or 3 primary component of Ш. The original numerical evidence is given in [20]; see [33] and [276] for additional computations.

Evidence 16.5.2. Exercise 5.4 says that isogenous elliptic curves have the same number of points modulo p for all primes p, and thus they have the same L-series. (One must also check that the factors for primes of bad reduction agree.) Consequently, if (C.16.5b) is true, then the quantity

$$\frac{\Omega \#\text{Ш}(E/\mathbb{Q}) R(E/\mathbb{Q}) \prod_p c_p}{\#E_{\text{tors}}(\mathbb{Q})^2}$$

must be an isogeny invariant. This was proved by Cassels [40] and extended to abelian varieties by Tate, in both cases under the assumption that Ш is finite. It is worth noting that *none* of the individual terms in the product need be the same for isogenous curves.

Evidence 16.5.3. Coates and Wiles [49] showed that if $E/\mathbb{Q}$ has complex multiplication and $E(\mathbb{Q})$ is infinite, then $L_E(1) = 0$. See also [6] and [214].

Evidence 16.5.4. Greenberg [98] showed that if $E/\mathbb{Q}$ has complex multiplication and $\operatorname{ord}_{s=1} L_E(s)$ is odd, then either $\operatorname{rank} E(\mathbb{Q}) \geq 1$, or else $\text{Ш}(E/\mathbb{Q})[p^\infty]$ is infinite for a set of primes p of density $\frac{1}{2}$. This last possibility is unlikely, to say the least. See also [211].

Evidence 16.5.5. In the case that $L_E(1) = 0$, Gross and Zagier [101] proved a limit formula relating $L'_E(1)$ to the canonical height of a point in $E(\mathbb{Q})$ called a *Heegner point*. Rubin [215] for elliptic curves with complex multiplication and Kolyvagin [130] for elliptic curves over $\mathbb{Q}$ developed the theory of *Euler systems* and used

---

[2]This conjecture on the order of vanishing of $L_E(s)$ is one of the $1,000,000 Millennium Prize problems; see www.claymath.org/millennium/.

it to prove that III is finite and that (C.16.5a) is true in many cases where $L_E(s)$ vanishes to order at most 1. In particular, the results of Kolyvagin and Gross–Zagier imply the following cases of (C.16.5a) for elliptic curves defined over $\mathbb{Q}$:

$$L_E(1) \neq 0 \quad \Longrightarrow \quad \operatorname{rank} E(\mathbb{Q}) = 0.$$
$$L_E(1) = 0 \text{ and } L'_E(1) \neq 0 \quad \Longrightarrow \quad \operatorname{rank} E(\mathbb{Q}) = 1.$$

In many cases (C.16.5b) is also known, at least up to a (square) rational factor. We remark that Gross and Zagier originally proved their theorems under the assumption that $E/\mathbb{Q}$ is modular, but their results are now unconditional because (C.16.4) tells us that all elliptic curves over $\mathbb{Q}$ are modular.

References. [6], [20], [21], [40], [49], [61], [62], [63], [64], [98], [100], [101], [102], [130], [197], [198], [199], [200], [211], [214], [215], [216], [239], [247], [248], [249], [276], [306], [309], [312].

C.17 Duality Theory

In (X §4) we discussed the bilinear pairing on the Shafarevich–Tate group. There is a complementary duality theorem in the local case.

Theorem 17.1. (Tate [281], [286]) *Let $v \in M_K$ be a nonarchimedean absolute value, and let E/K_v be an elliptic curve. There exists a nondegenerate bilinear pairing*

$$\langle \,\cdot\,, \,\cdot\, \rangle : E(K_v) \times \operatorname{WC}(E/K_v) \longrightarrow \mathbb{Q}/\mathbb{Z}.$$

More precisely, if we give $E(K_v)$ the v-adic topology and $\operatorname{WC}(E/K_v)$ the discrete topology, then $\langle \,\cdot\,, \,\cdot\, \rangle$ induces a duality of locally compact groups, i.e., the pairing $\langle \,\cdot\,, \,\cdot\, \rangle$ is continuous, every continuous homomorphism $E(K_v) \to \mathbb{Q}/\mathbb{Z}$ has the form $\langle \,\cdot\,, \xi \rangle$ for some $\xi \in \operatorname{WC}(E/K_v)$, and similarly every continuous homomorphism $\operatorname{WC}(E/K_v) \to \mathbb{Q}/\mathbb{Z}$ has the form $\langle P, \,\cdot\, \rangle$ for some $P \in E(K_v)$.

See Exercise 10.24 for a construction of the Tate pairing.

The global duality theory is not quite as satisfactory because it is not known in general whether the Shafarevich–Tate group can have divisible elements.

Theorem 17.2. (Cassels [38], Tate [281]) *Let E/K be an elliptic curve. There exists an alternating bilinear pairing*

$$\operatorname{III}(E/K) \times \operatorname{III}(E/K) \longrightarrow \mathbb{Q}/\mathbb{Z}$$

whose kernel on either side is precisely the group of divisible elements of III.

Corollary 17.2.1. *If $\operatorname{III}(E/K)$ is finite, then its order is a perfect square. More generally, the same is true for any p-primary component $\operatorname{III}(E/K)[p^\infty]$.*

References. [38], [177], [238], [286], [281].

C.18 Local Height Functions

Néron, in his original construction of canonical height functions [194], first constructed a local height pairing at each absolute value $v \in M_K$. He then formed the (global) canonical height by taking the sum of the local heights. A nice exposition of this local theory for elliptic curves was given by Tate [280] and published in [135]. The theory of local height functions is important in the study of the more delicate properties of the canonical height; see for example [101], [113], or [254]. It is also useful for numerical computation of the canonical height of points on elliptic curves.

Proposition 18.1. *Let $v \in M_K$, and let E/K_v be an elliptic curve given by a Weierstrass equation*

$$E : y^2 + a_1 xy + a_3 y = x63 + a_2 x^2 + a_4 x + a_6.$$

There is a unique function

$$\lambda_v : E(K_v) \smallsetminus \{O\} \longrightarrow \mathbb{R},$$

called the local height function *for E at v, with the following properties:*
  (i) *λ_v is continuous for the v-adic topology on $E(K_v)$ and the usual topology on $\mathbb{R}$.*
  (ii) *The limit*

$$\lim_{P \to O} \lambda_v(P) + \frac{1}{2} v(x(P))$$

  exists, where $P \in E(K_v)$ and $P \to O$ in the v-adic topology.
  (iii) *Let Δ be the discriminant of the Weierstrass equation. Then for all $P \in E(K_v)$ with $[2]P \neq O$ we have*

$$\lambda_v([2]P) = 4\lambda_v(P) + v(2y(P) + a_1 x(P) + a_3) - \frac{1}{4} v(\Delta).$$

Further, property (iii) *may be replaced by the "quasi-parallelogram law"*
(iii′) *For all $P, Q \in E(K_v)$ with $P, Q, P \pm Q \neq O$,*

$$\lambda_v(P + Q) + \lambda_v(P - Q) = 2\lambda_v(P) + 2\lambda_v(Q) + v(x(P) - x(Q)) - \frac{1}{6} v(\Delta).$$

Remark 18.1.1. The function λ_v in (C.18.1) does not depend on the choice of Weierstrass equation for E because conditions (i)–(iii) are invariant under change of coordinates.

Theorem 18.2. *Let E/K be an elliptic curve. Then for all points $P \in E(K) \smallsetminus \{O\}$, the canonical height $\hat{h}(P)$ is given by*

$$\hat{h}(P) = \frac{1}{[K : \mathbb{Q}]} \sum_{v \in M_K} n_v \lambda_v(P).$$

There are explicit formulas for the local height in all cases, but we are content to give the following statement.

Theorem 18.3. *Let E/K be an elliptic curve and let $v \in M_K$.*

(a) Case I. *v archimedean.*

Choose a lattice $\Lambda \subset \mathbb{C}$ and an isomorphism $E(\bar{K}_v) \cong \mathbb{C}/\Lambda$. Let $\sigma(z, \Lambda)$ be the Weierstrass σ-function and let $\Delta(\Lambda)$ and $\eta : \mathbb{C} \to \mathbb{R}$ be as in Exercises 6.6 and 6.4, respectively. If $P \in E(K_v)$ corresponds to $z \in \mathbb{C}/\Lambda$, then

$$\lambda_v(P) = -\log\left|\Delta(\Lambda)^{1/12} e^{-z\eta(z)/2} \sigma(z, \Lambda)\right|_v.$$

(b) Case II. *v nonarchimedean and $P \in E_0(K)$.*

Let x and y be coordinate functions on a minimal Weierstrass equation for E at v. Then

$$\lambda_v(P) = \max\left\{-\frac{1}{2}v(x(P)), 0\right\} + \frac{1}{12}v(\Delta).$$

(c) Case III. *v nonarchimedean, E has split multiplicative reduction at v, and $P \notin E_0(K)$.*

Fix an isomorphism $E(K_v)/E_0(K_v) \cong \mathbb{Z}/N\mathbb{Z}$, where $N = -\operatorname{ord}_v(j(E))$. (See (VII.6.1).) Suppose that with this identification, the point $P \in E(K_v)$ maps to $n \in \mathbb{Z}/N\mathbb{Z}$ with $1 \le n \le N - 1$. Then

$$\lambda_v(P) = -\frac{1}{2}B_2\left(\frac{n}{N}\right)v(j(E)),$$

where $B_2(T) = T^2 - T + \frac{1}{6}$ is the second Bernoulli polynomial.

Remark 18.3.1. There are also formulas for $\lambda_v(P)$ in the nonsplit multiplicative and additive reduction cases. See [263] or [266, exercises 6.7, 6.8] for details, but note that in any case we can always apply (C.18.3) after replacing K by a finite extension (VII.5.4c).

References. [135], [194], [263], [266]. For a reformulation and generalization in terms of arithmetic intersection theory, see for example [44] or [83].

C.19 The Image of Galois

Let E/K be an elliptic curve defined over a number field, and let ℓ be a prime. Many of the arithmetic properties of E are determined by the ℓ-adic representation

$$\rho_\ell : G_{\bar{K}/K} \longrightarrow \operatorname{Aut}(T_\ell(E)).$$

We state two fundamental results about ρ_ℓ.

Theorem 19.1. (Serre [237], [231]) *Assume that E does not have complex multiplication.*

(a) *For all primes ℓ, the image of ρ_ℓ is of finite index in $\operatorname{Aut}(T_\ell(E))$.*

(b) *For all but finitely many primes ℓ, the image of ρ_ℓ is equal to $\operatorname{Aut}(T_\ell(E))$.*

Theorem 19.2. (Faltings [82], [84]) *Let E/K and E'/K be elliptic curves. Then the natural map*

$$\operatorname{Hom}_K(E, E') \otimes \mathbb{Z}_\ell \longrightarrow \operatorname{Hom}_K(T_\ell(E), T'_\ell(E))$$

is an isomorphism. (Here the right-hand side is the group of $\mathbb{Z}_\ell$-linear homomorphisms from $T_\ell(E)$ to $T_\ell(E')$ that commute with the action of $G_{\bar{K}/K}$. We proved injectivity in (III.7.4); the real difficulty lies in showing that the map is surjective.)

References. [82], [84], [231], [237].

C.20   Function Fields and Specialization Theorems

Let V/K be a variety defined over a number field. We consider elliptic curves defined over the function field $K(V)$. These are curves given by Weierstrass equations

$$E : y^2 + a_1 xy + a_3 y = x^3 + a_2 x^2 + a_4 x + a_6 \qquad \text{with } a_1, \dots, a_6 \in K(V).$$

For almost all points $t \in V$, i.e., points outside of some proper subvariety of V, all of the functions $a_1, \dots, a_6$ are defined at t. This allows us to define the *specialization of E at t* to be the elliptic curve

$$E : y^2 + a_1(t)xy + a_3(t)y = x^3 + a_2(t)x^2 + a_4(t)x + a_6(t).$$

Similarly, if $P = (x, y) \in E(K(V))$, then the functions $x, y \in K(V)$ are defined for almost all $t \in V$, which allows us to *specialize P* to a point

$$P_t = (x(t), y(t)) \in E_t.$$

It turns out to be true, although we do not prove it, that the group $E(K(V))$ is finitely generated, i.e., the Mordell–Weil theorem holds in this function field setting. Thus by choosing a finite set of generators for $E(K(V))$, for almost all $t \in V$ we can define a specialization homomorphism

$$\sigma_t : E(K(V)) \longrightarrow E_t.$$

We observe that if $t \in V(K)$, then E_t is defined over K and the image of σ_t lies in $E_t(K)$. Néron used a generalization of Hilbert's irreducibility theorem to prove that the specialization homomorphism is frequently injective.

Theorem 20.1. (Néron [192]) *Let K be a number field, and let E be an elliptic curve defined over the function field $K(\mathbb{P}^n)$. Then there are infinitely many points $t \in \mathbb{P}^n(K)$ such that the specialization homomorphism*

$$\sigma_t : E(K(\mathbb{P}^n)) \longrightarrow E_t(K)$$

is injective.

PROOF. More precisely, Néron proves that the set of t for which σ_t is noninjective forms a "thin" set. See Néron's original paper [192], or the exposition in [139, Chapter 9] or [236, §11.1]. $\qquad\square$

Corollary 20.1.1. *There exist infinitely many elliptic curves $E/\mathbb{Q}$ such that $E(\mathbb{Q})$ has rank at least* 10.

PROOF. Néron [192] originally used (C.20.1) to find families of curves $E/\mathbb{Q}$ of rank 9 and 10. Subsequently, others have constructed families of rank up to 19; see for example [76, 85, 188]. $\qquad\square$

Remark 20.2. The way that Néron uses (C.20.1) to produce curves $E/\mathbb{Q}$ of moderately large rank is to find an elliptic curve over $\mathbb{Q}(T_1, \ldots, T_n)$ having large rank and then specialize. For example, taking $n = 18$, let $C/\mathbb{Q}(T_1, \ldots, T_{18})$ be the cubic curve passing through the nine points $(T_1, T_2), \ldots, (T_{17}, T_{18})$. It is not hard to show that the Jacobian of C has rank (at least) 9 over $\mathbb{Q}(T_1, \ldots, T_{18})$. With some additional work, Néron found infinitely many curves over $\mathbb{Q}$ of rank 10. Working within these families, an ingenious search method due to Mestre [171, 172] often enables one to find specific curves $E/\mathbb{Q}$ of even higher rank; see (VIII §10).

Remark 20.2.1. Masser [161] applied ideas from transcendence theory to extend and strengthen Néron's result.

In the case that the variety V is a curve, Néron's theorem (C.20.1) may be significantly improved.

Theorem 20.3. (Silverman [255]) *Let K be a number field, let C/K be a curve, and let E be an elliptic curve defined over the function field $K(C)$. Assume that E is nonconstant, i.e., $j(E) \notin K$. Then the specialization map*

$$\sigma_t : E\big(K(C)\big) \longrightarrow E_t$$

is well-defined and injective for all but finitely many points $t \in C(K)$. More generally, the set of points $t \in C(\bar{K})$ for which σ_t is not injective is a set of bounded height.

PROOF. See [139, Chapter 12], [256], or [266, III.11.4]. $\qquad\square$

The Birch–Swinnerton-Dyer conjecture (C.16.5) says that rank of the Mordell–Weil group $E(K)$ is determined by the reductions $\tilde{E}(\mathbb{F}_{\mathfrak{p}})$ of E modulo the various primes of K. A conjecture of Nagao gives a similar statement for the Mordell–Weil group of elliptic curves over function fields. For simplicity, we state the conjecture over $K(\mathbb{P}^1)$; see [213] for the general formulation.

Conjecture 20.4. (Nagao [189]) *Let K be a number field and let E be an elliptic curve defined over the function field $K(\mathbb{P}^1)$. Assume that $j(E) \notin K$. For each prime $\mathfrak{p}$ of K, let*

$$A_{\mathfrak{p}}(E) = \frac{1}{N_{K/\mathbb{Q}}\,\mathfrak{p}} \sum_{t \in \mathbb{P}^1(\mathbb{F}_{\mathfrak{p}})} a_{\mathfrak{p}}(\tilde{E}_t/\mathbb{F}_{\mathfrak{p}}).$$

In other words, $A_{\mathfrak{p}}$ is the average value of

$$a_{\mathfrak{p}}(\tilde{E}_t/\mathbb{F}_{\mathfrak{p}}) = \mathrm{N}_{K/\mathbb{Q}}\,\mathfrak{p} + 1 - \#\tilde{E}_t(\mathbb{F}_{\mathfrak{p}})$$

as we reduce modulo $\mathfrak{p}$ and specialize E at all of the points $t \in \mathbb{P}^1(\mathbb{F}_{\mathfrak{p}})$. Then

$$\lim_{X \to \infty} \frac{1}{X} \sum_{\mathrm{N}_{K/\mathbb{Q}}\,\mathfrak{p} \leq X} -A_{\mathfrak{p}}(E) \log \mathrm{N}_{K/\mathbb{Q}}\,\mathfrak{p} = \mathrm{rank}\, E(K(\mathbb{P}^1)).$$

Remark 20.4.1. If $A(T), B(T) \in K[T]$ are polynomials satisfying

$$\deg A(T) \leq 3 \quad \text{and} \quad \deg B(T) \leq 5,$$

then Rosen and Silverman [213] prove that (C.20.4) is true for the family of elliptic curves

$$y^2 = x^3 + A(T)x + B(T).$$

(The constraints on A and B imply that E is a rational elliptic surface.) More generally, they show that (C.20.4) is a consequence of a deep conjecture of Tate that relates the L-series of a variety to its geometry.

References. [76], [85], [139], [161], [188], [189], [192], [213], [256], [260], [284].

C.21   Variation of a_p and the Sato–Tate Conjecture

Let E/K be an elliptic curve defined over a number field. As in (C §16), we reduce E modulo finite places v and set

$$a_v = q_v + 1 - \#\tilde{E}_v(k_v),$$

where k_v is the residue field and $q_v = \#k_v$ is the norm of v. We know from (V.1.1) that $|a_v| \leq 2\sqrt{q_v}$, and we ask how the a_v values vary as we vary v. In order to normalize matters, it is natural to look at the ratios $a_v/2\sqrt{q_v}$. Since these ratios are between -1 and 1, we write them as

$$\frac{a_v}{2\sqrt{q_v}} = \cos\theta_v \quad \text{with} \quad 0 \leq \theta_v \leq \pi,$$

and we study the distribution of the angles θ_v.

    If E has complex multiplication, then it is not hard to prove that the θ_v values are uniformly distributed in the interval $[0, \pi]$. The non-CM case is covered by a deep conjecture of Sato and Tate

Conjecture 21.1. (Sato–Tate [287]) *With notation as above, assume that E does not have complex multiplication. Then the set of angles $\{\theta_v\}$ is equidistributed with respect to the measure $\frac{2}{\pi}\sin^2\theta\,d\theta$. In other words, for any $0 \leq \alpha \leq \beta \leq \pi$,*

$$\lim_{X \to \infty} \frac{\#\{v : q_v \leq X \text{ and } \alpha \leq \theta_v \leq \beta\}}{\#\{v : q_v \leq X\}} = \frac{2}{\pi}\int_\alpha^\beta \sin^2\theta\,d\theta.$$

Building on the groundbreaking methods devised by Wiles to prove the modularity conjecture (C.13.6), Clozel, Harris, Shepherd-Barron, and Taylor have mounted attacks on the Sato–Tate conjecture, leading to Taylor's proof of the conjecture for elliptic curves with nonintegral j-invariant

Theorem 21.2. *Let $E/\mathbb{Q}$ be an elliptic curve with nonintegral j-invariant. Then the Sato–Tate conjecture (C.21.1) is true for E.*

PROOF. [290]; see also [47] and [110]. $\square$

Remark 21.3. The $\sin^2 \theta$ distribution appearing in (C.21.1) may seem somewhat mysterious. It comes from Haar measure on the unitary group $\mathrm{SU}_2(\mathbb{C})$. To see how, we fix a prime ℓ and let

$$\rho : G_{\bar{K}/K} \longrightarrow \mathrm{Aut}\big(T_\ell(E)\big) \otimes_{\mathbb{Z}_\ell} \mathbb{Q}_\ell \cong \mathrm{GL}_2(\mathbb{Q}_\ell)$$

be the ℓ-adic representation attached to E. For each finite place $v \in M_K$ of good reduction for E, we choose a Frobenius element $\phi_v \in G_{\bar{K}/K}$. The matrix $\rho(\phi_v)$ is determined up to conjugation, so in particular its trace and norm are well-defined. Indeed, the trace and norm are integers,

$$\mathrm{Trace}\,\rho(\phi_v) = a_v \qquad \text{and} \qquad \mathrm{Norm}\,\rho(\phi_v) = q_v.$$

Thus $\rho(\phi_v)$ is conjugate to a matrix in $\mathrm{GL}_2(\bar{\mathbb{Q}})$. Further, Hasse's theorem (V.1.1) tells us that the characteristic polynomial of the matrix $\rho(\phi_v)/\sqrt{q_v}$,

$$\det\left(X - \frac{\rho(\phi_v)}{\sqrt{q_v}}\right) = X^2 - \frac{a_v}{\sqrt{q_v}}X + 1,$$

has complex roots of norm 1, so $\rho(\phi_v)/\sqrt{q_v}$ is actually conjugate to a matrix in $\mathrm{SU}_2(\mathbb{C})$. We thus get a well-defined map

$$M_K^0 \longrightarrow \mathrm{SU}_2(\mathbb{C})/(\text{conjugacy}), \qquad v \longmapsto \text{conjugacy class of } \rho(\phi_v)/\sqrt{q_v}.$$

(We may need to exclude finitely many elements of M_K^0.) It is an exercise to check that the trace map defines an isomorphism

$$\mathrm{Trace} : \mathrm{SU}_2(\mathbb{C})/(\text{conjugacy}) \xrightarrow{\ \sim\ } [-2, 2], \qquad A \longmapsto \mathrm{Trace}(A).$$

The compact group $\mathrm{SU}_2(\mathbb{C})$ has an invariant (Haar) measure μ that is unique if we normalize it to satisfy $\mu\big(\mathrm{SU}_2(\mathbb{C})\big) = 1$. One can check that the pushforward of μ to $[-2, 2]$ is given by $\frac{1}{2\pi}\sqrt{4 - x^2}\,dx$, i.e., for any continuous function on $[-2, 2]$ we have

$$\int_{\mathrm{SU}_2(\mathbb{C})} f \circ \mathrm{Trace}\,d\mu = \int_{-2}^{2} f(x) \cdot \frac{1}{2\pi}\sqrt{4 - x^2}\,dx.$$

The Sato–Tate conjecture asserts that the matrices $\rho(\phi_v)/\sqrt{q_v}$ are equidistributed in the quotient group $\mathrm{SU}_2(\mathbb{C})/(\text{conjugacy})$ relative to the measure induced from normalized Haar measure on $\mathrm{SU}_2(\mathbb{C})$. Moving everything to the interval $[-2, 2]$, this means that the trace values $\mathrm{Trace}\big(\rho(\phi_v)/\sqrt{q_v}\big)$ are equidistributed with respect to the measure $\frac{1}{2\pi}\sqrt{4 - x^2}$. Now the substitution $x = 2\cos\theta$ transforms the $\mathrm{SU}_2(\mathbb{C})$-equidistribution assertion into the formula given in (C.21.1).

The Sato–Tate conjecture deals with a fixed elliptic curve over a number field and counts the number of points modulo v for varying primes v. It is also natural to fix a finite field k and study the number of points in $E(k)$ as E varies over all isomorphism classes of elliptic curves, as in the following result.

Theorem 21.4. (Birch [18]) *Fix a prime p, and let $\mathbf{E}_p$ denote the set of $\mathbb{F}_p$-isomorphism classes of elliptic curves defined over $\mathbb{F}_p$. For each $E \in \mathbf{E}_p$, let*

$$a_p(E) = p + 1 - \#E(\mathbb{F}_p) \qquad and \qquad \cos\theta_p(E) = a_p(E)/2\sqrt{p}.$$

Then for all $0 \le \alpha \le \beta \le \pi$,

$$\lim_{p \to \infty} \frac{\#\{E \in \mathbf{E}_p : \alpha \le \theta_p(E) \le \beta\}}{\#\mathbf{E}_p} = \frac{2}{\pi} \int_\alpha^\beta \sin^2 \theta \, d\theta.$$

References. [18], [47], [110], [287], [290].

Notes on Exercises

Many of the exercises in this book are standard results that were not included in the text due to lack of space, while others are special cases of results that appear in the literature. The following notes thus serve two purposes. They are an attempt by the author to give credit for theorems that appear in the exercises, and they provide an aid for the reader who wishes to delve more deeply into some aspect of the theory. However, since any attempt to assign credit is bound to be incomplete in some respects, the author herewith tenders his apologies to all who feel that they have been slighted.

Except for an occasional computational problem (and for Exercise 3.16), we have not included solutions, or even hints. Indeed, since it is hoped that this book will lead the student on into the realm of active mathematics, the benefits of working without aid clearly outweigh any advantage that might be gained by having solutions readily available.

CHAPTER I

(1.1)   (a) $B(A^3 - 27B) = 0$.   (b) $4A^3 + 27B^2 = 0$.
(1.2)   (a) $(0,0)$. (b) $(0,0)$. (c) $(0,0)$. (d) $(0,0,1)$.
(1.3)   [111, I.5.1]
(1.5)   (b) $P_3 = (-8/9, 109/27)$
(1.7)   (b) $\psi = [Y, X]$.   (c) No.

CHAPTER II

(2.1)   [8, Proposition 9.2]
(2.4)   (b) [136, Lemma, Page 7]
(2.5)   [111, II.6.10.1] and [111, IV.1.3.5]
(2.6)   This volume (III §3).
(2.9)   This example is due to Hurwitz [117]. See also [41, §22].
(2.11)   This proof of Weil reciprocity is due to E. Kani.

CHAPTER III

(3.4) $P_2 = -[2]P_1 + P_3$, $P_4 = P_1 - P_3$, $P_5 = -[2]P_1$, $P_6 = -P_1 + [2]P_3$,
$P_7 = [3]P_1 - P_3$, $P_8 = -[4]P_1 + [3]P_3$
(3.6)   [111, IV.3.2b]
(3.7)   [36], [41, Lemma 7.2], [135, II Theorem 2.1]

(3.8)   (b) This volume (VI §6).
(3.9)   [210, II.1.24], [210, II.2.9]
(3.13)   [186, II §7 page 66]
(3.16)   (c) Since the publication of the first edition, this question has generated more inquiries to the author than any other question in the book, so we break our stated policy and provide a proof sketch. Choose points $P', Q', R \in E$ satisfying

$$[m]P' = P, \qquad [m]Q' = Q, \qquad P' \neq \pm Q', \qquad [2]R = P' - Q'.$$

Define divisors D_P and D_Q and choose functions $f_P, f_Q, g_P, g_Q \in \bar{K}(E)$ satsifying

$$D_P = (P) - (O), \qquad\qquad D_Q = (Q + [m]R) - ([m]R),$$
$$\mathrm{div}(f_P) = mD_P, \qquad\qquad \mathrm{div}(f_Q) = mD_Q,$$
$$f_P \circ [m] = g_P^m, \qquad\qquad f_Q \circ [m] = g_Q^m.$$

We claim that the following two functions in $\bar{K}(E)$ are constant when viewed as functions of X:

(i) $\dfrac{g_P(X + Q' + R)g_Q(X)}{g_P(X + R)g_Q(X + P')}.$ (ii) $\displaystyle\prod_{i=0}^{m-1} g_Q\big(X + [i]Q'\big).$

It suffices to show that the divisors of (i) and (ii) are 0, which is easily verified using

$$\mathrm{div}(g_P) = \sum_{T \in E[m]} (T + P') - (T), \qquad \mathrm{div}(g_Q) = \sum_{T \in E[m]} (T + Q' + R) - (T + R).$$

We now compute

$$
\begin{aligned}
\tilde{e}_m(P, Q) &= \frac{f_P(D_Q)}{f_Q(D_P)} && \text{definition of } \tilde{e}_m, \\[2mm]
&= \frac{f_P(Q + [m]R)f_Q(O)}{f_P([m]R)f_Q(P)} && \text{definition of } D_P \text{ and } D_Q, \\[2mm]
&= \frac{f_P([m]Q' + [m]R)f_Q([m]O)}{f_P([m]R)f_Q([m]P')} && \text{definition of } P' \text{ and } Q', \\[2mm]
&= \left(\frac{g_P(Q' + R)g_Q(O)}{g_P(R)g_Q(P')}\right)^m && \text{since } f_P \circ [m] = g_P^m \text{ and } f_Q \circ [m] = g_Q^m, \\[2mm]
&= \prod_{i=0}^{m-1} \frac{g_P(R + [i+1]Q')g_Q([i]Q')}{g_P(R + [i]Q')g_Q(P' + [i]Q')} && \text{since (i) is constant,} \\[2mm]
&= \frac{g_P(R + [m]Q')}{g_P(R)} \prod_{i=0}^{m-1} \frac{g_Q([i]Q')}{g_Q(P' + [i]Q')} && \text{since the product telescopes,} \\[2mm]
&= \frac{g_P(R + Q)}{g_P(R)} && \text{since (ii) is constant,} \\[2mm]
&= e_m(P, Q) && \text{definition of } e_m.
\end{aligned}
$$

(3.18)   (d) [60]
(3.20)   [249, Proposition 4.11]
(3.21)   [111, IV §4]
(3.23)   This volume (A.1.3). Poonen remarks that one can do (c) and (d) first, then prove (a) by finding an α such that the Deuring equation is nonsingular and has the same j-invariant as the given E.
(3.26)   (a) This volume (XI.7.1).
(3.27)   (e), (f) [267, 6.52]
(3.29)   This problem was suggested by David Masser.
(3.31)   This exercise was devised by Michael Rosen, René Schoof, and the author.
(3.32)   This volume (V.2.3.1).
(3.33)   This exercise was devised by Michael Rosen and the author.
(3.34)–(3.36) Elliptic divisibility sequences were first investigated by Ward [303]; see also [252, 271, 272].

CHAPTER IV

(4.1)   (a) [93, I §3 Proposition 1]
(4.2)   (b) [112, Theorem I.6.1]
(4.4)   [93, IV §2 Theorem 2]

CHAPTER V

(5.3)   [111, C.4.1]
(5.4)   (a) Due to F.K. Schmidt. See [41, Lemma 15.1].   (b) [282]
(5.8)   Due to Deuring. See [186, Section 22, Theorem on page 217].
(5.10)   (c), (d) solutions provided by Jaap Top.

(c)	$y^2 = x^3 - x$	$y^2 = x^3 - x + 1$	$y^2 = x^3 - x - 1$	$y^2 = x^3 + x$
$\#E(\mathbb{F}_3)$	4	7	1	4

(d)	$y^2 + y = x^3$	$y^2 + y = x^3 + x$	$y^2 + y = x^3 + x + 1$	
$\#E(\mathbb{F}_2)$	3	5	1	

(e), (f) suggested by René Schoof.
(5.11)   This proof of a weak version of [234, §4.3] was suggested by Serre.
(5.14)   This volume (XI.6.2).
(5.16)   This problem was suggested by Jonathan Lubin.
(5.18)   This babystep–giantstep algorithm is due to Shanks; see (XI.5.2).

CHAPTER VI

(6.3)   (d) [310, Chapter XX, Misc. Exercise 33]
(6.4)   (a)–(e) [3, Chapter 7, §3.2], [310, Chapter XX], [135, Chapter I,§6]
(6.8)   For more information about complex multiplication and class field theory, see (C §11) and the references listed there.
(6.11)–(6.13) The literature on elliptic integrals is vast. A nice summary may be found in [310, Chapter XXII].

(6.14)   [52]

(6.16)   This is due to Ward [303].

CHAPTER VII

(7.2)   [283, §3]

(7.4)   [283, §4]

(7.9)   [239, Theorem 2 and Corollaries]

(7.10)   For three proofs of this fact, see [266, II §6 and V.6.3]. For more on complex multiplication, see (C §11) and the references listed there.

(7.11)   This volume (A.1.4).

(7.13)   The observation that the ECDLP is trivial for anomalous curves is due to Semaev [228], Satoh–Araki [218], and Smart [269]. See this volume (XI.6.5).

CHAPTER VIII

(8.2)   This volume (X.6.1b).

(8.3)   This problem was suggested by D. Rohrlich.

(8.6)   [139, Chapter 3, Proposition 1.1]

(8.7)   [220]

(8.9)   [139, page 54]

(8.11)   [41, Theorem 17.2]

(8.12)   One example of each group allowed by (VIII.7.5).

(8.14)   (a,b) [241]   (d) [259]

(8.15)   Due to Tate. See [198].

(8.17)   (c) [254]   (d) [201]

(8.18)   Due to Dem'janenko and Zimmer. See [138] and [316]. For sharper bounds, see [55] and [265].

(8.21)   [196]

(8.23)   [196]

CHAPTER IX

(9.3)   (b) [153]   (d) [257]

(9.4)   [264]

(9.5)   [148]

(9.6)   Due to A. Thue.

(9.7)   [237, IV §2]

(9.8)   [139, Chapter 5, §7]

(9.10)   [56]

(9.11)   [185, Chapter 26]

(9.13)   This argument appears in an unpublished letter from Tate to Serre. One can also do (c) and (d) directly; see [185, Chapter 27, Theorem 2].
          (d) $2 \cdot 3 = 1 \cdot 2 \cdot 3$ and $14 \cdot 15 = 5 \cdot 6 \cdot 7$

(9.18)   The 14 integral points on E with $y > 0$ are $P_1 = (-12, 19)$, $P_2 = (-10, 33)$, $P_3 = (-4, 45)$, $P_4 = (3, 46)$, $P_5 = (8, 51)$, $P_6 = (18, 89)$, $P_7 = (60, 467)$, $P_8 = (71, 600)$, $P_9 = (80, 717)$, $P_{10} = (170, 2217)$,

$P_{11} = (183, 2476)$, $P_{12} = (698, 18441)$, $P_{13} = (9278, 893679)$, and $P_{14} = (129968, 46854861)$. The latter 10 of these are given by the linear combinations $P_5 = -P_2 - P_4$, $P_6 = -P_2 - P_3$, $P_7 = P_2 + P_3 + P_4$, $P_8 = -P_1 - P_2$, $P_9 = P_1 - P_3$, $P_{10} = P_3 - P_4$, $P_{11} = P_2 - P_3$, $P_{12} = P_1 - P_2$, $P_{13} = P_1 + [2]P_2 + P_3 + P_4$, $P_{14} = -P_2 + [2]P_3 - P_4$.

CHAPTER X

(10.2) [307]

(10.3) This is due to Châtelet. See [41, Theorem 11.1]

(10.4) (c) [41, Corollary to Lemma 10.3]

(10.6) This is due to Lang; see [144].

(10.8) [144]

(10.9) (a) [185, Chapter 16, Theorem 6]

(10.10) [37]

(10.11) (c,d) Due to Lang–Tate and Shafarevich. See [41, Lemma 12.2].

       (e) [39]

       (f) [37]

       (g) [149]

(10.19) (d) Due to Fueter [94].

(10.21) [266, V.5.2]

(10.24) [177], [286], [281].

(10.23) Due to Neumann [195] and Setzer [240]. A minimal Weierstrass equation for E_a is $y^2 + xy = x^3 + \frac{1}{4}(a-1)x^2 - x$.

CHAPTER XI

(11.1) (a) [23](195, 9) = (485, 573).   (b) [3211](2898, 439) = (243, 1875).

(11.2) See [22, §IV.3], [51, §15.1], or [116, Exercise 5.23]. The trick in (c) is due to Solinas.

       (d-i) $349 = 2^1 - 2^3 - 2^6 - 2^8 + 2^{10}$

       (d-ii) $9337 = 2^1 - 2^4 + 2^8 + 2^{11} + 2^{14}$

       (d-iii) $38728 = 2^4 + 2^7 - 2^9 - 2^{12} + 2^{14} + 2^{16}$

       (d-iv) $8379483273489 = 2^1 + 2^5 + 2^9 - 2^{12} - 2^{15} + 2^{22} - 2^{33} + 2^{38} - 2^{40} + 2^{44}$

(11.3) (d-i) $931 = -1 + \tau^2 + \tau^{10} + \tau^{14} - \tau^{17} - \tau^{19} - \tau^{21}$

       (d-ii) $32755 = -1 + \tau^2 + \tau^4 + \tau^6 + \tau^8 + \tau^{15} - \tau^{17} + \tau^{19} - \tau^{22} + \tau^{28} - \tau^{31}$

       (d-iii) $82793729188 = \tau^2 + \tau^8 - \tau^{10} - \tau^{12} + \tau^{15} + \tau^{18} + \tau^{20} - \tau^{24} - \tau^{27} + \tau^{30} - \tau^{34} + \tau^{36} - \tau^{40} + \tau^{44} + \tau^{46} - \tau^{48} + \tau^{50} - \tau^{52} + \tau^{55} + \tau^{58} + \tau^{61} + \tau^{68} - \tau^{71} - \tau^{73}$

(11.7) (a) $[5!]P$ gives $589 = 19 \cdot 31$.

       (b) $[7!]P$ gives $26167 = 191 \cdot 137$.

       (c) $[11!]P$ gives $1386493 = 1069 \cdot 1297$.

       (d) $[29!]P$ gives $28102844557 = 117763 \cdot 238639$.

(11.9) This algorithm is due to Pohlig and Hellman [205].

       (a) See [116, Theorem 2.32].

       (b) See [116, Proposition 2.34].

(11.11) This encryption scheme is due to Abdalla, Bellare, and Rogaway [1].

(11.5)   (b) For 50 people, the probability is 97%.

   For 23 people, the probability is 50.7%.

   (c) $c = 1.177$ gives probability 50%.

   $c = 5.256$ gives probability $1 - 10^{-6}$.

(11.12)

	$x_{S_i} \in A$	$x_{S_i} \in B$	$x_{S_i} \in C$
$x_{f(S_i)} \in A$	$\gamma_i + 2$	$2\gamma_i + 1$	$\gamma_i + 1$
$x_{f(S_i)} \in B$	$2\gamma_i + 2$	$4\gamma_i$	$2\gamma_i$
$x_{f(S_i)} \in C$	$\gamma_i + 1$	$2\gamma_i$	$\gamma_i + 1$

Value of γ_{i+1}

	$x_{S_i} \in A$	$x_{S_i} \in B$	$x_{S_i} \in C$
$x_{f(S_i)} \in A$	δ_i	$2\delta_i$	$\delta_i + 1$
$x_{f(S_i)} \in B$	$2\delta_i$	$4\delta_i$	$2\delta_i + 2$
$x_{f(S_i)} \in C$	$\delta_i + 1$	$2\delta_i + 1$	$\delta_i + 2$

Value of δ_{i+1}

(11.13) (a) $Q = [198]P$ (b) $Q = [6062]P$ (c) $Q = [62354]P$

(11.14) (b,c) Due to Teske [292]; see also [293].

(11.15) (b) $P' = (56, 11593)$, $Q' = (54, 12553)$.

   (c) $\log_E\big([137]P'\big) \equiv z\big([137]P'\big) \equiv 111 \cdot 137 \pmod{137^2}$,

   $\log_E\big([137]Q'\big) \equiv z\big([137]Q'\big) \equiv 65 \cdot 137 \pmod{137^2}$.

   (d) $m \equiv 111^{-1} \cdot 65 \equiv 66 \pmod{137}$.

(11.16) (a) Take $S = (0, 36)$. Then

$$f_{P'}(Q'+S) = 326, \ f_{P'}(S) = 523, \ f_{Q'}(P'-S) = 483, \ f_{Q'}(-S) = 576,$$

so $e_5(P', Q') = (326/523)/(483/576) = 512$.

   (b) $P' = 3P$ and $Q' = 4Q$.

List of Notation

$\bar{K}$	an algebraic closure of K, 1
$G_{\bar{K}/K}$	the Galois group of $\bar{K}/K$, 1
$\mathbb{A}^n$	affine n space, 1
$\mathbb{A}^n(K)$	the set of K rational points of $\mathbb{A}^n$, 1
P^σ	the action of $\sigma \in G_{\bar{K}/K}$ on the point P, 2
$\bar{K}[X]$	the polynomial ring $\bar{K}[X_1, \ldots, X_n]$, 2
V_I	subset of $\mathbb{A}^n$ associated to the ideal I, 2
$I(V)$	ideal associated to the algebraic set V, 2
$V(K)$	the set of K rational points of V, 2
$K[V]$	affine coordinate ring of V/K, 3
$K(V)$	function field of V/K, 3
$\dim(V)$	the dimension of V, 4
M_P	ideal associated to the point P, 5
$\bar{K}[V]_P$	local ring of V at P, 6
$\mathbb{P}^n$	projective n space, 6
$\mathbb{P}^n(K)$	set of K rational points of $\mathbb{P}^n$, 6
$K(P)$	minimal field of definition of the point P, 6
V_I	subset of $\mathbb{P}^n$ associated to the homogeneous ideal I, 7
$I(V)$	ideal associated to the projective algebraic set V, 7
$V(K)$	set of K-rational points of the projective variety V, 7
$\bar{V}$	projective closure of the algebraic set V, 9
$\dim(V)$	the dimension of the projective variety V, 10
$K(V)$	function field of the projective variety V/K, 10
$\bar{K}[V]_P$	local ring of the projective variety V at P, 11
ord_P	normalized valuation on $\bar{K}[C]_P$ and on $\bar{K}(C)$, 17
ϕ^*	map of function fields induced by rational map of curves, 20
$\deg \phi$	degree of the map ϕ, 21
$\deg_s \phi$	separable degree of the map ϕ, 21
$\deg_i \phi$	inseparable degree of the map ϕ, 21
ϕ_*	map of function fields induced by rational map of curves, 21
$e_\phi(P)$	ramification index of the map ϕ, 23
$f^{(q)}$	polynomial with coefficients raised to the qth power, 25
$C^{(q)}$	curve obtained by raising coefficients to the qth power, 25
$\mathrm{Div}(C)$	divisor group of a curve, 27
$\deg D$	degree of a divisor, 27
$\mathrm{Div}^0(C)$	group of divisors of degree 0 on a curve, 27
$\mathrm{Div}_K(C)$	group of divisors defined over K, 27

467

$\hat{\mathbb{G}}_a$	formal additive group, 121
$\hat{\mathbb{G}}_m$	formal multiplicative group, 121
$\hat{E}$	the formal group associated to the elliptic curve E, 121
$[m]$	multiplication-by-m map on a formal group, 121
$\mathcal{M}$	maximal ideal of the (local) ring R, 123
k	residue field of the (local) ring R, 123
$\mathcal{F}$	a formal group with formal group law $F(X, Y)$, 123
$\mathcal{F}(\mathcal{M})$	the group associated to $\mathcal{F}/R$, 123
$\omega(T)$	invariant differential on a formal group, 125
$\log_{\mathcal{F}}$	formal logarithm of the formal group $\mathcal{F}$, 127
$\exp_{\mathcal{F}}$	formal exponential of the formal group $\mathcal{F}$, 127
$Z(V/\mathbb{F}_q; T)$	the zeta function of the variety V over the finite field $\mathbb{F}_q$, 140
$\mathbb{C}(\Lambda)$	the field of elliptic functions for the lattice Λ, 161
$\sum_{w \in \mathbb{C}/\Lambda}$	sum over a fundamental parallelogram for Λ, 162
$\mathrm{Div}(\mathbb{C}/\Lambda)$	the divisor group of $\mathbb{C}/\Lambda$, 164
$\mathrm{Div}^0(\mathbb{C}/\Lambda)$	group of divisors of degree 0 on $\mathbb{C}/\Lambda$, 164
$\mathrm{div}(f)$	divisor of the elliptic function f, 164
sum	summation map$\mathrm{Div}^0(\mathbb{C}/\Lambda) \to \mathbb{C}/\Lambda$, 164
$\wp$	Weierstrass $\wp$-function, 165
G_{2k}	Eisenstein series of weight $2k$, 165
σ	Weierstrass σ-function, 167
g_2, g_3	Eisenstein series $60G_4$ and $140G_6$, 169
$\zeta(z)$	Weierstrass ζ-function, 178
$\eta(\omega)$	quasiperiod associated to the period ω, 179
$K(k), T(k)$	complete elliptic integrals, 181
K	a complete local field (Ch. VII), 185
v	discrete valuation on the field K (Ch. VII), 185
R	ring of integers of K (Ch. VII), 185
R^*	unit group of R (Ch. VII), 185
$\mathcal{M}$	maximal ideal of R (Ch. VII), 185
π	a uniformizer for R (Ch. VII), 185
k	residue field of R (Ch. VII), 185
$\tilde{}$	reduction modulo π, 187
$\tilde{E}$	reduction of the elliptic curve E modulo π, 187
$\tilde{P}$	reduction of the point P modulo π, 187
$E_0(K)$	set of points of $E(K)$ with nonsingular reduction, 188
$E_1(K)$	kernel of reduction modulo π, 188
K^{nr}	maximal unramified extension of K, 194
I_v	inertia subgroup of $G_{\bar{K}/K}$ for the valuation v, 194
K	a number field (Ch. VIII–X), 207
M_K	a complete set of inequivalent absolute values on K (Ch. VIII–X), 207
M_K^∞	the archimedean absolute values in K (Ch. VIII–X), 207
M_K^0	the nonarchimedean absolute values in K (Ch. VIII–X), 207
$v(x)$	$= -\log |x|_v$, for an absolute value $v \in M_K$ (Ch. VIII–X), 207
ord_v	normalized valuation for $v \in M_K^0$ (Ch. VIII–X), 207
R	the ring of integers of K (Ch. VIII–X), 208
R^*	the unit group of R (Ch. VIII–X), 208
K_v	the completion of K at v (Ch. VIII–X), 208

R_v	the ring of integers of K_v (Ch. VIII–X),	208
$\mathcal{M}_v$	the maximal ideal of R_v (Ch. VIII–X),	208
k_v	the residue field of R_v (Ch. VIII–X),	208
κ	the Kummer pairing $E(K) \times G_{\bar{K}/K} \to E[m]$,	209
$\tilde{E}_v/k_v$	the reduction of E modulo $\mathcal{M}_v$,	211
R_S	ring of S-integers,	213
$A[m]$	the m-torsion subgroup of the abelian group A,	216
H	height of a rational number,	220
h_x	height function on an elliptic curve over $\mathbb{Q}$,	220
$M_\mathbb{Q}$	the set of standard absolute values on $\mathbb{Q}$,	225
M_K	the set of standard absolute values on the number field K,	225
n_v	the local degree of the absolute value v,	225
$H_K(P)$	multiplicative height of P relative to K,	226
H	absolute multiplicative height,	227
$H(x)$	abbreviation for $H([x, 1])$,	230
$H_K(x)$	abbreviation for $H_K([x, 1])$,	230
$O(1)$	a bounded function,	234
h	absolute logarithmic height on projective space,	234
h_f	height function on an elliptic curve relative to f,	235
$\mathcal{D}_{E/K}$	minimal discriminant ideal of E/K,	243
$\mathfrak{p}_v$	prime ideal of R associated to v,	243
$\bar{\mathfrak{a}}_{E/K}$	Weierstrass class of E/K,	244
$\hat{h}$	canonical height function $E(\bar{K}) \to \mathbb{R}$,	248
$\langle \cdot , \cdot \rangle$	the canonical height pairing $E(\bar{K}) \times E(\bar{K}) \to \mathbb{R}$,	252
$R_{E/K}$	elliptic regulator of E/K,	253
N_E	the conductor of the elliptic curve E,	256
R_S	the ring of S-integers of K,	269
R_S^*	the unit group of R_S,	269
d_v	a v-adic distance function,	273
$m(\alpha)$	size of the unit α relative to a given basis for R_S^*,	287
$K(S, m)$	subgroup of $K^*/(K^*)^m$ with $\mathrm{ord}_v(b) \equiv 0 \pmod m$ for all $v \notin S$,	311
$\mathrm{Isom}(C)$	the group of isomorphisms from a curve C to itself,	318
$\mathrm{Isom}_K(C)$	the subgroup of $\mathrm{Isom}(C)$ of isomorphisms defined over K,	318
$\mathrm{Twist}(C/K)$	the set of twists of the curve C,	318
μ, ν	"addition" and "subtraction" maps on a homogeneous space,	322
$\mathrm{WC}(E/K)$	the Weil–Châtelet group of E/K,	324
sum	the summation map on the Picard group of a homogeneous space,	329
$S^{(\phi)}(E/K)$	the Selmer group of E/K for the isogeny ϕ,	332
$Ш(E/K)$	the Shafarevich–Tate group of E/K,	332
$S^{(m,n)}(E/K)$	the image of $S^{(m^n)}(E/K)$ in $S^{(m)}(E/K)$,	340
$\mathrm{Twist}\big((E, O)/K\big)$	twists of E, isomorphic to $H^1\big(G_{\bar{K}/K}, \mathrm{Aut}(E)\big)$,	342
$\nu(n)$	the number of distinct prime divisors of n,	346
$(a\mid b)$	Legendre symbol,	353
$\langle P, Q \rangle$	modified Weil pairing on $E[N]$,	390
τ	the Tate–Lichtenbaum pairing,	397
M^G	the submodule of M fixed by G,	415
$H^0(G, M)$	the 0th cohomology of the G-module M,	415
$C^1(G, M)$	group of 1-cochains from G to M,	416

References

[1] M. Abdalla, M. Bellare, and P. Rogaway. The oracle Diffie-Hellman assumptions and an analysis of DHIES. In *Topics in cryptology—CT-RSA 2001 (San Francisco, CA)*, volume 2020 of *Lecture Notes in Comput. Sci.*, pages 143–158. Springer, Berlin, 2001.

[2] D. Abramovich. Formal finiteness and the torsion conjecture on elliptic curves. A footnote to a paper: "Rational torsion of prime order in elliptic curves over number fields" [Astérisque No. 228 (1995), 3, 81–100] by S. Kamienny and B. Mazur. *Astérisque*, (228):3, 5–17, 1995. Columbia University Number Theory Seminar (New York, 1992).

[3] L. V. Ahlfors. *Complex analysis*. McGraw-Hill Book Co., New York, third edition, 1978. An introduction to the theory of analytic functions of one complex variable, International Series in Pure and Applied Mathematics.

[4] T. M. Apostol. *Introduction to analytic number theory*. Springer-Verlag, New York, 1976. Undergraduate Texts in Mathematics.

[5] T. M. Apostol. *Modular functions and Dirichlet series in number theory*, volume 41 of *Graduate Texts in Mathematics*. Springer-Verlag, New York, second edition, 1990.

[6] N. Arthaud. On Birch and Swinnerton-Dyer's conjecture for elliptic curves with complex multiplication. I. *Compositio Math.*, 37(2):209–232, 1978.

[7] E. Artin. *Galois theory*. Dover Publications Inc., Mineola, NY, second edition, 1998. Edited and with a supplemental chapter by Arthur N. Milgram.

[8] M. F. Atiyah and I. G. Macdonald. *Introduction to commutative algebra*. Addison-Wesley Publishing Co., Reading, Mass.–London–Don Mills, Ont., 1969.

[9] M. F. Atiyah and C. T. C. Wall. Cohomology of groups. In *Algebraic Number Theory (Proc. Instructional Conf., Brighton, 1965)*, pages 94–115. Thompson, Washington, D.C., 1967.

[10] A. O. L. Atkin and F. Morain. Elliptic curves and primality proving. *Math. Comp.*, 61(203):29–68, 1993.

[11] A. Baker. *Transcendental number theory*. Cambridge Mathematical Library. Cambridge University Press, Cambridge, second edition, 1990.

[12] A. Baker and J. Coates. Integer points on curves of genus 1. *Proc. Cambridge Philos. Soc.*, 67:595–602, 1970.

[13] R. Balasubramanian and N. Koblitz. The improbability that an elliptic curve has subexponential discrete log problem under the Menezes-Okamoto-Vanstone algorithm. *J. Cryptology*, 11(2):141–145, 1998.

[14] A. F. Beardon. *Iteration of Rational Functions*, volume 132 of *Graduate Texts in Mathematics*. Springer-Verlag, New York, 1991. Complex analytic dynamical systems.

[15] E. Bekyel. The density of elliptic curves having a global minimal Weierstrass equation. *J. Number Theory*, 109(1):41–58, 2004.

[16] D. Bernstein and T. Lange. Faster addition and doubling on elliptic curves. In *Advances in cryptology—ASIACRYPT 2007*, volume 4833 of *Lecture Notes in Comput. Sci.*, pages 29–50. Springer, Berlin, 2007.

[17] B. J. Birch. Cyclotomic fields and Kummer extensions. In *Algebraic Number Theory (Proc. Instructional Conf., Brighton, 1965)*, pages 85–93. Thompson, Washington, D.C., 1967.

[18] B. J. Birch. How the number of points of an elliptic curve over a fixed prime field varies. *J. London Math. Soc.*, 43:57–60, 1968.

[19] B. J. Birch and W. Kuyk, editors. *Modular functions of one variable. IV.* Springer-Verlag, Berlin, 1975. Lecture Notes in Mathematics, Vol. 476.

[20] B. J. Birch and H. P. F. Swinnerton-Dyer. Notes on elliptic curves. I. *J. Reine Angew. Math.*, 212:7–25, 1963.

[21] B. J. Birch and H. P. F. Swinnerton-Dyer. Elliptic curves and modular functions. In *Modular functions of one variable, IV (Proc. Internat. Summer School, Univ. Antwerp, Antwerp, 1972)*, pages 2–32. Lecture Notes in Math., Vol. 476. Springer, Berlin, 1975.

[22] I. F. Blake, G. Seroussi, and N. P. Smart. *Elliptic curves in cryptography*, volume 265 of *London Mathematical Society Lecture Note Series*. Cambridge University Press, Cambridge, 2000. Reprint of the 1999 original.

[23] D. Boneh and M. Franklin. Identity-based encryption from the Weil pairing. In *Advances in Cryptology—CRYPTO 2001 (Santa Barbara, CA)*, volume 2139 of *Lecture Notes in Comput. Sci.*, pages 213–229. Springer, Berlin, 2001.

[24] D. Boneh, B. Lynn, and H. Shacham. Short signatures from the Weil pairing. In *Advances in cryptology—ASIACRYPT 2001 (Gold Coast)*, volume 2248 of *Lecture Notes in Comput. Sci.*, pages 514–532. Springer, Berlin, 2001.

[25] A. I. Borevich and I. R. Shafarevich. *Number theory.* Translated from the Russian by Newcomb Greenleaf. Pure and Applied Mathematics, Vol. 20. Academic Press, New York, 1966.

[26] A. Bremner. On the equation $Y^2 = X(X^2 + p)$. In *Number theory and applications (Banff, AB, 1988)*, volume 265 of *NATO Adv. Sci. Inst. Ser. C Math. Phys. Sci.*, pages 3–22. Kluwer Acad. Publ., Dordrecht, 1989.

[27] A. Bremner and J. W. S. Cassels. On the equation $Y^2 = X(X^2 + p)$. *Math. Comp.*, 42(165):257–264, 1984.

[28] C. Breuil, B. Conrad, F. Diamond, and R. Taylor. On the modularity of elliptic curves over **Q**: wild 3-adic exercises. *J. Amer. Math. Soc.*, 14(4):843–939 (electronic), 2001.

[29] F. Brezing and A. Weng. Elliptic curves suitable for pairing based cryptography. *Des. Codes Cryptogr.*, 37(1):133–141, 2005.

[30] M. L. Brown. Note on supersingular primes of elliptic curves over **Q**. *Bull. London Math. Soc.*, 20(4):293–296, 1988.

[31] W. D. Brownawell and D. W. Masser. Vanishing sums in function fields. *Math. Proc. Cambridge Philos. Soc.*, 100(3):427–434, 1986.

[32] Y. Bugeaud. Bounds for the solutions of superelliptic equations. *Compositio Math.*, 107(2):187–219, 1997.

[33] J. P. Buhler, B. H. Gross, and D. B. Zagier. On the conjecture of Birch and Swinnerton-Dyer for an elliptic curve of rank 3. *Math. Comp.*, 44(170):473–481, 1985.

[34] E. R. Canfield, P. Erdős, and C. Pomerance. On a problem of Oppenheim concerning "factorisatio numerorum." *J. Number Theory*, 17(1):1–28, 1983.

[35] H. Carayol. Sur les représentations galoisiennes modulo l attachées aux formes modulaires. *Duke Math. J.*, 59(3):785–801, 1989.

[36] J. W. S. Cassels. A note on the division values of $\wp(u)$. *Proc. Cambridge Philos. Soc.*, 45:167–172, 1949.

[37] J. W. S. Cassels. Arithmetic on curves of genus 1. III. The Tate-Šafarevič and Selmer groups. *Proc. London Math. Soc. (3)*, 12:259–296, 1962.

[38] J. W. S. Cassels. Arithmetic on curves of genus 1. IV. Proof of the Hauptvermutung. *J. Reine Angew. Math.*, 211:95–112, 1962.

[39] J. W. S. Cassels. Arithmetic on curves of genus 1. V. Two counterexamples. *J. London Math. Soc.*, 38:244–248, 1963.

[40] J. W. S. Cassels. Arithmetic on curves of genus 1. VIII. On conjectures of Birch and Swinnerton-Dyer. *J. Reine Angew. Math.*, 217:180–199, 1965.

[41] J. W. S. Cassels. Diophantine equations with special reference to elliptic curves. *J. London Math. Soc.*, 41:193–291, 1966.

[42] J. W. S. Cassels. Global fields. In *Algebraic Number Theory (Proc. Instructional Conf., Brighton, 1965)*, pages 42–84. Thompson, Washington, D.C., 1967.

[43] J. W. S. Cassels. *Lectures on elliptic curves*, volume 24 of *London Mathematical Society Student Texts*. Cambridge University Press, Cambridge, 1991.

[44] T. Chinburg. An introduction to Arakelov intersection theory. In *Arithmetic geometry (Storrs, Conn., 1984)*, pages 289–307. Springer, New York, 1986.

[45] D. V. Chudnovsky and G. V. Chudnovsky. Padé approximations and Diophantine geometry. *Proc. Nat. Acad. Sci. U.S.A.*, 82(8):2212–2216, 1985.

[46] C. H. Clemens. *A scrapbook of complex curve theory*, volume 55 of *Graduate Studies in Mathematics*. American Mathematical Society, Providence, RI, second edition, 2003.

[47] L. Clozel, M. Harris, and R. Taylor. Automorphy for some l-adic lifts of automorphic mod l representations. 2007. IHES Publ. Math., submitted.

[48] J. Coates. Construction of rational functions on a curve. *Proc. Cambridge Philos. Soc.*, 68:105–123, 1970.

[49] J. Coates and A. Wiles. On the conjecture of Birch and Swinnerton-Dyer. *Invent. Math.*, 39(3):223–251, 1977.

[50] H. Cohen. *A Course in Computational Algebraic Number Theory*, volume 138 of *Graduate Texts in Mathematics*. Springer-Verlag, Berlin, 1993.

[51] H. Cohen, G. Frey, R. Avanzi, C. Doche, T. Lange, K. Nguyen, and F. Vercauteren, editors. *Handbook of Elliptic and Hyperelliptic Curve Cryptography*. Discrete Mathematics and Its Applications (Boca Raton). Chapman & Hall/CRC, Boca Raton, FL, 2006.

[52] D. A. Cox. The arithmetic-geometric mean of Gauss. *Enseign. Math. (2)*, 30(3-4):275–330, 1984.

[53] J. Cremona. *Elliptic Curve Data.* http://sage.math.washington.edu/cremona/index.html, http://www.math.utexas.edu/users/tornaria/cnt/cremona.html.

[54] J. E. Cremona. *Algorithms for modular elliptic curves*. Cambridge University Press, Cambridge, second edition, 1997. available free online at www.warwick.ac.uk/staff/J.E.Cremona/book/fulltext/index.html.

[55] J. E. Cremona, M. Prickett, and S. Siksek. Height difference bounds for elliptic curves over number fields. *J. Number Theory*, 116(1):42–68, 2006.

[56] L. V. Danilov. The Diophantine equation $x^3 - y^2 = k$ and a conjecture of M. Hall. *Mat. Zametki*, 32(3):273–275, 425, 1982. English translation: *Math. Notes Acad. Sci. USSR* **32** (1982), no. 3–4, 617–618 (1983).

[57] H. Davenport. On $f^3(t) - g^2(t)$. *Norske Vid. Selsk. Forh. (Trondheim)*, 38:86–87, 1965.

[58] S. David. Minorations de formes linéaires de logarithmes elliptiques. *Mém. Soc. Math. France (N.S.)*, (62):iv+143, 1995.

[59] B. M. M. de Weger. *Algorithms for Diophantine equations*, volume 65 of *CWI Tract*. Stichting Mathematisch Centrum Centrum voor Wiskunde en Informatica, Amsterdam, 1989.

[60] M. Deuring. Die Typen der Multiplikatorenringe elliptischer Funktionenkörper. *Abh. Math. Sem. Hansischen Univ.*, 14:197–272, 1941.

[61] M. Deuring. Die Zetafunktion einer algebraischen Kurve vom Geschlechte Eins. *Nachr. Akad. Wiss. Göttingen. Math.-Phys. Kl. Math.-Phys.-Chem. Abt.*, 1953:85–94, 1953.

[62] M. Deuring. Die Zetafunktion einer algebraischen Kurve vom Geschlechte Eins. II. *Nachr. Akad. Wiss. Göttingen. Math.-Phys. Kl. IIa.*, 1955:13–42, 1955.

[63] M. Deuring. Die Zetafunktion einer algebraischen Kurve vom Geschlechte Eins. III. *Nachr. Akad. Wiss. Göttingen. Math.-Phys. Kl. IIa.*, 1956:37–76, 1956.

[64] M. Deuring. Die Zetafunktion einer algebraischen Kurve vom Geschlechte Eins. IV. *Nachr. Akad. Wiss. Göttingen. Math.-Phys. Kl. IIa.*, 1957:55–80, 1957.

[65] W. Diffie and M. E. Hellman. New directions in cryptography. *IEEE Trans. Information Theory*, IT-22(6):644–654, 1976.

[66] L. Dirichlet. Über den biquadratischen Charakter der Zahl "Zwei." *J. Reine Angew. Math.*, 57:187–188, 1860.

[67] Z. Djabri, E. F. Schaefer, and N. P. Smart. Computing the p-Selmer group of an elliptic curve. *Trans. Amer. Math. Soc.*, 352(12):5583–5597, 2000.

[68] D. S. Dummit and R. M. Foote. *Abstract algebra*. John Wiley & Sons Inc., Hoboken, NJ, third edition, 2004.

[69] R. Dupont, A. Enge, and F. Morain. Building curves with arbitrary small MOV degree over finite prime fields. *J. Cryptology*, 18(2):79–89, 2005.

[70] B. Dwork. On the rationality of the zeta function of an algebraic variety. *Amer. J. Math.*, 82:631–648, 1960.

[71] H. M. Edwards. A normal form for elliptic curves. *Bull. Amer. Math. Soc. (N.S.)*, 44(3):393–422 (electronic), 2007.

[72] M. Eichler. Quaternäre quadratische Formen und die Riemannsche Vermutung für die Kongruenzzetafunktion. *Arch. Math.*, 5:355–366, 1954.

[73] D. Eisenbud. *Commutative algebra*, volume 150 of *Graduate Texts in Mathematics*. Springer-Verlag, New York, 1995. With a view toward algebraic geometry.

[74] T. ElGamal. A public key cryptosystem and a signature scheme based on discrete logarithms. *IEEE Trans. Inform. Theory*, 31(4):469–472, 1985.

[75] N. Elkies. List of integers x, y with $x < 10^{18}$, $0 < |x^3 - y^2| < x^{1/2}$. www.math.harvard.edu/~elkies/hall.html.

[76] N. Elkies. $\mathbb{Z}^{28}$ in $E(\mathbb{Q})$. Number Theory Listserver, May 2006.

[77] N. D. Elkies. The existence of infinitely many supersingular primes for every elliptic curve over $\mathbb{Q}$. *Invent. Math.*, 89(3):561–567, 1987.

[78] N. D. Elkies. Distribution of supersingular primes. *Astérisque*, (198-200):127–132 (1992), 1991. Journées Arithmétiques, 1989 (Luminy, 1989).

[79] N. D. Elkies. Elliptic and modular curves over finite fields and related computational issues. In *Computational perspectives on number theory (Chicago, IL, 1995)*, volume 7 of *AMS/IP Stud. Adv. Math.*, pages 21–76. Amer. Math. Soc., Providence, RI, 1998.

[80] J.-H. Evertse. On equations in S-units and the Thue-Mahler equation. *Invent. Math.*, 75(3):561–584, 1984.

[81] J.-H. Evertse and J. H. Silverman. Uniform bounds for the number of solutions to $Y^n = f(X)$. *Math. Proc. Cambridge Philos. Soc.*, 100(2):237–248, 1986.

[82] G. Faltings. Endlichkeitssätze für abelsche Varietäten über Zahlkörpern. *Invent. Math.*, 73(3):349–366, 1983.

[83] G. Faltings. Calculus on arithmetic surfaces. *Ann. of Math. (2)*, 119(2):387–424, 1984.

[84] G. Faltings. Finiteness theorems for abelian varieties over number fields. In *Arithmetic geometry (Storrs, Conn., 1984)*, pages 9–27. Springer, New York, 1986. Translated from the German original [Invent. Math. **73** (1983), no. 3, 349–366; ibid. **75** (1984), no. 2, 381; MR 85g:11026ab] by Edward Shipz.

[85] S. Fermigier. Une courbe elliptique définie sur **Q** de rang $\geq$ 22. *Acta Arith.*, 82(4):359–363, 1997.

[86] E. V. Flynn and C. Grattoni. Descent via isogeny on elliptic curves with large rational torsion subgroups. *J. Symbolic Comput.*, 43(4):293–303, 2008.

[87] D. Freeman. Constructing pairing-friendly elliptic curves with embedding degree 10. In *Algorithmic number theory*, volume 4076 of *Lecture Notes in Comput. Sci.*, pages 452–465. Springer, Berlin, 2006.

[88] G. Frey. Links between stable elliptic curves and certain Diophantine equations. *Ann. Univ. Sarav. Ser. Math.*, 1(1):iv+40, 1986.

[89] G. Frey. Elliptic curves and solutions of $A - B = C$. In *Séminaire de Théorie des Nombres, Paris 1985–86*, volume 71 of *Progr. Math.*, pages 39–51. Birkhäuser Boston, Boston, MA, 1987.

[90] G. Frey. Links between solutions of $A - B = C$ and elliptic curves. In *Number theory (Ulm, 1987)*, volume 1380 of *Lecture Notes in Math.*, pages 31–62. Springer, New York, 1989.

[91] G. Frey and H.-G. Rück. A remark concerning m-divisibility and the discrete logarithm problem in the divisor class group of curves. *Math. Comp.*, 62:865–874, 1994.

[92] A. Fröhlich. Local fields. In *Algebraic Number Theory (Proc. Instructional Conf., Brighton, 1965)*, pages 1–41. Thompson, Washington, D.C., 1967.

[93] A. Fröhlich. *Formal groups*. Lecture Notes in Mathematics, No. 74. Springer-Verlag, Berlin, 1968.

[94] R. Fueter. Ueber kubische diophantische Gleichungen. *Comment. Math. Helv.*, 2(1):69–89, 1930.

[95] W. Fulton. *Algebraic curves*. Advanced Book Classics. Addison-Wesley Publishing Company Advanced Book Program, Redwood City, CA, 1989. An introduction to algebraic geometry, Notes written with the collaboration of Richard Weiss, Reprint of 1969 original.

[96] J. Gebel, A. Pethő, and H. G. Zimmer. Computing integral points on elliptic curves. *Acta Arith.*, 68(2):171–192, 1994.

[97] S. Goldwasser and J. Kilian. Almost all primes can be quickly certified. In *STOC '86: Proceedings of the eighteenth annual ACM symposium on Theory of computing*, pages 316–329, New York, 1986. ACM.

[98] R. Greenberg. On the Birch and Swinnerton-Dyer conjecture. *Invent. Math.*, 72(2):241–265, 1983.

[99] P. Griffiths and J. Harris. *Principles of algebraic geometry*. Wiley Classics Library. John Wiley & Sons Inc., New York, 1994. Reprint of the 1978 original.

[100] B. Gross, W. Kohnen, and D. Zagier. Heegner points and derivatives of L-series. II. *Math. Ann.*, 278(1-4):497–562, 1987.

[101] B. Gross and D. Zagier. Points de Heegner et dérivées de fonctions *L. C. R. Acad. Sci. Paris Sér. I Math.*, 297(2):85–87, 1983.

[102] B. H. Gross and D. B. Zagier. Heegner points and derivatives of *L*-series. *Invent. Math.*, 84(2):225–320, 1986.

[103] R. Gross. A note on Roth's theorem. *J. Number Theory*, 36:127–132, 1990.

[104] R. Gross and J. Silverman. *S*-integer points on elliptic curves. *Pacific J. Math.*, 167(2):263–288, 1995.

[105] K. Gruenberg. Profinite groups. In *Algebraic Number Theory (Proc. Instructional Conf., Brighton, 1965)*, pages 116–127. Thompson, Washington, D.C., 1967.

[106] M. Hall, Jr. The Diophantine equation $x^3 - y^2 = k$. In *Computers in number theory (Proc. Sci. Res. Council Atlas Sympos. No. 2, Oxford, 1969)*, pages 173–198. Academic Press, London, 1971.

[107] D. Hankerson, A. Menezes, and S. Vanstone. *Guide to elliptic curve cryptography*. Springer Professional Computing. Springer-Verlag, New York, 2004.

[108] G. H. Hardy and E. M. Wright. *An introduction to the theory of numbers*. The Clarendon Press Oxford University Press, New York, fifth edition, 1979.

[109] J. Harris. *Algebraic geometry*, volume 133 of *Graduate Texts in Mathematics*. Springer-Verlag, New York, 1992. A first course.

[110] M. Harris, N. Shepherd-Barron, and R. Taylor. A family of Calabi-Yau varieties and potential automorphy. *Ann. of Math. (2)*. to appear.

[111] R. Hartshorne. *Algebraic geometry*. Springer-Verlag, New York, 1977. Graduate Texts in Mathematics, No. 52.

[112] M. Hazewinkel. *Formal groups and applications*, volume 78 of *Pure and Applied Mathematics*. Academic Press Inc. [Harcourt Brace Jovanovich Publishers], New York, 1978.

[113] M. Hindry and J. H. Silverman. The canonical height and integral points on elliptic curves. *Invent. Math.*, 93(2):419–450, 1988.

[114] M. Hindry and J. H. Silverman. *Diophantine geometry*, volume 201 of *Graduate Texts in Mathematics*. Springer-Verlag, New York, 2000. An introduction.

[115] G. Hochschild and J.-P. Serre. Cohomology of group extensions. *Trans. Amer. Math. Soc.*, 74:110–134, 1953.

[116] J. Hoffstein, J. Pipher, and J. H. Silverman. *An introduction to mathematical cryptography*. Undergraduate Texts in Mathematics. Springer-Verlag, New York, 2008.

[117] A. Hurwitz. Über ternäre diophantische Gleichungen dritten Grades. *Vierteljahrschrift d. Naturf. Ges. Zürich*, 62:207–229, 1917.

[118] D. Husemöller. *Elliptic curves*, volume 111 of *Graduate Texts in Mathematics*. Springer-Verlag, New York, second edition, 2004. With appendices by Otto Forster, Ruth Lawrence and Stefan Theisen.

[119] J.-I. Igusa. Class number of a definite quaternion with prime discriminant. *Proc. Nat. Acad. Sci. U.S.A.*, 44:312–314, 1958.

[120] A. Joux. A one round protocol for tripartite Diffie-Hellman. In *Algorithmic number theory (Leiden, 2000)*, volume 1838 of *Lecture Notes in Comput. Sci.*, pages 385–393. Springer, Berlin, 2000.

[121] S. Kamienny. Torsion points on elliptic curves and *q*-coefficients of modular forms. *Invent. Math.*, 109(2):221–229, 1992.

[122] S. Kamienny and B. Mazur. Rational torsion of prime order in elliptic curves over number fields. *Astérisque*, (228):3, 81–100, 1995. With an appendix by A. Granville, Columbia University Number Theory Seminar (New York, 1992).

[123] N. M. Katz. An overview of Deligne's proof of the Riemann hypothesis for varieties over finite fields. In *Mathematical developments arising from Hilbert problems (Proc. Sympos. Pure Math., Vol. XXVIII, Northern Illinois Univ., De Kalb, Ill., 1974)*, pages 275–305. Amer. Math. Soc., Providence, R.I., 1976.

[124] N. M. Katz and B. Mazur. *Arithmetic moduli of elliptic curves*, volume 108 of *Annals of Mathematics Studies*. Princeton University Press, Princeton, NJ, 1985.

[125] M. A. Kenku. On the number of $\mathbb{Q}$-isomorphism classes of elliptic curves in each $\mathbb{Q}$-isogeny class. *J. Number Theory*, 15(2):199–202, 1982.

[126] J.-H. Kim, R. Montenegro, Y. Peres, and P. Tetali. A birthday paradox for Markov chains, with an optimal bound for collision in Pollard rho for discrete logarithm. In *Algorithmic number theory*, volume 5011 of *Lecture Notes in Comput. Sci.*, pages 402–415. Springer, Berlin, 2008.

[127] A. W. Knapp. *Elliptic curves*, volume 40 of *Mathematical Notes*. Princeton University Press, Princeton, NJ, 1992.

[128] N. Koblitz. Elliptic curve cryptosystems. *Math. Comp.*, 48(177):203–209, 1987.

[129] N. Koblitz. *Introduction to elliptic curves and modular forms*, volume 97 of *Graduate Texts in Mathematics*. Springer-Verlag, New York, second edition, 1993.

[130] V. A. Kolyvagin. Finiteness of $E(\mathbb{Q})$ and $\text{III}(E, \mathbb{Q})$ for a subclass of Weil curves. *Izv. Akad. Nauk SSSR Ser. Mat.*, 52(3):522–540, 670–671, 1988.

[131] S. V. Kotov and L. A. Trelina. S-ganze Punkte auf elliptischen Kurven. *J. Reine Angew. Math.*, 306:28–41, 1979.

[132] D. S. Kubert. Universal bounds on the torsion of elliptic curves. *Proc. London Math. Soc. (3)*, 33(2):193–237, 1976.

[133] E. Kunz. *Introduction to plane algebraic curves*. Birkhäuser Boston Inc., Boston, MA, 2005. Translated from the 1991 German edition by Richard G. Belshoff.

[134] M. Lal, M. F. Jones, and W. J. Blundon. Numerical solutions of the Diophantine equation $y^3 - x^2 = k$. *Math. Comp.*, 20:322–325, 1966.

[135] S. Lang. *Elliptic curves: Diophantine analysis*, volume 231 of *Grundlehren der Mathematischen Wissenschaften [Fundamental Principles of Mathematical Sciences]*. Springer-Verlag, Berlin, 1978.

[136] S. Lang. *Introduction to algebraic and abelian functions*, volume 89 of *Graduate Texts in Mathematics*. Springer-Verlag, New York, second edition, 1982.

[137] S. Lang. *Complex multiplication*, volume 255 of *Grundlehren der Mathematischen Wissenschaften [Fundamental Principles of Mathematical Sciences]*. Springer-Verlag, New York, 1983.

[138] S. Lang. Conjectured Diophantine estimates on elliptic curves. In *Arithmetic and geometry, Vol. I*, volume 35 of *Progr. Math.*, pages 155–171. Birkhäuser Boston, Boston, MA, 1983.

[139] S. Lang. *Fundamentals of Diophantine geometry*. Springer-Verlag, New York, 1983.

[140] S. Lang. *Elliptic functions*, volume 112 of *Graduate Texts in Mathematics*. Springer-Verlag, New York, second edition, 1987. With an appendix by J. Tate.

[141] S. Lang. *Number theory III*, volume 60 of *Encyclopedia of Mathematical Sciences*. Springer-Verlag, Berlin, 1991.

[142] S. Lang. *Algebraic number theory*, volume 110 of *Graduate Texts in Mathematics*. Springer-Verlag, New York, second edition, 1994.

[143] S. Lang. *Algebra*, volume 211 of *Graduate Texts in Mathematics*. Springer-Verlag, New York, third edition, 2002.

[144] S. Lang and J. Tate. Principal homogeneous spaces over abelian varieties. *Amer. J. Math.*, 80:659–684, 1958.

[145] S. Lang and H. Trotter. *Frobenius distributions in* GL_2*-extensions.* Springer-Verlag, Berlin, 1976. Distribution of Frobenius automorphisms in GL_2-extensions of the rational numbers, Lecture Notes in Mathematics, Vol. 504.

[146] M. Laska. An algorithm for finding a minimal Weierstrass equation for an elliptic curve. *Math. Comp.*, 38(157):257–260, 1982.

[147] M. Laska. *Elliptic curves over number fields with prescribed reduction type.* Aspects of Mathematics, E4. Friedr. Vieweg & Sohn, Braunschweig, 1983.

[148] D. J. Lewis and K. Mahler. On the representation of integers by binary forms. *Acta Arith.*, 6:333–363, 1960/1961.

[149] S. Lichtenbaum. The period-index problem for elliptic curves. *Amer. J. Math.*, 90:1209–1223, 1968.

[150] C.-E. Lind. Untersuchungen über die rationalen Punkte der ebenen kubischen Kurven vom Geschlecht Eins. *Thesis, University of Uppsala,*, 1940:97, 1940.

[151] J. Liouville. Sur des classes très-étendues de quantités dont la irrationalles algébriques. *C. R. Acad. Paris*, 18:883–885 and 910–911, 1844.

[152] E. Lutz. Sur l'equation $y^2 = x^3 - ax - b$ dans les corps p-adic. *J. Reine Angew. Math.*, 177:237–247, 1937.

[153] K. Mahler. On the lattice points on curves of genus 1. *Proc. London Math. Soc. (3)*, 39:431–466, 1935.

[154] J. I. Manin. The Hasse-Witt matrix of an algebraic curve. *Izv. Akad. Nauk SSSR Ser. Mat.*, 25:153–172, 1961.

[155] J. I. Manin. The p-torsion of elliptic curves is uniformly bounded. *Izv. Akad. Nauk SSSR Ser. Mat.*, 33:459–465, 1969.

[156] J. I. Manin. Cyclotomic fields and modular curves. *Uspehi Mat. Nauk*, 26(6(162)):7–71, 1971. English translation: Russian Math. Surveys 26 (1971), no. 6, 7–78.

[157] R. C. Mason. The hyperelliptic equation over function fields. *Math. Proc. Cambridge Philos. Soc.*, 93(2):219–230, 1983.

[158] R. C. Mason. Norm form equations. I. *J. Number Theory*, 22(2):190–207, 1986.

[159] D. Masser. *Elliptic functions and transcendence.* Springer-Verlag, Berlin, 1975. Lecture Notes in Mathematics, Vol. 437.

[160] D. Masser and G. Wüstholz. Isogeny estimates for abelian varieties, and finiteness theorems. *Ann. of Math. (2)*, 137(3):459–472, 1993.

[161] D. W. Masser. Specializations of finitely generated subgroups of abelian varieties. *Trans. Amer. Math. Soc.*, 311(1):413–424, 1989.

[162] D. W. Masser and G. Wüstholz. Fields of large transcendence degree generated by values of elliptic functions. *Invent. Math.*, 72(3):407–464, 1983.

[163] D. W. Masser and G. Wüstholz. Estimating isogenies on elliptic curves. *Invent. Math.*, 100(1):1–24, 1990.

[164] H. Matsumura. *Commutative algebra*, volume 56 of *Mathematics Lecture Note Series*. Benjamin/Cummings Publishing Co., Inc., Reading, Mass., second edition, 1980.

[165] B. Mazur. Modular curves and the Eisenstein ideal. *Inst. Hautes Études Sci. Publ. Math.*, (47):33–186 (1978), 1977.

[166] B. Mazur. Rational isogenies of prime degree (with an appendix by D. Goldfeld). *Invent. Math.*, 44(2):129–162, 1978.

[167] H. McKean and V. Moll. *Elliptic curves.* Cambridge University Press, Cambridge, 1997. Function theory, geometry, arithmetic.

[168] A. J. Menezes, T. Okamoto, and S. A. Vanstone. Reducing elliptic curve logarithms to logarithms in a finite field. *IEEE Trans. Inform. Theory*, 39(5):1639–1646, 1993.

[169] A. J. Menezes, P. C. van Oorschot, and S. A. Vanstone. *Handbook of Applied Cryptography.* CRC Press Series on Discrete Mathematics and Its Applications. CRC Press, Boca Raton, FL, 1997.

[170] L. Merel. Bornes pour la torsion des courbes elliptiques sur les corps de nombres. *Invent. Math.*, 124(1-3):437–449, 1996.

[171] J.-F. Mestre. Construction d'une courbe elliptique de rang ≥ 12. *C. R. Acad. Sci. Paris Sér. I Math.*, 295(12):643–644, 1982.

[172] J.-F. Mestre. Courbes elliptiques et formules explicites. In *Seminar on number theory, Paris 1981–82 (Paris, 1981/1982)*, volume 38 of *Progr. Math.*, pages 179–187. Birkhäuser Boston, Boston, MA, 1983.

[173] M. Mignotte. Quelques remarques sur l'approximation rationnelle des nombres algébriques. *J. Reine Angew. Math.*, 268/269:341–347, 1974. Collection of articles dedicated to Helmut Hasse on his seventy-fifth birthday, II.

[174] S. D. Miller and R. Venkatesan. Spectral analysis of Pollard rho collisions. In *Algorithmic number theory*, volume 4076 of *Lecture Notes in Comput. Sci.*, pages 573–581. Springer, Berlin, 2006.

[175] S. D. Miller and R. Venkatesan. Non-degeneracy of Pollard rho collisions, 2008. arXiv:0808.0469.

[176] V. S. Miller. Use of elliptic curves in cryptography. In *Advances in Cryptology—CRYPTO '85 (Santa Barbara, Calif., 1985)*, volume 218 of *Lecture Notes in Comput. Sci.*, pages 417–426. Springer, Berlin, 1986.

[177] J. S. Milne. *Arithmetic duality theorems*, volume 1 of *Perspectives in Mathematics*. Academic Press Inc., Boston, MA, 1986.

[178] J. S. Milne. *Elliptic curves.* BookSurge Publishers, Charleston, SC, 2006.

[179] J. Milnor. On Lattès maps. ArXiv:math.DS/0402147, Stony Brook IMS Preprint #2004/01.

[180] R. Miranda. *Algebraic curves and Riemann surfaces*, volume 5 of *Graduate Studies in Mathematics*. American Mathematical Society, Providence, RI, 1995.

[181] A. Miyaji, M. Nakabayashi, and S. Takano. Characterization of elliptic curve traces under FR-reduction. In *Information security and cryptology—ICISC 2000 (Seoul)*, volume 2015 of *Lecture Notes in Comput. Sci.*, pages 90–108. Springer, Berlin, 2001.

[182] P. Monsky. Three constructions of rational points on $Y^2 = X^3 \pm NX$. *Math. Z.*, 209(3):445–462, 1992.

[183] F. Morain. Building cyclic elliptic curves modulo large primes. In *Advances in cryptology—EUROCRYPT '91 (Brighton, 1991)*, volume 547 of *Lecture Notes in Comput. Sci.*, pagès 328–336. Springer, Berlin, 1991.

[184] L. J. Mordell. The diophantine equation $x^4 + my^4 = z^2$.. *Quart. J. Math. Oxford Ser. (2)*, 18:1–6, 1967.

[185] L. J. Mordell. *Diophantine equations.* Pure and Applied Mathematics, Vol. 30. Academic Press, London, 1969.

[186] D. Mumford. *Abelian varieties.* Tata Institute of Fundamental Research Studies in Mathematics, No. 5. Published for the Tata Institute of Fundamental Research, Bombay, 1970.

[187] D. Mumford, J. Fogarty, and F. Kirwan. *Geometric invariant theory*, volume 34 of *Ergebnisse der Mathematik und ihrer Grenzgebiete (2) [Results in Mathematics and Related Areas (2)]*. Springer-Verlag, Berlin, third edition, 1994.

[188] K.-I. Nagao. Construction of high-rank elliptic curves. *Kobe J. Math.*, 11(2):211–219, 1994.

[189] K.-I. Nagao. $\mathbb{Q}(T)$-rank of elliptic curves and certain limit coming from the local points. *Manuscripta Math.*, 92(1):13–32, 1997. With an appendix by Nobuhiko Ishida, Tsuneo Ishikawa and the author.

[190] T. Nagell. Solution de quelque problèmes dans la théorie arithmétique des cubiques planes du premier genre. *Wid. Akad. Skrifter Oslo I*, 1935. Nr. 1.

[191] NBS–DSS. Digital Signature Standard (DSS). FIPS Publication 186-2, National Bureau of Standards, 2000. http://csrc.nist.gov/publications/ PubsFIPS.html.

[192] A. Néron. Problèmes arithmétiques et géométriques rattachés à la notion de rang d'une courbe algébrique dans un corps. *Bull. Soc. Math. France*, 80:101–166, 1952.

[193] A. Néron. Modèles minimaux des variétés abéliennes sur les corps locaux et globaux. *Inst. Hautes Études Sci. Publ.Math. No.*, 21:128, 1964.

[194] A. Néron. Quasi-fonctions et hauteurs sur les variétés abéliennes. *Ann. of Math. (2)*, 82:249–331, 1965.

[195] O. Neumann. Elliptische Kurven mit vorgeschriebenem Reduktionsverhalten. I. *Math. Nachr.*, 49:107–123, 1971.

[196] J. Oesterlé. Nouvelles approches du "théorème" de Fermat. *Astérisque*, (161-162):Exp. No. 694, 4, 165–186 (1989), 1988. Séminaire Bourbaki, Vol. 1987/88.

[197] A. Ogg. *Modular forms and Dirichlet series*. W. A. Benjamin, Inc., New York-Amsterdam, 1969.

[198] A. P. Ogg. Abelian curves of 2-power conductor. *Proc. Cambridge Philos. Soc.*, 62:143–148, 1966.

[199] A. P. Ogg. Abelian curves of small conductor. *J. Reine Angew. Math.*, 226:204–215, 1967.

[200] A. P. Ogg. Elliptic curves and wild ramification. *Amer. J. Math.*, 89:1–21, 1967.

[201] L. D. Olson. Torsion points on elliptic curves with given j-invariant. *Manuscripta Math.*, 16(2):145–150, 1975.

[202] *PARI/GP*, 2005. http://pari.math.u-bordeaux.fr/.

[203] A. N. Paršin. Algebraic curves over function fields. I. *Izv. Akad. Nauk SSSR Ser. Mat.*, 32:1191–1219, 1968.

[204] R. G. E. Pinch. Elliptic curves with good reduction away from 2. *Math. Proc. Cambridge Philos. Soc.*, 96(1):25–38, 1984.

[205] S. C. Pohlig and M. E. Hellman. An improved algorithm for computing logarithms over GF(p) and its cryptographic significance. *IEEE Trans. Information Theory*, IT-24(1):106–110, 1978.

[206] J. M. Pollard. Monte Carlo methods for index computation (mod p). *Math. Comp.*, 32(143):918–924, 1978.

[207] H. Reichardt. Einige im Kleinen überall lösbare, im Grossen unlösbare diophantische Gleichungen. *J. Reine Angew. Math.*, 184:12–18, 1942.

[208] K. A. Ribet. On modular representations of $\mathrm{Gal}(\overline{\mathbf{Q}}/\mathbf{Q})$ arising from modular forms. *Invent. Math.*, 100(2):431–476, 1990.

[209] R. L. Rivest, A. Shamir, and L. Adleman. A method for obtaining digital signatures and public-key cryptosystems. *Comm. ACM*, 21(2):120–126, 1978.

[210] A. Robert. *Elliptic curves*. Springer-Verlag, Berlin, 1973. Notes from postgraduate lectures given in Lausanne 1971/72, Lecture Notes in Mathematics, Vol. 326.

[211] D. E. Rohrlich. On L-functions of elliptic curves and anticyclotomic towers. *Invent. Math.*, 75(3):383–408, 1984.

[212] P. Roquette. *Analytic theory of elliptic functions over local fields*. Hamburger Mathematische Einzelschriften (N.F.), Heft 1. Vandenhoeck & Ruprecht, Göttingen, 1970.

[213] M. Rosen and J. H. Silverman. On the rank of an elliptic surface. *Invent. Math.*, 133(1):43–67, 1998.

[214] K. Rubin. Elliptic curves with complex multiplication and the conjecture of Birch and Swinnerton-Dyer. *Invent. Math.*, 64(3):455–470, 1981.

[215] K. Rubin. Tate-Shafarevich groups and *L*-functions of elliptic curves with complex multiplication. *Invent. Math.*, 89(3):527–559, 1987.

[216] K. Rubin. The "main conjectures" of Iwasawa theory for imaginary quadratic fields. *Invent. Math.*, 103(1):25–68, 1991.

[217] T. Saito. Conductor, discriminant, and the Noether formula of arithmetic surfaces. *Duke Math. J.*, 57(1):151–173, 1988.

[218] T. Satoh and K. Araki. Fermat quotients and the polynomial time discrete log algorithm for anomalous elliptic curves. *Comment. Math. Univ. St. Paul.*, 47(1):81–92, 1998.

[219] E. F. Schaefer and M. Stoll. How to do a *p*-descent on an elliptic curve. *Trans. Amer. Math. Soc.*, 356(3):1209–1231 (electronic), 2004.

[220] S. H. Schanuel. Heights in number fields. *Bull. Soc. Math. France*, 107(4):433–449, 1979.

[221] W. M. Schmidt. *Diophantine approximation*, volume 785 of *Lecture Notes in Mathematics*. Springer, Berlin, 1980.

[222] S. Schmitt and H. G. Zimmer. *Elliptic curves*, volume 31 of *de Gruyter Studies in Mathematics*. Walter de Gruyter & Co., Berlin, 2003. A computational approach, With an appendix by Attila Pethő.

[223] R. Schoof. Elliptic curves over finite fields and the computation of square roots mod *p*. *Math. Comp.*, 44(170):483–494, 1985.

[224] R. Schoof. Counting points on elliptic curves over finite fields. *J. Théor. Nombres Bordeaux*, 7(1):219–254, 1995. Les Dix-huitièmes Journées Arithmétiques (Bordeaux, 1993).

[225] E. S. Selmer. The Diophantine equation $ax^3 + by^3 + cz^3 = 0$. *Acta Math.*, 85:203–362 (1 plate), 1951.

[226] E. S. Selmer. A conjecture concerning rational points on cubic curves. *Math. Scand.*, 2:49–54, 1954.

[227] E. S. Selmer. The diophantine equation $ax^3 + by^3 + cz^3 = 0$. Completion of the tables. *Acta Math.*, 92:191–197, 1954.

[228] I. A. Semaev. Evaluation of discrete logarithms in a group of *p*-torsion points of an elliptic curve in characteristic *p*. *Math. Comp.*, 67(221):353–356, 1998.

[229] J.-P. Serre. Géométrie algébrique et géométrie analytique. *Ann. Inst. Fourier, Grenoble*, 6:1–42, 1955–1956.

[230] J.-P. Serre. Complex multiplication. In *Algebraic Number Theory (Proc. Instructional Conf., Brighton, 1965)*, pages 292–296. Thompson, Washington, D.C., 1967.

[231] J.-P. Serre. Propriétés galoisiennes des points d'ordre fini des courbes elliptiques. *Invent. Math.*, 15(4):259–331, 1972.

[232] J.-P. Serre. *A course in arithmetic*. Springer-Verlag, New York, 1973. Translated from the French, Graduate Texts in Mathematics, No. 7.

[233] J.-P. Serre. *Local fields*, volume 67 of *Graduate Texts in Mathematics*. Springer-Verlag, New York, 1979. Translated from the French by Marvin Jay Greenberg.

[234] J.-P. Serre. Quelques applications du théorème de densité de Chebotarev. *Inst. Hautes Études Sci. Publ. Math.*, (54):323–401, 1981.

[235] J.-P. Serre. Sur les représentations modulaires de degré 2 de $\mathrm{Gal}(\overline{\mathbf{Q}}/\mathbf{Q})$. *Duke Math. J.*, 54(1):179–230, 1987.

[236] J.-P. Serre. *Lectures on the Mordell-Weil theorem.* Aspects of Mathematics. Friedr. Vieweg & Sohn, Braunschweig, third edition, 1997. Translated from the French and edited by Martin Brown from notes by Michel Waldschmidt, With a foreword by Brown and Serre.

[237] J.-P. Serre. *Abelian l-adic representations and elliptic curves,* volume 7 of *Research Notes in Mathematics.* A K Peters Ltd., Wellesley, MA, 1998. With the collaboration of Willem Kuyk and John Labute, Revised reprint of the 1968 original.

[238] J.-P. Serre. *Galois cohomology.* Springer Monographs in Mathematics. Springer-Verlag, Berlin, english edition, 2002. Translated from the French by Patrick Ion and revised by the author.

[239] J.-P. Serre and J. Tate. Good reduction of abelian varieties. *Ann. of Math. (2),* 88:492–517, 1968.

[240] B. Setzer. Elliptic curves of prime conductor. *J. London Math. Soc. (2),* 10:367–378, 1975.

[241] B. Setzer. Elliptic curves over complex quadratic fields. *Pacific J. Math.,* 74(1):235–250, 1978.

[242] I. R. Shafarevich. Algebraic number fields. In *Proc. Int. Cong. (Stockholm 1962),* pages 25–39. American Mathematical Society, Providence, R.I., 1963. Amer. Math. Soc. Transl., Series 2, Vol. 31.

[243] I. R. Shafarevich. *Basic algebraic geometry.* Springer-Verlag, Berlin, study edition, 1977. Translated from the Russian by K. A. Hirsch, Revised printing of Grundlehren der mathematischen Wissenschaften, Vol. 213, 1974.

[244] I. R. Shafarevich and J. Tate. The rank of elliptic curves. In *Amer. Math. Soc. Transl.,* volume 8, pages 917–920. Amer. Math. Soc., 1967.

[245] A. Shamir. Identity-based cryptosystems and signature schemes. In *Advances in Cryptology (Santa Barbara, Calif., 1984),* volume 196 of *Lecture Notes in Comput. Sci.,* pages 47–53. Springer, Berlin, 1985.

[246] G. Shimura. Correspondances modulaires et les fonctions ζ de courbes algébriques. *J. Math. Soc. Japan,* 10:1–28, 1958.

[247] G. Shimura. On elliptic curves with complex multiplication as factors of the Jacobians of modular function fields. *Nagoya Math. J.,* 43:199–208, 1971.

[248] G. Shimura. On the zeta-function of an abelian variety with complex multiplication. *Ann. of Math. (2),* 94:504–533, 1971.

[249] G. Shimura. *Introduction to the arithmetic theory of automorphic functions,* volume 11 of *Publications of the Mathematical Society of Japan.* Princeton University Press, Princeton, NJ, 1994. Reprint of the 1971 original, Kano Memorial Lectures, 1.

[250] G. Shimura and Y. Taniyama. *Complex multiplication of abelian varieties and its applications to number theory,* volume 6 of *Publications of the Mathematical Society of Japan.* The Mathematical Society of Japan, Tokyo, 1961.

[251] T. Shioda. An explicit algorithm for computing the Picard number of certain algebraic surfaces. *Amer. J. Math.,* 108(2):415–432, 1986.

[252] R. Shipsey. *Elliptic divisibility sequences.* PhD thesis, Goldsmith's College (University of London), 2000.

[253] V. Shoup. Lower bounds for discrete logarithms and related problems. In *Advances in cryptology—EUROCRYPT '97 (Konstanz),* volume 1233 of *Lecture Notes in Comput. Sci.,* pages 256–266. Springer, Berlin, 1997. updated version at www.shoup.net/papers/dlbounds1.pdf.

[254] J. H. Silverman. Lower bound for the canonical height on elliptic curves. *Duke Math. J.,* 48(3):633–648, 1981.

[255] J. H. Silverman. *The Néron–Tate height on elliptic curves*. PhD thesis, Harvard University, 1981.

[256] J. H. Silverman. Heights and the specialization map for families of abelian varieties. *J. Reine Angew. Math.*, 342:197–211, 1983.

[257] J. H. Silverman. Integer points on curves of genus 1. *J. London Math. Soc. (2)*, 28(1):1–7, 1983.

[258] J. H. Silverman. The S-unit equation over function fields. *Math. Proc. Cambridge Philos. Soc.*, 95(1):3–4, 1984.

[259] J. H. Silverman. Weierstrass equations and the minimal discriminant of an elliptic curve. *Mathematika*, 31(2):245–251 (1985), 1984.

[260] J. H. Silverman. Divisibility of the specialization map for families of elliptic curves. *Amer. J. Math.*, 107(3):555–565, 1985.

[261] J. H. Silverman. Arithmetic distance functions and height functions in Diophantine geometry. *Math. Ann.*, 279(2):193–216, 1987.

[262] J. H. Silverman. A quantitative version of Siegel's theorem: integral points on elliptic curves and Catalan curves. *J. Reine Angew. Math.*, 378:60–100, 1987.

[263] J. H. Silverman. Computing heights on elliptic curves. *Math. Comp.*, 51(183):339–358, 1988.

[264] J. H. Silverman. Wieferich's criterion and the abc-conjecture. *J. Number Theory*, 30(2):226–237, 1988.

[265] J. H. Silverman. The difference between the Weil height and the canonical height on elliptic curves. *Math. Comp.*, 55(192):723–743, 1990.

[266] J. H. Silverman. *Advanced topics in the arithmetic of elliptic curves*, volume 151 of *Graduate Texts in Mathematics*. Springer-Verlag, New York, 1994.

[267] J. H. Silverman. *The arithmetic of dynamical systems*, volume 241 of *Graduate Texts in Mathematics*. Springer, New York, 2007.

[268] N. P. Smart. S-integral points on elliptic curves. *Math. Proc. Cambridge Philos. Soc.*, 116(3):391–399, 1994.

[269] N. P. Smart. The discrete logarithm problem on elliptic curves of trace one. *J. Cryptology*, 12(3):193–196, 1999.

[270] K. Stange. The Tate pairing via elliptic nets. In *Pairing Based Cryptography*, Lecture Notes in Comput. Sci. Springer, 2007.

[271] K. Stange. *Elliptic Nets and Elliptic Curves*. PhD thesis, Brown University, 2008.

[272] K. Stange. Elliptic nets and elliptic curves, 2008. `arXiv:0710.1316v2`.

[273] H. M. Stark. Effective estimates of solutions of some Diophantine equations. *Acta Arith.*, 24:251–259, 1973.

[274] W. Stein. *The Modular Forms Database.* `http://modular.fas.harvard.edu/Tables`.

[275] W. Stein. *Sage Mathematics Software*, 2007. `http://www.sagemath.org`.

[276] N. M. Stephens. The Diophantine equation $X^3 + Y^3 = DZ^3$ and the conjectures of Birch and Swinnerton-Dyer. *J. Reine Angew. Math.*, 231:121–162, 1968.

[277] D. R. Stinson. *Cryptography: Theory and Practice*. CRC Press Series on Discrete Mathematics and Its Applications. Chapman & Hall/CRC, Boca Raton, FL, 2002.

[278] W. W. Stothers. Polynomial identities and Hauptmoduln. *Quart. J. Math. Oxford Ser. (2)*, 32(127):349–370, 1981.

[279] R. J. Stroeker and N. Tzanakis. Solving elliptic Diophantine equations by estimating linear forms in elliptic logarithms. *Acta Arith.*, 67(2):177–196, 1994.

[280] J. Tate. Letter to J.-P. Serre, 1968.

[281] J. Tate. Duality theorems in Galois cohomology over number fields. In *Proc. Internat. Congr. Mathematicians (Stockholm, 1962)*, pages 288–295. Inst. Mittag-Leffler, Djursholm, 1963.

[282] J. Tate. Endomorphisms of abelian varieties over finite fields. *Invent. Math.*, 2:134–144, 1966.

[283] J. Tate. Algorithm for determining the type of a singular fiber in an elliptic pencil. In *Modular functions of one variable, IV (Proc. Internat. Summer School, Univ. Antwerp, Antwerp, 1972)*, pages 33–52. Lecture Notes in Math., Vol. 476. Springer, Berlin, 1975.

[284] J. Tate. Variation of the canonical height of a point depending on a parameter. *Amer. J. Math.*, 105(1):287–294, 1983.

[285] J. Tate. A review of non-Archimedean elliptic functions. In *Elliptic curves, modular forms, & Fermat's last theorem (Hong Kong, 1993)*, Ser. Number Theory, I, pages 162–184. Int. Press, Cambridge, MA, 1995.

[286] J. Tate. WC-groups over p-adic fields. In *Séminaire Bourbaki, Vol. 4 (1957/58)*, pages Exp. No. 156, 265–277. Soc. Math. France, Paris, 1995.

[287] J. T. Tate. Algebraic cycles and poles of zeta functions. In *Arithmetical Algebraic Geometry (Proc. Conf. Purdue Univ., 1963)*, pages 93–110. Harper & Row, New York, 1965.

[288] J. T. Tate. Global class field theory. In *Algebraic Number Theory (Proc. Instructional Conf., Brighton, 1965)*, pages 162–203. Thompson, Washington, D.C., 1967.

[289] J. T. Tate. The arithmetic of elliptic curves. *Invent. Math.*, 23:179–206, 1974.

[290] R. Taylor. Automorphy for some l-adic lifts of automorphic mod l representations. II. *Inst. Hautes Études Sci. Publ. Math.* submitted.

[291] R. Taylor and A. Wiles. Ring-theoretic properties of certain Hecke algebras. *Ann. of Math. (2)*, 141(3):553–572, 1995.

[292] E. Teske. A space efficient algorithm for group structure computation. *Math. Comp.*, 67(224):1637–1663, 1998.

[293] E. Teske. Speeding up Pollard's rho method for computing discrete logarithms. In *Algorithmic Number Theory (Portland, OR, 1998)*, volume 1423 of *Lecture Notes in Comput. Sci.*, pages 541–554. Springer, Berlin, 1998.

[294] E. Teske. Square-root algorithms for the discrete logarithm problem (a survey). In *Public-Key Cryptography and Computational Number Theory (Warsaw, 2000)*, pages 283–301. de Gruyter, Berlin, 2001.

[295] D. Ulmer. Elliptic curves with large rank over function fields. *Ann. of Math. (2)*, 155(1):295–315, 2002.

[296] B. L. van der Waerden. *Algebra. Vols. I and II*. Springer-Verlag, New York, 1991. Based in part on lectures by E. Artin and E. Noether, Translated from the seventh German edition by Fred Blum and John R. Schulenberger.

[297] J. Vélu. Isogénies entre courbes elliptiques. *C. R. Acad. Sci. Paris Sér. A-B*, 273:A238–A241, 1971.

[298] P. Vojta. A higher-dimensional Mordell conjecture. In *Arithmetic geometry (Storrs, Conn., 1984)*, pages 341–353. Springer, New York, 1986.

[299] P. Vojta. Siegel's theorem in the compact case. *Ann. of Math. (2)*, 133(3):509–548, 1991.

[300] J. F. Voloch. Diagonal equations over function fields. *Bol. Soc. Brasil. Mat.*, 16(2):29–39, 1985.

[301] P. M. Voutier. An upper bound for the size of integral solutions to $Y^m = f(X)$. *J. Number Theory*, 53(2):247–271, 1995.

[302] R. J. Walker. *Algebraic curves*. Springer-Verlag, New York, 1978. Reprint of the 1950 edition.

[303] M. Ward. Memoir on elliptic divisibility sequences. *Amer. J. Math.*, 70:31–74, 1948.

[304] L. C. Washington. *Elliptic curves*. Discrete Mathematics and Its Applications (Boca Raton). Chapman & Hall/CRC, Boca Raton, FL, second edition, 2008. Number theory and cryptography.

[305] A. Weil. Numbers of solutions of equations in finite fields. *Bull. Amer. Math. Soc.*, 55:497–508, 1949.

[306] A. Weil. Jacobi sums as "Grössencharaktere." *Trans. Amer. Math. Soc.*, 73:487–495, 1952.

[307] A. Weil. On algebraic groups and homogeneous spaces. *Amer. J. Math.*, 77:493–512, 1955.

[308] A. Weil. Über die Bestimmung Dirichletscher Reihen durch Funktionalgleichungen. *Math. Ann.*, 168:149–156, 1967.

[309] A. Weil. *Dirichlet Series and Automorphic Forms*, volume 189 of *Lecture Notes in Mathematics*. Springer-Verlag, 1971.

[310] E. T. Whittaker and G. N. Watson. *A course of modern analysis*. Cambridge Mathematical Library. Cambridge University Press, Cambridge, 1996. Reprint of the fourth (1927) edition.

[311] A. Wiles. Modular elliptic curves and Fermat's last theorem. *Ann. of Math. (2)*, 141(3):443–551, 1995.

[312] A. Wiles. The Birch and Swinnerton-Dyer conjecture. In *The millennium prize problems*, pages 31–41. Clay Math. Inst., Cambridge, MA, 2006.

[313] G. Wüstholz. Recent progress in transcendence theory. In *Number theory, Noordwijkerhout 1983 (Noordwijkerhout, 1983)*, volume 1068 of *Lecture Notes in Math.*, pages 280–296. Springer, Berlin, 1984.

[314] G. Wüstholz. Multiplicity estimates on group varieties. *Ann. of Math. (2)*, 129(3):471–500, 1989.

[315] D. Zagier. Large integral points on elliptic curves. *Math. Comp.*, 48(177):425–436, 1987.

[316] H. G. Zimmer. On the difference of the Weil height and the Néron-Tate height. *Math. Z.*, 147(1):35–51, 1976.

[317] K. Zsigmondy. Zur Theorie der Potenzreste. *Monatsh. Math.*, 3:265–284, 1892.

Index